International Conference on Eritrean Studies (ICES2025)

International and Regional Cooperation for Sustainable Development

January 4-6, 2025

Selected Works

Volume I

HDRI PUBLISHERS

ices2024asmara@gmail.com

Cover Design Yosief Aregahgn and Rita Berane

Text Design Rita Berhane

ISBN - 978-99948-2-020-7

Printed by Sabur Printing Services
Asmara - Eritrea

Content

EDUCATION AND LANGUAGE

LAW

POLITICS AND REGIONAL AFFAIRS

Keynote Speech

Madame Chairperson, Dr. Halima Mohammed

Honourable Government and Senior PFDJ Officials

Excellencies Ambassadors, Members of the Diplomatic Corps and UN International Organizations

Distinguished Participants and Invited Guests

Ladies and Gentlemen

Let me first congratulate all the Members of the Organizing Committee – both from the academic and relevant government institutions here at home as well as from our Communities in the Diaspora – for their commendable work in making this event a reality.

The research papers – on a broad spectrum of themes – slated for discussion during the next three days will, needless to emphasize, immensely enrich development-oriented academic research in the country.

They will also spur and crystallize vital networking among key national as well as foreign experts and stakeholders in all domains of vital national interest in the period ahead.

Furthermore, another invaluable spinoff of this Conference will be its impact and contribution in debunking and rectifying the stifling, negative narrative, on Eritrea. This has been pursued, for almost two decades now without let-up, by major legacy media outlets and associated think thanks.

The optimism must of course be tempered with a dose of realism.

As we all know, the defamatory campaigns against Eritrea never stemmed from paucity of information or from good-faith media misperception.

I do not wish to waste your time here to delve into anecdotal incidents. But there are so many instances in which reports and documentaries by visiting journalists were suppressed, or willfully altered, by senior editors presumably because, or under the implausible ruse, that they were "too good to be true".

The fact is the negative media campaigns – the dogged portrayal of Eritrea in the most pejorative terms – were and remain, closely intertwined with the agendas of extraneous powers, who have long opted to pursue policies of regime-change for their own geopolitical interests and calculus.

Let me now revert to highlight, in very broad terms, Eritrea's perspectives on current and unfolding international and regional trends as well as its domestic development policies and priorities.

As we all agree, these are very turbulent and trying times; by all historical metrics. The global security structure is literally under unprecedented and dangerous stress.

The spiral of inexorable escalation in Ukraine; mushrooming wars in Gaza and the greater Middle East; are fraught, perhaps for the first time after the Second World War, with precipitating a nuclear Armageddon; irrational and improbable as this doomsday scenario may have seemed in the past decades; or even a few years ago.

The world has of course seen calamities of monstrous proportions in earlier times. Due to the sheer magnitude of the destruction and human losses that they inculcated, the First and Second World Wars may represent the nadir or lowest point of humanity's inexcusable failures. But slavery, colonialism and the multiple proxy wars that have, and continue to rage, especially in the Global South were, and are no less tolerable in terms of humane and ethical considerations.

Indeed, if our so-called civilized, technologically-advanced, post-industrial world has been spared from another global conflagration in the past eighty years, this was not due to some higher moral calling, statesmanship, or wisdom. It was not based on a judicious architecture of global governance. Sadly, the fragile peace was, and remains predicated, on a raw and crude balance of terror. Due to what is termed as Mutually Assured Destruction or MAD as the acronym cynically suggests.

The nuclear powers did not enjoy the privilege of success in annihilating their adversaries through pre-emptive first strikes. It was this binary balance of terror that constituted – for the last eighty years – the sole and primary basis of deterrence and fragile peace.

Tactical and political brinkmanship and posturing aside, there are worrying signs now that this paradigm might shift for the worst. Some, worst-case analysists are even invoking what is known as the Thucydides Trap to emphasize historical precedents for the high probability of the outbreak of war in this period of global transition of hegemony.

Innate and systemic aberrations of the so-called rules-based international order are not limited, unfortunately, to the architecture of global security. Degrees of magnitude and intensity aside, the global legal and economic architectures are also dithering from similar and profound anomalies.

In legal terms, the UN remains emasculated and marginalized in global affairs of vital impact and substance. As amplified in the tragic situation in Gaza, the six-decades-old unilateral sanctions on Cuba, and other similar cases, the Resolutions of the UN General Assembly remain, invariably, toothless.

As it happens, the moral voice of the global community that should have reinforced our ethical compass has been routinely rendered irrelevant and reduced to sheer exercises in futile, Public Relations, stints.

Similarly, the UN Security Council, which has exclusive purview on grave matters of regional and international security, seems to have forfeited its legal authority on various instances due to structural and systemic anomalies and loopholes.

The latest and most appalling instance was not only its failure to impose a ceasefire on Gaza. What is more astounding is its attempts to rationalize non-adherence to this resolution – which elicits punitive action in accordance with Chapter VII of the UN Charter – as optional and dependent on the discretionary political will of the culpable party.

And in general, structural unfairness embedded in the veto power; routine political horse-trading in the adoption of resolutions; bloc politics and overriding interests of the major powers have always rendered the UNSC as a partisan and politicized entity; not the legally flawless and morally authoritative body cherished by the vast majority of humanity.

The architecture of global economic governance – which is a vast subject and beyond the scope of this brief presentation – has been equally rendered dysfunctional in the duplicitous, rules-based international order on account of various parameters.

The modus operandi of the Bretton Woods Institutions, heavily influenced as they are by value-laden ideological biases and political conditionalities, have vastly reduced their efficacy in providing transformative and effective development support – through concessional loans and grants – to countries in the Global South.

Gross and widening gaps of wealth between the rich North and underdeveloped countries in the Global South as well as the huge chasm of income within these countries – especially in the post-industrial North in which less than 0.1% of the population own 90% of the national wealth – are not only morally repugnant but also recipes and fertile grounds for breeding intractable instability and chaos.

And these days, the inherent absurdity in the global economic governance architecture seems to have scaled another height. Sovereign wealth and other assets of an adversarial country are considered as legitimate ransom. They can indeed be arbitrarily weaponized for imposing political solutions in flagrant breach of all accepted economic orthodoxies and the nominal rules of the game.

All these aberrations – in the security, legal and economic global architectures of governance – are occurring at a time when technological advancement has achieved unprecedented progress in all its dimensions.

The disconnect between reality and what is possible and desirable, remains immensely and absurdly huge.

Humanity can indeed work out a viable and sustainable global system that can advance human dignity under a common Global Social Contract of compassion, fairness, and solidarity.

This is simple common sense. It is what is arguably cherished by the vast majority of the peoples of the world.

Let us be more clear: the choice is not necessarily between antagonistic, dogmatic, conventional ideologies. Nor is it between a uni-polar, bi-polar or multi-polar systems of embedded and inevitable confrontation.

The choice is for a compassionate and humane architecture of global governance under transparent, fair, and equitable rules of the game.

These precepts were succinctly articulated in President Isaias's remarks at the 15th BRICS Summit, held in South Africa in August 2023, when he argued for concerted action to rectify the flaws and deficits of the prevailing global governance architecture whose defining features were and remain, dysfunctional, non-inclusive and unfair, rules and regulations. The President went on to say", and I quote, "...the lofty aspirations of humanity as a whole is for a just and fair global order where justice and the rule of law prevail; where nations and peoples forge meaningful and symmetric ties of cooperation and partnership on the basis of respect for national independence and sovereignty; for societies anchored on compassion and social justice."

These are some of the challenges – and the list is far longer than what has been presented above in skeletal format – that true and dispassionate academic research has to grapple with. Hopefully, some of these challenges will be broached in the papers that will be presented in the Conference to constitute topics of intense and vibrant discussion.

Let me now revert to the African continent and our region in particular.

The developmental and security challenges and trends in the African continent as a whole and the wider Horn of Africa, Nile Basin and Red Sea neighborhoods in particular, can only be gauged against the backdrop, and within the constraints and parameters of, the fragile global governance architecture outlined above.

Africa's daunting challenges are too evident to merit much elaboration. Indeed, as the grim statistics amplify – Africa's share of world trade stands at a paltry 2%, while extreme poverty affects over 460 million of its population. Few exception notwithstanding, Africa continues to export raw-materials to the global market at lowest prices; it remains overburdened and paralyzed by debt; and, it continues to be bedeviled by perennial internal conflicts.

These malaises are counter-balanced by substantial positive attributes that can propel Africa on the path of rapid and sustainable development once the underlying problems are fully addressed and resolved.

Africa is reportedly endowed with 60 percent of global natural resources, especially significant reserves of critical minerals vital for global green transition. It is blessed with a young, dynamic and educated population as well as large Diaspora whose expertise can be tapped. Africa has also bright prospects in reducing – at a fast pace – the knowledge gap with the developed North if it pursues the right mix of policies on its human capital.

But its options and policy choices for domestic development cannot be seen in isolation from the collective and purposeful measures that it has to pursue to assert its rightful place in the global power pecking order.

Indeed, as Eritrea has consistently advocated in the various Summits of Partnership between Africa and its prospective interlocutors, Africa needs to accomplish, first and foremost, its homework with full diligence to ensure that the rules of engagement ultimately become symmetric and mutually beneficial. This will require sober and meticulous work and consensus building that transcends nominal and cosmetic results such as inconsequential membership at the G20 or allocation of permanent seats for a couple of African countries at the UN Security Council.

Closer to home, the situation in our immediate neighborhood is not different, in substantive terms, from the continental reality.

Naturally, and for reasons of history, geographic proximity, pragmatic calculus, and other cogent considerations, Eritrea's policy perspectives on continental African cooperation and partnership are primarily predicated on a modular approach. This policy perspective was first articulated at the OAU Summit held in Tunis in 1994.

In Eritrea's views, while the aspirations for fostering continent-wide institution of African effective cooperation is laudable in terms of broad abstract principles, realism dictates that this is implemented – over a long span of decades – through the establishment and consolidation of regional building blocks.

In the Horn of Africa, IGAD was revitalized in 1995 to include broader scope of economic cooperation and integration as well as Inter-State conflict resolutions through the expansion of its original mandate that was confined to combating drought.

But all these ideals remain – to date – mere aspirations that could not crystallize into visible and tangible programmes of action.

Furthermore, institutionalized ethnic and religious exclusivity and polarization continue to remain the Achilles heel in the region accounting for cyclical internal conflicts that have, and continue to inculcate, huge loss of life and physical destruction.

Debilitating dependency on, and entrenched tendencies and proclivities to act as surrogates of global and regional hegemons, have also encumbered independent and home-grown development policies and options. Ill-advised ambitions of territorial aggrandizement and expansion, often induced by overriding external agendas, have and continue to constitute avoidable factors of additional regional destabilization; thereby compromising the national and regional interests of the nations and peoples of the Horn of Africa.

In a nutshell, and without sounding in any way prescriptive, there are compelling arguments for the region to revisit and map-out appropriate and functional National Social Contracts that foster and bolster internal unity and cohesion. Within this framework, regional cooperation can be seriously pursued with purposeful and incremental programmes and timelines on the basis of respect of each other's sovereignty and territorial integrity.

Let me now revert to GOE's domestic development approaches and policy perspectives within the broad global and regional ecosystems outlined above.

Eritrea's endowments are substantial by all standards. The blue economy consisting of considerable marine resources; locational advantages; ports as well as pristine beaches and islands can be leveraged to catalyze and expedite rapid economic growth under conditions of congenial regional peace and stability.

The extractive industry; agriculture; revived manufacturing geared on educated human capital and knowledge-intensive subsectors enmeshed in global networks and value-chains; are other prospective sectors that have huge potential.

And above all, Eritrea's fundamental policy precepts that imbue highest priority to investment in human capital will concretely guarantee qualitative dividends for the longer term. This sober approach is underpinned by the profound conviction that singular focus and dependence on finite resources – which have to be exploited judiciously for reasons of sustainability and to preserve what must be preserved for posterity – has limited viability and is ultimately fraught with irreparable distortions.

The policy precept is also closely intertwined with Eritrea's over-arching commitment to Social Justice in all its manifestations and dimensions. Free education from Kindergarten to the tertiary level, and provision of highly subsidized health services – including waiver of all payments for those without any means when necessary – so as to ascertain equal opportunities and a level-playing field; greater focus on deprived areas; narrowing the urban-rural gap in terms of allocation of budgetary resources for physical and social infrastructure; are pursued rigorously to advance this policy commitment.

It must be emphasized here that Eritrea's Social Contract, hinged as it is on comprehensive Social Justice, does not stem from a humanistic philosophical and political convictions alone. Embedded in the philosophy is the keen and implicit tribute to, and recognition of, Stakeholder rights; the additional acknowledgement that Eritrea is a nation whose citizens have all paid heavy and almost unparalleled sacrifices to assert and defend their inalienable human and national rights. As such, they deserve their proper share in the national pie.

But in spite of these purposeful policy perspectives, Eritrea's post-independence developmental achievements have not been commensurate with its relentless efforts in the past three decades or in terms of its aspirations and latent potentialities.

Imposed wars, unwarranted sanctions, and related adversities have compounded the generic global and regional hurdles described above. Let me also emphasize that instead of making reparations for the damage incurred for almost ten years until 2018 by the unwarranted UNSC sanctions, the US and the EU have slammed new and unilateral sanctions against Eritrea's defense and security institutions in the past three years. The raft of sanctions imposed by the Biden Administration target the Eritrean Defense Forces; the PFDJ; and their respective Institutions including the Army Chief of Staff and the Head of PFDJ Economic Affairs while the EU's coordinated measures are focused at Eritrea's National Security Office. These deplorable acts, essentially concocted to weaken and downgrade Eritrea's economic and defense capabilities, only expose their entrenched policies of hostility.

All these cogent realities amplify the interlinked and symbiotic global and regional challenges that Eritrea will continue to grapple with in its focused and full-fledged pursuit of national development and growth.

In this context, it is gratifying to note that the Conference will address all these inter-related themes in greater depth in its proceedings in the next three days.

Let me conclude by wishing you the best of success in all your proceedings

I thank you.

H.E. Mr. Yemane Gebremeskel,
Minister of Information

Closing Speech

Madame chairperson, Dr. Halima Mohammed, Honourable Ministers, Senior PFDJ and Government Officials,

Excellencies Ambassadors, members of the Diplomatic Community and UN International Organizations,

Distinguished Participants, and Invited Guests,

Ladies and Gentlemen,

First and foremost, allow me to extend my heartfelt congratulations to the organizing committee, whose unparalleled dedication and meticulous planning have made this event possible. Their collaborative efforts, bridging academic institutions, government bodies, and the Eritrean diaspora, and the seamless orchestration of this event, are, without doubt, an exemplar of what can be achieved when diverse stakeholders align their aspirations towards a common goal.

As we draw the curtains on this august International Conference on Eritrean Studies, it becomes our solemn obligation to reflect, synthesize, and duly articulate the profound significance of the deliberations we have collectively engaged in over these past three days. This gathering has not merely been an academic exercise; rather, it has epitomized the convergence of scholarly rigor, policy-oriented discourse, and the unassailable ethos of intellectual pursuit that underscores Eritrea's resolute commitment to knowledge-driven development.

In line with the conference theme – Regional and International Cooperation for Sustainable Development – this dialogue has underscored the importance of collaboration, extending beyond Eritrea's borders into the realms of regional and global partnerships aimed at achieving lasting, inclusive, and equitable growth.

The past three days have provided us with an extraordinary platform to engage with a diverse range of research papers and discourses that span themes integral to Eritrea's developmental trajectory, historical narrative, and global positioning. These contributions have illuminated critical pathways to fostering a deeper understanding of Eritrea's unique narrative and exploring solutions to the complex challenges that define our time. The depth and breadth of the discourse presented are further proof of the vibrant intellectual community that Eritrea proudly sustains, both within its borders and across the globe. But more importantly, these discussions have also highlighted the power of international collaboration in tackling shared challenges, emphasizing the role of cross-border cooperation in ensuring that progress is sustainable and equitable for all.

The corpus of research papers presented during this conference, traversing a wide array of themes, have illuminated critical issues, fostered rigorous debate, and, most importantly, laid the groundwork for future inquiries that will undoubtedly enrich the corpus of knowledge on Eritrea and its multifaceted realities. The contributions have not only expanded the horizons of academic discourse but have also provided actionable insights for policymakers, practitioners, and stakeholders invested in Eritrea's development trajectory. And, as we have seen, the impact of these discussions extends far beyond academic spaces. It serves as a call to action for the strengthening of regional and international cooperation, encouraging the alignment of policies and strategies that catalyze meaningful transformation.

This event has been instrumental in challenging and deconstructing the reductive narratives that have hitherto obfuscated Eritrea's agency, aspirations, and achievements. For too long, certain sectors of the global discourse have promulgated superficial caricatures that belie the rich socio-political complexities of this nation. Through the empirically grounded research and nuanced debates presented herein, we have laid the cornerstone for a more balanced and equitable understanding of Eritrea – one that accords due recognition to its developmental philosophy, its resilient people, and its strategic aspirations.

While it is unrealistic to expect that these efforts will immediately alter deeply entrenched positions, their incremental impact in broadening awareness and fostering informed discourse cannot be overstated. The dissemination of the proceedings from this conference will undoubtedly serve as a valuable resource for academics, policymakers, and the general public alike, amplifying its reach and resonance.

It is equally important to underscore the catalytic role this conference has played in fostering networks of collaboration among scholars, experts, and institutions. Such synergies are indispensable in addressing the complex and interconnected challenges of our time. By facilitating dialogue across disciplines, geographies, and perspectives, this forum has contributed to the cultivation of a robust intellectual ecosystem that will continue to yield dividends long after the conclusion of this event.

Our deliberations have not unfolded in isolation; rather, as highlighted in the opening statement by Minister Yemane, have occurred against the backdrop of an increasingly fractious global landscape marked by escalating conflicts, economic inequities, and the erosion of multilateral frameworks. These dynamics have rendered the principles of cooperation and sustainable development more pertinent than ever.

As underscored in the conference, Eritrea's developmental philosophy is firmly rooted in the principles of self-reliance, social justice, and sustainable growth. This approach far from being a mere reaction to external exigencies is reflective of the nation's historical experiences and collective aspirations. Eritrea's emphasis on human capital development, exemplified by investments in education and healthcare, attest to its recognition that its greatest resource lies in its people. By prioritizing inclusivity and equity, Eritrea is building a society that is not only resilient in the face of adversity but is also capable of harnessing its potential for transformative change, where no one is left behind.

Similarly significant, the presentations and discussions have also illuminated the immense opportunities that Eritrea possesses, from its strategic geographic position to its untapped resources in the blue economy, agriculture, and extractive industries. However, realizing this potential requires a conductive regional and international conditions. Peace, stability, and mutual respect must form the foundation of meaningful partnerships within the Horn of Africa and beyond.

It is equally pertinent to acknowledge that the African continent continues to grapple with the malignant effects of tribal and ethnic conflicts and external interference, which exacerbate socio-political fractures. The legacy of colonialism, coupled with the entrenchment of structural dependencies, has deepened these challenges, undermining the continent's ability to forge unified progress. Additionally, external meddling often inflames local disputes, perpetuating cycles of instability and division.

These realities underscore the necessity for African nations to assert greater agency in addressing internal dynamics while resisting external forces that seek to exploit vulnerabilities.

Concurrently, the broader Global South remains marginalized in the global governance architecture, facing systemic poverty, inequities, and the exploitative paradigms of an imbalanced economic order. The inequitable distribution of resources and the perpetuation of debt traps hinder developmental strides. This confluence of challenges underscores the urgency of fostering solidarity and advocating for transformative change that dismantles these entrenched inequities and builds an inclusive global order rooted in fairness and mutual respect.

The challenges confronting Eritrea, the African continent, and the wider Global South are, as we have repeatedly observed, deeply intertwined with the broader anomalies of the global governance architecture. The marginalization of African voices in international decision-making processes, the inequitable distribution and opportunities, and the perpetuation of exploitative economic paradigms are but a few manifestations of these systemic injustices. Yet, amidst these challenges, there is also a growing recognition of Africa's potential and agency.

In this regard, Eritrea's advocacy for modular regional integration, as articulated in its longstanding policy perspectives, offers a pragmatic framework for fostering cooperation and addressing shared challenges. By prioritizing functional and incremental approaches to integration, grounded in the principles of sovereignty and mutual benefit, Eritrea seeks to contribute to the establishment of a stable and prosperous regional order. This vision, however, necessitates a collective commitment to addressing the root causes of conflict and underdevelopment, including the pernicious legacies of colonialism, dependency, and external interferences.

The dialogues during this conference have reinforced the imperative for Africa to assert its rightful place in the global order. This entails not only addressing the structural impediments to development but also leveraging the continent's immense resources, both human and natural, to drive transformative change. The pursuit of these objectives must be guided by a clear and unified vision, underpinned by a commitment to self-determination, solidarity, and innovative policy-making.

In this context, Eritrea's experiences and perspectives offer valuable insights for the broader African and Global South community. Its emphasis on resilience, self-reliance, and people-centered development serves as a powerful reminder that meaningful progress is achievable even in the face of formidable challenges. Moreover, Eritrea's principled stance on issues of sovereignty and non-alignment underscores the importance of maintaining an independent and constructive approach to international engagement.

As we reflect on the myriad discussions held over these past three days, it is evident that this conference has not only served as a platform for intellectual exchange but has also ignited a renewed sense of purpose among its participants. The diverse range of topics addressed-from governance and socio-economic development to environmental sustainability and cultural heritage-reflects the multidimensionality of Eritrea's reality and complexity of the challenges it confronts. Equally, these discussions have highlighted the resilience of the Eritrean people, whose unwavering commitment to progress continues to defy adversity.

The importance of collaboration cannot be overstated in this endeavor. As the conference has demonstrated, the collective expertise of academics, practitioners, and policymakers is indispensable in addressing the intricate and multifaceted issues of our time. The networks forged and strengthened during this event will undoubtedly serve as a foundation for future initiatives aimed at advancing Eritrea's development and enhancing its global engagement.

Moreover, this conference has underscored the necessity of bridging the gap between academic research and policy execution. The insights generated through rigorous scholarship must inform decision-making processes at all levels, ensuring that policies are grounded in evidence and attuned to the needs of the people. This iterative process of learning and adaptation is crucial for achieving sustainable and inclusive development.

As we bring this conference to its conclusion, it is essential to recognize that our discussions represent not an end but a beginning. The onus lies upon each of us to ensure that the knowledge and insights gained here are translated into tangible actions that benefit not only Eritrea but also the broader global community.

This enquires ongoing collaboration among all stakeholders, including academics, policymakers, practitioners, and national organizations. It will also necessitate a commitment to continuous learning, critical reflection, and the pursuit of excellence in all endeavors.

Let us continue to challenge misconceptions, advocate for justice and fairness, and contribute to the development of a more equitable and sustainable world.

In closing, let me reiterate my profound gratitude to all those who have contributed to the success of this conference. To our esteemed researchers and academics, I encourage you to persist in pushing the boundaries of inquiry and contributing to the eclectic array of Eritrean studies.

To our policymakers and practitioners, I urge you to draw upon the wealth of knowledge generated here to inform decisions that will shape the future of our nation and region. And to all participants, I extend my deepest gratitude for your engagement, insights, and commitment to advancing the goals of this conference.

To the organizers and support staff, your tireless efforts behind the scenes have ensured that this conference has been both seamless and impactful.

Thank you and I wish you all safe travels and continued success in your respective endeavors.

H.E. Mr. Osman Saleh,
Minister of Foreign Affairs

Introduction

The International Conferences on Eritrean Studies (ICES)2025) builds upon the foundations laid by its predecessor, the International Conference on Eritrean Studies (ICES 2016). Both conferences strive to revitalize the ethos of Eritrean studies by embracing interdisciplinary research across all scientific fields. As long as the research adheres to scientific methods and addresses matters relevant to Eritrea—whether at the local, regional, or global level—it is considered eligible, irrespective of whether it belongs to the humanities, natural sciences, or social sciences. Studies that utilize fundamental scientific tools and methodologies, regardless of their discipline, are welcomed. While the Eritrean studies have traditionally focused on the Humanities and Social Sciences, the last two conferences adopted a more inclusive approach. They added a new dimension to the identity of Eritrean Studies by broadening its scope and enriching its academic texture.

The Origin of Eritrean Studies

The history of academic conference on Eritrean studies dates back to the era of the Eritrean armed struggle. In January 1979, a group of Eritrean academics and international scholars supportive of Eritrean people`s cause organized the first Eritrea-specific international symposium in London under the theme "Behind the Wars in Eritrea" (Davidson et al. 1980: 7-9). This symposium brought together renowned historians, analysts, and activists with first-hand experience in the Eritrean armed struggle. The works presented at the symposium focused on three main issues: the right of the Eritrean people to nationhood and self-determination, the social revolution initiated by the Eritrean People`s Liberation Front (EPLF), and the condemnation of Soviet support for Ethiopian military offensives. By exposing the humanitarian consequences of the war, the symposium`s underlying message was a call for the international community to pay serious attention to and address the Eritrean armed struggle as it would any other African liberation movement.

The symposium injected new energy into the Eritrean armed struggle, contributing to the pursuit of a durable solution. Such initiatives added value to the broader field of the struggle, which was predominantly military in nature. By articulating and reinforcing the narrative of the Eritrean armed struggle through academic and intellectual channels, the symposium strengthened the political wing of the Eritrean liberation effort. This marked a turning point, amplifying the political and academic dimensions of the struggle. Having diverse audience, including participants from Eritrea, Europe, North America, African and Ethiopia, the conference provided the Eritrean causes with significant international visibility. This, in turn, broadened the dissemination of the narrative surrounding the Eritrean armed struggle, reaching various observers and increasing its global circulation.

Another milestone in Eritrean Studies was 1983's conference, organized by the Eritrean armed struggle in collaboration with Eritrean academics. This event marked the beginning of regularized conferences on Eritrean studies under the auspices of the Research and Information Center of Eritrea (RICE). Building on this foundation, RICE organized another conference from July 26-28, 1988, which featured 28 papers presented by EPLF members, Eritrean diaspora scholars, and African and European academics (RICE 1988). Unlike the previous conferences—which had focused primarily on advocating for the cause of the armed struggle—this gathering was futuristic, shifting its emphasis to post-independence economic modernization. The papers sought to conceptualize the theoretical and practical challenges Eritrea might face after independence, bridging historical context with future aspirations by exploring four thematic areas. These are economic modernization (development models, industrialization), research traditions (avoiding epistemic dependence on colonial framework), legal systems (balancing indigenous and modern practices), and technological transfer (lessons from colonial and wartime advancements).

This conference reflected a forward-looking approach. It anticipated a challenging nation-building process while reaffirming confidence in the inevitability of Eritrea`s independence. In tandem to this mission, while the development economics papers proposed industrial strategies and economic models tailored to Eritrea`s post-war reconstruction, in the papers about decolonizing knowledge, scholars highlighted the need to break free from inherited research paradigms, emphasizing locally grounded scholarship. Similarly, the papers on legal framework centered their discussion on blending indigenous Eritrean justice system with modern legal principles. Whereas the papers on technological transfer, presenters evaluated past technological adaptations (e.g. during colonial rule and the liberation struggle), and their potential applications in a sovereign Eritrea.

The 1988 conference embodied a futurist orientation, operating on the premise that Eritrean independence was inevitable. By focusing on post-liberation governance, the gathering implicitly affirmed this certainty while urging the leadership to prepare for the challenges of statehood. Essentially, it presented a crucial argument: the long-term success of the liberation movement would be measured not by military victory alone, but by achievements in political and economic fronts in post-independence. Without progress in these areas, independence would merely make Eritrea conform to Africa`s postcolonial struggles rather than transcend them.

In a nutshell, the three-armed struggle conferences addressed multiple dimensions simultaneously. While they powerfully validated the Eritrean people`s right to self-determination, their greater legacy lay in institutionalizing Eritrean Studies as a formal discipline. They served dual purposes: amplifying the liberation narrative while systematically developing an academic framework for Eritrean scholarship. This represented a decisive break from colonial-era knowledge production, which had been carefully constructed to sustain European hegemony. By creating independent platform for scholarship, the conferences enabled Eritreans to reclaim their narrative from both Ethiopian occupation and colonial historiography.

Post-independence Conferencing

The July 22-24, 1991 conferences at the University of Asmara marked a historic transition—the first scholarly gathering in liberated Eritrea, convened barely two months after independence. Organized by the EPLF`s Department of Economic Planning and Coordination in collaboration with diaspora academics (Eritreans for Peace and Democracy), its theme "Yesterday in Liberation Struggle and Today in National Construction" reflected the urgent pivot from resistance to reconstruction. The Conference`s subthemes encompassed the full spectrum of nation-building the leadership was sailing ahead. Having seven themes, Macroeconomic framework, Human Resources Development, Agricultural environment, Finance and Banking, Trade and Industry, natural Resources Development and Construction, and Technology and Information system, this conference was an elaborated version of the previous conference (GebreHiwet 1993).

This unprecedented assembly brought together policymakers, practitioners and scholars to present forty papers analyzing post-independence challenges. Its unique character derived from being the first intellectual platform on Eritrean soil where the visions of the liberation struggle could be translated into concrete development plans, initiating the crucial work of transforming revolutionary ideals into governance realities. The papers provided a comprehensive analysis of the economic challenges facing newly independent Eritrea. Building upon but significantly expanding the scope of previous conferences, these studies offered detailed policy recommendations tailored to the war-torn economic inherited by the EPLF. With particular emphasis on agricultural development as the cornerstone of economic revival, the papers established the fundamental orientation for Eritrea economic policies, articulated the nation`s collective economic aspirations, objectively assessed the country`s economic realities, and proposed development strategies at both macro and micro levels. Understandably, this body of work served as a crucial reference point for the Provisional Government of Eritrea, providing evidence-based policy recommendations during the formative period of state building.

Marking 10th Anniversary of the Eritrean independence, the Eritrean Studies Association (ESA) convened a follow-up conference from July 23-26, 2001 under the theme "Independent Eritrea: Lessons and Prospects." This gathering of 150 scholarly papers systematically evaluated the nation`s post-independence experience across six key areas: Constitutional and Legal Development, Economic Performances and Evaluation, Women`s Status in Society, Healthcare System Progress, Media's role in Democratization, and Education and Human Capital Development (Abraham 2001: 2). These post-independence conferences represented a significant evolution in focus that shifted from political liberation to economic development and transitioned its nature from theoretical proposals to policy evaluations. In other words, while the 1991 conference established development frameworks, the 2001 conference assessed implementation outcomes.

More profoundly, these academic gatherings advanced the decolonization of Eritrean knowledge production by challenging colonial-era narratives, recovering marginalized historical knowledge, asserting Eritrean epistemic autonomy and expanding the boundaries of national scholarship. Thus, they served dual purposes—addressing immediate policy needs while simultaneously liberating Eritrea Studies from the constraints of colonial intellectual paradigms, representing a significant leap forward in the development of authentic national scholarship.

The ICES

The International Conference on Eritrean Studies (ICES2016 and ICES2025) epitomize the latest evolution of a long tradition of Eritrean scholarship, building upon the historical foundations laid by previous gatherings. These conferences mark significant milestones in making the Eritrean Studies a robust academic discipline.

Held from July 20-22, 2016 in Asmara and organized by the National Higher Education and Research Institute (NHERI), ICES2016 represented a pivotal reconstitution of Eritrean Studies. The conference emerged in response to two key observations, i.e. growing global interest in Eritrean affairs and increasing polarization within the existing scholarship. The event attracted remarkable international participation, with 240 abstract submissions from scholars in 26 countries, out of which 140 were presented (ICES 2016). This overwhelming response demonstrated both the global relevance of Eritrean Studies and the urgent need for a balanced, scholarly platform. ICES2016, successfully established such a forum, enabling rigorous, critical examination of diverse perspectives and paradigms related to Eritrea.

Seeking to advance the dialogue, ICES2025, which was held from January 4-6, 2025, adopted the theme "Regional and International Cooperation for Sustainable Development." The conference set to meet objectives such as fostering open academic discourse among Eritrean Studies scholarly community, promoting innovative research and knowledge production, and showcasing Eritrea`s research capabilities and state-building experiences.

The conference received an unprecedented 300 paper submissions spanning all scientific disciplines from researchers across Eritrea, Africa, Asia, Europe, and North America. After rigorous blind peer review, 152 papers were selected for presentation through various formats mainly plenary sessions, panels, parallel sessions, roundtable discussions, workshops and poster presentations. These sessions facilitated in-depth exploration of diverse topics, from regional geopolitics to scientific advancements, reflecting the conference`s multidisciplinary approach. The conference attracted distinguished scholars and emerging researchers alike with several significant outcomes. It demonstrated the success of Eritrea`s 30-years educational investment, as 70 per cent of Eritrean presenters were products of this system. It showcased the growing depth and breadth of Eritrean scholarship and highlighted the increasing international collaboration in Eritrean studies.

The ICES conferences have fundamentally reshaped the identity of Eritrean Studies by introducing an expansion beyond traditional social sciences and humanities to embrace STEM fields, transcending from a single-discipline focus to a multidisciplinary approach, and promoting rigorous academic standards through peer review process. These transformations are responses to three critical factors evident in the contemporary academia locally and globally. Firstly, the disciplinary boundary has become blurred. Contemporary academic realities have significantly eroded the traditional demarcation between natural and social sciences . While natural sciences fundamentally examine the physical world, their ultimate focus on improving human welfare has created substantial overlap with social sciences' concerns. This convergence has rendered obsolete the rigid separation that once characterized scholarly inquiry. Although integrating these domains challenges long-established academic traditions, recent decades have witnessed remarkable methodological innovations facilitating meaningful interdisciplinary collaborations.

Secondly, there is inherent subjectivity across all fields. All research, regardless of discipline, exists within socio-historical contexts that shape its priorities and interpretations. While social sciences openly acknowledge their ideological dimensions, natural sciences are equally susceptible to value-laden influence in setting research agenda, funding allocations, interpretation frameworks. This recognition has accelerated cross-disciplinary approaches, particularly in development studies where technical solutions require social implementation contexts. Thirdly, there is lack of systematic coordination among different Eritrean studies activities. The ICES initiative represents a crucial reassertion of authentic Eritrean scholarship. For various reasons, home-based, mainstream Eritrean studies had become relatively inactive, creating a vacuum filled by politically motivated actors who weaponized academic platforms for activism. The danger of this breed of platforms allowed a prioritized decontextualized critique that focuses on what is missing rather than what is happening. In answering the question, it employs cherry-picking methodologies, and disregards rigorous field work research. In contrast, the ICES is an attempt to liberate the Eritrean studies recent tradition by promoting, evidence-based, methodologically rigor, balanced and on the-ground research and multidisciplinary integrated scholarship.

In a nutshell, the ICES conferences represented both continuity with Eritrea`s scholarly traditions, and innovations in their methodological and conceptual approaches. By maintaining academic rigor while embracing interdisciplinary perspectives, these gatherings have positioned Eritrean studies as a dynamic, evolving field capable of addressing both national priorities and global academic discourses. While maintaining academic rigor, the ICES2025 also attempted to address international misperceptions about Eritrea, correct global narrative and ensure strategic communication of Eritrean realities. Moreover, the strong representation of Eritrea-based scholars at the ICES2025 conference demonstrated the nation`s thirty-years investment in human capital development.

Now the Eritrean studies stands at a critical juncture which will be determined by the continuum of the conferences. However, its robust existence does not only depend on conferencing but how far can it adopt characters such as methodological diversity, interdisciplinary approaches, balanced local and global perspectives and rigorous evidence-based scholarship. So long as it does strive to internalize those characters, it will surely maintain its core mission of authentically representing Eritrean realities and contributing to national development.

In line to the objective of the conference, the conference proceedings are compiled into two volumes, broadly categorized by discipline—social sciences and natural sciences—although some papers overlap and span across both fields. In total, 58 papers are published: Volume I, focused on social sciences, contains 31 papers, while volume II, which primarily includes natural science contributions, contains 27 papers. Selecting the papers was extensive and time-consuming task. At the end of the conference, presenters were invited to revise their submissions based on feedback received during their presentations. Following this call 76 papers were submitted. Each paper underwent blind review by two independent reviewers. Despite some delays and logistical challenges in exchanging manuscripts during the review process, approximately 97% of the papers were successfully reviewed and assessed on a five-point scale: rejected, accepted with major revisions, accepted with minor revisions, and accepted without revisions.

After receiving outcomes, authors were asked to revise and resubmit their papers accordingly. Through ongoing communication and coordination, 55 out of the 76 (72%) submitted papers ultimately reached publication. Factors contributing to the exclusion of the remaining papers included insufficient quality, non-compliance with reviewers' comments, and changes in authors` decisions. While the primary aim of these proceedings is to disseminate the scientific findings shared at the conference to a wider audience, considerable effort has been made to ensure the content is accessible and readable. Nevertheless, variations exist across disciplines in language proficiency, referencing styles, and interpretation of data and facts. Responsibility for any errors in these elements rest solely with the individual authors.

Reference

Abraham, Tedros, "Independent Eritrea: Lessons and Prospects: ESA, Conference Here Concludes," Eritrean Profile, January 28, 2001.

Davidson, Basil, Lionel Cliff and Bereket Habte Selassie, Behind the War in Eritrea, (Nottingham: Spokesman, 1980).

International Conference on Eritrean Studies 20-22 July 2016, Proceedings, Volume II.

Research and Information Centre of Eritrea (RICE), Proceedings of the First Conference on Research and Development, 1988.

GebreHiwet Tesfagiorgis, Emergent Eritrea: Challenges of Economic Development (Trenton: The Red Sea Press, 1993).

ICES 2025

Introduction

The International Conference on Eritrean Studies (ICES) 2025, held from January 4-6, 2025, at the Hotel Asmara Palace, was a pivotal event that built upon the foundational work of the ICES 2016. Conceived as a strategic national undertaking, the conference was designed to advance academic discourse on Eritrea across a multitude of disciplines.

The conference was underpinned by several key objectives. These included; establishing a forum for constructive dialogue and critical analysis, facilitating the exchange of academic ideas and fostering knowledge production, and cultivating a global community of scholars to bridge geographical and disciplinary divides. A crucial national objective was to amplify Eritrean voices and narratives to promote a more nuanced understanding of the country's historical trajectory, current realities, and future aspirations. The conference also aimed to deepen the understanding of Eritrea's role in regional peace and stability.

The scale of ICES 2025 underscored its significance, convening over 500 scholars and professionals. Participants included; high-level government officials, members of the diplomatic community, and heads of United Nations agencies in Eritrea. The international reach was notable, with participants from a wide array of countries, including Germany, Italy, the United Kingdom, China, Russia, the United States, and Canada. The conference's intellectual breadth was equally impressive, structured around nineteen broad themes that spanned disciplines from Agriculture and Marine Sciences to Regional and International Affairs, indicating a holistic approach to Eritrean studies.

ICES 2025 was largely assessed as a successful endeavor, providing a valuable platform for practitioners and emerging scholars to engage with renowned experts. The event successfully navigated a dual mandate by both genuinely fostering academic inquiry and simultaneously functioning as a strategic platform for Eritrea to articulate its national narrative and development aspirations on a global stage. This strategic intent was underscored by the presence of high-level government officials alongside scholars, indicating that the academic output was designed to inform and support national policy and image-building efforts. The conference also served as a vital platform for showcasing Eritrea's rich cultural heritage through oral poetry and ethnomusicology.

Comprehensive Overview of Conference Themes and Topics

The International Conference on Eritrean Studies (ICES) 2025 showcased an impressive intellectual scope, organized under nineteen primary themes designed to offer a holistic understanding of Eritrea's multifaceted realities.

- **Agriculture & Marine Sciences:** Discussions focused on food security and climate-smart agricultural practices in East Africa.
- **Archaeology & Heritage:** Sessions highlighted the discovery of ancient knowledge, the potential of untapped sites, and the process of archiving and salvaging heritage documents globally.
- **BRICS & International Relations:** This theme addressed contemporary geo-

political developments, including the expansion of BRICS nations and the trend toward de-dollarization.

- **Diaspora Studies:** This area explored the vital role of the Eritrean diaspora in nation-building and the challenges they face, such as international sanctions.
- **Economics & Development:** Discussions encompassed economic integration, sustainable development, and the economic implications of global shifts.
- **Energy, Environment & Climate Change:** Presentations covered alternative energy resources, digital tools in the energy sector, and strategies for environmental conservation.
- **Education:** This theme was extensively covered, with sessions on social justice in education and challenges to ensuring access, equity, and quality at all levels.
- **History:** Discussions covered the complexities of writing Eritrean historiography, the history of the armed struggle, and the contributions of prominent writers.
- **Information and Communication Technologies (ICT):** Topics included digital development, the potential of big data, telemedicine, and the use of drones in agriculture.
- **Medicine & Health Science:** Presentations covered critical areas such as medication safety, maternal and child health, and the social determinants of health. Evidence-based research from these sessions is poised to inform policy and practice within Eritrea's health service system.
- **Politics:** This theme included discussions on political power in conflicts and the past and contemporary socio-economic and political dynamics of Eritrea and its neighbors.
- **Regional & International Affairs:** Topics addressed included regional and global cooperation for sustainable development, peace and security in the Horn of Africa, and the pursuit of just and constructive international relations.

The conference employed a variety of session formats, including keynote lectures, plenary sessions, panel sessions, parallel sessions, round-table discussions, workshops, and poster presentations, to facilitate comprehensive engagement. The extensive diversity of themes, particularly the strong emphasis on science, technology, environment, and health alongside traditional humanities, indicates a comprehensive and forward-looking national development agenda.

Key Highlights and Insights from Sessions

Keynote addresses emphasized the critical role of science and technology in national development. Prof. Ghebrebrhan Ogubazghi highlighted the importance of mathematical modelling for seasonal rainfall forecasts and the need for disaster mitigation strategies. Prof. Girma Bitsuamlak focused on the necessity of resilient infrastructure, the potential of renewable energy, and the adoption of sustainable design practices. Prof. Solomon Tesfamariam addressed the urgent need for effective management of Massawa's built heritage. These keynotes demonstrated a deliberate effort to align top-tier academic expertise with immediate national priorities like food security, climate resilience, and sustainable urban development.

Plenary sessions delved into broader geopolitical and historical contexts. Prof. Francesco Perticone challenged the term "globalization," proposing "globality" to describe a system manipulated by powerful nations. Dr. Redie Bereketeab addressed the governance challenges in the Horn of Africa, linking instability to the legacy of colonialism and incomplete nation-state formation.

The historiography plenary revealed a concerted effort to critically examine and shape Eritrea's national historical narrative. Mr. Zemhret Yohannes discussed the challenges of writing the history of the armed struggle with objectivity, while Mr. Alemseghed Tesfai underscored the critical need for a comprehensive Eritrean historiography to counter distorted narratives.

Panel discussions offered focused examinations of contemporary issues. In the "Science & Development" panel, Prof. Solomon Tesfaldet presented research on affordable electrochemical biosensors for drug detection, and Dr. Abraham Bahre proposed establishing local production facilities for sterile pharmaceuticals to reduce import reliance. Mr. Dawit Gebremichael explored the potential of information technology, big data, telemedicine, and drones to transform the Red Sea economy.

Conference Outcomes and Impact

The conference provided valuable insights across diverse sectors. The application of science and digital tools sparked lively debates, inspiring development planners in Eritrea. Sessions on archaeology and heritage unveiled ancient knowledge and highlighted the need to salvage Eritrean heritage documents globally. The quality of natural sciences papers was generally exceptional, and social sciences sessions, particularly those on Eritrea-Ethiopia relations and regional integration, were energetic and popular.

Roundtable discussions regarding ensuring access, equity, and quality education, garnered significant attention, prompting diverse opinions on the state of education in Eritrea. Another round table reviewed the role of women in national development, highlighting both their achievements and persistent challenges, such as limiting stereotypes and a sharp drop in female enrollment in higher education.

Feedback from attendees, collected via surveys and interviews, highlighted that some moderators exhibited inadequate time management and that the attempt to cover too many significant topics in a limited timeframe was perceived to have undermined the purpose of robust idea exchange. Inconsistency in paper quality and overlapping presentation times were also noted as challenges.

The conference was described as a historic event that satisfied both participants and the broader public. A key takeaway was the necessity of institutionalizing ICES to consistently identify research gaps related to Eritrea, thereby supporting national development plans and policy enhancements. The overall impressions from participants were positive, with a strong consensus on the importance of maintaining the conference's strengths, such as the diverse participation from experienced and junior researchers, as well as academic and non-academic representatives.

Future Recommendations

The organizing committee put forth several key recommendations to ensure the continuity and future enhancement of Eritrean studies conferences. The core recommendations focus on establishing a permanent framework for Eritrean studies.

Establish an Independent ICES Institution: The current reliance on an ad-hoc committee is seen as unsustainable. An independent institution could provide a stable and dedicated platform for sustained research and recurring conferences, advancing Eritrea's national interests.

Integrate ICES with Existing Research Institutes: Merging the conference with existing research entities, such as the Strategic Studies Center, would leverage existing resources and expertise, enhancing efficiency and impact.

Align ICES with the National Higher Education and Research Institution (NHERI): This proposal aims to embed the conference within the national research ecosystem by affiliating it with NHERI. This alignment would ensure that research remains a national priority and that the conference continues to thrive as a key academic event.

These distinct recommendations collectively suggest a strategic recognition of the imperative for long-term sustainability and institutionalization of Eritrean studies. This shift from ad-hoc events to a more structured, permanent framework is explicitly tied to broader national identity and foreign policy objectives. The commitment to preparing conference proceedings and hosting ongoing workshops further highlights a dedication to knowledge dissemination and continuous academic activity beyond the main event.

Key Takeaways from the ICES 2025 Conference

The International Conference on Eritrean Studies (ICES) 2025 held from January 4-6, 2025, in Asmara, was a landmark event that successfully fulfilled a dual mandate: upholding academic rigor while strategically projecting Eritrea's national narrative and development aspirations on a global stage. The conference served as a vital platform for fostering dialogue and global collaboration on research related to Eritrean peace-building and sustainable development. The event's intellectual breadth was notable, spanning nineteen themes from archaeology to artificial intelligence.

A key outcome was the identification of a strategic, long-term vision to move beyond an ad-hoc event model. A crucial suggestion was to institutionalize ICES as a regular occurrence every 2-3 years to consistently identify research gaps and support national development plans and policy enhancements. Recommendations for achieving this include establishing an independent ICES institution, integrating it with existing research institutes like the Strategic Studies Center, or aligning it with the National Higher Education and Research Institution (NHERI).

The conference demonstrated Eritrea's growing research capacity and its intellectual engagement with both national and international scholarly agendas. The depth and breadth of topics, expertise of speakers, and networking opportunities were all highly praised as relevant to Eritrea's knowledge-driven development agenda. The commitment to publishing proceedings and hosting ongoing workshops further solidifies the intent to build a sustained and impactful scholarly ecosystem for Eritrean studies.

Conclusion

The ICES 2025 conference was a successful and significant endeavor, demonstrating Eritrea's commitment to fostering academic discourse and leveraging scholarly research for sustainable development. The broad thematic coverage, encompassing both humanities and critical STEM fields, underscored a holistic approach to national development and a search for evidence-based solutions. The strategic recommendations for institutionalizing ICES are critical for ensuring its long-term continuity and enhancing its impact. These proposed structures aim to transition the conference into a permanent fixture within Eritrea's academic landscape, fostering a sustained research environment and consistently contributing to national interests and state-building efforts. The commitment to publishing proceedings and hosting ongoing workshops further solidifies the intent to build a vibrant and impactful scholarly ecosystem for Eritrean studies.

Archeology and Heritage

Asmara: Decentring Modern Heritage for a Planetary Age

Edward Denison

Architecture and Global Modernities
The Bartlett School of Architecture / UCL

Email: e.denison@ucl.ac.uk

Abstract

This paper explores the themes of just and constructive international engagements through cultural heritage and sustainable development in the context of Eritrea, building on Asmara's pioneering role in the promotion, protection, and contestation of modern heritage on the African continent. Building on the experience of Asmara's successful UNESCO World Heritage nomination in 2017, Africa's first and currently only site inscribed solely as 'modern heritage', this paper critiques Eurocentric notions of 'value' and the impact this has on the ascription not only to the tangible and intangible legacies of our recent collective pasts, but also on its application in non-European territories, whose modern encounters – whether positive or negative – are comparatively unrecognised on the World Heritage List. Asmara's example is particularly instructive, not only because of its architecturally conspicuous aesthetic and historical connections to coloniality and structures of power from Europe in the first half of the twentieth century, but also through its intra-African, trans-continental and extra-planetary experiences throughout the second half of the century. While the former reflects a type of contested heritage that is often mis- and under-represented on global registers, the latter offers novel ways of widening the conceptualisation of modern heritage beyond the European frame and ascribing it value. The paper argues that to recognise the true value of modern heritage in Asmara, and by extension Africa, it is necessary to decentre and reframe the conceptualisation of modern heritage by looking beyond the urban and the situated and by adopting an extractive and extra-planetary perspective for a planetary age. This wider and more plural approach supports the aims of the global collaborative, MoHoA (Modern Heritage of Africa / Modern Heritage in the Anthropocene); an equitable network of practitioners, academics, and publics engaged in the decentring of heritage theory and practice.

Keywords: modern heritage; Anthropocene; Africa; decentring; planetary.

Introduction

Responding to the conference's overarching agenda of global cooperation and sustainable development, and specifically to the complementary themes of cultural heritage preservation and development[1] and just and constructive international engagements, this paper builds on

1 As defined by the UN's Sustainable Development Goal 11 (Make cities and human settlements inclusive, safe, resilient and sustainable) and specifically Target 11.4 (Strengthen efforts to protect and safeguard the world's cultural and natural heritage), and associated targets, including 4.A in Education (to ensure that all learners acquire the knowledge and skills needed to promote sustainable development, including, among others, through education for sustainable development and sustainable lifestyles, human rights, gender equality, promotion of a culture of peace and non-violence, global citizenship and appreciation of cultural diversity and of culture's contribution to sustainable development) and

research spanning 25 years and critically reflects on the experiences leading up to and beyond Asmara's successful UNESCO World Heritage nomination in 2017. Inscribed[2] under Criteria 2[3] and 4[4] , 'Asmara – A Modernist African City,' was Eritrea's first World Heritage Site and Africa's first (and still only) solely 'modern' site. The rejection by UNESCO's Advisory Body, ICOMOS, [5] of Criterion 3[6] which sought to highlight the central role of Eritrean experiences in the creation of modernist Asmara expressed through constructional skills, labour, and materials has been the focus of critique in previous papers. These include 'The Value of Others: Modern Heritage and Historiographic Inequity' (Denison, 2022) and the Guest Editorials (Denison and Vawda, 2022 & 2023) of the two special editions of Curator: The Museum Journal published after the two MoHoA conferences, Modern Heritage of Africa (2021) and Modern Heritage in the Anthropocene (2022).Citing the Nomination Dossier submitted to UNESCO by the State Party of Eritrea in January 2016, the justification for nomination under Criterion 3 stated:

> As a well-preserved total urban landscape and modern capital of Eritrea, Asmara bears exceptional testimony to the universal aspiration for and attainment of national self-determination founded on the protracted development of indigenous cultural and political consciousness through multiple encounters with regional civilisations and colonial experiences. Asmara was once a centre of established cultural traditions and trading networks spanning the region, before becoming the locus of uniquely varied foreign agendas of global significance before and during the Second World War and throughout the Cold War … Asmara's complex encounter with modernity throughout the twentieth century is embodied in a modernist city outstanding not merely in urban planning and architectural terms, but also, crucially, in expanding the parameters of conventional modernist history.[7]

Developing this critique, this paper seeks to widen the spatial, thematic, temporal and conceptual framing of modernity by focussing attention on the under-researched topic of Asmara's experiences after Italian colonisation, during the nation's complex and protracted path to Independence in 1991. Particular attention is paid to how the United States of America, followed by the Soviet Union, exploited Eritrea's geographic, territorial, and political conditions for technological and geo-political gain and how the material legacies of these encounters have left their mark on Asmara's wider urban landscape. These marks, scars even, and the histories they reflect help to construct this expanded field of modernist history, elevating it literally and figuratively from the grounded, bounded, and physical terrain of architecture and urban planning to the invisible, infinite and immaterial realms of the inter-planetary. The underlying thesis therefore argues that the meaning and value of Asmara as a modernist African city can only be fully realised by adopting an expanded spatial, temporal and conceptual perspective that transcends constraining categorisations often used to

2 To get a site inscribed on the UNESCO World Heritage List, the State Party must prove 'outstanding universal value' under a minimum of one of ten criteria (six of these are intended for cultural sites and four are for natural sites).

3 To exhibit an important interchange of human values, over a span of time or within a cultural area of the world, on developments in architecture or technology, monumental arts, town-planning or landscape design.

4 To be an outstanding example of a type of building, architectural or technological ensemble or landscape which illustrates (a) significant stage(s) in human history.

5 International Council on Monuments and Sites. www.icomos.org

6 To bear a unique or at least exceptional testimony to a cultural tradition or to a civilization which is living or which has disappeared.

7 Asmara Heritage Project (2016), Asmara – Africa's Modernist City, UNESCO Nomination Dossier. 291.

define modern heritage, such as architecture, planning, the city, and the colonial, and instead presents an entire landscape of modernity extending to the extraterrestrial in which the definition and deployment of these terms are more fluid and contestable.

The paper concentrates on the experiences that support an expanded field of enquiry for modernist history and, by extension, modern heritage. In doing so, it owes a dept to the outstanding research by Harry Fisher, especially his MA studies titled 'The dependency of the Children of Monday: a study of the indigenous modernities in and around Asmara's Kagnew Station'.[8] It also draws into question current approaches to the assessment of modern sites and attempts to further support the central thesis of Asmara's nomination for a more plural conceptualisation of modern heritage not only in the context of Asmara's, but also beyond Europe, geographically, culturally, and conceptually. In aligning with sustainable agendas more broadly, an important aspect of this research confronting the post-colonial condition is to better understand and acknowledge the traumas inflicted on the people and the biosphere in the name of modernity and progress. It attempts to add weight to the proposition that Asmara's modern heritage, and the 'values' ascribed to it,[9] whether favourable or adverse, were not concentrated only in the Italian colonial period and ceasing with the defeat of Fascist Italy in 1941, but rather extended well beyond the 1940s, and up to Independence in 1991.

By critiquing Western notions of heritage 'value' and the impact of its ascription not only to the tangible and intangible legacies of our collective pasts, but also to its application in non-Western territories, where modern encounters are comparatively unrecognised on the World Heritage List, this paper employs Asmara's example to expose the Eurocentricity of how modern sites are assessed and valorised, and the consequences of this partiality. Beyond Western milieu, as evidenced in the case of 'Asmara—A Modernist African City,' modern heritage sites invariably possess aesthetic, cultural and historical connections to coloniality and structures of power from Europe. The effect is to privilege European experiences and values over those of others, especially local actors, limiting and even denying their experiences and therefore the true value of the heritage site.

This wider reading of Asmara's modern heritage, far from diminishing the recognised and recognisable cultural assets from the Italian era as decolonial critics might argue, would instead enrich them by adding cultural context and content within and beyond the temporal limits defined by a singular colonial encounter. It would also contribute to the important function of acknowledging the trauma that accompanies territorial occupation and colonial violence and its enduring impact racially, spatially, and environmentally. Cultural heritage is not a zero-sum process in which the recognition of one type of value diminishes another, but, like knowledge, is a process enriched by plurality. To deny or discourage this is to enable and encourage what I call 'historiographical inequity' – the deliberate and systemic denial of the historical experiences of the disempowered to sustain the cultural ascendancy of those in power.

Confronting this historical prejudice has been one of the main aims of the global collaborative, MoHoA (Modern Heritage of Africa / Modern Heritage in the Anthropocene),

8 Harry Fisher, The dependency of the Children of Monday: a study of the indigenous modernities in and around Asmara's Kagnew Station, essay submitted for Multiple Modernities Architecture, MA Architectural History, The Bartlett School of Architecture, UCL, 2023.

9 It is important to note here that heritage 'value' does not necessarily carry a positive or celebratory connotation, but can also apply to 'difficult', 'dark' or 'negative' heritage.

an equitable research network of practitioners, academics, and publics engaged in the decentring of heritage theory and practice (www.mohoa.org). One of the main outputs of MoHoA (www.mohoa.org) has been the *Cape Town Document on Modern Heritage*, a radical policy proposal presented to UNESCO advocating the reconceptualization of 'modern' heritage that is relieved of its Eurocentric, homogenous, universalising, developmental, and colonial associations, and is instead reframed as an inherently plural and planetary phenomenon.

Colonial Reverberations

Following Italy's surrender to the Allies in Asmara on 1 April 1941, it was not only the city of Asmara, which would later gain global recognition for its modernist architecture and urban planning, that got turned over to British administrators, but also the industrial, technological, and military sites beyond the city centre. As with Asmara's more recognisable modernist heritage assets and their stated attributes, some of these sites and their later extensions maintained these same characteristics defined by a 'wide-ranging encounter with modernity on a global scale, encompassing political ideology, technological development, cultural interaction and upheaval, and prolific artistic expression.'[10]

Most notable after the Second World War was the intersection of pioneering technological developments and Asmara's role in a growing constellation of global connections that quickly surpassed our planetary limits and extended into the extra-planetary realm. This expanded field of modernist experiences – from the planetary to the extra-planetary – recasts and enriches our understanding of modern heritage and Africa's role in this process.

For Asmara, the seeds of this expanded field were sown during the Second World War in the outskirts of the city's southern suburbs where the Italians established a naval radio relay station, Radio Marina. Asmara's auspicious location at the junction of three continents, 15 degrees latitude north of the equator, and its high altitude (nearly 2,500m above sea level) was perfect for sending and receiving long-distance radio signals. Echoing colonial practices everywhere, sites of territorial expropriation provided fertile grounds for accelerated technological development due to their exigency in serving the wider colonial agenda and to their expedience as sites of experimentation. These combined characteristics of technological need and experimentation are increasingly well-documented in varied colonial contexts, from Japan's South Manchuria Railway[11] to France's nuclear testing in the Algerian desert.[12] The same was also true of Italy in Eritrea before Radio Marina was established. Major investments were made not only in road and rail, but also in the world's longest cable car linking the port of Massawa with Asmara with a continuous cycle of 1,500 cars over 110km.

For Radio Marina, the acquisition of a site in Asmara's southern suburbs marked the start of a growing network of quasi-colonial territorial possessions that would eventually occupy multiple sites on Asmara's periphery in service to multiple extraterritorial powers and at multiple scales, from the terrestrial to the extraterrestrial. Following Italy's surrender in April 1941, the British Military Administration (BMA) took control of the three steel transmitter towers that comprised Radio Marina. This lasted only until 1943 when the Americans, who were now fully engaged in the war, acquired the facility on behalf of the United States War Department (USWD) and rebranded it Asmara Barracks, within Tract A, which also

10 Asmara Heritage Project (2016), Asmara – Africa's Modernist City, UNESCO Nomination Dossier. 31.

11 For further reading, see Denison and Ren, 'Ultra-Modernism – Architecture and Modernity in Manchuria'. (Hong Kong University Press, 2017).

12 For further reading, see Henni, 'Colonial Toxicity: Rehearsing French Radioactive Architecture and Landscape in the Sahara'. (If I Can't Dance, I Don't Want To Be Part Of Your Revolution / Framer Framed, 2024).

included the site of the American Embassy, half a kilometre south of the city centre. On 15 September 1945, this became the home of The Second Signal Service Battalion, redesignated the Army Security Agency (ASA).

Tests were conducted by the US Army on the technological potential of the 1000-watt AM radio transmitter by sending signals to Washington DC. The results proved so successful, the USWD ordered the expansion of operations. Over the summer, two officers and 44 enlisted men began training in the US National Security Agency (NSA) and United States Army signals intelligence and electronic warfare facility at Vint Hill Farms Station, Virginia (USA). By the end of the year, Radio Marina was staffed by four U.S. Army officers and 50 enlisted employees, with additional transmitter and receiver towers erected outside the city limits as part of an emerging web of global communications infrastructure linking Washington with New Delhi (India) and Tehran (Iran). Operations for these technologically advanced facilities were constructed in underground bunkers made of 10-inch bombproof concrete and functioned throughout the Second World War, after which they were deactivated and some of the equipment shipped back to the US.[13]

Kagnew Station

American disinterest in Africa was an enduring mark of the post-war era. Military aid to Sub-Saharan Africa between 1946 and 1973 accounted for 0.4% of the '$62.1 billion of U.S. foreign military assistance doled out worldwide during the period', the bulk of which went to Ethiopia and Zaire.[14] According to the Historical Office of the Office of the Secretary of Defense, 'Ethiopia received $181.6 million, mostly to allow continued access to maintain its only major military installation in the region, the Kagnew Station.'[15]

Kagnew Station became the appellation for the US Army base in Asmara, officially titled the 'United States Army Radio Station: Kagnew Station'. Although Kagnew Station comprised multiple sites within and beyond the city, the name became synonymous with the army base later established on the western edge of the city directly south of the former Italian fort and designated Tract E, later called 'New Kagnew Station'. Despite its American occupation, Kagnew Station derived its name from the Ethiopian word meaning 'to bring into harmony' and 'to bring order out of chaos'. Steeped in local modern history, the editor of the *Kagnew Gazelle*, a monthly newspaper printed at Kagnew Station, explained in the first edition:[16]

> Kagnew first appeared in Ethiopian history during the Battle of Adowa in 1896. Kagnew was the name of an Ethiopian general's horse that, riderless, galloped towards the attacking Italians heartening the Ethiopians into repulsing them. Successive attacks by the Ethiopian ... Jn 1936 when Ethiopia was again invaded by the Italians, an airplane named Kagnew flown by an Ethiopian, inflicted considerable damage before being destroyed ... In 1951, the Kagnew Battalion was sent to Korea as a part of Ethiopia's support of the United Nations.[17]

The US maintained modest operations in Asmara after the war, but the Korean War

13 John Rasmuson. (1973). A History of Kagnew Station and American Forces in Eritrea. Il Poligraphico: Asmara. 40-42.

14 Joel C. Christenson, Anthony R. Crain, and Richard A. Hunt. (Undated.) 'The Collapse of the Portuguese Empire and the Arrival of the Cold War in Africa,' Volume 8, Chapter 18. The Decline of Détente: Elliot Richardson, James Schlesinger, and Donald Rumsfeld, 1973-1977 (Vol. VIII). U.S. Government Publishing Office. 2.

15 Ibid.

16 In June 1963 the paper went bi-monthly.

17 Rasmuson. 48-49.

(1950-1953) reanimated their interest in and need for global communications facilities in the region. Coinciding with Eritrea's federation with Ethiopia in 1952, new sites were acquired for transmitter and receiver towers in a series of 'Tracts' beyond the western perimeter of Asmara (Fig.1). On 22 May 1953, a Base Rights Agreement was signed between the US and the Ethiopian Governments, permitting the US Army to lease the site as a base for 25 years.

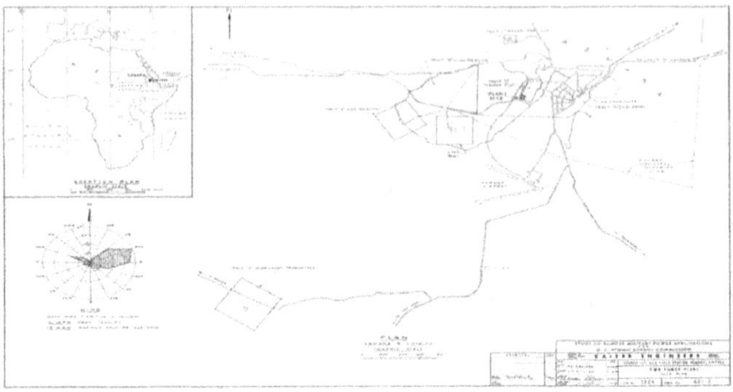

Fig 1. Map by Kaiser Engineers (1960) showing the successively acquired 'Tracts' (A-E) that comprised Kagnew Station.

Tract E, known by the acronym CINTIA during Italian occupation, had been used by the Italians as a site for 'heavy equipment and vehicle assembly complex and motor pool until 1942', which the British continued to use until 1952 'when it was abandoned'.[18] A pattern of use, or abuse, for this neglected site on the edge of the city was beginning to form that reflected different kinds of colonial trauma that remain evident to this day in the physical and spatial scars that mark this modern urban landscape. After the British abandoned the site, the US Army used it for training in mortar flares and a blank firing range before realising its potential as a site for a larger base. A formal lease agreement was signed by the US Army on 21 February 1955 and $8m was budgeted to convert it into a new army base by 1956.[19] Tract E would later be plugged into Eritrea's existing networks of modern infrastructure with connections to the main road to Massawa and a spur track to the Massawa to Keren railway line (see Fig. 1).

Rather than being a single compound, Kagnew Station grew to become a constellation of sites overlaid onto the civilian structure of Asmara that comprised one of the largest and most sophisticated US Army bases in Africa throughout the post-war period. By the late 1960s, Kagnew comprised 10 tracts covering a total area of 3,400 acres and supported: an atomic energy detection facility; two Army Security Agency stations; a US Navy communications station; a US Army strategic communications Defense Communications Agency facility;[20] a communications facility for the Diplomatic Telecommunications Service; the US Consulate General in Asmara; and the Stonehouse facility. At its height it was responsible for over 6,000 service personnel and their dependents. Facilities included a school, chapel, outpatient clinic, laundry, housing, gymnasium, boxing ring, 10-lane bowling alley, shooting range, racing track, miniature golf course, three night clubs (Officers' Open Mess, Top Five and Oasis), a newspaper, radio station, and the world's first Army-operated television studios broadcasting 11-hours a day.[21]

21 KANU-TV was one of the first occupants of the Tract E site, broadcasting from 1 December 1955.

Atomic Asmara

Kagnew Station's rapid growth exerted unsustainable pressures on Asmara's urban and natural environments. Water scarcity was omnipresent. With the city's storage capacity in localised reservoirs around 528m gallons, Kagnew Station's daily consumption of 145,000 gallons imposed a permanent pressure on the local population. Power was also unreliable, with municipal supply not sufficient for the projected demands of new technological equipment such as radar. Asmara's experience was not isolated, as the reach and demands of US power expanded throughout the post-war era. Consequently, on 17 August 1959, the US Atomic Energy Commission contracted Kaiser Engineers to conduct 'an evaluation of the construction and operation of nuclear power plants at several designated military installations where increased power generating capabilities of 5 to 40 mwe may be required by the Government for the period 1963 through 1970'.[22] The selected sites were Okinawa (Ryukyus Island); Guam (Mariana Island); Thule (Greenland); NORAD (USA); Asmara (Eritrea); Super Sage (USA); Nike-Zues (USA); Inchon (Korea); Atlantic Barge (Atlantic Ocean); and McMurdo Sound (Antarctica), while the following three sites were discarded: Pacific Barge (Pacific Ocean); Panama, (Canal Zone); and Hokkaido (Japan) (Fig.2).

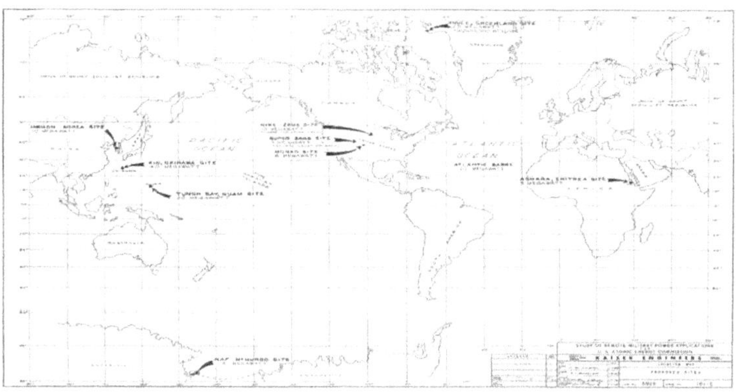

Fig.2. Study of Remote Military Power Applications for U.S. Atomic Energy Commission, Kaiser Engineers (1960), showing Asmara as one of 10 potential sites globally. Credit: Kaiser Engineers.

Owing to power shortages at Kagnew, 'The Fourth U. S. ASA Field Station, Asmara, Eritrea, was one of the sites selected for further study. Power requirements for this location were established at 5 mwe (firm) in 1963.'[23] In 1960, Kaiser Engineers produced a detailed report for the design of a 5MW nuclear power station to be constructed on the southwest corner of Tract E, including detailed drawings of the reactor (Fig.3-6). This site was specifically chosen because of its central location in the constellation of tracts dotted around Asmara and within a 15-mile radius. 'With the exception of the nuclear reactor, no special construction problems [were] anticipated'. [24] The proposed reactor was a pressurised water reactor, designated SM-2. The reactor, comprising a matrix of highly enriched uranium oxide (UO2), was

22 Kaiser Engineers, Study of Remote Military Power Applications Report No.6 Fourth U.S. ASA Field Station, Asmara, Eritrea, for United States Atomic Energy Commission, New York Operations Office, New York. January 1960 (Revised July 1960), Henry J Kaiser Company, Oakland, California. 1.

23 Ibid.

24 Ibid. 2.

to be housed in a containment vessel over 19m high and 8m in diameter, fabricated in steel plate up to 3.2cm thick. The fuel plates, containing '0.08 grams of boron-10 as a burnable poison in the form of 'natural' zirconium diboride (ZrB2)',[25] would need replacing every 16 months. Radioactive waste was to be stored on site in a 5,000-gallon stainless steel tank and spent fuel rods dispensed into on-site reinforced concrete pits. Liquid waste was to be dispensed into the sewer and gaseous waste emitted through a stack beside the containment vessel. Six radiation monitors were designed to measure high radiation levels from 'the main steam line, de-mineralizers, vent stack, containment vessel vent and spent fuel pit'.[26]

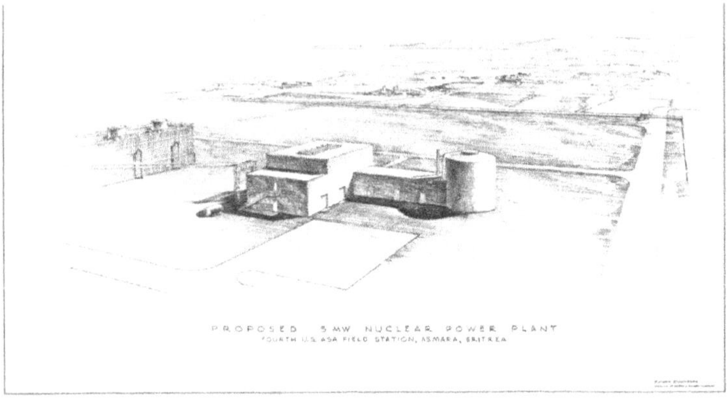

Fig. 3. Proposed 5MW Nuclear Power Plant at Asmara, by Kaiser Engineers (1960). Credit: Kaiser Engineers.

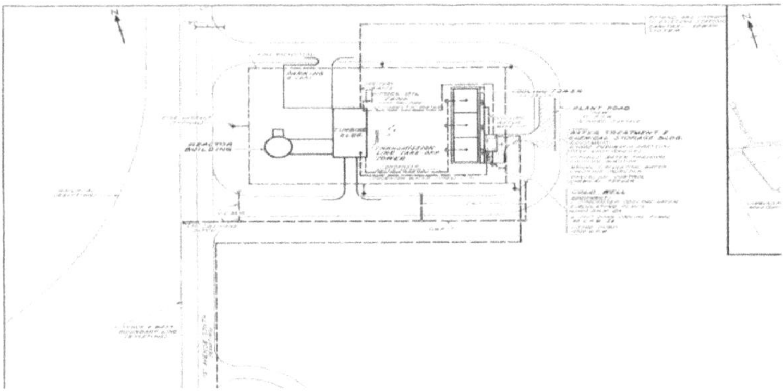

Fig. 4. Plot Plan of the Proposed 5MW Nuclear Power Plant at Asmara, by Kaiser Engineers (1960). Credit: Kaiser Engineers.

25 Ibid. 46.
26 Ibid. 31.

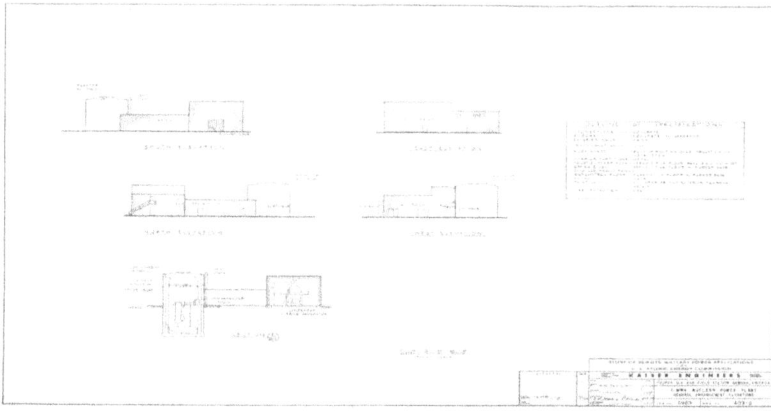

Fig. 5. Elevations and Sections of the Proposed 5MW Nuclear Power Plant at Asmara, by Kaiser Engineers (1960). Credit: Kaiser Engineers.

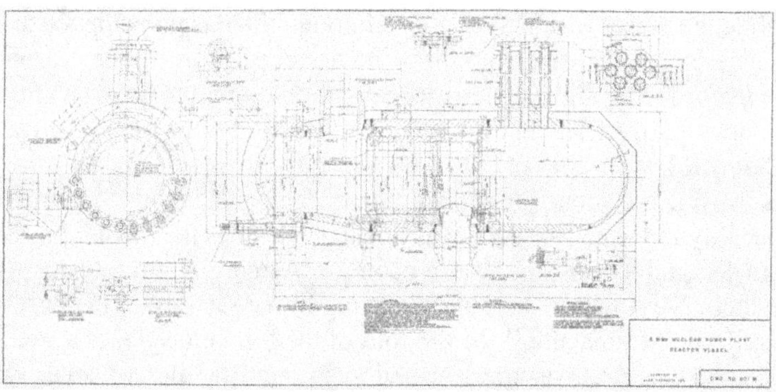

Fig. 6. Sections of the Reactor Vessel in the proposed 5MW Nuclear Power Plant at Asmara, by Kaiser Engineers (1960). Credit: Kaiser Engineers.

The design of Asmara's proposed nuclear reactor was a larger and improved version of the SM-1 operating in Fort Belvoir, Virginia (USA), [27] which 'despite having no experience building nuclear reactors, the Army and its contractor completed in only 18 months using standard Corps of Engineers contracting and management procedures.'[28] The feasibility of Asmara's nuclear reactor was made possible by a $3,500,00 collaboration between The United States Atomic Energy Commission and the Department of Defense who commissioned Alco Products, Inc. to develop and design a small scale reactor for deployment overseas before 1 September 1959. The reactor would therefore be purchased and manufactured in the US before being shipped to Eritrea.

Construction was expected to be carried out by US contractors with the support of local workers. In their report, Kaiser Engineers painted a disparaging picture of local workers[29] that contrasted with the experiences of the US marine salvage expert, Commander Edward Ellsberg, hired in 1941 to put the port of Massawa back into working order after the Italians had destroyed it upon their retreat. 'Eritreans in Massawa did their bit to win the war', wrote Ellsberg, adding somewhat prophetically, 'it should not be forgotten when some day around some table in the United Nations delegates meet to decide the future of Eritrea.'[30]

30 Edward Ellsberg. (1946). Under the Red Sea Sun. (Dodd, Mead & Co., New York). 176.

By contrast, Kaiser's report offers an insight into the state of the construction industry in Eritrea in the late-1950s and the long shadow cast by colonialism and its racial laws that prevented Eritreans from participating in, less still having agency in, the construction trades, as well as reinforcing the racist attitudes prevalent in white American and European culture at that time. Besides the inevitable fact that Eritreans were 'not acquainted with American methods of construction' or 'U. S. specifications and standards', 'English, Danish, Italian and Israeli contractors' were said to be responsible for carrying out the larger construction projects in Eritrea and 'with the exception of brick, sand and gravel and some cement products, building materials are generally imported from European, Asian and some African sources.'[31] Even the Italian craftsmen whose skills and labour has since been credited for the construction of modernist Asmara were reported 'to be versed solely in hand construction methods' and therefore 'a technical project, such as the construction of a nuclear power plant, would have to rely heavily on imported U. S. supervision and European labor in the more skilled trades.'[32]

In their report, Kaiser Engineers offered a cost comparison between a nuclear power plant and a conventional diesel plant. The estimated total cost of the nuclear plant was $5,800,000, while the diesel was $2,915,000. Although there is no evidence to prove conclusively the grounds for the decision to opt for the diesel plant, it can be reasonably assumed that cost was a major factor. Although it was never built, it would have been the first nuclear power plant to operate in Africa and the second of its kind to be deployed by the US Army anywhere in the world.

Besides power, the most pressing problem facing Kagnew Station throughout the 1950s and especially in the early-1960s was housing. Besides the on-site barracks, the 'insatiable American demand for housing' was a boon for 'Asmara's entrepreneurs [who] took advantage of the lucrative situation' and built housing in the vicinity of the bases, in particular Tract E and the Radio Marina site.[33] At the foot of the hill below Tract E, the southern suburbs of Tiravolo and Gaggiret was transformed by the proliferation of small villas contained within garden compounds, the architecture a distinctively flamboyant angular development on the more austere orthogonal early-modernism found in the centre of town (Figs. 7-10).

31 Kaiser Engineers. 15.
32 Ibid. 16.
33 Rasmuson. 53.

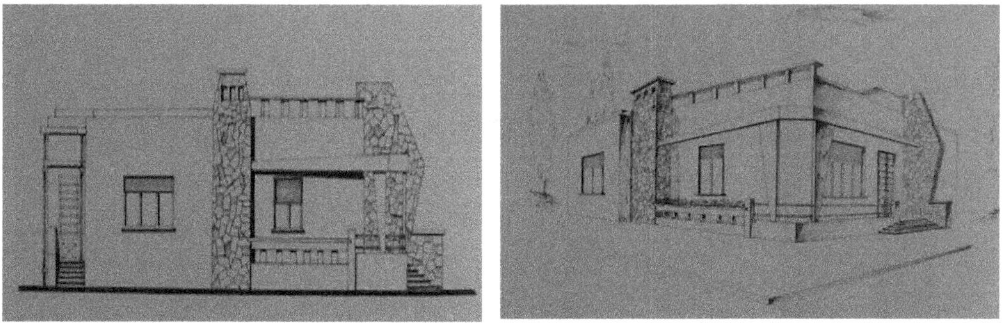

Figs 7-10. Elevations and perspective views of proposed villas and apartments in Gaggiret (top tow) designed by Augusto Alfano and in Tiravolo (bottom row) designed by Alberto Amighini in the 1960s. Credit: Municipality of Asmara Archives.

Stonehouse

Kagnew Station reached its peak in the mid-1960s with the addition of a new facility that would transform its function and its role in the modern history in and beyond this world. In January 1964, the US Department of Defense announced the construction of Stonehouse, a deep space listening station that would join a string of similar facilities in Turkey, Pakistan and Japan that, combined with sites in North America, encircled the globe to give the US unbroken radio coverage of outer space. Stonehouse was located on the old Radio Marina transmitter site, northwest of the city centre. In 1961, after the old transmitters had been relocated to a new site to the southwest of the city, Tract D, the old site was converted into Kagnew Station's primary recreational facility, harbouring a rifle and pistol range, a skeet range and 'the Afro-American Racing Club's banked-dirt oval track.'[34] However, despite serving a large and restless population of US military personnel, these facilities were sacrificed to serve the loftier ambitions of the US Intelligence and the burgeoning space programme.

In 1960, the National Security Agency (NSA) had established the Space Surveillance Signals Intelligence (SIGINT) Program, which immediately began searching for ideal sites to locate its deep space communications facilities. The first were part of the BANKHEAD facilities, which established sites in Peshawar (Pakistan), Chitose (Japan) and, later, Sinop (Turkey), designed to track launches from Soviet space and military sites. In early 1963, a facility in Asmara was being proposed to 'collect and exploit data transmissions from Soviet lunar spacecraft before they could be obtained from NASA's own lunar exploration program. The data would be of great value in the Apollo manned lunar landing program'[35] and later served as a tracking station for the Apollo space missions.[36] NASA concluded: 'The proposed NSA facility at Asmara should be accelerated. The NASA is willing to assist the NSA in this regard, if desired by the NSA.'[37]

In April 1964, $40m was assigned to the development of the equipment needed by the Space Surveillance Signit Program (Phase 1) at Stonehouse. The 46m antenna and 26m rotatable parabolic dish were prefabricated in sections in the US and shipped to Massawa, from where they had to be transported by road to Asmara, using cranes to lift them over the railway bridges (Fig.11). The equipment was installed and operational on 17 May 1965, their futuristic forms providing 'a visual juxtaposition of the modern world against the ancient Ethiopian plateau.'[38] (Fig.12).

38 Nixon. 30.

Fig. 11. Sections of the support structures for the Stonehouse's antennae and the rotatable parabolic dish being lifted over the Massawa-Asmara railway line on the Massawa Road (c.1964). Credit: Public Affairs Office, United States Army Garrison, Kagnew Station.

Fig. 12. Priests and other onlookers outside the Stonehouse deep space monitoring facility. Credit: Patrick Striplin.

Stonehouse had various functions, but its main purpose was a deep space monitoring facility to track Soviet satellites and space probes (Fig.13). Asmara was ideally located for this newfound role in the burgeoning planetary communications race. Stonehouse was well suited not only because of the relatively low electronic noise and high altitude that made it ideal for sending and receiving radio signals, but it was also located due south of the Soviet deep space and planetary radar station at Yevpatoria (Ukraine) (Fig.14), enabling the US to listen to all Soviet communications with their spacecraft. Through Stonehouse, the US was therefore able to get a comprehensive understanding of the Soviet lunar and planetary programmes, including communications with probes sent to Mars and Venus.

Fig.13. The 46m antennae and 26m rotatable parabolic dish at Stonehouse, Asmara. Credit: The US National Security Archive.

Fig.14. Former Soviet deep space and planetary radar station at Yevpatoria (Ukraine). Credit: Wiki-commons.

Stonehouse can be seen functionally, technologically and spatially, as an extractive extension of the Italian colonial occupation, characterised by the exploitation of others' land in service to a foreign power. It is a condition that has marked Asmara, and by extension Eritrea, for centuries, given their respective locations at the intersection of three continents. As occurred throughout history, the arrival of one foreign power was inevitably followed by their departure and the arrival of the next. This was true of Kagnew Station, although America's departure was accelerated by a combination of local and global events.

Although the Pentagon had planned to withdraw from Eritrea between 1976 and 1978, reduced defence spending after the Vietnam War and the growing counterinsurgency by the Eritrean Liberation Front (ELF) against Ethiopian forces around Asmara in early 1975, hastened the withdrawal. The ELF had been able to take advantage of the instability caused by the overthrow of Emperor Haile Selassie's government and the monarchy by the Communist Derg in late 1974. Wary of this threat, the aging Emperor had visited President Nixon on 15 May 1973 to seek military support for his government. His pleas were partially answered, with the US promising F-5E fighters, M-60 tanks, Sidewinder missiles, and Vulcan air defence systems, only some of which was ever delivered. Events in Ethiopia and Eritrea eventually overtook the sluggish supply of military aid. On 4 March 1975, Stonehouse ceased operations, in the same month the Derg declared Ethiopia a socialist state. On 12 December the ELF attacked the Navy Transmitter site at Tract D, capturing two Americans and destroying the transmitter tower. The Departments of State and Defense began planning for the orderly decommissioning of Kagnew Station over the coming years. However, the deployment of new satellites that made land-based communications facilities largely redundant and the worsening situation in Eritrea by 1977 meant that US began winding down the facility. In early 1977, they began evacuating remaining staff and dependents and destroying intelligence equipment that was too costly to ship back to the US, including the KW-26, KG-13 and KW-7 cryptographic equipment for tele-typewriters (Fig.14)[39] On 29 April 1977, the last Americans left and Kagnew Station closed.

Fig.14: Destruction of classified documents and equipment at the foot of the 46m antennae at Stonehouse prior to the US Army's eventual withdrawal from Kagnew Station in 1977. Credit: US National Security Archive.

Throughout the late 1970s and 1980s, Eritrea experienced some of the most violent and brutal phases of the Struggle, as the Soviet Union poured military hardware and personnel into the region in support of the Communist Derg regime. One witness to this barbarous campaign was the Ukrainian writer Serhiy Synhaivsky, formerly a military translator in Ethiopia from 1984 to 1987. His experiences during this tour of duty informed and inspired his novel Road to Asmara (2016, orig. Дорога на Асмару), in which the protagonist 'realized that he has become part of a complex and cynical game, that pumps weapons to a bloody regime.'[40] It is no coincidence that Synhaivsky, like the Australian author Thomas Keneally before him, inscribed Asmara in the title of his book with an aspiration prefix. [41]

Although relatively under-researched, especially in terms of heritage, the legacies of this terrible period of Eritrea's intra-African colonial conflict can be read in Asmara's urban landscape. Nowhere is this more striking than in the Tank Cemetery, a vast collection of Soviet-era hardware amassed and piled up on the land surrounding Kagnew Station's former Tract E (Figs.15-16). This monument to one of the darkest periods of Eritrea's modern history embodies the spatial reverberations from successive and interwoven phases of different forms of colonialism. More than the modernist architecture for which Asmara earned its titled on the UNESCO World Heritage List, the Tank Cemetery more completely encapsulates the range of experiences of modernity through colonial occupation, appropriation, and violence and their legacies etched in spatial trauma. It is no coincidence, that this monstrous scrapheap lies within a few metres of a former Italian machine workshop that was inherited by the British and converted into a firing range by the Americans before being developed into one of the most sophisticated army barracks in the world. A similar fate can be observed spatio-functionally at the sites of other Tracts, where the shadows of former extraterritorial areas are cast long into the future (Fig.17).

Fig.15: The snaking outer perimeter wall of the Tank Cemetery constructed of tanks, armoured personnel carriers and other military hardware from Eritrea's Struggle for Independence. Credit: Edward Denison 2022.

Fig.16: Armoured personnel carriers and other military hardware from Eritrea's Struggle for Independence stacked up inside the Tank Cemetery on Asmara's western outskirts. Credit: Edward Denison 2022.

Fig.16: The central hinge bracket of the 26m rotatable parabolic dish at Stonehouse, now in a transport depot outside Asmara.

Conclusion

This paper has attempted to present a case for studying the modernist heritage site of Asmara as an expanded field of enquiry that transcends conventional categorisations and designations based on architecture and urban planning to accommodate the wide range of other factors that make the city – as a landscape of modernity – an outstanding, indeed exceptional, example of modern heritage. By presenting Asmara as a landscape of modernity that transcend the planetary, the paper argues that these wider experiences should be acknowledged when seeking to ascertain the true value of its modern heritage.

It is hoped that this example serves to support the adoption of the *Cape Town Document on Modern Heritage* (2022) developed by the global collaborative MoHoA (Modern Heritage of Africa / Modern Heritage in the Anthropocene), a decentring and restitutive framework for an equitable and sustainable approach to the theory and practice of modern heritage in a planetary age.[42] In line with the aims of the *Cape Town Document on Modern Heritage*, it calls:

> … for a reframing of how modern heritage is defined, identified, conceived, valorised, protected and safeguarded. The Document responds to the paradox that modern heritage is simultaneously of modernity and yet threatened by its consequences. By acknowledging the role of inequitable and unsustainable practices that gave rise to and often characterised modernity—slavery, imperialism, colonialism, and fossil fuel dependence—the *Cape Town Document on Modern Heritage* is a decentring and restitutive framework aimed at sustaining modern heritage for future generations in a more cohesive, inclusive and equitable way.[43]

By adopting an expanded perspective, it is possible to acknowledge the city's proud and often difficult history as a site of pioneering technologically experimentation that, although enabled by colonial and quasi-colonial violence, connected Asmara to the cosmos and in so doing gave Asmara a role in the subsequent global communications revolution that we now take for granted. As the unknown author of one declassified US National Security Agency Report wrote: 'Advanced systems [of communication] of today and the future evolved from lessons learned at Asmara.' [44]

---

42 For a more detailed account of the draft Cape Town Document on Modern Heritage, see the special edition of Curator dedicated to the first conference of Modern Heritage of Africa / Modern Heritage in the Anthropocene (MoHoA), Modern Heritage of Africa, hosted by the University of Cape Town in 2021: Shahid Vawda and Edward Denison, "Towards the Cape Town Document on Modern Heritage," and "Co-Curating the Cape Town Document on Modern Heritage," Curator: The Museum Journal 65, no. 3 (July 2022): 485–496; 497–508., resp.

43 MoHoA, Preamble, The Cape Town Document on Modern Heritage, draft text agreed after the Expert Meeting hosted at Liverpool University, October 29, 2022.

44 Author anonymised. STONEHOUSE: The first collected of [redacted] signals. Declassified report by the United States' National Security Agency (NSA), released on 18 September 2007, FOIA #7319.

30 years of geo-palaeontological research in northern Danakil depression of Eritrea: The Late Neogene Buia and Engel Ela - Ramud basins

Rook L.[1], Martinez-Navarro B.[2,3,4], Medin T.[5]

[1]*Dipartimento di Scienze della Terra, Paleo[Fab]Lab, Università di Firenze, Italy*
[2] *IPHES-CERCA, Institut Català de Paleoecologia Humana i Evoluciò Social, Tarragona, Spain*
[3] *Area de Prehistoria, Universitat Rovira i Virgili (URV), Tarragona, Spain*
[4]*ICREA, Barcelona, Spain*
[5]*Commission of Culture and Sports (Eritrea), Asmara, Eritrea*

Abstract

We summarise 30 years of research in the Late Neogene of the northern Danakil Depression of Eritrea (Buia and Engel Ela-Ramud Basins). These research activities began in 1994 at the Buia Basin as a joint project between the Asmara Department of Mines (Eritrea) and the Earth Sciences Department of the University of Florence (Italy). It was mainly focused on geological surveys.

The first surveys carried (late 1994–early 1995) resulted in the recognition of a Pleistocene sedimentary succession, some tens of kilometres south of the Buia village, rich in fossils and stone tools. Interest in the area increased tremendously with the discovery in late 1995 of an almost complete fossil human calvarium. This gave a much larger focus to the project and requested enlarging the collaborations to a wider spectrum of disciplines and institutions. On the Eritrean side, the renewed project had (and has) main counterparts in the National Museum of Eritrea, in the Eritrea Commission of Culture and Sports, and in the Northern Red Sea Regional Museum of Massawa.

With the same goals as the Buia Project and in collaboration with it, an independent Eritrean–Spanish Project began in 2012 to study the Early Pleistocene basin of Engel Ela-Ramud, located 30 km south of Buia. It started with the association between the National Museum of Eritrea and the Catalan Institute of Human Paleoecology and Social Evolution of Tarragona (Spain). Since 2016, the Eritrean Commission of Culture and Sports and the Northern Red Sea Regional Museum of Massawa have also been involved in this research.

In addition to the mentioned scientific outputs, both projects have acted over the years as cooperative platforms for sustainable development. This demonstrates how culture and research initiatives can have immediate returns in terms of sustainable development.

Keywords: Fossil vertebrates; Fossil *Homo*; Stone tools; Early Pleistocene; Sustainable Development.

Introduction

The search for hominid remains is one of the objectives of several international research teams. However, this type of research activity is difficult and seldom results in new findings, which become precious due to their rarity. Based on relatively few and, in most cases, extremely fragmentary remains, the available modern technologies allow researchers to derive crucial information for reconstructing phases and patterns of human evolution. The reconstruction of palaeoenvironments and the study of Palaeolithic artefacts (in most cases associated with fossil remains) provide important information for evaluating the cultural placement of hominids and for understanding the factors that drove hominid dispersal across the African continent and over into Eurasia.

The collaborations between the authors of this contribution provide mutual benefits for unique research projects involving these lines of research. The contributors include an international collaborative research group active in Eritrea, who have, to their credit, both palaeoanthropological and palaeontological findings in the Danakil depression of Eritrea. Thanks to these results, the Eritrean-Italian and Eritrean-Spanish teams can be aligned with other teams from the United States, United Kingdom, Germany and France, with a long tradition in research on palaeoanthropological sites and their geological context. Our team includes specific and different expertise on the study of physical and (palaeo-) biological environment in several areas of east Africa, on chronologic placement of sedimentary successions and geologic events, on palaeontology and palaeoanthropology and on Palaeolithic archaeology.

The present contribution is intended to offer an account of the history of this international group over 30 years of geo-palaeontological research in the Danakil depression of Eritrea. It first introduces an overview of the international framework aimed at placing these activities into a wider international context.

Geo-Palaeontological and Palaeoantrhopologiocal Research: Why in Eritrea?

Prior to the 1990s, Eritrea was unexplored (or under-explored) regarding its Plio-Pleistocene palaeontology and palaeoanthropology. This could be considered surprising, especially due to the country's geographic location in proximity to regions like Ethiopia or Kenya, which are well known for their celebrated palaeoanthropological records. Thanks to our activities, the region of the northern Danakil depression has been identified as an extremely promising area for Plio-Pleistocene geo-palaeontological and palaeoanthropological research.

East Africa is the region where most studies on early hominisation, on Pliocene and Pleistocene peopling and on early bio- and cultural human evolution were initiated and developed. From this region, the main phases of the so-called 'out-of-Africa' dispersal of our ancestors started and possibly were repeated over time. The latest Pliocene to Early Pleistocene interval represents a timespan when crucial evolutionary phases in human evolution occurred, but the palaeoanthropological record of this timespan remains relatively scarce.

The last decades have seen remarkable new findings across Europe (Italy, Spain) or at the gates of Europe (Georgia) that have yielded clear new evidence of a human dispersal out of the African continent that occurred since the Early Pleistocene, at about 1.8 Ma. Remarkable in this respect are the 1.8 Ma human remains and Palaeolithic artefacts from Dmanisi (Georgia). These new data allow the drawing of a new scenario to understand early human dispersal(s) out of Africa towards Eurasia that incorporates a wider context of vertebrate fauna dispersal (and palaeoecological and palaeoclimatological evolution).

In this context, the 1995 discovery in the Buia sedimentary basin (the Dandiero valley and the northern Danakil Depression) of an excellently preserved calvarium dated approximately 1 My, referred to as *Homo* aff. *H. erectus*, is of extreme relevance. During the Early Pleistocene, up until one million years BP, the *Homo* genus is represented in North-Eastern Central Africa by some specimens, commonly referred to as *H. erectus/ergaster* (or *H. erectus* sensu lato), which seem characterised by a relevant morphological and anthropometric variation.

Buia represents the northernmost evidence of the East African Plio-Pleistocene palaeoanthropological record. (Buia is located about 500 km Nord to Daka, in Ethiopia, where an almost coeval cranial finding is documented.) Buia thus represents a key area for the study of the timing and patterning of human dispersal pathways out of the African continent. North of Buia, the first Plio-Pleistocene evidence available is at Ubeidiya and Gesher Benot Ya'aqov, in the southernmost portion of the Eurasian continent in Israel. Here, the Levantine corridor is well known for the occurrence of mammal faunas and hominids dispersing out of Africa towards Eurasia. In fact, the documentation of human activity in these Pleistocene Israel sites is evident proof of the actual dispersal routes from the African continent throughout the Dead Sea graben. Acceleration in the vertical tectonism, which produced the present physiography of the region, began slightly before humans had started their way northwards out of Africa. The first human dispersal began around 1.8 Ma, coinciding with the intense climatic and environmental changes that started during the Plio-Pleistocene transition.

The most likely *Homo erectus* dispersal routes out of the African continent are traceable along the northernmost East African Rift (including the Danakil Depression and the Buia area) throughout northeastern Sudan and the Red Sea coastal regions until the Dead Sea and the Levantine corridor. These are actually the northernmost sector of a longer pathway that starts from Tanzania and extends across Kenya and Ethiopia and can be detected along the east African rift by the widely known documentation of celebrated archaeological and palaeoanthropological sites. As a matter of fact, the morpho-tectonics and climatic–environmental conditions that favoured northward dispersion occur noticeably within the complex of African rifts. The African rifts are characterised by frequent tectonic activity due to the sinking of the rift floor and uplift of the margins. The occurrence of transverse tectonic structures and/or volcanic structures accentuates the environmental differentiation within each basin and determines the high environmental variability and, consequently, favours the increase in biodiversity. The continuous uplift of rift shoulders creates orographic barriers favouring the condensation of humid air masses. Morphological differences in height and intense atmospheric precipitation thus cause erosive processes that result in rapid sedimentation of detrital successions and have good preservation potential for fossils.

Sedimentary Basins and Palaeoanthropological and Archaeological Sites Along the East African Rift from Tanzania to Eritrea and Beyond

These types of morpho-tectonic and climatic-environmental conditions are typically found along the entire East African Rift, from Tanzania to Eritrea. Several hominid or Palaeolithic sites are known within these rift sedimentary basins. Here, we summarise the main characteristics of these basins, from Tanzania to Eritrea, to provide a regional reference frame.

In this context, the southernmost basin is Olduvai. The celebrated Olduvai Gorge has yielded a number of hominid remains and abundant Palaeolithic industries. Beds I to IV cover the Plio-Pleistocene transition until about 0.7 Ma. Bed I is made up of a fluvio-lacustrine succession rich in large mammals (including hominids) and birds. It transitions into a clay–

sandy succession (Bed II), representing the depositional phases of rivers that, after flowing across wide savannahs, flowed into the lake. Bed III, characterised by reddish clay sands and conglomerates, testifies to the occurrence of a climatic–environmental change towards more arid conditions and playa lakes. Bed IV, again, represents a depositional environment characterised by water abundance, with rivers and lakes. Moving northwards to the area of the Turkana basin (northern Kenya), the Early Pleistocene times are documented by sedimentary successions (rich in palaeolithic industries) deposited in shallow lake environments and then pass to marginal lakes and aeolic deposits.

The Omo basin, in southern Ethiopia, is located in the southern part of the Main Ethiopian Rift and is characterised by a series of semi-grabens filled by fluvial, lacustrine and fan-delta deposits, with frequent interbedded volcanic layers. This area is known for hominids, including *Homo erectus*, and widespread mode I and mode II Palaeolithic industries. Further north, within the Main Ethiopian Rift, similar depositional conditions are reported at Gadeb and Kesem-Kebena. At these sites, the occurrence of water bodies (rivers and lakes) also seems to be a crucial aspect for human occupation in the Early Pleistocene. In fact, the high-standing level of water in the lakes and the widespread volcanism along the lowermost (topographic) rift parts may be responsible for the marginal distribution (with respect to the Rift axis) of some sites, such as Konso-Gardula and Melka-Kunturè.

Further north, in the Afar area, several palaeoanthropological sites are reported in the literature. Among them, we stress here the importance of the deposit outcropping at Daka/Bouri (the middle Awash valley). There, within the deltaic sand belonging to the Dakanihylo Member (Bouri Formation), a level dating back to 1 My yielded a *Homo erectus* calvarium that was associated with Oldowan (mode I) and Acheulean (mode II) Palaeolithic industries. The Bouri mammal assemblage, dominated by bovids, seems indicative of open savannah spaces with available water bodies.

This short review of the Plio-Pleistocene East African sedimentary basins ends with a mention of the first basins to yield palaeoanthropological evidence found as we move northwards from East Africa. These are the Ubeidiya and Gesher Benot Ya'aqov sites, found in Israel along the Levantine corridor. Here, the geological and palaeoenvironmental contexts were, by and large, comparable to those occurring in the African Rift, as they were also located within a rift depression in a marginal lake context.

Located in an intermediate position, north of the Ethiopian sites and south of the Israeli ones, the Buia and Engel Ela-Ramud basins play a crucial role in depicting the framework of Plio-Pleistocene human dispersal and evolutionary histories.

The Buia and Engel Ela-Ramud Basins: Their Research History

The first 10 years of the Buia project (1994 to 2004)

The fieldwork research we are covering in this contribution began in late 1994 as a joint project between the Asmara Department of Mines and the Department of Earth Sciences of the University of Florence. This first initiative (led by Dr. Alem Kibreab of the Asmara Department of Mines and by Prof. Ernesto Abbate from the University of Florence) was mainly focused on geological surveys and regional geology (*The Danakil Horst and adjoining area project*). From late 1994 to early 1995, the first field survey led to the identification of a Plio-Pleistocene fossiliferous succession (mammals and other vertebrates) located 20 km south of the Buia village in the northern Danakil (Afar) depression (Figure 1).

Figure 1: Landscape view allowing to appreciate the extension of outcrops in the Buia area

Interest in this area increased with the discovery in late 1995 (during the second field campaign in the area) of an almost complete human calvarium, as well as several sites with widespread and abundant Oldowan and Acheulean stone tools. Since the discovery of the fossil mammals and palaeoanthropological and archaeological remains, the project has become of much wider interest and has since been enlarged to include further collaborations, such as the National Museum of Eritrea (directed by Dr. Yosief Libsekal) and the Archaeology & Anthropology, and Biology Departments of the University of Asmara, as well as other European researchers and institutions within a 'new' project (the *Buia Research Project*). At that time, the project was approved by the Research Committee of Asmara University (whose chairperson changed over time, being first led by Dr. Azieb Okbagìhebriel, then by Dr. Ghebrehabram Oqubazaghi and then by Dr. Zemenfes Tsighe).

The Buia project activities were not limited to developing research but were also aimed at providing adequate structure and capacity building. Overall, the activity of the Buia project, which gave equal importance to developing structure and capacity building and research, was considered pivotal for future research projects in Eritrea and for the management of the palaeontological, palaeoanthropological and archaeological heritage of this country. Venues within the Eritrea National Museum were refurbished by the project to set up a lab oratory for the storage and preparation of fossils (including a safe for the conservation of unique specimens, such as the human remains) (Figure 2). An important point to stress is that the entire preparation and restoration of fossil material (cleaning, hardening and preservative treatments, moulding and casting, etc.) has been realised in Asmara thanks to the refurbished laboratory and to scholarships supporting the training of Eritrean Museum personnel (the late Habtom Kashai) in Italy.

Figure 2 Lorenzo Rook handling the Buia human calvarium to the Ministry of Culture, Minister Osman Salih, and Yosief Libsekal, in the venues of the palaeontological laboratory at the Eritrean National Museum (November 2006).

Under the energetic and strong organisational skills of Prof. Abbate, the aspect of capacity building was one of the main purposes of the project. One of the talented young geologists of the Asmara Department of Mines (Dr. Tewelde Medhin Tecle) was granted a PhD scholarship at an Italian university, and a Buia field school was organised and has been functioning for a number of seasons (from 2002 onwards). Notably, some of the students who attended the first Buia field school were in charge as graduate assistants in the following years.

Some of the difficulties encountered during the 'first phase' of the Buia project were related to competition with researchers external to the project who were trying to access the Buia research areas. Among these, we recall some celebrated institutions, including the Institute of Human Origins (IHO; founded and directed by Donald Johanson, who was possibly the most famous worldwide-known palaeoanthropologist as the discoverer of the *Australopithecus* skeleton known as 'Lucy' in the 1970s) and the Stony Brook University in New York (with the palaeoanthropologist Randal Susmann). All of these attempts (despite the related difficulties that affected the normal development of the project) were unsuccessful and were brought to the attention of the international community for the unethical behaviour of the institutions (cfr. Dalton 2002a,b).

The Buia basin sedimentary succession is more than 500 meters thick and crops out into the Eritrean Danakil depression, about 500 km north of middle Awash. It covers a time span from the late Early Pleistocene to the early Middle Pleistocene (approximately from 1.2 to 0.6 Ma). The basin includes fluviatile sands, lacustrine clays and alluvial gravels arranged in two lacustrine transgressive–regressive cycles. The Buia humans and associated vertebrate fauna lived in savannah areas close to lakes and rivers, where permanent water resources were available and abundant. The Palaeolithic technocomplexes at Buia are dominated by Acheulean hand-axe artefacts (mode II) and are comparable with the documentation of most East African Rift sites, especially Bodo (Ethiopia) and Olorgesailie (Kenya).

Since the first year, the research activity has been intense. The results published in scientific journals (Abbate et al., 1998; Sagri et al., 1998; Ferretti et al., 2003) have received much interest. Several presentations were also welcomed by the scientific community at international conferences: The IV 'Dual Congress' of the International Association for the Study of Human Palaeontology / International Association of Human Biologists (Sun City / Johannesburg, South Africa, June 1998); the 68th Annual Meeting of the American Association of Physical Anthropologists (Columbus, USA, April 1999); the 70th Annual Meeting of the American Association of Physical Anthropologists (Kansas City, USA, March 2001); the Annual Meeting of the Paleoanthropological Society (Denver, USA, March 2002); the 71st Annual Meeting of the American Association of Physical Anthropologists (Buffalo, USA, April 2002); the international conference on 'The Role of Research in Nation Building' (Asmara, Eritrea, June 2002); the 72nd Annual Meeting American Association of Physical Anthropologists (Tempe, USA, April 2003); the International Symposium on Human Origins and Environmental Backgrounds (Kyoto, Japan, March 2003); and the 32° International Geological Congress (Firenze, Italy, August 2004).

The last event was probably the most important occasion, as it presented the results of the project to the widest international audience. A topical symposium on 'Hominin evolution across environmental change' was organised within this congress, and on that occasion, the special volume titled *A Step Towards Human Origins. The Buia Homo one-million-years ago in the Eritrean Danakil Depression (East Africa)* (edited by Ernesto Abbate, Beraki

Woldehaimanot, Yosief Libsekal, Tewelde Medin Tecle & Lorenzo Rook) was published and distributed worldwide. The volume included several articles dealing with different aspects of the project, from geology to palaeontology, archaeology and palaeoanthropology (Abbate et al., 2004; Albianelli & Napoleone, 2004; Bigazzi et al., 2004; Delfino et al., 2004; Fiore et al., 2004; Macchiarelli et al., 2004; Martínez-Navarro et al., 2004; Martini et al., 2004). Within the different initiatives organised within the 32nd International Geological Congress was an itinerary exhibition. It was presented in Florence at the opening of the congress and was moved to Asmara later on and put on display in the venues of the Eritrea National Museum for the dissemination and popularisation of activities and results of the project.

The second phase of the Buia project (2004 to 2014)

The second phase of the project history has been characterised and driven by two main aspects: one related to the scientific focus of the research and the other to the management and coordination of the research. During these years, the research activity has been focused on the deepening studies and research covering the enormous amount of data collected during the previous years, including the description of new human remains found after the discovery of the Buia calvarium in 1995 (Bondioli et al., 2006). Much work has also been devoted to the description of the palaeontological record (Martínez-Navarro et al., 2007; Rook et al., 2010, 2012), including the description of a new species from the area, named *Bos buiaensis* (Martínez-Navarro et al., 2010), and to deepening the geological context (Ghinassi et al., 2009, 2012; Abbate et al., 2012b; Papini et al., 2014).

The coordination of the research was characterised by a major shift that influenced the development of the project. With the retirement of Prof. Ernesto Abbate in 2010, the coordination of the project from the Italian side moved from the University of Florence to the Sapienza University of Rome under the responsibility of Prof. Alfredo Coppa.

Figure 3: Ernesto Abbate (right) and Tsegai Medin (left) at the Buia Homo site (November 2010).

The latest phases of the project, which were still coordinated by Prof. Abbate, were also devoted to the investigation in areas outside the Buia basin. In the Mendefera region, new fossiliferous areas were recognised (Abbate et al., 2012a). Since 2016, during the post-Prof. Abbate coordination phase, field research was carried out under the supervision and authorisation of the Eritrea Commission of Culture and Sports (Figure 3). The studies were conducted on a non-regular basis, with a somewhat more intense phase in the 2010–2013 time span. This was possible mainly thanks to specific financial support from the Italian Ministry of University and Research for a national project coordinated by Prof. Massimiliano Ghinassi of Padua University. This led to some important new discoveries, including a new *Homo*-bearing site in an area of the basin, specifically the Mulhuli-Amo site (Zanolli et al., 2014) (Figure 4).

Despite the reduced field activity, the Buia project research never stopped but continued to produce a number of scientific contributions on the geological survey and mapping (Ghinassi et al., 2015, 2022; Sani et al., 2017; Scarciglia et al., 2018), on palaeontology (Medin et al., 2015, 2022; Delfino et al., 2018; Martínez-Navarro et al., 2022; Pandolfi et al., 2023), and on palaeoanthropology (Bruner et al., 206; Hammond et al., 2018).

Figure 4 The Buia research team at the Homo site (November 2010).

Expanding the research area: the Engel Ela-Ramud project (2012 to the present)

With the same goals, and in collaboration with the Buia Project, an independent Eritrean–Spanish Project began in 2012 to study the Early Pleistocene archaeo-palaeontological basin of Engel Ela-Ramud, located 30 km south of Buia and close to the Eritrean-Ethiopian border. Except for a short first exploratory survey by the team of Prof. Abbate, no research had previously been carried out. This new project is directed by Tsegai Medin (who obtained his scientific training in Spain at the Institut Català de Paleoecologia Humana i Evolució Social/ Universitat Rovira I Virgili, in Tarragona, funded by successive Erasmus Mundus grants and Atapuerca Foundation Pre-Doc and Post-Doc Grants) and by Bienvenido Martínez-Navarro, as well as by Eudald Carbonell, who was co-director of the Project until 2019 (Figure 5). To date, 10 field seasons have been carried out, with a total of 120 effective working days in the field (unpublished reports, Reports Engel Ela-Ramud Project years 2012, 2013, 2014, 2016, 2017, 2018, 2019, 2020, 2022 and 2023).

Figure 5: Eudald Carbonell (left) and Bienvenido Martinez Navarro (right) with Mohamed Habona (centre) showing a large handaxe found at Ganabá site in 2014).

Since the project started, its general objective has been to record the palaeontological and prehistoric heritage to understand the potential of the basin. To do this, an intense survey of an area of approximately 50 km^2 around the Ramud Camp has been carried out. We have insisted on geological work involving studying the stratigraphic series, advancing the chronological dating and conducting a survey around the entire Basin.

The stratigraphic sequence is more than 200 m thick. It is tilted towards the west, highly tectonised and dominated by the presence of abundant normal faults. Sedimentologically, the sequence was deposited in a deltaic palaeoenvironment, with the source area located to the west in the Eritrean Highlands. It is composed of conglomerates at the base, superimposed by sandy-type strata with a tendency to become finer grained towards the top, where abundant large vertebrate fossil remains and Acheulean lithic artefacts were found. The stratigraphic sequence is discordantly closed by a package of conglomerates in the western sector of the basin and by volcanic rock (ignimbrite) in the eastern one. During the middle and late Pleistocene, accelerated erosion occurred, fitting a new hydrographic network that extended downwards to 150 meters in the Ramud sector camp. Various outcrops with Acheulean stone industry and fauna of large vertebrates in situ have been identified.

To pinpoint the chronology of the series, we have applied different dating techniques, such as ESR, luminescence, palaeomagnetism and biochronology. This has enabled us to fit all the stratigraphic sequences into a time span between 1.8 and 0.7 million years, which corresponds to approximately the second half of the Early Pleistocene.

Figure 6 A buffalo metatarsal found in situ at Gameré-Sasaktoli in 2018.

During the intense survey, we found important areas significantly rich in fossil bones and lithic Acheulian artefacts. Most of the fossils appear on the surface, in the bottoms of the gorges, where they have fallen by gravity from the upper strata. They consist of relatively abundant fossil bone fragments and teeth of elephants, rhinos, equids, hippos, suids (different species of pigs), giraffids, bovids (buffaloes and different species of large, medium and small antelopes) and a few carnivores, as well as crocodiles and turtles. Sometimes they appear *in situ* on the slopes of the ravines, such as a buffalo metatarsal from Gameré-Sasaktoli (Figure 6) or a tree trunk from Dibakole-Diaritana. Large Acheulian tools are also abundant at different sites in the basin, but small-format Acheulian tools in quartz are the most abundant. Those from the Sasaktoli sector are peculiar, as an association of 180 lithic pieces was found in the 2014 season in a fallen block, but well located stratigraphically. The raw material used to make these tools was mainly quartz, although volcanic rocks were also used. Associations of flakes and carving remains are common (Martínez-Navarro et al., 2016a, 2016b, 2017).

At the sites of Ado Quauleh, Gameré, Luba Gadhi 1 and 2, and Erau 5, we have conducted several systematic excavations in places where we found fossil fauna and lithic artefacts in situ. The most important excavations were at Ado Quauleh, Gameré and Luba Gadhi 2-Gallardo:

- Ado Quauleh: This site was excavated from 2014 to 2019 and provided more than 100 small-format lithic artefacts made of quartz, one coprolite, and a few bone fragments (including one giraffid cranium, one identifiable tooth of a small bovid, and four unidentifiable bone fragments).
- Gameré: In the 2017 season, 139 small carved quartz pieces and 2 indeterminate bone fragments were found on a surface area of 3.5 m^2.
- Luba Gadhi II-Gallardo: 60 m^2 were excavated from 2018 to 2023, and 89 large-format lithic artefacts were recorded, together with rich and diversified vertebrate remains, which included plate fragments of *Elephas recki* ssp., a tooth of a rhino (*Ceratotherium simum*), abundant teeth and bone fragments of *Hippopotamus* cf. *gorgops*, several teeth of three species of pigs (*Metridichoerus modestus*, *Metridichoerus compactus* and *Kolpochoerus majus*), a mandibular fragment of Hippotragini indet. (probably *Oryx* sp.), a distal end of a humerus corresponding to a monkey (Cercopithecidae indet.), several teeth and bone fragments of *Crocodylus* cf. *nyloticus*, and bone fragments of a turtle. The most abundant species were hippos and crocodiles, indicating that the site was generated in a water-dominant environment (Figure 7).

Figure 7 The excavation at Luba Gadhi II-Gallardo in 2020.

After 13 years of research and 10 field seasons at the Engel Ela-Ramud basin, we can say that this is a new complex site with a very good potential palaeontological and archaeological record that will better inform us about the human evolution in this area of the northern Afar region. In the near future, the excavation of these sites will provide interesting new ideas regarding the palaeoenvironments in which humans evolved and from where they dispersed outside Africa during the Early Pleistocene.

Concluding

The importance of the Buia and Engel Ela-Ramud projects and their deep interconnection is confirmed by the recent publication of a joint synthesis in the pivotal book published by Springer International titled *Handbook of Pleistocene Archaeology of Africa: Hominin behaviour, geography, and chronology*. In that book, an entire chapter (Abbate et al., 2023) is dedicated to the important sites in the Buia and Engel Ela-Ramud basins.

The teams of the Buia and the Engel Ela-Ramud Projects consist of the people and the territory itself to learn about and protect the natural and cultural heritage. Over the years, both projects have acted as cooperative platforms for sustainable development. The synergistic actions aimed at developing and disseminating research findings and at involving local communities are fully central to some of the objectives of the 2030 Agenda for Sustainable Development, particularly goals #4 (Quality education), #8 (Decent work and economic growth), #11 (Sustainable cities and communities), #13 (Climate action) and #15 (Life on Land). This demonstrates how culture and research initiatives can have an immediate return in terms of sustainable development. Among the recent examples of actions developed by these projects are lectures to university students and field schools, carried out in conjunction with research activities, as well as interactions with local people who are active in local museums and communities, with the aim of enhancing the value of their geo-palaeontological heritage and local geodiversity (Figure 8).

Figure 8 Meeting with Mohamed Habona and the Engel Ela Council (2018 field campaign).

References

Aalad and Mulhuli-Amo (Buia), Eritrean Danakil: micromorphology and microstructure. *Journal of Human Evolution*, 74: 96-113.

Abbate E., Albianelli A., Azzaroli A., Benvenuti M., Tesfamariam B., Bruni P., Cipriani N., Clarke R.J., Ficcarelli G., Macchiarelli R., Napoleone G., Papini M., Rook L., Sagri M., Tecle T.M., Torre D., Villa I. (1998). A one-million-year-old *Homo* cranium from the Danakil (Afar) depression of Eritrea. *Nature* 393: 458-460.

Abbate E., Beraky W., Bruni P., Falorni P., Papini M., Sagri M., Simret G., Tewolde M.T. (2004). Geology of the *Homo*-bearing Pleistocene Dandiero basin (Buia region, Eritrean Danakil depression). *Rivista Italiana di Paleontologia e Stratigrafia*, 110(supplement): 5-34.

Abbate E., Bondioli L., Collina C., Coppa A., Delfino M., Ghinassi M., Libsekal Y., Macchiarelli R., Martínez-Navarro B., Martini F., Medin T., Oms Lobet O., Papini M., Rook L., Sagri M., Zanolli C. (2023). Buia, Eritrea. In: Beyin A., Wright D., Wilkins J., & Olszewskion D. (eds) *Handbook of Pleistocene Archaeology of Africa: Hominin behavior, geography, and chronology*: 243-261. Springer International Publishing AG.

Abbate E., Bruni P., Coppa A., Aria D., Ferretti M.P., Libsekal Y., Rook L., Sagri M. (2012a). A new Oligo-Miocene mammal-bearing site from a sedimentary intarcalation in the Trap Basalts of central Eritrea. *Riv. It. Paleont. Strat.*, 118(3): 545-550.

Abbate E., Bruni P., Landucci F., Pellicanò G. (2012b). Unusual ichnofossils in *Homo erectus*-bearing beds of the Pleistocene Lake deposits in central-eastern Eritrea, East Africa. *Palaios* 27: 97-104.

Albianelli A., Napoleone G. (2004). Magnetostratigraphy of the *Homo*-bearing Pleistocene Dandiero basin (Danakil depression, Eritrea). *Rivista Italiana di Paleontologia e Stratigrafia*, 110(supplement): 35-44.

Bigazzi G., Balestrieri M.L., Norelli P., Oddone M., Tewolde M.T. (2004). Fission-track dating of a tephra layer in the Alat formation of the Dandiero Group (Danakil depression, Eritrea). *Rivista Italiana di Paleontologia e Stratigrafia*, 110(supplement): 45-49.

Bondioli L., Coppa A., Frayer D.W., Libsekal Y., Rook L., Macchiarelli R. (2006). A one-million-year-old human pubic symphysis. *Journal of Human Evolution*, 50: 479-483.

Bruner E., Bondioli L., Coppa A., Frayer D.W., Holloway R.L., Libsekal Y., Medin T., Rook L., Macchiarelli R. (2016). The endocast of the one-million-year-old human cranium from Buia (UA 31), Danakil Eritrea. *American Journal of Physical Anthropology*, 160: 458-468.

Dalton R. (2002a). Survey overlap impedes fossil hunters' study. Nature, 415: 463.

Dalton R. (2002b). Hints of bone bounties rile fossil hunters. Nature, 416: 356.

Delfino M., Candilio F., Carnevale G., Coppa A., Medin T., Pavia M., Rook L., Urciuoli A., Villa A. (2018) The early Pleistocene vertebrate fauna of Mulhuli-Amo (Buia area, Danakil Depression, Eritrea). *Bollettino della Società Paleontologica Italiana*, 57: 27-44.

Delfino M., Segid A., Yosief D., Shoshani J., Rook L., Libsekal Y. (2004). Fossil reptiles from the Pleistocene *Homo*-bearing locality of Buia (Eritrea, Northern Danakil depression) *Rivista Italiana di Paleontologia e Stratigrafia*, 110(supplement): 51-60.

Ferretti M.P., Ficcarelli G., Libsekal Y., Tecle T.M., Rook L. (2003). Fossil elephants from Buia (northern Afar Depression, Eritrea) with remarks on the systematics of *Elephas recki* (Proboscidea, Elephantidae). *Journal of Vertebrate Paleontology*, 23: 244-257.

Fiore, I., Bondioli, L., Coppa, A., Macchiarelli, R., Russom R., Kashay H., Solomon T., Rook L., Libsekal Y. (2004). Taphonomic analysis of the late Early Pleistocene bone remains from Buia (Dandiero Basin, Danakil Depression, Eritrea): evidence for large mammal and reptile butchering. *Rivista Italiana di Paleontologia e Stratigrafia*, 110(supplement): 89-97.

Ghinassi M., Billi P., Libsekal Y., Papini M., Rook L. (2012). Inferring fluvial morphodynamics and overbank flow control from 3D outcrop sections of a Pleistocene point bar, Dandiero basin, Eritrea. *Journal of Sedimentary Research*, 83: 1065-1083.

Ghinassi M., Libsekal Y., Papini M., Rook L. (2009). Palaeoenvironments of the Buia *Homo* site: high-resolution facies analysis and non-marine sequence stratigraphy in the Alat formation (Pleistocene Dandiero basin, Danakil depression, Eritrea). *Paleogeography Paleoclimatology Paleoecology*, 280: 415-431.

Ghinassi M., Oms O., Papini M., Scarciglia F., Carnevale G., Sani F., Rook L., Delfino M., Pavia M., Libsekal Y., Bondioli L., Coppa A., Frayer D.W., Macchiarelli R. (2015). An integrated study of the *Homo*-bearing Aalat stratigraphic section (Eritrea): an expanded continental record at the Early-Middle Pleistocene transition. *Journal of African Earth Sciences*, 112: 163-185.

Ghinassi M., Papini M., Rook L., Oms O., Sani F. (2022). Pleistocene landscape variability recorded in the *Homo*-bearing Dandiero basin (Eritrea). In: Billi P. (ed.) *Landscapes and Landforms of the Horn of Africa. Eritrea Djibouti, Somalia*: 155-167. Springer Nature.

Hammond A.S., Almécija S., Libsekal Y., Rook L., Macchiarelli R. (2018). A partial *Homo* pelvis from the Early Pleistocene of Eritrea. *Journal of Human Evolution*, 123: 109-128.

Macchiarelli R., Bondioli L., Chech M., Coppa A., Fiore I., Russom.R, Vecchi F, Libsekal Y, Rook L. (2004). The late Early Pleistocene human remains from Buia, Danakil depression, Eritrea. *Rivista Italiana di Paleontologia e Stratigrafia*, 110(supplement): 133-144.

Martínez-Navarro B., Carbonell E., Medin T., Oms O., Rodríguez-Álvarez X.P., Libsekal Y., Ros-Montoya S., Araia D., Tsehale L., Libsekal Y. (2016a). The Engel Ela - Ramod Basin: A new Plio-Pleistocene Archaeo - Paleopaleontological site in Eritrea. ESHE Conference, Alcalá de Henares (Madrid, Spain), 14-17 Sept. 2016. Abstracts: 154.

Martínez-Navarro B., Carbonell E., Medin T., Oms O., Rodríguez-Álvarez X.P., Ros-Montoya S., Araia D., Zerai A., Ghirmay M., Yohannes H., Mehare H., Libsekal Y. (2017). A new Neogene-Quaternary archeo-paleontological site in Eritrea: The Engel Ela-Ramud Basin. III Jornadas de Prehistoria Africana. Tarragona, 18-19 March, 2017.

Martínez-Navarro B., Medin T., Oms O., Rodríguez-Álvarez X.P., Libsekal Y., Ros-Montoya S., Araia D., Tsehale L., Rook L., Carbonell E. (2016b). The Eritreo-Spanish Project: A four-years Geo - Archaeo - Paleopaleontological field research at the Engel Ela - Ramod Basin (Eritrean Danakil Depression). International Conference of Eritrean Studies: The Way Forward. (Poster).

Martínez-Navarro B., Pandolfi L., Tsegai Medin, Yosieph Libsekal, Ghinassi M., Papini M. & Rook L. (2022). The ontogenetic patter of *Hippopotamus gorgops* revealed by a juvenile cranium from the one-million-years-old paleoanthropological site of Buia (Eritrea). *Comptes Rendus Palevol*, 21: 157-173.

Martínez-Navarro B., Perez Claros J.A., Palombo M.R., Rook L., Palmqvist P. (2007). The African buffalo *Pelorovis oldowayensis* and the origin of the genus Bos. *Quaternary Research*, 68: 220-226.

Martínez-Navarro B., Rook L., Papini M., Libsekal Y. (2010). A new species of bull from the Early Pleistocene paleoanthropological site of Buia (Eritrea): parallelism on the dispersal of the genus Bos and the Acheulian culture. *Quaternary International*, 212:169-175.

Martínez-Navarro B., Rook L., Segid A., Yosief D., Ferretti M.P., Shoshani J., Tecle T.M., Libsekal Y. (2004). The large fossil mammals from Buia (Eritrea): systematics, Biochronology and Paleoenvironments. *Rivista Italiana di Paleontologia e Stratigrafia*, 110(supplement): 61-88.

Martini F., Libsekal Y., Filippi O., Ghebreher A., Kashay H., Kiros A., Martino G., Okubatsion D., Segid A., Solomon T., Teka Z., Yosief D., Yamane S. (2004). Characterization of lithic complexes from Buia (Dandiero basin, Danakil depression, Eritrea). *Rivista Italiana di Paleontologia e Stratigrafia*, 110(supplement): 99-132.

Medin T., Martínez-Navarro B., Ferretti M.P., Ghinassi M., Papini M., Libsekal Y., Rook L. (2022). Mammal Paleoecology from the late Early Pleistocene sites of the Buia Basin (Eritrea), with emphasis on the suid record. Chapter 16. In: Reynolds S.C., Bobe R. (eds.) *African Paleoecology and Human Evolution*: 187-196. Cambridge University Press.

Medin T., Martínez-Navarro B., Rivals F., Libsekal Y., Rook L. (2015). The late Early Pleistocene suid remains from the paleoanthropological site of Buia (Eritrea): systematics, biochronology and eco-geographical context. *Paleogeography Paleoclimatology Paleoecology*, 431: 26-42.

Papini M., Ghinassi M., Libsekal Y., Rook L. (2014). Facies associations of the northern Dandiero basin (Danakil depression, Eritrea, including the Pleistocene Buia *Homo* site). *Journal of Maps*, 10: 126-135.

Rook L., Ghinassi M., Carnevale G., Delfino M., Pavia M., Bondioli L., Candilio F., Coppa A., Martínez-Navarro B., Medin T., Papini M., Zanolli C., Libsekal Y. (2012). Stratigraphic context and paleoenvironmental significance of minor taxa (Pisces, Reptilia, Aves, Rodentia) from the late Early Pleistocene palaeoanthropological site of Buia (Eritrea). *Journal of Human Evolution*, 64: 83-92.

Rook L., Ghinassi M., Libsekal Y., Martínez-Navarro B., Medin T., Papini M. (2010). Stratigraphic context and taxonomic assessment of the large cercopithecoid (Primates, Mammalia) from the late Early Pleistocene palaeoanthropological site of Buia (Eritrea). *Journal of Human Evolution*, 59: 692-697.

Sagri M., Abbate E., Azzaroli A., Balestireri M.L., Benvenuti M., Bruni P., Fazzuoli M., Ficcarelli G., Marcucci M., Papini M., Pavia G., Reale V., Rook L., Tecle T.M. (1998). New data on the Jurassic and Neogene sedimentation in the Danakil Horst and Northern Afar Depression, Eritrea. *Mémoires du Muséum national d'Histoire naturelle,* 177: 193-214.

Sani F., Ghinassi M., Papini M., Oms O., Finotello A. (2017). Evolution of the northern tip of Afar triangle: inferences from the Quaternary succession of the Dandiero - Massawa area (Eritrea). *Tectonophysics*, 717: 339-357.

Scarciglia F., Mercatante G., Fondevilla V., Anadón P., Oms O., Donato P., Agnini C., Papini M., Rook L., Ghinassi M. (2018). Pleistocene paleosol development and paleoenvironmental dynamics in East Africa: a multiproxy record from the Homo-bearing Aalat pedostratigraphic succession, Dandiero Basin (Eritrea). *Quaternary Science Reviews*, 191: 275-298.

Zanolli C., Bondioli L., Coppa A., Dean C.M., Bayle P., Candilio F., Capuani S., Dreossi D., Fiore I., Frayer D.W., Libsekal Y., Mancini L., Rook L., Medin Tekle T., Tuniz C., Macchiarelli R. (2014). The late Early Pleistocene human dental remains from Uadi

Adulis (Eritrea): Sustainable Archaeology in the Horn of Africa

Tsegai Medin,[1] Abraham Zerai,[1] Serena Massa,[2]

Marco Bocciolone,[3] Susanna Bortolotto,[3] Nelly Cattaneo[3]

[1] *Commission of Culture and Sports of Eritrea - Archaeological Heritage Research Branch*

[2] *Research Centre on Eastern Desert, Università Cattolica di Milano, Politecnico di Milano*

[3] *Politecnico di Milano*

Abstract

The Eritrean-Italian Adulis project was born in 2011, after the will of the Eritrean Authorities. In its broader frame Adulis project is methodologically dictated by landscape archaeology. The interest of the research is aimed at investigating all the phases of this historical and cultural landscape to understand the context in which the ancient community found and managed resources for its long-term survival and prosperity, building a broad commercial and cultural networks and system of relationships and social identity of which archaeology reads the material traces that have been preserved up to present time.

The project has been and is carried on following these main strategies: participation of the local communities; stratigraphic investigation; conservation alongside excavation; historic and present landscape analyses; land use and risk assessment: hydrology, farming techniques, renewable energy; planning of a permanent, sustainable research centre within the archaeological site; designing a sustainable management plan.

The present paper outlines how, by achieving its specific objectives, the project contributes to designing new strategies of heritage sustainable management, enhancing cultural values and highlighting the potential of past civilizations to support problem solving for the present society. This will be the p remise to suggest viable alternatives for territorial planning and growth, creating better lives for the people, while preserving the environment and its resources.

Keywords: *Adulis, Eritrean Archaeological Heritage, Sustainable Archaeological and Natural Park, Preservation.*

Introduction

Since 2011, an Eritrean-Italian research has been conducted on the Adulis archaeological site (Eritrea), an important emporium of the Horn of Africa, a place inhabited since at least the 2nd millennium BC, when it was the principal destination of the pharaonic expeditions to the Land of Punt.[1] The project is led by the Eritrean Commission of Culture and Sports and by the Research Centre on the Eastern Desert, in collaboration with the National Museum of Eritrea, the Northern Red Sea Regional Museum of Massawa, Politecnico di Milano and other Italian universities.[2]

1 BARD et al. 2021, pp. 241, 590.

2 The Adulis Project scientific research is directed by Prof. Serena Massa (Università Cattolica del Sacro Cuore di Milano) and Dr. Tsegai Medin (Archaeological Heritage Research Branch of the commission of Culture and Sports of Eritrea), in collaboration with the Department of Culture and Sports of the Northern Red Sea Region. The project is financed 90% by the government of the State of Eritrea and with a

In addition to scientific research, the objectives of which will be illustrated below, one of the main aims of the project is to create an archaeological park.

We owe to the joint Eritrea - Italian Adulis Project for bringing ancient Adulis, among the vast Eritrean Archaeological Heritage, back to light, more than a hundred years after the most important research carried out on the site by different travellers and amateur archaeologists starting the late mid of the 19th century. Their intuition was based on their deep understanding of the role of the African continent in the formation of the best-known ancient civilizations, such as the Egyptian one. The Castiglioni brothers, together with their Eritrean counterparts, with this new research project, aimed at demonstrating with archaeological evidence the relationship between Egypt and the Land of Punt, a relationship that we know from the sources to have existed at least since the second millennium BC.

And in fact, the Red Sea, with its coasts and inlands, represents an essential geographical and cultural space for a thorough understanding of the contacts and strategic relations between the Mediterranean, Africa, Arabia and the Indian Ocean, which have built over the long term the extraordinary cultural wealth of current Eritrea and its people.

The geographic position of the site, at the connection of ancient international maritime and regional terrestrial routes, was a key factor for the flourishing of Adulis. The research project, aimed at the rediscovery, study and valorisation of the archaeological remains of the ancient town, has also the intent to create the first "Sustainable Archaeological and Natural Park" in Sub-Saharan Africa.

Introducing the concept of Sustainability in an archaeological research project has meant considering the manifold meanings that this term has been integrating in the past years. They can be all included in the idea that the site and its context are so intertwined, that the addressing the future of the site means addressing holistically the future of a wider system of relations at different levels: cultural (how the understanding of the site and of its civilization can enrich today's population? How can it promote the commitment of stakeholders in its enhancement?) environmental (what can we learn from the past use of natural resources? How can the improvement of natural environment contribute to the safeguard of the cultural site and *vice versa*?), scientific (how can we find a proper balance between the needs of archaeological research, which is destructive, and the preservation of the site? How to address the transfer of scientific knowledge to a wide audience of non-experts?), management approach (which strategies should be designed to ensure a constant maintenance and improvement of the site?) etc.

The objectives have been designed together with decision makers and in dialogue with the local communities to plan the excavation and conservation fieldworks. Any plan must be in close relationship with the farmed and natural environment, giving high attention

scientific support from Ce.R.D.O and logistical assistance from Piccini Group. It has also received contributions from the Italian Ministry of Foreign Affairs and International Cooperation since 2012, from AICS (Agenzia Italiana per la Cooperazione allo Sviluppo) within the project VITAE in 2020 and is supported by ISMEO (Associazione Internazionale di Studi sul Mediterraneo e l'Oriente).

Università Cattolica del Sacro Cuore di Milano, and the contribution of Università degli Studi di Padova. Other collaborations, limited in time, have been carried out with Museo di Rovereto, Centro di Geotecnologie (Università di Siena), Università degli Studi L'Orientale di Napoli, Pontificio Istituto di Archeologia Cristiana, Università degli Studi dell'Insubria di Varese.

Our warmest thanks to the Foro Sub zone Administration, the people of Zula and Afta, and the Commissioner of Culture and Sports, Ambassador Zemede Tekle, to Piccini Group, Mr. Tedros Berhane and to all the Eritrean and Italian colleagues and workers.

to the study and valorisation of traditional hydrogeological and agronomic knowledge, as a strategic asset to planning the sustainable development of the area. Specific needs have been expressed by the inhabitants of the villages, neighbouring the ancient site, to the archaeological project, which will have to deal also with improved accessibility of road links and services creating economic development without threatening the cultural and natural heritage, part of which consists in the traditional knowledge of resource management.

The project in fact proposes a model of research and valorisation of the cultural landscape in the Horn of Africa, based on the study and enhancement of the ancient holistic approach as a key element to sustainability. The research is conducted with an interdisciplinary method that integrates the latest advanced techniques of remote sensing, geoarchaeology, bioarchaeology, archaeometry, 3D modelling and computational models.

Such objectives need long-term archaeological research and dedicated strategies to obtain a comprehensive understanding of the territorial context where the archaeological site is located. The creation of a sustainable park where the evidence of the discovery is publicly shared can be pivotal for the development of the local community, going along with the empowerment of specialized skills to manage the site, the artefacts and the future park.

Fig. 1 – The results of archaeological excavations in the south-eastern part of Adulis.

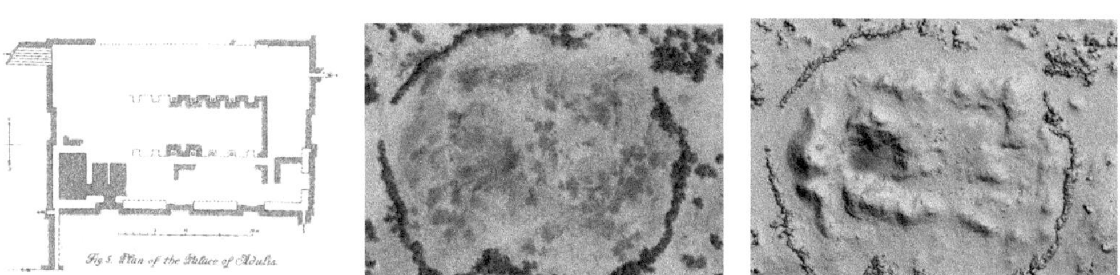

Fig. 2 – Bibliographical research, aerophotogrammetry and 3D modelling supporting archaeological understanding: the case of the so-called "Sundström Palace" in Adulis.

Fig. 3 – 3D accurate documentation of archaeological finds in collaboration with the National Museum of Asmara.

The New Research (2011-2024)

The evidence brought to light at Adulis during the annual field surveys and excavation campaigns since 2011,[3] enabled us to understand the ground-breaking research potential of the site in relation to the origin and development of African Horn civilization, its place in the construction of local identities and the strategic relationship between the Mediterranean, Africa, Arabia and Indian Ocean. In the common opinion, the development of the port town of Adulis is strictly connected with the flourishing of Aksum, the capital of the Aksumite Empire, located on the Ethiopian Highlands,[4] at an altitude of 2.200 meters and at a distance of about ten caravan days from Adulis, gate to the sea for the metropolis. Nevertheless, remains of an earlier settlement of huts have been excavated in the southwestern area of the site, with material culture dated to the II millennium BC,[5] when this portion of the coast was part of the Land of Punt, the objective of expeditions by the Pharaohs to find precious and exotic items, such as ivory, ebony, aromatic resins and animals,[6] therefore long before the rising of Aksumite kingdom.

Adulis, centre of extra-regional exchanges in the Pharaonic era, already possessed at that time some of the founding characteristics of an urban reality which, currently, represent the oldest evidence of the rise of urbanism in the Horn of Africa.[7]

It is probable, by reaching deeper archaeological levels in future research, that earlier documents of human frequentation/settlement could be recovered, like those found in the coastal environment of the Bay of Zula, where artefacts datable to the Middle and Late Stone Age document the presence of human groups capable of exploiting the marine and coastal resources of an ideal habitat for living. Almost 15 years of fieldwork have also enabled us to refine our understanding of the current social system and cultural landscape, which the archaeological site is part of.

The excavation surveys, carried out annually except in cases of force majeure due to extreme weather conditions or epidemics,[8] have made it possible to extend the topographical knowledge of the town and to mark out with precise chronological sequences linked to the stratigraphies and the finds the events between the 1st/2nd and 7th/8th centuries and beyond.

The most important novelties concern: the recognition of settlement phases datable to the 1st-3rd century AD, characterised by stone architecture and a material culture that, alongside local production, includes imports from Egypt, the Mediterranean, the northern Red Sea and the Persian Gulf;[9] the early dating of the Northern Urban Church to the sec-

9 ZAZZARO & MANZO 2012; MASSA 2021; ZERAI et al. 2022.

ond half of the 4th - first half of the 5th century, offering for the first time a stratigraphic chronological reference for the Aksumite early Christian churches, evidence that constitutes a valuable confirmation of the testimony of the textual sources on the early Christianization of the Aksumite kingdom. The analysis of the construction techniques by Susanna Bortolotto provides further evidence for the recognition of different and later construction techniques of the Cathedral and the Eastern Church. [10]

Aerial surveys and surface research have made it possible to recognise elements of fortification that had never been identified by previous research, with consequences for the general comprehension of the origin of urbanism in the Northern Horn of Africa, as well as to better define the extent of the remains outcropping in the topsoil. A number of burials from the Christian period and others from the late Muslim period have been excavated, which partially bridge the gap between the ancient settlement and the cemetery use of the site by the Zula and Afta communities.[11]

The study of the numerous finds has brought great novelties, relevant not only to Adulis but to the Red Sea in general.[12]

Methodology: Landscape Archaeology and Holistic Approach

The theoretical premises of the scientific research underlying the Adulis project are those dictated by landscape archaeology, denoting the term landscape in its broad sense of cultural expression, i.e., the result of the interaction between man and the environment. The interest of the research is therefore aimed at investigating all the phases of this historical landscape, in order to understand the context in which the ancient community found and managed the resources for its survival and prosperity in the long term, building a system of relationships and social identity of which archaeology reads the material traces that have been preserved up to our time. It is a dynamic system, which can be read with the method of archaeological stratigraphy, recognizing its environmental, productive and social elements, within a systemic approach and a multidisciplinary methodology involving, alongside the archaeological and historic ones, natural sciences, mathematic and statistics. [13]

This holistic and systemic approach, as well as being central to knowledge, is the basis of a vision of public archaeology that places at its centre the economic and social values of the community that currently resides in the area surrounding the archaeological site. This is in fact the heir of a cultural tradition that is one of the key factors for a sustainable safeguard of the landscape. [14]

All the more so in contexts hard for livelihood, such as the case-study illustrated here, a semi-arid environment in which wisdom in water management is as vital today as two thousand years ago, at the time when the splendid urban civilization of Adulis flourished, in climatic conditions probably not so dissimilar to the current ones. [15]

---

10 MASSA 2017, MASSA 2019; BORTOLOTTO et al. (in press).

11 LARENTIS 2021

12 ZERAI et al. 2020; MASSA & GORLA 2024

13 BROGIOLO & CHAVARRIA ARNAU 2019, pp. 104-105.

14 As recognized and ratified globally since the 1992 Rio Conference, also known as the 'Planet Earth Summit'. Following the Conference, world conventions, international organizations such as the United Nations, the FAO, UNESCO, the European Union, promote projects for the protection of environmental resources and for the creation of a new ethical economy and a technology capable of considering the wisdom of ancient practices.

15 BEYN & BAR-YOSEF MAYER 2018, p. 26.

Traditional knowledge[16] is an elaborate and often multipurpose system, part of an integral approach between society, culture and economy and grounded on the idea of the world based on the careful management of local resources.[17] The close and intimate bond existing in traditional cultures between man, natural environment and the universe, permeates techniques, spaces of sacredness and defines the identity of the communities.[18] This is the intangible component of the heritage that runs the greatest risk of being lost if not properly investigated alongside its more easily recognizable material evidence.

In public archaeology [19], the community plays a central role in the process of knowledge, protection and enhancement of cultural heritage, considered as a reflection and expression of its constantly evolving values, beliefs, knowledge and traditions[20] . It is a different way of doing archaeology in which "the energies invested in research and in the growth of knowledge favour processes of participation, social cohesion, valorisation and sustainable economic development"[21] , with the ultimate goal of proposing sustainable solutions to the current problems,[22] among which the scarcity of water resources is certainly one of the most serious.

As noted by several voices, there are no regulatory legal tools or codified guidelines for the implementation of participatory archaeology projects, but in the case of Eritrea there are two elements that have facilitated this approach. The first is the local culture itself, which is based on a vision of the world whose values are defined by relationships, not only between individuals who make up the community, but also with the environment. From this derives a strong sense of cohesion and sharing of the community members, which also includes the process of knowledge and research, and which contributes, in the end, to the quality of life.[23]

The second facilitating factor in the Adulis Project is due to the methodology that has always characterized the ethnographic and archaeological research work at the origin of the project, i.e., respectful, inclusive and participatory.

16 As defined by the Science and Technology Committee set up by the Convention for the fight against desertification: www.unccd.int.

17 www.droughtmanagement.info/literature/UNCCD_promotion_of_traditional_knowledge_2003.pdf

18 LAUREANO 2001, p. 80.

19 Participatory research in archaeology. Legal issues and good practises, "European Journal of Post-Classical Archaeologies", n. 9, 2019: NUCCIOTTI et al. 2019; DRAGONI & CERQUETTI 2019. In these texts a critical review of the multiple aspects connected with Public Archaeology is analysed, starting from the definition itself and of the key concept of community versus Public, experts, the "bottom up" and "top-down approaches", legal and ethical issues, social implications of the interpretation of the results, democratization of knowledge, collaboration, with the conclusion that both are necessary and an equilibrium between needs

20 VOLPE 2017, p. 48.

21 VOLPE 2019, p. 10.

22 In line with the more democratic and open approach to cultural heritage adopted by a series of international Conventions and Recommendations promoted by UNESCO and the Council of Europe, which consider the population not only the recipient of research and projects to enhance the heritage but also an actor actively participant to its management, in the optics of an economic and social development: from the Convention for the Protection of the World Cultural and Natural Heritage (Paris, 1972), to that for the Protection of the Archaeological Heritage (Valletta, 1992), up to that on the Value of Cultural Heritage for Society (Faro, 2005).
The latter convention is fundamental because it promotes the involvement of all actors in the research and conservation of heritage, closely linking cultural heritage to human rights. BROGIOLO & CHAVARRIA ARNAU 2019, p. 104, with previous references.

23 See further about African Relational Evaluation Approaches.

More than ten centuries separate the Islamic graves from the monumental, early Christian churches that belong to the Byzantine phase of Adulis, a completely foreign architecture in the eyes of the local communities, but full of meaning from the point of view of understanding of the past. It is therefore appropriate to ask ourselves whether our academic value of the past has the same meaning for those who now live in places where, from natural catastrophic causes and following the Islamic conquest, a drastic change in the cultural landscape took place.

The Adulis project has to be acknowledged as an opportunity that has been offered to the Italian team of experts, not only for scientific achievements but also for the possibility to develop a multidisciplinary, inclusive and dynamic research approach, within a context of valuable human relationships. In this context, we can consider Public Archaeology, both as a theoretical/methodological framework and a mitigation tool, as the design of any future activity that undertakes an evaluation of local needs.

In the Adulis case-study there are different layers of meaning of the word *heritage* and, of course, many others not formalized in official definitions. As Laura-Jane Smith claims, *"there is a dominant Western discourse about heritage, which [she] terms the 'authorized heritage discourse' that works to naturalize a range of assumptions about the nature and meaning of heritage".* For the team working in Adulis it has been important to refer to the definition of *heritage* that Eritrea has shaped and formally defined in the national law of 2015. The *"Cultural and Natural Heritage Proclamation"* n. 177 affirms (art. 2.1.g): *"Cultural Heritage means any tangible or intangible resource, which is the product of human creativity and labour in the discernible historical times, describing and witnessing to such creativity and labour because of its scientific, archaeological, historical, cultural, artistic, architectural or aesthetic value or content ultimately bearing the identity and/or collective memory of peoples or communities"* and further (art. 3 f): *"the objectives of this Proclamation shall be to: [...] c) protect Cultural and Natural Heritage against all forms of damage; f) empower and encourage the general population to nurture and conserve heritage resources and their cultural and indigenous values".*

The Proclamation provides a broad and inclusive definition of heritage and can be considered the achievement of a long process of debate on the nature and the role Eritrean culture, started within the Eritrean People Liberation Front cultural committees.

Towards the Design of an Archaeological Park

The points of interest of the Adulis site are not limited to the ongoing excavations. In fact, there are traces of the past excavations by the following expedition: the British Museum in 1868, the Italian archaeologist Roberto Paribeni in 1906, the Swedish missionary Richard Sundström's in 1906 and the French archaeologist Francis Anfray in the 1960s. The excavations and the surveys on Adulis also tell the story of archaeology as an evolving disciplinary approach to the ancient world and as a methodological corpus for learning from it. In designing the Archaeological Park, it is important to consider this 'disciplinary' history as a theme of the visitor's itinerary, as it explains aspects of the current state of the site, the different destinations of the finds, and the quantity and quality of information that different methods of excavation and archaeological investigation can yield.

Related to the past excavations are the mounds, several metres high, created near the excavations by the deposit of earthwork material. There are also outcrops of walls in the archaeological area, brought to light by rainfall, which enrich the mosaic of information on the layout of the buildings and the walls' characteristics. Finally, there are other mounds, mainly of basalt and schist slabs, originating from collapsed structures, as well as little *cumuli* of pottery sherds with concentrations in specific parts of the site from the top of which one can view the various excavation sectors and also the natural environment.

The site constitutes an interesting 'research workshop' for multiple aspects. First of all, the built heritage of stone architecture is of great interest — the site holds specimens of internationally recognized importance — in a vast area, most of which has still to be uncovered from thick layers of alluvial deposits. The ongoing activities for the future realization of an archaeological and natural park are multidisciplinary, based on a holistic approach as a key element for sustainability and aim at constituting an example that can be replicated in relation to the adopted training methods in research, conservation, safeguard of archaeological sites and cultural landscape in the Horn of Africa.

The settlements near the archaeological area of resident communities in traditional villages (Afta and Zula) are considered pivotal and supporting their livelihood is important to avoid depopulation; the respect for the local traditions of local communities is paramount: visitor itineraries as well as being designed to make the results of the archaeological research clear, must also protect the burial areas between the centre of the site and its south-eastern edges, where the tombs of Scheick Mahmoud, Bet Khalifa, and their respective family groups are located; these areas are still in use by the Zula and Afta communities, and their sacredness must be maintained. The high potential of the plain of Zula is considered, by designing possible measures to trigger a transition towards equitable and sustainable prosperity from the environmental and cultural points of view, integrating local knowledge with international experience in agriculture/breeding and agroforestry in semi-arid contexts. The paramount challenge is to define a suitable master plan that clearly defines the limits of the park, the permitted anthropic expansion, the borders of the farmed fields, pasture areas and compatible side transit corridors specifically for caravans with livestock.

Fig. 5 – The landscape around the archaeological site.

The development of experimental guidelines for a preservation/maintenance programme is another important challenge to contrast the effects of persisting extreme environmental conditions and hazardous events in the area (rising temperatures, flowing water and gullies; strong gusts of *khamsin* winds; infesting vegetation), occurring with increasing intensity occasionally and/or cyclically and putting pressure on both the excavated areas and the structures over time; these best practices could be replicated in other similar contexts.

The peculiarities of the cultural/rural landscape and the panoramic natural/environmental landscape are of great interest, with the possibility of seeing views of the upland and sea from the site, as well as glimpses of local micro and macrofauna (birds, hyenas, ostriches, hares, etc.).

Eventually the opportunity to use activities in the field (reconnaissance, survey, excavation, conservation, etc.) is important as they can structure the lifelong learning practises of a new generation of researchers in Eritrea, and of collaborators in the local communities, who will guarantee not only the management of the process in the long term, but the protection and enhancement of the site as well.

Fig. 6 – On field activities regarding geophysical investigation, conservation and archaeological excavation and documentation.

Fig. 7 - Processing of ceramic finds on field, cataloguing and selecting samples for archaeometric analyses.

Alongside studies and research on the archaeological area, assessments are being developed on the context, in particular on the environmental risks exacerbated by the ongoing climate change are being drawn up; strategies for their mitigation have been identified, particularly with regard to hydrological and agro-environmental hazards and what they entails in terms of conservation of the archaeological area, including the risk to the communities of Foro, Zula and Afta and the uncertainty of the resources required for agro-pastoral activities. For the plain of Zula a valuable and comprehensive risk assessment was carried out by Eritrean analysts and scholars in 2018.

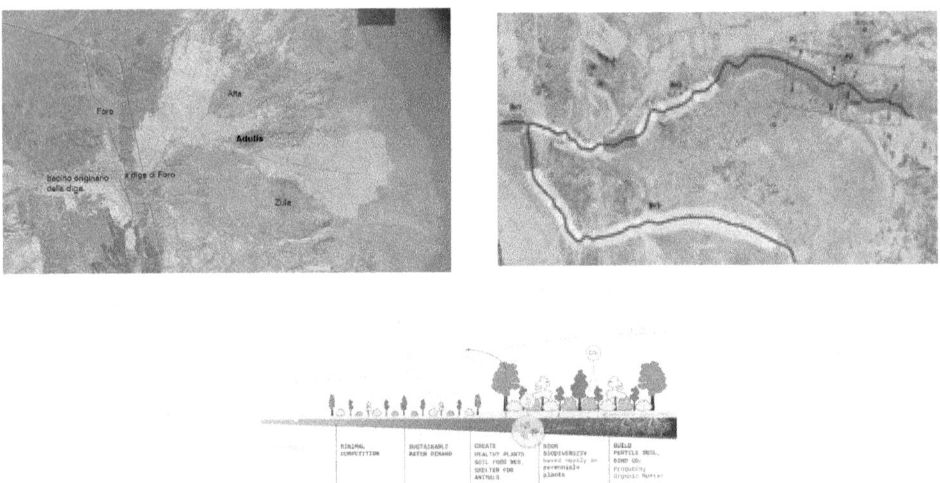

Fig. 8 – Overview on the 2024 hydrogeologic and environmental assessment and possible mitigation measures

It is well known that socio-economic aspects are a fundamental component in the formulation of an adequate management plan: here, in particular, the development of visitor circuits must be integrated with the agropastoral development of the area, at the level of the communities which are in charge of the management of the irrigation system of the Zula plain as well as of the constant monitoring and emergency interventions on the archaeological area. Agro-climatic and environmental hazards, exacerbated by soil erosion, biodiversity depletion and climate change, are threatening local communities and the site, therefore the research is also addressing community-based land regeneration strategies able to reduce the direct effects of floods and windstorms on the villages, on the farmed lands and on the archaeological site and to improve the livelihood of the local communities.

The maintenance and *in situ* conservation of monuments and archaeological emergencies is one of the main objectives of the management, protection and enhancement of the heritage of Adulis.

Before describing the operative strategies for the conservation site, it is appropriate to introduce the multiple factors to be considered:

- assessment of the specificities of each situation (building, excavation sector etc.);
- congruence with the diagnostic analysis and respect for the materials and construction techniques of the historic buildings;
- compliance with minimum intervention and maximum reversibility;
- awareness of the limits of the intervention and of the ageing processes over time and its durability; possible degree of optimization, upgradability; relative cost-effectiveness of the intervention and its strategic evaluation.
- activation of documentation procedures of the conditions preceding the intervention and subsequent monitoring to evaluate the merits and possible defects.

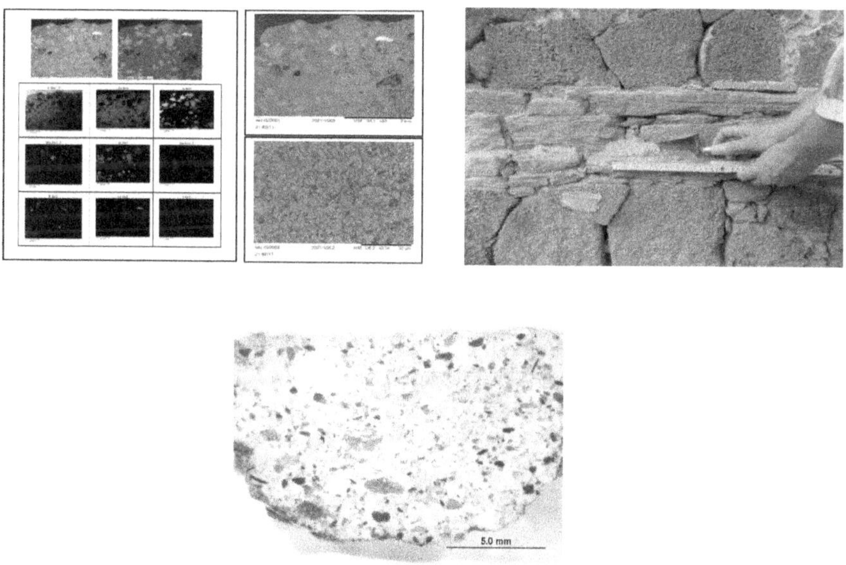

Fig. 9 – Sampling ancient mortar for diagnostics in conservation activities: from documentation to lab analysis.

Fig. 10 – Excavation protection at the end of the annual fieldwork.

Moreover, the actions must respect a protocol that implies three main points: the new interventions must clearly distinguish themselves from the original parts of the structures; the use of locally available resources must be favoured; the conservation results must minimize maintenance work.

Operations in the conservation of ruined architecture of Adulis complied with the points indicated above in the awareness, on the one hand of securing the archaeological ruins and handing them down to posterity, on the other hand, of being able to grasp the opportunity to train restoration technicians in the field. At each mission opening, the work team is called upon to initiate assessments of the current state of the sites and undertake maintenance and conservation activities.

Conclusions

The multidisciplinary methodological approach used to define the strategic conservation plan for the Adulis archaeological site aims to provide a range of integrated solutions for sustainable development, the exploitation and enjoyment of cultural heritage and a better quality of life in the neighbouring villages of Afta and Zula through the activation of innovative lines, high-profile research and training related to the realization of the Adulis Sustainable Archaeological and Natural Park. A key strategy for site management is community involvement in sharing the future land conservation plan (large area) and in partnerships for preventive maintenance and conservation with sustainable raw materials. Thanks to the existing relationships with local authorities, the project will increase the impact of interventions on the territory strengthening its status and offering the opportunity to activate new lines of action, considering its improvement replicability and scalability in other contexts.

References

BARD *et al.* 2021: Bard K. A., R. Fattovich R., A. Manzo A., *Harbor of the Pharahos to the Land of Punt II. Archaeological Investigations at Mersa/Wadi Gawasis, Egypt, 2006-2011*, Napoli 2021.

BEYN & BAR-YOSEF MAYER 2018: Beyn A., Bar-Yosef Mayer D., "Prehistoric Settlements on the Red Sea Coast of Eritrea", in Sarathi A. (Ed.), *Early Maritime Cultures in East Africa and the Western Indian Ocean,* Oxford: Archaeopress Access Archaeology, 2018, pp. 1-32.

BORTOLOTTO *et al.* 2013: Bortolotto S, Cattaneo N, Massa S., "Adulis, una città-porto tra Mediterraneo e Oceano Indiano: archeologia e restauro", in *Conservazione e valorizzazione dei siti archeologici: approcci scientifici e problemi di metodo*, 29° Convegno Internazionale Scienza e Beni culturali, Bressanone 9-12 luglio 2013, Venezia, pp. 923-934.

BORTOLOTTO et al. 2021: Bortolotto S., Cattaneo N., Massa S., "Seasonal watercourses as a source of wealth and a cause of destruction: the water management in Adulis (Eritrea) in antiquity and today", *LARHYSS Journal P-ISSN 1112-3680/E-ISSN 2521-9782,* vol. 47, 2021, pp. 25-38.

BORTOLOTTO et al. (in press): Bortolotto, S. Cattaneo, N., Massa, S., Medin, T., Zerai, A. (in press), *At the Origins of the Built Environment in the Horn Africa. The Case Study of Adulis*, L'Erma di Bretschneider, Roma.

BROGIOLO & CHAVARRIA ARNAU 2019: Brogiolo G.P., Chavarria Arnau A. (eds). *Participatory Research in Archaeology. Legal Issues and Good Practices, PCA Studies 9*, Mantova, 2019, 61-76.

CASTIGLIONI et al. 1995: Castiglioni,Al., Castiglioni An., Salza A., *Madre Africa ritorno alle origini.* Milano, Mondadori Editore, 1995.

CASTIGLIONI et al. 2018a: Castiglioni A., Bortolotto S., Cattaneo N., Massa S., "Past landscapes to shape future societies: Research in the Horn of Africa", *Journal of Universities and international development Cooperation, 1*, 2018, pp. 12-25.

CASTIGLIONI et al. 2018b: Castiglioni Al., Castiglioni An., Gebreyesus Y., Massa S., Manzo A., Giostra C., Bortolotto S., "Archaeological research at Adulis. The Eritrean-Italian joint project 2011-2015". In Tsighe, Z., S.M. Idris, Mesfu Asfaha, Y., Woldeab Andemariam, S., Kifle Taddese, R., Ogubazghi, G. (Eds.), *International Conference on Eritrean Studies Proceedings, 2*, (pp. 895-916), Asmara, National Higher Education and Research Institute, 2018.

CASTILLO MENA 2019: Castillo Mena A., "Participative processes in cultural heritage management. Methodology and critical results based on experiences within the Spanish World Heritage context", in Brogiolo G.P., Chavarria Arnau A. (Eds.), *Participatory Research in Archaeology. Legal Issues and Good Practices, PCA Studies 9*, Mantova, 2019, 61-76.

CATTANEO & MASSA 2020: Cattaneo N., Massa S., "Adulis (Eritrea) criticità e peculiarità di un sito complesso nel corno d'Africa", *Archeologia e Calcolatori, 31.2*, 2020, pp. 49-59.

DRAGONI & CERQUETTI 2019: Dragoni P., Cerquetti M., *L'archeologia pubblica prima e dopo l'archeologia pubblica.* Il capitale culturale. Studies on the Value of Cultural Heritage Supplementi 09, 2019.

FATTOVICH 2008: Fattovich R., "The Archaeology of Ancient Eritrea by Peter R. Schmidt, Matthew C. Curtis, Zelalem Teka", *The International Journal of African Historical Studies, 41*, pp. 346-348.

GREEN & HUNTER 1995: Green H., Hunter C., *Tourism and the environment: a sustainable relationship?* London: Routledge, 1995.

HARROWER et al. 2019: Harrower M.J., Dumitru I.A., Perlingieri C., Nathan S., Zerue K., Lamont J.L., Bausi A., Sweida J.L., Bongers J.L., Woldekiros H.S., Poolman L.A., Pohl C.M., Brandt S.A., Peterson E.A., "Beta Samati: discovery and excavation of an Aksumite town", *Antiquity, 93(372)*, 2019, pp. 1534-1552.

LAUREANO 2001: Laureano P., *Atlante d'acqua.* Torino: Bollati Boringhieri, 2001

LARENTIS 2021: Larentis O., "A Muslim Grave: First Evidence of the Ancient Islamic Funerary Practice in AA.VV., For an Archaeology of Religious Identity in Adulis and in the Northern Red Sea Region of Eritrea", *Journal of African Archaeology*, 19, 2021, pp. 33-39.

MANZO 2010: Manzo A., "Adulis before Aksum? Possible 2nd and 1st Millennium BC Evidence from the Site of the Ancient Port", in Lusini G. (Ed.), *Current Trends in Eritrean Studies* (p. 29-43). *Annali, 70.* Napoli: Università degli Studi di Napoli "L'Orientale", 2010.

MANZO 2017: Manzo A., "Bi3w Pwnt in the archaeological record Preliminary results and perspective of research", in Micheli I. (Ed.), *Cultural and Linguistic Transition explored Proceedings of the ATrA closing workshop Trieste* (p. 87-108). Trieste: EUT, 2017.

MASSA 2017: Massa S., "La prima chiesa di Adulis. Le origini della cristianità nel Corno d'Africa alla luce delle testimonianze archeologiche", *Rivista di Archeologia Cristiana, 93*, 2017, pp. 411-455.

MASSA 2019: Massa, S., "The Christianisation of Adulis in Light of the Material Evidence", in Manzo A., Zazzaro C., De Falco D. J. (Eds.), *Stories of Globalisation: The Red Sea and the Persian Gulf from Late Prehistory to Early Modernity, Selected Papers of red Sea Project VII*, Brill, Leiden, 2019, pp. 314-334.

MASSA 2021: Massa S., "Alcuni reperti diagnostici", in Castiglia G. *et al.*, "Un nuovo complesso episcopale nel Corno d'Africa di età tardo antica? Gli scavi del Pontificio Istituto di Archeologia Cristiana ad Adulis (Eritrea): 2018–2020", *Mitteilungen zur Christlichen Archäologie*, 27, 2021, pp. 41-46.

MASSA & GORLA 2024: Massa S., Gorla D., "Mediterranean goods from the excavation contexts of Adulis (Eritrea)", 33rd Congress of the Rei Cretariæ Romanæ Fautores Leiden, the Netherlands, 15-22 September 2024.

NUCCIOTTI et al. 2019: Nucciotti M., Bonacchi C., Molducci C., *Archeologia Pubblica in Italia.* Firenze: University Press, 2019.

PARIBENI 1907: Paribeni R., "Ricerche nel luogo dell'antica Adulis (Colonia Eritrea)", *Monumenti Antichi, 18*, 1907, cc. 437-572.

RUSSOM 2018: Russom R., "Conservation challenges on Eritrean cultural heritage: A case of Qohaito and Matara archaeological sites", in Tsighe Z., Idris S.M., Mesfu Asfaha Y., Woldeab Andemariam S., Kifle Taddese R., Ogubazghi G. (Eds.), *International Conference on Eritrean Studies Proceedings*, Asmara: National Higher Education and Research Institute, 2018, pp. 881-894.

SCHMIDT et al. 2007: Schmidt P.R., Curtis M.C., Teka Z., *The Archaeology of Ancient Eritrea.* Asmara: The Red Sea Press, 2007.

SMITH 2006: Smith L., *Uses of Heritage*. New York: Routledge, 2006.

STATE OF ERITREA 1999: State of Eritrea - Ministry of Tourism, *National Tourism Development Plan (2000-2020)*, 1999, unpublished report.

VOLPE 2017: Volpe G., "Patrimoni culturali e comunita colpite dai terremoti", in De Bonis L., Giovagnoli M. (Eds.). *Territori fragili. Comunità, patrimonio, progetto. Scienze del Territorio, 7*, 2017, pp. 48-53.

VOLPE 2019: Volpe G., "Archeologia al futuro. Teoria e prassi dell'archeologia pubblica", in Dragoni P., Cerquetti M. (Eds.), *L'archeologia pubblica prima e dopo l'archeologia pubblica.* Il capitale culturale. Studies on the Value of Cultural Heritage Supplementi 09, 2019, pp. 9-23.

Vertical Archipelago Model and Mountain Adaptation of Saho Pastoralists in Qoḥayəto Plateau–Gulf of Zula Culture Area

Robel Haile Gebru
Department of Sociology and Anthropology
College of Business and Social Science, Adikeih

Abstract

The existence of multiple topographic and climatic (topo-climatic) units located at different altitudinal zones in mountainous regions creates a zonal complementarity, where communities inhabiting various topo-climatic units engage in different forms of exchange systems (Mura 1972). Vertical archipelago model was the conceptual and methodological tool invented by cultural ecologists to examine the patterns of resource utilization of mountain dwellers and their inter-community exchange systems throughout the altitudinal length of the mountain. Guillet et al. (1983) criticizes the fragmentary tendency of ecological studies in mountainous adaptation as being too particularistic based on distinct localities. Consequently, this research pioneers the application of vertical archipelago model and zonal complementarity of the Qoḥayəto-Zula Culture Area (with 2700 altitudinal range). The fieldwork was conducted between September 2014 and July 2015 among the Fäqatə Ḥarakə *mela*, which occupy the territory from Central Qoḥayəto to the Village of Zula in the Red Sea coast.

Four topo-climatic units are identified extending from the village of Dəgədəgəta in Qoḥayəto Plateau to the coastal village of Zula; *Highland, Mid Altitude, Coastal Lowland,* and *Marine* (mainly islands) units demarcated based on the ecological characteristics, and land use pattern of the inhabitants strongly defined by tribal affiliation and segmentary lineage social organization. Finally, this research concludes that a combination of ecological verticality and tribal horizontality are the governing principles of the mountain adaptation of the inhabitants.

Keywords: Cultural Ecology, *Mela*, Topo-Climate, Qoḥayəto-Zula, Saho.

Introduction

Cultural ecological studies on mountainous region contributed profoundly to the understanding of human-environment relations in contemporary and past contexts in several parts of the world: Near East (McCorriston and Hole 1991), Mesoamerica (Fox 1980), Yemen (Korotayev 1995), and the Andes (Buren 1996). The mountainous region adjoining the Red Sea Basin and the Horn of Africa, that also extends deep into interior East Africa, sets a similar physiographic and social condition suitable for cultural ecological studies of mountain adaptation. Recently, environment-based inquiry for the co-emergence of complex societies and economic diversification expanded to the southern Red Sea (Harrower et al. 2010). The Southern Red Sea attracts researchers due to the existence of diverse topo-climatic and ecological units, and proximity to the Nile Valley and Near East. The territorial expansion is accompanied with thematic diversity with new interest developing on the interplay of environment, landscape, social and family factors in the origin of complex societies and new economic regimes (Purrington 1984, Goring-Morris and Belfer-Cohen 2001).

Map 1: Location of Qoḥayəto-Zula Mountainous Region

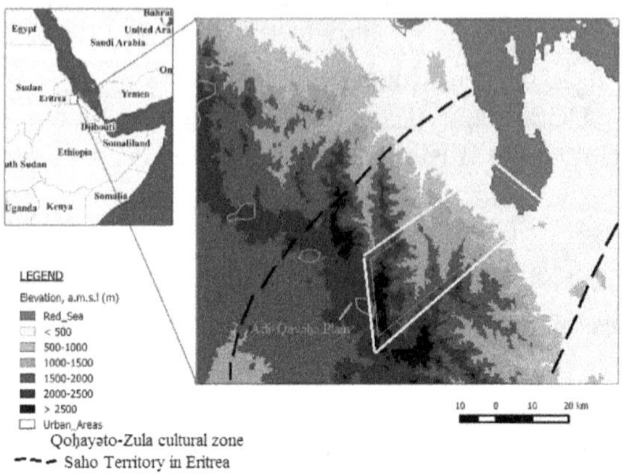

The topographic and climatic characteristics of the Qoḥayəto Plateau-Gulf of Zula (hereafter Qoḥayəto-Zula) region form mountainous terrain with a 2700 altitudinal stretch (Map 1). Well preserved archaeological and historical evidences, and ethnographic accounts indicate that the Qoḥayəto-Zula region experienced shared cultural past at least for the last three millennia. Despite the convincing material evidences, little textual and ethnohistorical accounts, no integrated cultural ecological study covering the entire mountainous zone, in a holistic way, has been carried out. Fragmentary archaeological research separately in the Plateau and the Gulf parts of the region were carried out in a highly localized scope: ʿÄdulisə (Peacock and Blue 2007; Samuel Yemane and Tsegai Medin 2008), Mäṭära (Anfray 1954), Kesəḳese (Curtis and Daniel Habtemichael 2008), and Qoḥayəto (Steffen Wenig 1997; CARP 2007; Yosief Lisbsekal 2008; Wenig and Curtis 2008, Robel Haile 2018 among many others).

None of these researches sought to explore the cultural ecological condition of the entire length of Qoḥayəto-Zula. This mountainous habitation needs to be studied in a holistic framework to comprehensively understand the adaptive mechanism taking place across the vertically stratified topo-climatic units (Purrington 1984). Guillet et al. (1983) criticizes the fragmentary tendency of cultural and ecological studies in mountainous regions for being too particularistic and localized, and when generalization occurs, it is just a micro-macrocosm nature. To create a holistic cultural ecological outline, a preliminary classification and detail documentation of activities is required.

The preliminary classification proposed in this paper is based on the cultural (lineage organization, and land use pattern), and topo-climatic (altitude and rainfall in particular) parameters. This article employed the vertical archipelago model to develop the preliminary classification and outline of the cultural ecology of Qoḥayəto-Zula region.[1] The vertical archipelago model discusses the existence of diverse ecological zones in a mountain region, and the exploitation pattern of these vertical archipelagoes by the inhabitants.

Study Area

Qoḥayəto Plateau forms the highest plateau in Eritrea with an elevation of 2700m above sea level (Map 1), and is found north of the Soira Mountain Range (3000 meter above sea level); see Plate 1. The Plateau is bounded by two major rivers that descend almost in parallel courses from the Plateau to the Red Sea; River Ḥädasə and River Komᶜäli. The Plateau covers a total area of about 32km^2 (CARP 2007), and has annual rainfall ranging from 500 to 700 mm (Wenig and Curtis 2008). The Plateau has a dry savanna vegetation cover dominated by small cactus shrubs and grasses. A few isolated pockets of even green and semi-evergreen trees are found in the Plateau including juniper (Juniperas procera), acacia, aloe, and some montane forest trees.

Qoḥayəto Plateau is inhabited by three Saho *mela*s (clans); ᶜÄsa Lesanə, Fäqatə Ḥarakə and Dasamo *mela*s from north to south sequentially. The ᶜÄsa Lesanə clan dwells in the northern part of the Plateau in villages such as Saro and Saro Gubi. The central part of the Plateau is occupied by the Fäqatə Ḥarakə *mela*, and has many villages compared to the northern and southern regions of the Plateau. Wäqäyəro, Mashurə, and Gurəbətiya, are among big settlements with more than forty families in each. The southern part of Qoḥayəto Plateau, connected by a tiny bridge of land between the villages of Donəgoləgolo and Masagole Zola, is inhabited by the Dasamo *mela*, and ᵓIgila is the biggest village in this part of the Plateau.

Members of each *mela* dwell in their respective clan territory of the Plateau, and each tribal territory covers from the Plateau to the coast in a quasi-parallel orientation (Fig. 1). Onc of the unique features of the settlement pattern of the Saho speaking communities, compared to other ethno-linguistic groups, is that majority of the *mela*s (clans) and *kišo*s (tribes) possess a tribal territory forming a narrow strip of east-west oriented land extending from the Plateau to the Gulf. This settlement pattern and territorial organization ensures access to multiple topo-climatic units (resources), and to flexibly practice diverse subsistence activities.

Fig. 1: Sketch of the topo-climatic verticality and tribal horizontality of Qoḥayəto-Zula

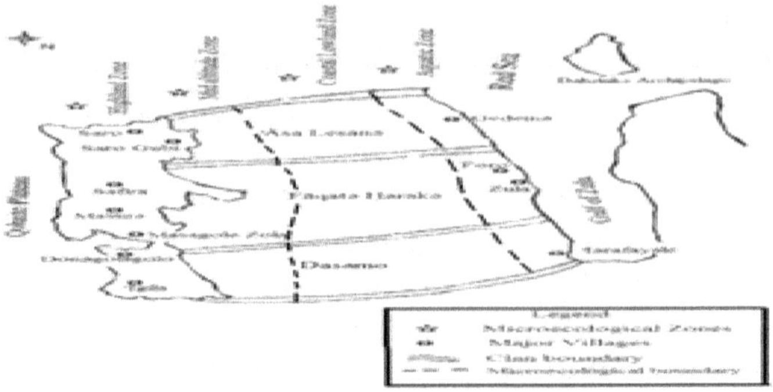

Segmentary Lineage Organization is the primary social organization system of the inhabitants, and accordingly land is owned at *kišo* (tribe) level. All members of the *kišo* are entitled to a usufruct right throughout the tribal territory. Although land is owned at *kišo* level, the tribal land is practically further sub divided into internal clans. The *kišo* territory of the Minifärä stretches from central part of Qoḥaito Plateau to the vicinity of the Gulf of Zula, and the sub tribes designate their sub-tribal territory within the general *kišo* territory.

The tribal and clan territorial organization is the key factor with respect to the application of vertical archipelago model. There is no uniform lineage-territory relationship that characterizes all the topo-climatic units. It varies from one unit to the other. In the Plateau and coastal lowland unit, the clans have bounded territory, and villages are inhabited exclusively by distinct lineage group. In the Mid Altitude topo-climatic unit, land is owned at *kišo* level, and villages are interspersed. In Minifärä *kišo*, the *mela*s under this *kišo* have separate territory in the Plateau. In the Mid Altitude unit (Nabagadä Valley) however the villages of these *mela*s are found interspersed. Hence, the variable lineage-territory relationship across the different topo-climatic units demonstrates the mutual relationship between the cultural and ecological domains.

Theoretical and Methodological Background

Vertical archipelago model, proposed by Murra (1972), was developed to understand the cultural ecological pattern of mountain adaptation, and the inter-ecological zone relationship among several ethnic groups in the Andes. This model classifies the whole altitudinal range of a mountainous region into several topographic and climatic sub-units. According to Mary Buren (1996), Murra was highly influenced by the structural functionalism school of anthropology and Karl Polanyi's work on exchange among pre-capitalistic societies. Understanding the exchange system in pre-capitalist mode of economy is one of the essential components of this model. Comparison of ethnographic data in Africa and Polynesia gave Murra insight on how kinship and political organization contribute to maintain the equilibrium of a cultural system (Buren 1996).

Vertical archipelago model is inherently linked with the notion of zonal complementarity, which explains how the inhabitants of the vertically organized ecological archipelagoes complement their subsistence needs and other requirements by engaging in different forms of exchange among them. The altitudinal zonation create heterogeneous groups (Stanish 1989), which complement their needs by inter-ethnic or inter-ecological zone exchange networks.

Depending on the number of ecological zone exploited by one particular society, Guillet et al. (1983) identified general and specialized strategies (types) of exploitation. The general (mixed) strategy refers to economic systems where a particular population engages in utilizing several ecological zones found in its mountainous terrain. Secondly, specialized strategy means the total dependency of the inhabitants in one ecological zone, and exchanging the surplus product in regional trade network in order to acquire products from other ecological zones. The former reduces risk, while the latter exposes the inhabitants into a greater risk, which is why in most mountainous habitations societies tend to mix their subsistence base in broad range subsistence activities.

Application of the original vertical archipelago model in this research requires slight modification to explore additional variables applicable to Qoḥayəto-Zula context. Most vertical archipelago applications tend to focus on the heterogeneous nature of the mountain dwellers and the resulting relationship among them. In Qoḥayəto-Zula mountainous region however the dwellers are homogenous in terms of subsistence pattern, ethnicity and territorial organization pattern. Moreover, the settlement pattern is characterized not only by the ecological verticality but also by clan/tribal horizontality (Fig. 1). Tribal horizontality refers to the fact that the territory of the three clans is horizontally divided. Hence, Qoḥayəto-Zula is defined by dual cultural ecological organization derived from the topographic and climatic verticality, and cultural (tribal) horizontality. This dual organization is the unique cultural ecological characteristics of the Qohaito-Zula area. Thus, the main objective of this research is to outline the topo-climatic units and create classification based on cultural and ecological variables.

Classification of the topo-climatic units in the study area modifies the highland-lowland dichotomy commonly practiced in some mountain studies as well as ethnological classification in the region. Nadel (1944) for instance classified Saho in to Saho ᵓAgani (lowland), and Gubi (highland), and such simplistic classification reduced the complexity of the human-nature relationship. The new cultural ecological classification proposed in this research is based on several interrelated parameters. Some of the parameters are social organization with special focus on lineage units and territoriality, distribution of village, permanence of settlements, means of income, land tenure, terrain, soil type, and family and lineage group structure. Ultimately, four topo-climatic units were identified; Highland, Mid Altitude, Coastal Lowland and Marine units. Murra and several scholars of vertical archipelago model often ignored the marine (island) zone in their cultural ecological classification, while this research covers it significantly because of its considerable input to the livelihood of the inhabitants.

Methodologically the research uses primarily ethnographic observation carried out among the inhabitants of Qoḥayəto-Zula area. Little (1999) distinguished two major areas of anthropological research on environment; ecological anthropology and anthropology of environmentalism. The former uses ecological methods to study interrelationship of human groups with their environment while the latter uses ethnographic methodologies to study environmentalism as a type of human action (Little 1999). This research falls to the second category and sought to examine the Saho clan's activities in their territory.

The data collection was carried out by intermittent field visits between September 2014 and July 2015. The data collection techniques were combination of participatory observation, interview, oral sources and literature survey. Since the research constitutes a multisite ethnography on large geographical area, the researcher had to supplement the direct observation method by other data collection techniques. In the Highland and Mid Altitude topo-climatic units, participant observation and interview were carried out, while in the other two topo-climatic units interviewing seasonal migrants and secondary sources encompassed the data gathering method. A final note on the methodology is that the fieldwork was carried out among the Fäqatə Ḥarakə *mela*, which occupy the territory from Central Qoḥayəto to the Gulf of Zula. They have occupied the central Qoḥayəto Plateau since around 16th CAD (according to clear genealogical data).

Topo-Climatic Units of Qoḥayəto-Zula Culture Area

Four topo-climatic units are identified in the territory of the Fäqatə Ḥarakə *mela* extending from the village of Dəgədəgəta in Qoḥayəto Plateau to the coastal village of Zula; *Highland, Mid Altitude, Coastal Lowland,* and *Marine* (coast and islands) zones. The terminology used in this research is 'topo-climatic' rather than 'micro-ecology' due to the absence of detail ecological data of the whole clan (tribal) territory. The notion of topo-climatic means that the classification of the vertical units (zonation) of the study area is predominantly based on the topographic and climatic characteristics affecting the exploitation pattern of the inhabitants.

The preliminary classification of the topo-climatic units is based on the ecological (topographic and climatic characteristics), lineage organization, and land use practices of the inhabitants. Despite the diversity in climatic and topographic setting, the resources in most topo-climatic zones are seasonal in availability. The swapping rainy seasons between the two ends of the clan territory enables the inhabitants to engage in optimal exploitation by seasonally migrating into the topo-climatic unit where rainfall is available. Below is discussed the four topo-climatic units in detail.

Highland

This topo-climatic unit encompasses the Qoḥayəto Plateau and its surrounding slope. Qoḥayəto Plateau has a dry savanna climatic condition with rainy season in summer ranging from June to August. Qoḥayəto Plateau receives an average annual rainfall of 500mm (MoLWE 2005). The other three topo-climatic units do not get rainfall during summer, but get flashflood from the Plateau. Due to high erosion rate, poor vegetation cover, and rocky geomorphology, the Plateau has thin soil profile of loamy and sandstone type, which compared to other types, renders less productive soil.

Summer cultivation[2], June to August, is the predominant economic activity in this topo-climatic unit. The inhabitants cultivate cereal crops such as wheat and barley. Wheat and barley are the most widely cultivated crops in the Plateau because the local bread and porridge are prepared from these grains. In addition to the family diet, the crops yield stalks used as fodder. Compared to the legumes, wheat and barley are less resistant crops to fluctuations in the rainfall distribution. Cultivating both crops requires a uniform distribution of rain throughout the summer, and minor irregularity of rain results in poor or no harvest.

Supplementary crops cultivated in the Plateau include chick-pea, lentils, flaxseed, and

2 The introduction of agriculture in the modern period among the Fäqatə Ḥarakə *mela* is attributed to Qadi ʿÄbədäla, who was a local chief of the *mela* during the Italian colonial period. He invited his close friend, probably ally, Dägəyatə Zahəlayə from a neighboring village called Täk̠onədaʿe. In the event, on June 13 Geez calendar (of an unknown year), Dägəyatə Zahəlayə brought fifty pairs of plough oxen and fifty sets of plough tools to train Qadi ʿÄbədäla men ploughing, and according to local informants this marked the first ever acquisition of plough tools in the recent history of the *mela*. Due to the absence of written sources, however, it is not clear when the introduction of agriculture took place. Some informants say that Qadi ʿÄbədäla's friendship with Dägəyatə Zahəlayə of Täk̠onədaʿe became intimate after both were invited to Rome to attend ceremonial proceedings by the then Italian Emperor. From the inscription written on Qadi ʿÄbədäla's medal awarded in his visit to Rome, it was in 1911 that he visited Rome. The second landmark reference to deduce the time of the event is the death of Qadi ʿÄbədäla, which according to his grandson Maḥəmudə ʿÄbədu, was around 1930 (+/-3). Thus putting the two pieces of information together, we could conclude that the introduction of agriculture to modern day inhabitants of Qoḥayəto Plateau was between 1911 and 1930.

beans. These legume crops require less labor, amount and regularity of rain, and the sowing period is at the terminus of the rainy period. This makes cultivating legumes easier for the inhabitants by accurately predicting the end of the rainy days. Legumes can grow and ripe merely by the water stored in the thin soil layer, and the low evaporation rate in the Plateau enhances longer storage of humidity in the subsurface. Despite relatively easy productivity, legumes are not widely consumed by the inhabitants, and often are sold in the nearby towns for a reasonable price, because pulses are usually in high demand by the neighboring Təgərəña society for food, condiments, and ceremonial ingredients.

Plate 1: General view of Qoḥayəto Plateau

Plate 2: Basketry as an alternative source of income

Plate 3: Topography of the Mid Altitude

Plate 4: Gärädafə Plain

Plate 5: River bank cultivation at Mid Altitude.

Plate 6: Farmland Preparation in Coastal Lowlands

Table 1: Types and Number of Livestock in Some Villages in the Highlands *(Data collected in 2015)*

Sn	Village/ cluster	No. Family	Sample	Family Detail	Livestock Type		
					Cattle	Sheep	Goat
1	Anədalə	2	2	Family 1	-	50	15
				Family 2	2	40	-
2	Märäbaʿe	9	6	Family 1	5-8	25-30	5-8
				Family 2	2-3	-	-
				Family 3	5-8	-	-
				Family 4	-	5-10	-
				Family 5	-	-	-
				Family 6	-	-	-
3	ᵓAfuma	12	6	Family 1	1-2	-	5-8
				Family 2	4-6	35-40	-
				Family 3	-	2-5	-
				Family 4	-	2-5	-
				Family 5	2-4	-	-
				Family 6	-	-	-
4	Bozo	15	6	Family 1	5-6	25-35	60-80
				Family 2	-	5-10	5-10
				Family 3	-	5-10	5-10
				Family 4	1-2	5-10	-
				Family 5	-	3-5	-
				Family 6	-	-	-
5	ᵓAbəʿa	27	6	Family 1	2-5	-	150-200
				Family 2	7-10	30-40	-
				Family 3	15-20	20-25	-
				Family 4	2-4	10-15	5-7
				Family 5	5-10	20-25	10-15
				Family 6	-	-	-
6	Gurubətiya	>40	6	Family 1	10-15	15-20	-
				Family 2	-	15-20	-
				Family 3	10-15	30-40	40-50
				Family 4	6-10	10-21	21-28
				Family 5	10-15	14-15	-
				Family 6	61-50	14-16	-
7	Qələʿuz		6	Family 1	-	20-30	13-21
				Family 2	13	-	20-31
				Family 3	20-30	-	-
				Family 4	14-16	-	-
				Family 5	-	-	-
				Family 6	13-14	-	-

Pastoralism makes the second most important subsistence activity in the Plateau. Majority of the families in the highland keep varieties of small and large livestock for different purposes. Cattle, sheep/goat, chicken, and donkey, and seldom camel are raised and kept for plough, dairy supply, commercialization, pack animals and long distance traveling. The average herd size per family is 3-4 cattle, 10-15 sheep/goat, 2 donkeys and a set of chicken (about 20). Families can keep this herd size sedentarily in the Plateau perennially. But, if the herd size exceeds the average size, the family needs to migrate in winter, November to March, to the coastal plains for pasture and water. Table 1 presents a detailed data of livestock types and number in some villages in Central Qoḥayəto.

In autumn and winter, two types of seasonal migrations are common in the Plateau. Transhumance is the first type of seasonal migration, where some members of a family migrate to the other topo-climatic unit for pasture and water. The second type of seasonal migration is non-pastoral in nature and shorter in duration. In this mode of migration, the male adult members of a family go to work in the cash crop fields in the valley at the foot the Plateau. Villages like ʕəšəka located in the Mid Altitude area are famous for cash crop cultivation; vegetables, fruits and tobacco.

Besides summer cultivation and pastoralism, families who live in the Plateau engage in diverse forms of economic activities to supplement their income. Craftworks, such as basketry and bead works, and petty trade including firewood collection, charcoal production, apiculture, and small scale poultry are frequently practiced by women (Plate 2). Wage labor in construction works, stone quarrying, food for work projects, retail shops, and civil works (teachers, local administration personnel, site guards etc) are among common jobs in the Plateau. The diversification of subsistence activities increases the sedentary nature of settlement in the Plateau.

Mid Altitude Unit

This topo-climatic unit is located at the Mid Altitude region between the Plateau and the coastal plain with an average elevation of 1000 to 1500 m above sea level. The terrain is characterized by deep valley, gorges, steep slopes, meandering river beds and scattered patches of flat terrain (Plate 3). These different topographic areas are used for different purposes; the banks of the rivers for seasonal cultivation; the hilltops for settlement; and the slopes for grazing. Unlike the Plateau, the Mid Altitude unit has thick soil profile of alluvial and fluvial deposits that are found throughout the foot of the slopes and banks of the river.

The Mid Altitude topo-climatic unit is characterized by hot and dry climatic conditions with annual rainfall below 500mm (MoLWE 2005). Since this zone is located at the foot of the slopes, where the tributaries enter their middle course, the area receives adequate flashflood from the Plateau. The slow velocity and broader banks of the streams enable the farmers to use river diversion techniques to water their fields. Normally, the rivers flow in two periods of a year; briefly in spring (March-April) and throughout summer (June-August). Long cycle crops, such as maize and sorghum, are widely cultivated in this area with little fruits and vegetables as cash crop supplements.

The rugged topographic nature of this topo-climatic unit results in low population. A village is common composed of a cluster of few families. Villages are located near the banks of the streams, and inhabited on semi-sedentary or sedentary bases. Rarely, there are isolated patches of flatland where population is relatively dense. Gärädafə Plain located at the Nabagadə Valley is a good example. The Nabagadə Valley will be discussed below in detail as a case study.

The Nabagadə Valley has about five major tributaries (streams); Bakura, Ba ʿewa, ʾÄrəbaba ʿe, Forohima and Madəhulo, and numerous smaller streams such as Säbulale. A relatively extensive land within the Nabagadə Valley is located at Gärädafə Plain, which is estimated to be 4km long and 1.5km wide (Plate 4). Despite the semi-dry climate of the area, the Plain has relatively fertile soil, and hence is excellent for drought resistant crops such as maize and millet.

Compared to the nearby regions, the Nabagadə Valley has high population density, and during my fieldwork 112 households (both nuclear and extended types) dwell in the banks and adjacent hills. Unlike the other topo-climatic units, the agricultural land in Gärädafə Plain is owned by minimal lineage groups instead of individual family.

Secondly, the land tenure systems in Gärädafə Plain is not based on communal (tribal) ownership. It is rather owned on hereditary basis by the minimal lineage groups. This hereditary ownership of land sheds an interesting clue to the beginning of agriculture in this area. In Ma ʿetarəta Village, for instance, the hereditary agricultural fields continued for four generations, and this could be estimated to have taken place around 75 years ago.

Table 2: Sample of Villages in Nabagadə Valley

	Village	Families	No. of Agricultural Plots	Perimeter of each plot	Watering Stream	Owner Clan
1	ʿAsa ʾAbəsali Dage	12	36	20m x 6m	Nabagadə	Nare, ʾIgila,
2	Saganəto	13	29	20m x 5m	Forohima/ Nabagadə	Gaʾaso/Dasamo
3	ʾAməbo	15	24	20m x 5m	Forohima	Dasamo
4	Madəhukuro	16	23	19m x 6m	Nabagadə	Fäqatə Harakə
5	Maʿetarəta	18	9	20m x 5m	Forohima	Dasamo
6	ʾIbahə Gibitə	8	40	30m x 5m	ʾArəbabaʿe / Nabagadə	Fäqatə Harakə
7	Forohima	13	16	20m x 5m	Forohima	Gaʾaso

This account corresponds significantly with the chronology of the *Ziyari Tena*[3] ritual initiated by Qadi ʿÄbədäla as a commemorative event of the beginning of agriculture. Qadi ʿÄbədäla was the tribal chief of the Fäqatə Harakə during Italian colonial period, who in collaboration with Dägəyatə Zahəlayə of Täkonəda ʿe started cultivation, and the ritual became the commemoration for the event. According to colonial texts, Qadi ʿÄbədäla reigned between 1911 and 1930. This chronology fits very well to the estimation of four generation count of hereditary farmland in Gärädafə Plain. Moreover, the farmland share per family increasingly shrank across generations, and this could be a factor for diversification of subsistence activities, and a reason for the Valley-to-Plateau migration. Other factors that catalyze the migration include access to social services, overseas remittance sources, and drought.

Spate-irrigation in the banks of the tributaries is the cultivation technique in this topo-climatic unit (Plate 5). The river flows twice a year; March to April, and June to August. Spate irrigation requires the construction of diversion wall, canals, ditches, defensive wall, digging wells, and constant maintenance. Hence, it depends on communal labor mobilization among several families whose agricultural fields are located adjacent, and watered from the same tributary and canal. This constitutes corporate labor created among members of minimal lineage to construct and maintain the spate irrigation infrastructure.

Coastal Plains: Graze Land and Deltaic Cultivation

Coastal zone covers area from the lower frontier of the Mid Altitude all the way to the coast. Desert and semi-desert climatic conditions prevail in this area with less than 100mm annual rainfall (MoLWE 2005). Drought resistant deciduous trees are found, mostly acacia and cactus species. Most of the topographic feature of this zone is characterized by flat land with muddy and sandy soil types, wide banks, and small hills. A riverine vegetational type dominates the area with dense trees found near the banks, and the density of vegetation reduces as moving away from the banks. A homestead is located at an average distance of 100 or 200 meters from the nearest bank, and the people have to fetch drinking water from the wells dug on or near the river bed or nearby cistern, and water their herds there.

Rainy season in the coastal plain zone lasts from November to February. Due to low productivity and insufficient rain, agricultural activity is limited to few areas; delta and banks of river - sometimes near big settlements such as Foro. Foro is a coastal town located in the River Ḥädasə and River Komʿäli's deltas. The surrounding environment is deltaic with thick soil layer, flat terrain, and higher capacity to store water. Corns, vegetables and fruits are cultivated in the delta and banks during both rainy seasons of the Plateau and the coastal lowland.

Areas interior to the delta have low agricultural potentials, and often are used for grazing. Due to continuous renewal and recharging soil by the flood, the meadows grow abundant grasses and the tree regenerate leaves during the rainy season. The extensive meadows with small shrubs and grasses adequately support pastoralism, and most pastoral families practice transhumance form of pastoralism in this zone during winter (November to February). The herds are composed of predominantly small animals, goat and sheep, and small proportion of cattle and camel.

Marine: Coast and Island

As mentioned earlier, most Saho *mela*s have access to the sea, and this adds aquatic resources to their diverse ecological niches. The economic potentials of the marine resources include access to fishing ground, sea-shell and other mineral collection, salt fields, and long–short distance trade and transport. Fishery and marine exploitation however are the least developed sector in Eritrea, and had been declining till the 1990s. According to Tesfa G Gebremedhin (1996), the number of fishermen in the country over the last few decades were 8514 (1967), 5240 (1971), 54 (1978), 124 (1984), 550 (1987), and 2600 (1991). Factors that affected the extremely underdevelopment of the fishery industry were harsh environment, lack of auxiliary facilities such as ice plant, poor inland and maritime connectivity between small and main harbors, traditional technique of fishery, and conflict.

Nowadays, very few clan members practice subsistence fishing in the coast and island. Often men carry out fishing expedition in group organized under expedition master, usually the owner of the fishing boat. They collect several species of fishes, and the most widely collected fishes are grupper, snapper, emporer, baracuda and varieties of tuna. Women are mostly engaged in collecting snail and sea cucumber in the shore. The fishing period, where abundant marine resources become available, is in November and December. Usually the catches are sold to agents of the Ministry of Fishery or to a corporate association called EriFish, who later process, distribute and/or export the collection.

Discussion

The ecological diversity created by vertical archipelago of topo-climatic units in the Horn of Africa, represented by Qoḥayəto-Zula culture area, has the potential for developing and testing of anthropological theories regarding social ecological complexity and human-nature relationship. Michael Harrower et al. (2010) noted continental and global scales of explanation of transitions to agriculture frequently obscure sub-regional complexities particularly lesser known areas, and the Southern Red Sea margins are therefore well suited to clarify the diversity of trajectories towards agriculture. The Qoḥayəto-Zula area covers the shortest distance in the Horn of Africa between the coast and the highland, and the existence of four topo-climatic units in less than 100km thus makes it very appropriate region or such research.

The first discussion point refers to the dynamics of adaptation by colonizing four topo-climatic units seasonally, which enables flexible adaptation strategies for the three Saho *mela*s inhabiting Qoḥayəto-Zula. The striking aspect of this multi-ecological adaptation of these *mela*s proves to be unique in Eritrea, and probably in the Horn of Africa too. None of the other ethno-linguistic groups in Eritrea possesses such holistic ecological diversity that ranges from the Plateau to island ecologies. This shows that every Saho *mela/kišo* has adequate ecological resources at its disposal and cultural mechanism to exploit these diverse resources.

By outlining the multiplicity of topo-climatic units aligned vertically along the altitudinal variation, this research paper attempted to reformulate the highland-lowland dichotomy prevalent in anthropological and archaeological works. This refinement of altitudinal stratification increases our ability to understand the transition and dynamics of environmental diversity and emergence of new economic regimes by elaborating the intermediary economic relationship between the highland and the lowland. The Mid Altitude zone forms an intermediary cultural ecological nature in terms of land ownership, social organization, seasonal migration patterns and other social-ecological parameters.

The second discussion point deals with the impact of ecological diversity and holistic colonization of the topo-climatic units by each *mela* to the social life of the inhabitants especially in terms of social and political organization. Unlike the notion of complementarity of vertical archipelago model, where different ethnic groups engage in regional exchange networks, Qoḥayəto-Zula culture area shows a homogenous mountain adaptation by the Saho *kišo*s' to satisfy their economic need by occupying range of ecological zones. In the case of Andean societies, access to diverse resources was maintained by inter-ethnic exchange mechanisms among various communities dwelling in distinct ecological zones. Charles Stanish (1989) noted that the Andean vertical archipelagoes are characterized by a striking regional ethnic heterogeneity in space and time. The Saho *mela*s of Qoḥayəto Plateau on the other hand acquire diverse resources by seasonal occupation of the multi-ecological zones of their tribal territory, and this creates the homogeneity and self sufficiency of the Saho society at clan/tribal level.

Conclusion

This research advocates for the introduction of regional scale anthropological and archaeological study to generate a holistic understanding of the complexity and dynamics of Qoḥayəto-Zula (and beyond) cultural zone. As Nawata (2001) pointed out, most pastoral studies concentrate on terrestrial adaptation, which result on meager information on pastoralists' exploitation of coastal ecosystem. Thus, the multiplicity of ecological setting and the existence of four topo-climatic units in this culture area with widespread archaeological and historical material remains throughout the area offer wonderful setting for a social-ecological complexity (nonlinear relationship) and dynamics (resilience and change) study.

The research concludes with open question regarding the cause-effect relationship between the autonomous governance and the ecological homogeneity at *mela/kišo* level. This question provokes anthropological research to test models of cross cultural of exchange and economic interdependence and homogeneity of the *kišo*s occupying the Qoḥayəto-Zula cultural area. Similarly, a preliminary correlation between modern day multi-ecological adaptation between the Gulf and Plateau, and the nature of settlement pattern precisely matches the prehistoric pattern of settlement pattern. Currently, the bigger settlements are established at both ends of the culture area, while the two climatic units in between are extremely sparsely populated and occupied seasonally. Similarly ancient cities of ʿÄdulisə and Qoḥayəto overlap in carbon copy fashion with contemporary bigger settlements of Foro and Safira respectively. I believe social ecological oriented ethnoarchaeological and ethnographic approaches possess the potential to lead us into a profitable venture to explain this correlation.

References

Buren, Mary, 1996. Rethinking the Vertical Archipelago: Ethnicity, Exchange and History in the South Central Andes. American Anthropologist, New Series, 98: 338-351.

CARP (Cultural Asset Rehabilitation Project)2007. The Archaeological and Cultural Landscape of Qoḥayəto, Eritrea: Site Management and Implementation Plan.

Curtis, Matthew and Daniel Habtemichael, 2008. Mäṭära, Keskese, and the "Classical Period" Archaeology of the Akele Guzay Highlands: A Brief Overview. In (Schimidt, Peter and Zelalem Teka, eds). The Archaeology of Ancient Eritrea. pp 311-328, Red Sea Press, Inc, Trenton, NJ.

Fox, W John, 1980. Lowland to Highland Mexicanization Processes in Southern Mesoamerica. American Antiquity, 45, 1: 43-54

Goring-Morris,A. Nigel and Belfer-Cohen, Anna, 2001. Neolithization Processes in the Levant: The Outer Envelope. Current Anthropology, 52, S4: 195-208

Guillet, David, 1983. Towards Cultural Ecology of Mountains: The Central Andes and the Himalayas Compared. Current Anthropology, 24, 56- 574

Harrower, Michael, Corriston, M. Joy and D'Andrea, A. Catherine, 2010. General/Specific, Local/Global: Comparing the beginning of Agriculture in the Horn of Africa (Eritrea and Ethiopia) and Southwest Arabia (Yemen). American Antiquity, 75, 3: 452-472

Korotayev, Andrey, 1995. Some Trends and Factors of Evolution of the Sabaic Language and Sabaean Culture. Oxford University Press, Oxford.

MoLWE, (Ministry of Land, Water and Environment, State of Eritrea), 2005. Strategic Approach of Water Harvesting Using Roof Catchment: The Eritrean Experience.

Nadel, S. Frederick ,1944. *Races and Tribes of Eritrea.* British Military Administration, Asmara.

Nawata, Hiroshi, 2001. Coastal Resource Use by Camel Pastoralists: A Case Study of Gathering and Fishing Activities among the Beja in Eastern Sudan. Nilo-Ethiopian Studies, 7, 23-43.

Peacock, David and Blue, Lucy, 2007. The Ancient Red Sea Port ofᶜÄdulisə, Eritrea: Results of the British Expedition 2004-5. Oxbow Books, Oxford.

Purrington, L. Burton, 1984. The Context of Cultural Ecology in Highland Environments. In (Beaver, D. Patricia and Purrington, L. Burton, eds). Cultural Adaptation to Mountain Environments. pp 3-11. University of Georgia Press, Athens.

Robel Haile, 2018. The Social Archaeology of Cave Dwelling in Qoḥayəto Plateau. *Ethnorema,* 17: 54-71.

Samuel Yemane and Tsegai Medin, 2008. The Ancient Port City of ᶜÄdulisə. (in) Paleoanthropological and Archaeo-Historical Research on Eritrea. pp 25-34, National Museum of Eritrea: Asmara.

Schimidt, Peter and Zelalem Teka, 2008. The Archaeology of Ancient Eritrea. Red Sea Press, Inc, Trenton, NJ

Stanish, Charles, 1989. Household Archaeology: Testing models of zonal complementarity in the South Central Andes. American Anthropologist, New Series, 91: 7 – 24

Tesfa G. Gebremedhin, 1996. Beyond Survival: the Economic Challenges of Agriculture and Development in Post Independence Eritrea. Red Sea Press, Inc, Asmara.

Wenig, Steffen, 1997. German Fieldwork in Eritrea. Nyame Akuma, 48, pp 20-21

Wenig, Steffen and Curtis, Matthew, 2008. Qoḥayəto: An Ancient Highland Urban Center. In (eds) Schimidt, Peter and Zelalem Teka. The Archaeology of Ancient Eritrea. pp 287-300, Red Sea Press, Inc, Trenton, NJ.

Yosief Libsekal, 2008. The Development of Ancient Complex Society and Early Urban System of Qoḥayəto. In (Tekeste Melake, ed) Paleoanthropological and Archaeo-Historical Research on Eritrea. pp 35-44, National Museum of Eritrea, Asmara.

Architectural and Urban Heritage in Eritrea - A Conservation Approach

Medhanie Teklemariam

Asmara Heritage Office
Email: medhanetm@gmail.com

Abstract

Eritrea is situated in a strategic location in the Horn of Africa. Asmara is the capital city located in the center of the country on a highland plateau. The city was founded on a site where four ancient villages are settled. Its modern development began in 1889 with its occupation by Italy. The city became an experimental laboratory for the application of modern architecture and urban planning ideas in its geographical, cultural, and local contexts. The city possesses one of the largest concentrations of modernist architecture in the world. Asmara is an outstanding example of the interchange of cultural influences brought about by the encounter with modernity in an African context and bears exceptional testimony to the combination of modern urban planning and modernist architecture in the African region.

In many African cities and other regions, 20th-century architectural heritages have been massively destroyed and swept away from the historic urban landscape under the pretext of modernization, economic growth, and searching for a new identity. Eritrea's approach to conserving Asmara's colonial-era architecture represents a profoundly different attitude towards modern heritage conservation and its interpretation and treatment compared to many other post-independence urban environments. This continued conservation effort resulted in the inscription of Asmara: A Modernist City of Africa on the prestigious UNESCO- World Heritage List.

The paper aims to offer an overview of Asmara's urban conservation effort to safeguard its modernist architecture, although the methodology applied in its regional, cultural, and historical context is very specific. The approach that is being pursued in the Asmara urban conservation process is based on the lessons learned in the region and could be adopted as the best practices of 20th-century urban heritage conservation in the African region or beyond.

Keywords: Africa, architecture, conservation, heritage, urban

Introduction

Asmara is a capital city, which is located in the center of the country, on a highland plateau, 2300m above sea level. The first human settlement, according to the oral tradition, was established on the same land where the town is located now, between the XII and the XIII century. More precisely, it was located northeast of the town, in the same place where the Coptic cathedral is now situated. The four villages did not have a peaceful life. They were often exposed to cattle raids carried out by the inhabitants of the lowlands, migrating towards the mountains during the dry season, and by nomadic tribes which, from inland regions pushed on as far as the lowlands to exchange hides and dairy products. Besides stealing the livestock, they also captured all the heads of household and most of the able men from the four villages. As a result of this raid, the women gathered and held a council meeting. After several days they decided to merge the four villages into one on the site of Kidisti Mariam. The new village, which counted about six or seven hundred inhabitants, was called **ኣርባዕቱ፣ ኣስመሩ**, Arbatè Asmera. The name can be interpreted as *"four women agreed"* or "*the women put the four together*" (Leonardo Oriolo, 1998, pp39-44).

The main objective of the paper is to offer an overview of Asmara's urban conservation effort to safeguard its modernist architecture. It traces and analyzes the various modern urban heritage conservation projects carried out in the historic center of Asmara. Key lessons learned from the case study and the possibility that it could be replicated as one of the best practices on the conservation of 20th century urban heritage in Eritrea and beyond will be addressed.

Modern heritage conservation is a widely under-researched area at the national as well as global level. This paper intends to contribute to broadening our knowledge about modern heritage conservation. In addition, the results of the study can complement the knowledge about conservation issues, and motivate others to contribute similar studies on the subject.

This research explores Asmara as a case study through the qualitative method, by analyzing historical documents and various architectural and urban heritage projects and reports. It also undertakes selected comparative analysis to provide an overviews of five African cities´ modern heritage conservation practices, namely Addis Ababa, Nairobi, Dar es Salaam, Zanzibar, Kinshasa, and Lubumbashi.

Architecture and Urban Planning

The first attempt to organize the city of Asmara with a "building plan" dates back to 1894-96 (Figure 01). Engineer Captain Ernesto Cantoni had been behind the preparation of the first Asmara Master Plan in1894 (Godio Anna, 2008). Following this initiative, it was the two civilian Governors of Eritrea Ferdinando Martini (1897-1907) and Giuseppe Salvago Raggi (1907-1915) who were the driving forces behind the implementation of the city's town plan. With the purpose of improving the colony's civil organization, mechanisms were outlined whereby the cities and towns in Eritrea were to become worthy representatives of Italy's great cultural and architectural traditions (Greseleri G., Massaretti P.G., et al, pp. 23-47).

Engineer Odoardo Cavagnari developed the town planning concept of zoning that was later adopted in the planning of the urban layout of Asmara in 1913 (Godio Anna, 2008, p.41). Cavagnari's urban layout gave more attention to the racial segregation zoning between Italians and Eritreans, broader spaces and squares, offices, commercial and financial activities, as well as the geographic and topographic characteristics. The 1916 Master Plan illustrates that the neighborhoods were designed according to a rectangular grid pattern type of urban layout and zoning (Figure 02).

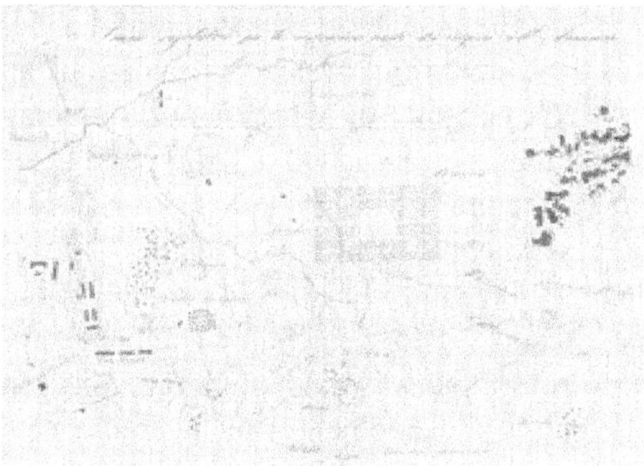

Figure 01. Asmara Master Plan, 1893

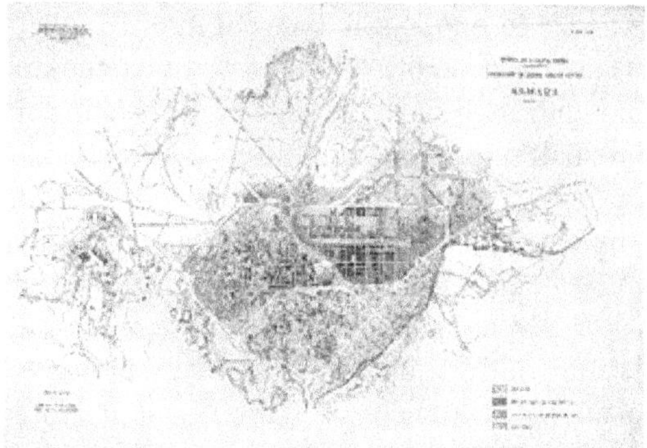

Figure 02. Asmara Master Plan, 1916

The modern layout of the colony's capital can be attributed to the construction and town planning program of Asmara's modern urban planning, which was prepared during 1935-1939 by a group of technicians led by Architect Vittorio Cafiero (Figure 03). Cafiero incorporated the street layout originally drawn up by Cavagnari into his new master plan, placing greater emphasis on the main arterial roads of Harnet Avenue (formerly Viale Mussolini and Viale Cordona) and the traffic problem (Godio A., 2008, pp.43-48). From 1935 to 1941, the Italians changed the face of Asmara with a system of new infrastructure and construction of thousands of new buildings. Asmara became known as "Piccola Roma", meaning little Rome. The unique urban setting, extensive road network with modern traffic control systems, wide tree-lined avenues, modern infrastructure, and the proliferation of Rationalist and other progressive forms of architecture made Asmara one of the most modern cities in Africa.

The urban plan drawn by Vittorio Cafiero could not be fully realized due to the fall of the Italian colonial government in 1941 by Allied Forces.

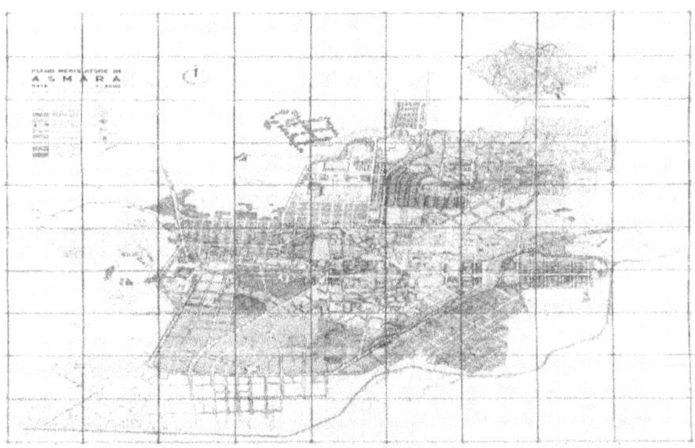

Figure 03. Asmara Master Plan Designed by Architect Vittorio Cafiero, 1938

In the following decades, Asmara continued to grow along the west, north and east directions. Until the early 1970's, Asmara had no updated master plan to guide urban development (Tecle Belula, 2017). In 1973, the Municipality of Asmara commissioned Studio Mezzedimi, an Italian consulting office, to prepare a new master plan. Arturo Mezzedimi's

Master Plan presented a detailed image of the postcolonial city. It was a comprehensive study plan based on the principles of modern urban planning to manage the expansion of the city that seemed out of control. In addition, the study of the Mezzedimi Master Plan undertook an in-depth analysis of the state of the existing city until the early 1970s, which included tracing urban development, and the economic, social and demographic situation. Even though some development activities were realized according to Mezzedimi's Master Plan, it was not fully implemented because a new Ethiopian Military Government (Derg Regime) came to power in 1974, and all development activities were halted.

Even though the 1970s master plan was too old and did not reflect the social, economic and political context of the city, the Mezzedimi Master Plan was used as an interim urban planning instrument by the city administration to guide development growth until a new master plan was implemented.

Following the Eritrea´s independence, the Municipality of Asmara established a Planning Committee in 1998 to outline urban development strategies for Asmara. After analyzing the urban problem, the committee recommended the need for preparing a new urban plan of the Greater Asmara Area. Accordingly, the Asmara Infrastructure Development Study was launched in 2004. The study comprised of preparing the Strategic Urban Development Plan (SUDP) and the Outline Urban Planning Regulation. The SUDP was developed for the Municipal authorities as a guiding planning strategy to coordinate and manage spatial development in the Greater Asmara Area. The SUDP envisaged a less ambitious, more economical, and more sustainable compact type of development, based on the fact that cities historically tend to evolve in cycles of spatial expansion, consolidation and densification of the urban structure. In addition, it defined strategies for preserving the historic character of the city (SUDP, 2006).

From an urban planning perspective, Asmara is not the outcome of a one-off urban plan, but rather the result of an innovative planning process encompassing colonial, postcolonial and regional planning and its application in an African context. The tailor-made urban plan integrated topographic features, local climate, local cultural conditions created by different ethnic and religious groups and used zoning principles to achieve racial segregation and functional organization. Mixed zone planning was not only used as functional organization of the city, but also as a meeting point between Italians, Eritreans and other nationalities. As a result, this space contributed by becoming "*the melting pot*" of the interchange of different cultures, the peaceful coexistence of religions such as the Orthodox, the Islamic, the Catholic and the Protestant. This exchange of religious and cultural values had a significant contribution on the development of architecture and intangible heritage in Asmara.

Asmara Modernist Architecture

Asmara's modernist architecture represents one of the most complete collections of its genre in the world. The existing stock of historical buildings displays a considerable diversity of styles and typologies. Before 1935, the architecture reflected various historical styles, including Classicism, Lombard, Romanesque, Renaissance, Neo-Baroque, Neo-Gothic, and Eclectic.

Figure 04. Asmara Theatre, 1919 Eclectic style

Figure05. Asmara Testing Centre, 1915 Neo-Baroque style

After 1935, following the invasion of Ethiopia, there was a swift transition to modernism, expressed diversely: from proto-modernist approaches like Novocento to Rationalism, Futurism, Monumentalism and even vernacular forms (Asmara UNESCO Nomination Dossier, 2016, p.56).

Figure 06. Cinema Impero, 1938 Art Deco Style

Figure 07. Fiat Tagliero, 1938, Futurist Style

In Europe architectural modernism was expressed with the new materials of glass and concrete, but in Eritrea, where such materials were comparatively expensive, many of the modernist buildings were constructed using local materials and made to look modern. Although reinforced concrete was available and frequently used, many buildings in Asmara used large quantities of local basaltic rocks rendered in lime plaster to give the appearance of concrete in order to create modern geometric forms befitting the modern machine age, albeit constructed using traditional materials. Asmara's unique architectural value does not only derive from the prevalence of Rationalism. The city also possesses other expressions of modernism. Collectively, Asmara's architectural character is therefore broad and aptly represents the transformation of early-20th-century architecture from tradition to modernism in Africa (Denison, E. Teklemariam M. and Abraha D., 2016, p.33).

Evolution of Modern Urban Heritage Conservation

Overview

During the struggle for liberation (1961-1991), few new buildings were constructed in Asmara. As a result, the city faced acute shortage of residential housing, commercial and other social services. The municipality of Asmara has been under immense pressure to improve living conditions, provision of commercial facilities, housing and modernization of infrastructural services. Since early 1990s, the Municipality of Asmara has encouraged investment and construction of new buildings, which had started throughout the city. The Asmara Municipality Technical Department granted permits to construct high-rise buildings in the city center. During this period, a few isolated high-rise buildings were built in different locations (Teklemariam, M. 2017, pp 47-49). Residents and professionals of the city noticed that these buildings were becoming strange and incompatible with the city's physical characters, and skyline, and were threatening its historic fabric. This gradual damage to Asmara's architectural heritage triggered discussions and debates that contributed as a wakeup call for professionals and the public to take appropriate measures to address the challenge before it became uncontrollable. In the mid-1990s, a heated debate arose on the rationale for the development of the high-rise buildings in the historic perimeter and on how to protect Asmara's modern architectural and urban heritage (CARP, 2003).

Asmara is regarded by Eritreans their national identity, pride and symbol. Asmarainos have a strong attachment and love for the architecture of their city, unlike many other parts of Africa where denigration of the colonial past took place. It can be said that Eritreans have successfully appropriated colonial buildings and reinvented the built environment according to their own aspirations. In this regard, Denison stated that 'Many ask why, in spite of all the colonial and racial indignities perpetrated by foreigners, Eritreans still seek to maintain the town plan and architectural heritage of their colonial oppressor, and a fascist one at that. The answer partly lies in the fabric of Eritrean society'. Gebremedhin Naigzy stated that the reason why Asmara's architecture should be preserved because 'Eritreans played a significant role, perhaps the main role in the building the city of Asmara. Had it not been for the labor of thousands of Eritreans, many whom perished without recognition, and equally many who labored under duress, Asmara would not have been built. Their memory should be preserved in equal measure with the Italian Architects and Engineers who designed the city'[1].

Italian colonization had its own dark history in Eritrea, nevertheless, Eritreans have no hatred for the colonial buildings, rather they have different attitude toward the built environment. Eritrea's approach to the conservation of colonial-era architecture in Asmara represents a profoundly different attitude towards modern heritage preservation and its interpretation and treatment compared with many other postcolonial urban environments (Denison E., Teklemariam M., and Abraha D., 2016). The preservation of Asmara's urban heritage cannot be imagined without the local sense of ownership and belonging.

Since independence, several conservation projects have been carried out to protect and conserve the city's outstanding architectural and urban heritage. Some of the key projects are described below:

1 Asmara, Africa's Secret Modernist City – Paper prepared for the African Perspectives: Dialogue on Urbanism and Architecture, TU, Delft, Naigzy Gebremehin, 2007, p18.

The Cultural Assets Rehabilitation Project (CARP)

The CARP was formally established in 1997 and concluded in 2007. CARP's mandate was to identify, preserve, and promote awareness of the diversity of Eritrea's cultural heritage, and it was engaged in several parallel programs across the country. The CARP project included archaeological sites, record management and museum development, and the recording of oral history and traditions. The built environment is one of CARP's key programs, of which the historic center of Asmara and Massawa were of priority concern. The CARP undertook many activities and results, namely:

- Delineation of the historic perimeter of Asmara,1998
- Inventory and identification of historical buildings, 2001
- Planning initiative for the historic perimeter of Asmara, 2003
- Publication of Asmara architectural books and map, 2003
- Preparation of tentative listing application for potential UNESCO of Asmara, 2004 and
- Promoting built environment of Eritrea includes Asmara, Massawa and Keren.

The National Heritage Program (NHP)

The NHP was initiated to safeguard the built environment and implement a series of capacity building and training programs on conservation at the municipal level (2008-2011). The program identified Eritrea's built heritage as an important cultural and economic asset and carried out the surveying and rehabilitation of Cinema Capitol and the central market (Mieda Eritrea). In addition, a series of capacity building and training programs and archival management system were carried out.

The Asmara Heritage Project (AHP)

The Asmara Heritage Project was formally established in March 2014 under the auspices of the Administration of Maekel Region (Municipality of Asmara). The AHP builds on successes and lessons learned from previous similar initiatives, namely the CARP (1997-2007) and National Heritage Program (2008-2011). The main objectives of AHP were the preparation of:

- UNESCO – World Heritage nomination dossier, 2016
- Integrated Management Plan, 2016
- Disaster Risk Management Framework, 2016
- Comprehensive survey of historical buildings, streets, public and open spaces, and archival documentations, 2014
- Detailed Urban Conservation Plan, and of the Planning Norms and Technical Regulations, 2023
- Promoting public awareness on architectural and urban heritage.

Comprehensive Inventory and Documentation

The Central Region Administration through the AHP had conducted a comprehensive inventory of the entire Historic Center. This is the first ever exercise that was carried out at the city and national level. The main goal of the inventories was to provide the Central Region Administration and other concerned authorities with a tool to control and monitor historical building conditions, as well as the overall physical and functional transformations which may affect the urban fabric of the historic center. The inventories were therefore intended to offer updated and comprehensive information, which would help establish the degrees of protection and to define the possible required interventions. This was meant to give both the public and the private sectors a clear and solid base to plan and design the work to be undertaken. Furthermore, the inventories were also aimed at serving as baseline information to develop Detailed Urban Conservation Plan for the World Heritage Property

A comprehensive inventory of buildings, infrastructures and open spaces in the World Heritage Property was conducted by the AHP during 2014 to 2016. The AHP collected the following data form the field survey: over 4,300 buildings, 257 roads, 38 open and public spaces, and scanned and catalogued over 80,000 technical documents from the Municipality's archive.

A heritage database was also developed to assist in the monitoring of the state of conservation. In particular, the information management system was developed to provide and retrieve information including condition of buildings, typology, construction technique, number of floors, use and function; architectural significances, categories of interventions such as preservation, renovation, new construction and demolition.

The database system also supports the analysis and dissemination of information to inform decision making processes. The ultimate goal of the comprehensive inventory is to serve as the basis for the preparation of the urban conservation master plan, as well as to support for the establishment of a good benchmark for monitoring the state of conservation and management of Asmara World Heritage Site (Teklemariam, M. 2018, pp. 42-57).

Asmara UNESCO World Heritage Inscription Process

Asmara was in the tentative list of UNESCO since 2005. To be inscribed on UNESCO World Heritage List, the full nomination dossier had to be prepared including the supporting documents such as the management and conservation plan. Therefore, the Central Region Administration in collaboration with the Ministry of Education and Commission of Culture and Sports initiated the Asmara UNESCO nomination dossier preparation in 2014.

After the compilation of all the data and extensive research studies for over two years, Eritrea submitted its first-ever application to UNESCO for inscription on the World Heritage List on 1st February 2016. The AHP prepared the 1,300-page Nomination Dossier in collaboration with professionals from a wide range of disciplines in Eritrea and overseas, with the support of many national and international government agencies. The advisory body to the World Heritage Committee - the International Council for Monuments and Sites (ICOMOS) carried out a vigorous evaluation of the nomination dossier for one and a half year, and it recommended for the inscription of Asmara. The UNESCO -World Heritage Committee at its 41st session held in Krakow, Poland, on 12th July 2017, declared Asmara: A Modernist City of Africa a World Heritage Site.

The Asmara World Heritage Site perimeter comprises the 'historic center' of the city, with an area of 470 hectares (Asmara Nomination dossier, 2016, pp14-17). Asmara met two criteria of UNESCO and was inscribed on the World Heritage List under the following two UNESCO criteria:

> ***Criterion (ii)***: Asmara, a Modernist City of Africa, represents an outstanding example of the transportation and materialization of ideas about planning in an African context and were used for functional and segregation purposes. The adaptation to the local context is reflected in the urban layout and functional zoning, and in the architectural

forms, which, although expressing a modernist and rationalist idiom, and exploiting modern materials and techniques, also relied on and borrowed heavily from the local morphologies, construction methods, materials, skills and labour. Asmara's creation and development contributed significantly to Eritrea's particular response to the tangible legacies of its colonial past. Despite the evidence of its colonial imprint, Asmara has incorporated into the Eritrean identity, acquiring important meaning during the struggle for self-determination that motivated early efforts for its protection.

Criterion (iv): Asmara's urban plan layout and character, in combining the orthogonal grid with radial street patterns, and picturesque elements integrating topographical features, taking into account local cultural conditions created by different ethnic and religious groups, and using the principles of zoning for achieving the racial segregation and functional organization, bear exceptional witness to the development of the new discipline of urban planning at the beginning of the 20th century and its application in African context, to serve the Italian colonial agenda. This hybrid plan, that combined the functional approach of the grid with picturesque and the creation of scenic spaces, vistas, civic plaza and monumental places, served the functional, civic, symbolic requirement for the colonial capital. The architecture of Asmara complements the plan and forms a coherent whole, although re-electing eclecticism and rationalist idioms, and is one of the most complete and intact collection of modernist architecture in the world.

In addition, Asmara had not only met the above two criteria but also satisfies the requirement of integrity and authenticity of the UNESCO Operational Guidelines.

Integrity: All the significant architectural structures and the original urban layout, including most of the characteristic features and public spaces, have been retained in their entirety. The site has also retained its historical, cultural, functional and architectural integrity with their elements largely intact and generally in good condition. Despite continuing developmental pressures, the establishment of the 'Historic Perimeter' around the center of the city since 2001 and a moratorium on new construction within this perimeter by the municipal authorities have safeguarded the site's integrity (Asmara Nomination dossier, pp346-347, 2016).

Authenticity: Asmara's combination of innovative town planning and modernist architecture in an African context represent important and early developmental phases of town planning and architectural modernism and therefore possess a high level of authenticity (Asmara Nomination dossier, pp 346-347, 2016).

Protection and Management: The Operational Guidelines indicate that: "All properties inscribed on the World Heritage List must have adequate long-term legislative, regulatory, institutional and/or traditional protection and management to ensure their safeguarding". In response to the management requirement for inscription, Eritrean government has already established ahead an appropriate institutional arrangement, i.e. the Asmara Heritage Project office in 2014.

Asmara inscription is the first World Heritage Site for the nation, and it is a pioneering Modern Heritage Site for the sub-Saharan region. The ICOMOS gap Analysis document (2005) revealed that the Africa region is mostly unrepresented on World Heritage List compared with other region such as Europe and North America, Latin America and Asia – in particular, in the Modern Heritage List category. In this respect, Asmara has contributed an important step, albeit small, in highlighting and redressing the underrepresentation of African Modern Heritage on the World Heritage List.

Asmara's inscription has also created many opportunities, however, there are several challenges on how to manage the World Heritage Site based on its Outstanding Universal Value (OUV) and transcend it to future generations. Bandarian, F. and Van Ores, R. state that a specific approach need to be developed for the management of change in the field of architecture, public space and uses of the existing buildings. In this respect, management plans should be seen as strategic documents linking the different areas of the management (Historic Urban Landscape- HUL, p.69).

Eritrea's conservation methodology is not merely on historical building preservation but rather an integrated approach toward conserving all the attributes including buildings, infrastructure, pubic and open spaces. In addition, Eritrea recognizes the potential of the HUL Approach and incorporated to its long-term conservation policies and management planning.

Management and Conservation Tools

World Heritage Sites are recognized as places of Outstanding Universal Value under the terms of the 1972 UNESCO Convention concerning the Protection of the World Cultural and Natural Heritage. State members of UNESCO presenting such sites are hence obliged to identify, protect, conserve, present and transmit them to future generations (UNESCO 1972, Article 4). As per UNESCO World Heritage Committee requirement the State Party owner of the World Heritage Property should prepare management plan and present to the World Heritage Center.

Management: - In Eritrea, the Asmara Integrated Management Plan (IMP) was prepared in 2016 to manage the World Heritage Site. The IMP is an important planning instrument, vital for the protection, use, maintenance and sustainable development of the World Heritage Site. Asmara´s IMP developed the vision, goal, objectives, strategies, short and medium-term action plans for the protection and conservation of the site. The primary aim of the management plan is the safeguarding of the Outstanding Universal Values and managing of the World Heritage Site (Asmara Integrated Management Plan, 2016).

Urban Conservation Plan: - One of the main challenges to the preservation of Asmara's historic center is the lack of proper conservation planning and legal instruments, which has hampered the Central Region Administration in preserving the rich architectural heritage. There have been several attempts to develop conservation planning guidelines for the historic center of Asmara during the past two decades; however, they were unable to produce concrete results in time.

The recent initiative to develop a conservation planning tool based on a comprehensive inventory of historic resources and archival documentation was conducted. Moreover, studies were undertaken by a multidisciplinary team, who carried out analyses of historical, physical, and socio-economic data. After six years extensive research, in 2023, two key strategic conservation documents were produced: the *Detailed Urban Conservation Plan (DUCP)* and the *Technical Implementation, Norms, and Regulations*.

The DUCP is a fundamental tool to achieve the final goal of protecting and enhancing cultural heritage on the one hand, and the sustainable development of architectural and cultural heritage on the other. Given its importance in supporting to sustain and enhance the significance of the World Heritage Site, the DUCP was developed in a participatory approach by fully involving all interested parties, particularly those responsible for managing, owning, or administering the site.

The DUCP is a document specifying in detail urban strategies and tools to be used for heritage protection and which at the same time respond to the needs of contemporary life. It contains legislations, rules, as well as conservation and monitoring plans (Technical Implementation Norms and Regulations, 2023, p.9). The Detailed Urban Conservation Plan (2023) was prepared based on the broader concept of the UNESCO- Historic Urban Landscape (HUL, 2011) Recommendation and principles. The HUL provides the basis for a comprehensive and integrated approach for the identification, assessment, conservation, and management of historic urban landscapes within an overall sustainable development framework (HUL definition 10).

The DUCP regulates the conservation and valorization of the historical modernist fabric, integrated with strategic urban areas. This perimeter is defined as the Integrated Historic Perimeter. The DUCP intends to contribute to the realization of the objectives of Cultural and Natural Heritage Proclamation No.177/2015 and other national and regional laws and regulations related to the provisions of this Plan (Technical Implementation Norms and Regulations, 2023, p.6).

Urban Heritage Conservation Practices in Africa

Post-independence, many African cities embarked on major urban and economic development schemes in the city center to demonstrate modernization and a new identity. To realize these ambitious programs, numerous historic cities have been destroyed and their modern heritage swept away from the historical urban landscapes. Following, we present a comparative analysis of selected African cities to provide a general overview of conservation policies and approaches. The selected cities are Addis Ababa, Nairobi, Dar es Salaam, and Kinshasa.

Addis Ababa, Ethiopia

The capital city Addis Ababa has experienced unprecedented urban development during the past three decades. As a result, massive urban transformation is taking place in the city today. The construction of high-rise buildings condominium housing blocks and road infrastructure are transforming the city dramatically. Architects Fasil Giorghis and Denis Gerard state what is happening now in reality today is large-scale renewal of buildings. The Arada with the market square and the Piazza, Imperial Gibbi, and Casa Inchis area are parts of Addis Ababa that need special urban heritage protection[2]. However, all the above professional recommendations were discarded, and many historical buildings in the city center have been demolished and replaced by new high–rise developments. Architect Fasil Giorghis stated 'I was very disappointed by this, as the period was an important part of our recent history and most of the early buildings of Addis Ababa were disintegrating in front of our eyes'. Furthermore, the bulldozing has sparked outcry from some of the city's most prominent architects and planners. Fasil Giorghis, chair of conservation of urban and architectural heritage at Addis Ababa University, has argued that the destruction reflects a fundamental "lack of consciousness, lack of awareness of the value of heritage". "Addis Ababa is beginning to look like an imitation of some other city," says Esther Sellassie Antohin, founder and director of Heritage Watch, which advocates for the restoration of urban heritage in Ethiopia. "There's a hollowness, there are disconnections, a disjointedness that presents a problem.

Nairobi, Kenya

In 1905, the British moved the capital of British East Africa from the coast (Mombassa) to the fertile and salubrious conditions on the high plateau (Nairobi), mirroring the Italians' similar strategy in 1900 when they moved the capital of the then nascent Italian East Africa from Massawa to Asmara. Early urban planning and subsequent architectural developments in Nairobi were similar to those in other British colonial cities in Africa (e.g. Harare, Lagos, Lusaka, Kampala, etc.). Although Nairobi was never the modernist city like Asmara was, the

2 The City and Its Architectural Heritage Addis Ababa, Giorghis, F. and Gerard D., P.382

largest distinction between the two cities in relation to preservation of Nairobi's architecture. Nairobi almost complete loss of the architecture that defined the early modern period – demolished largely to make way for subsequent wave of development before and in particular after independence in 1963.

Dar es Salaam, Tanzania

In 1867 the British commander Richard Bradshaw said of the new palace and town plan of Dar es Salaam: 'It is thought that this port will supersede Quiloa [Kilwa] as the domestic depot of the Sultan, the caravans from the interior altering their route for it on account of its proximity to Zanzibar. With the advent of the European powers (1885-1890), Dar es Salaam developed into a German Garden city, and Zanzibar was taken over by the British administration and turned into a typical segregated colonial city. The Stone Town Conservation Development Authority (STCD, 1984) has the responsibility of Conserving Stone Town and legal provision of declaring any area or town as conservation area. In 2007, the increasing economic pressure on heritage by developers and politicians alike caused the rejection of the Mainland's Preservation Act. The situation in 2013 is that entire blocks of the city center of Dar es Slaam, multiple early worthy buildings of the early twentieth century, are being demolished to make room for high-rise developments (Antoni S.Floker et al., 2017, pp74-76). Comparing Zanzibar Stone Town with Asmara, both have similarities in preserving their historic center. In contrary, Dar es Salaam demolished its architectural heritage and losses integrity and incomparable with Asmara conservation context.

Kinshasa, Democratic Republic of the Congo

Situated on the opposite bank of the Congo River to Brazzaville, Kinshasa was among the most important Belgian colonial cities. Founded by the British explorer Henry Stanley in 1881, Kinshasa was laid out according to the French system of quartiers and became the capital of the Belgian Congo in 1923. The city's urban planning was grander in scale and more ambitious than neighboring Brazzaville, but subsequent development and conflict throughout the second half of the twentieth century has seriously altered Kinshasa's original form and layout, causing the city to bear none of the integrity that Asmara possesses in its architectural and urban character.

The Belgian colonization produced immense infrastructure, the large part of which facilitated the economic exploitation of the territory's natural resources including in Kinshasa, Lubumbashi and Likasi. Urbanization occurred at different pace throughout the territory in the mid-1920s to 1950s. In the context of the recent building booms in Kinshasa and Lubumbashi, several examples of colonial architecture have been demolished or transformed without any restriction (John Lagae, 2017, pp 42-43).

Key lessons from other African cities

Many African cities have been experiencing massive transformation in their city center during the past five decades. Subsequently, they have been replacing their heritage withe new high-rise buildings to imitate other cities like Dubai, Singapore and Hong Kong, in search of a new identity. They have lost their sense of place "*genius loci*' and irreplaceable rich modern heritage. This destruction reflects ill-planned urban development policies and strategies and lack of awareness of urban heritage preservation. It is not to say that development should be stopped or frozen in the city center, but rather development should respect and take proper account of the existing built environment. Eritrea has learned from its own urban development mistakes implemented during the 1990s and from the experience of other African cities. After critically assessing various urban development policies and strategies, Eritrea has taken appropriate correction measures to conserve the architectural heritage of the 20th century.

Conclusions and Recommendations

Conclusion

The protection and conservation of architectural heritage as an expression of cultural identity is rare in African cities. In this respect, Eritrea has a unique experience in initiating and executing ambitious architectural and urban heritage projects that are sensitive to sub-Saharan Africa. Furthermore, in Africa, modern architectural heritage conservation has remained marginalized and under studied.

As a young nation, Eritrea has many socio-economic challenges and priorities. Nevertheless, the government has considered cultural heritage development as one of the key pillars of national development and has given appropriate attention. Eritrea has been pursuing a comprehensive and an integrated approach conservation, linked to national policies and UNESCO charters and Historic Urban Landscape recommendations. In particular, it has developed scientific methodologies on how to conserve modern architectural and urban heritage. As a testament to its successful efforts, it received UNESCO recognition in 2017 and the President Medal Award for the best research in 2016 from the Royal Institute of British Architects (RIBA). This is a unique and promising experience in the sub-Saharan African region.

Eritrea's success story is based on lessons and experiences acquired from the various cultural heritage projects implemented and the development coherent strategies towards the conservation of 20th-century urban heritage. Although so far this is an encouraging success story on the conservation of modern heritage in Eritrea, there remains much work to be done. Therefore, Eritrea's experience in modern heritage conservation can be one of the best practices to be replicated in other African cities and beyond.

Recommendations

Preventing the destruction of old buildings in the historic center of African cities requires creating a harmonious and integrated development policy - balancing between the heritage perseveration and development, through the Historic Urban Landscape Approach and harmonious urban development strategies.

The importance of extensive collection and compilation of data on architectural and urban heritage as basis for urban conservation planning. Develop urban conservation heritage policies and strategies at city or site-specific level.

Modern architectural and urban heritage conservation is an under-research area. Therefore, a need and urgency to carry out systematic research on the conservation of modern architectural and urban heritage in African cities.

A need for strong support and recognition by decision-makers as well as the sense of ownership of the city inhabitants towards the modern architectural heritage conservation are necessary.

African cities have many common values and similarities, the Eritrean experience could be borrowed and adapted to their context. Establish architectural and urban heritage practices exchange programs or networking among professionals including architects, engineers, urban planners, and cultural heritage experts.

References

AHP Compiled Sample Reports and Survey Forms (2015), Asmara Heritage Project, Department of Public Works Development.

Arturo Mezzedimi's Master Plan (1970).

Asmara UNESCO Nomination Dossier (2016), Asmara Heritage Project, Department of Public Works Development.

BANBARIN, F., and Van Ores, R. (2012). The Historic Urban Landscape: Managing Heritage in Urban Century, p.69.

BENEVOLO, L., (1971). History of Modern Architecture, Cambridge, Massachusetts

DENISON, E. Ren, G. and Gebremedhin N. (2003). Asmara – Africa's Secret Modernist City, Merrell Publishers.

DENISON, E. and Ren, G. et al (2024). Asmara a Guide to the Built Environment (pp. 10-14).

DENISON, E. Teklemariam M. and Abraha D. (2016). Africa's Modernist City (UNESCO World Heritage Nomination), *Journal of Architecture, Volume 22, Number 1, pp.11-53. Routledge publishers.*

Disaster Risk Management Framework (2016), Asmara Heritage Project, Department of Public Works Development.

Draft Final Detailed Urban Conservation Master Plan (2023), Asmara Heritage Project, Department of Public Works Development.

Draft Final Technical Implementation Norms and Regulations (2023), Asmara Heritage Project, Department of Public Works Development.

Fasil Giorghis and Denis Gerard (2019). The City and Its Architectural Heritage Addis Ababa.

FRAMPTON, K. (1985). Modern Architecture: a Critical History, Thames and Hudson, New York.

FULLER, Mia. (2007). Moderns Abroad: Architecture, Cities and Italian Imperialism. 2011. Italy's Colonial Futures: Colonial Inertia and Postcolonial Capital in Asmara." California Italian Studies, New York, Routledge.

GEBREMEHIN N, (2007). Asmara, Africa's Secret Modernist City – Paper prepared for the African Perspectives: Dialogue on Urbanism and Architecture, TU, Delft, p18.

GODIO A. (2008). Archiettura Italiana in Eritrea- Italian architecture in Eritrea, La Rosa publishers.

GRESLERI, G., Massaretti, P.G., and Zagnoni, S. (1993). Architettura Italiana d'Oltremare 1870–1940, Marsilio, Venezia.

Integrated Management Plan for Asmara (2016), Asmara Heritage Project, Department of Public Works Development.

Interim Report Responses for: Nomination of Asmara: Africa's Modernist City for inscription on the UNESCO World Heritage List (2017), pp. 25-28.

LAGAE, Johan (2004). "Rewriting Congo's Colonial Past: History, Memory, and Colonial Built Heritage in Lubumbashi, Democratic Republic of the Congo", in Repenser les limites: l'Architecture à travers l'space, le temps et les disciplines, Paris, INHA (Actes de Colloques).

LEONARDO Oriolo (1998). Asmara Style, pp. 39-44.

Outline Urban Planning Regulation (2005), Ministry of Public Works.

Planning Initiative for the Historic Perimeter of Asmara (2003) - Cultural Asset Rehabilitation Project.

Report on Archive System of Asmara Historical Buildings (2011), National Heritage Programme.

Socio-Economic Report (2015), Asmara Heritage Project, Department of Public Works Development.

Strategic Urban Development Plan (2006), Ministry of Public Works.

TECLE-MISGHINA, B. 2014. Asmara, an urban history. L'Architettura Delle Citta' the Journal of the Scientific Society Ludovico Quaroni UNESCO-Series #1 2014. Rome: Edizioni Nuova Cultura.

TEKLEMARIAM, M. (2018). "An Overview of the Asmara Heritage Project: An Integrated Conservation Approach" Urbanstica e architettura modern alla prova della contemporaneita, Susanna Bortlotto e Renzo Riboldazzi pp. 42-57.

TEKLEMARIAM, M. and DENISON, E (2017). "The Asmara Heritage Project- Heritage Preservation: Past: Present and Future" in Architecture in Asmara. Colonial Origin and Postcolonial Experiences, Volger, Peter and Stefan Graf. DOM publishers, Berlin, pp 245-253.

TEKLEMARIAM, M., (2017). "Eritrea architectural heritage conservation policies" in Time Frames: Conservation Policies for Twentieth–Century Architectural Heritage. Ugo CARIUGHI and Massimo Visione editors Routledge Publisher, pp. 47-49.

The World Heritage List Filling the Gaps – an Action for the Future (2005)

WORTHING, D. AND Bond, S. (2008). Managing Built Heritage: The Rule of Cultural Significance. Balckwell Publishing

هجرة الصحابة إلى الحبشة
التناقضات في الروايات الموثقة للهجرتين

الدكتور تاج الدين نورالدائم يوسف

المقدمة

ما تمت كتابته من تاريخ بخصوص الهجرتين الى ارض الحبشة، ليس يحتمل الخطأ والصواب فقط، انما تمت كتابة بعاطفة ورؤية إسلامية لسيرة ومسيرة بدايات الدين الإسلامي كهدف شامل يختزل كل الاحداث بخصوص الهجرتين. إضافة الى ان ما كتب من تاريخ، يعتمد على معرفه المؤرخين التامة بالجوانب المحلية عندهم، دون الربط بينها وبين ما يتوفر من معلومات إقليمية. لنجد ان جل ما كتب عن تاريخ ارض الحبشة وعن الهجرتين، تم بمنظور عربي خالص نقلا عن مصادر متكررة. مما جعل كل الكتب كأنها فصولا في كتاب واحد، تتشابه في السرد، والتشبيهات، وتصوير الاحداث. عدا ذلك فان التاريخ التوثيقي للروايات المذكورة رغم قلته، ذاب في تنعيم الالفاظ والسرد القصصي والخيالي للأحداث، مما يثير التشكيك بان وراء ذلك هدف او اهداف لها مرامي ويفتقد الربط بين تاريخ الهجرتين كحدث، وتاريخ المكان الذي هو أساس ومحور الحدث.

ان تاريخ ارض الحبشة ضارب في القدم وفصوله مرتبطة بإنسانه وحضارته من نواحي متعددة، منها الاجتماعي، السياسي، الديني، الاقتصادي والثقافي. غير ان المؤرخين العرب وبعض المستشرقين، عند التعرض لأصل سكان القرن الافريقي، يتم ربطهم بالأجناس الوافدة، او الهجرات. وربطه بالأديان. مثل الهجرات السامية التي يتناولها معظم المؤرخين على انه تحديد الاجناس نظريا، وليس على دراسة تاريخ الانسان على الأرض.

يتم تناول اهل بلاد الحبشة بنظرة لا تخلو من التعالي والعنصرية فيما يورده بعض المؤرخين العرب، عن علاقة الحبشة بالعرب بانها كانت (منذ ان عرف العرب حاجتهم الى الرقيق ليرعى إبلهم ويجلب نياقهم)[١]. وقد ذُكر ان النجاشي نفسه تعلم اللغة العربية لأنه كان قد اشتراه أحد التجار.[٢] تذكر بعض الروايات ان النجاشي عاش في بلاد العرب ورعى الغنم هناك.[٣]

هناك روايات سعت الى التقليل من شان الهجرتين، وأخرى تصب جام الغضب على الاحباش في ذلك الوقت، وتصفهم بأنهم كانوا يعاملون المسلمين بقسوة وبغض وكراهية.[٤] هذه الورقة العلمية تمت كخلاصة لبحث دؤوب دام ميدانيا، لأكثر من عقدين من الزمان. قضيتها في جمع الإشارات والملاحظات على الأرض والتوثيق الشفاهي للأحداث. لذلك قد تخلو الورقة من تعدد المراجع والمصادر.

الكلمات المفتاحية: التناقضات، الهجرة، الصحابة ارض الحبشة، النجاشي، احمد قراني

التناقضات

التناقضات في معجم لسان العرب (التناقض) هو اختلاف الشيء مع نفسه، بحيث يؤدي إثبات أحد الطرفين إلى نفي الآخر.

في معجم الوسيط: التناقض هو تعارض المعاني أو الأقوال أو الأفعال بحيث لا يمكن التوفيق بينها.

إذن، التناقض يعبر عن حالة من التعارض التام بين أمرين، بحيث لا يمكن الجمع بينهما منطقياً أو واقعياً.

يتمثل الاختلاف في تسجيل الحدث بين مؤرخ واخر في سياق سرد الأحداث التاريخية، او وجود اختلافات أو تعارض بين الروايات أو المصادر التي توثق لنفس الحدث. او عندما تقدم جهة معينة وصفًا أو تفسيرًا مختلفًا عن جهة أخرى، سواء من حيث التفاصيل، الأسباب، النتائج، أو حتى الشخصيات.

وأسباب التناقضات يعود في اغلب الأحيان الى التحيز، حيث يتم نقل الحدث من وجهة نظر الكاتب الأيدلوجية، الدينية، العرقية، القبلية، او الجهوية،.. الخ، مما يخلق معلومات تفتقد للإثباتات الميدانية المبنية على الحقائق والاثار على ارض الحدث، لذلك يلجا معظم الكتاب في الكتابات التاريخية الشفاهية الى السرد الخيالي، وتشبعه بالاوصاف والقصص التي يصعب على المتتبع التحقق من صحة معلوماتها.

مع مرور الوقت، تظهر وثائق جديدة تُغير النتائج عن المعلومات وسبل الاثبات، او تحقيق دراسات تبنى على النظم المتطورة الحديثة والتجارب المعملية. لذلك في معظم حالات التناقض. يعتمد المؤرخون عادة على مقارنة المصادر

وتحليلها نقديًا للوصول إلى الرواية او المعلومة الأقرب الى للحقيقة.

من هم الصحابة

كلمة «صحابي» مشتقة من الجذر العربي «صحب»، والذي يعني المرافقة أو الصحبة.

الصحابة كانوا جزءًا أساسيًا من المجتمع الإسلامي الأول، وكان لهم دور كبير في نشر تعاليم الإسلام وحمايته. كما أن النبي محمد (ص) كان يشير إليهم بأنهم «أصحابه، مما يعكس العلاقة القوية والوثيقة بينهم. كاتباع للرسول محمد (ص) وهم قلة امنوا بالدين الجديد[٥] في بداياته في مكة. وقد تعرّض العديد منهم للتعذيب والاضطهاد في بداية الدعوة الإسلامية، خاصة أولئك الذين لم تكن لهم مكانة اجتماعية قوية، أو حماية من قبائلهم لإجبارهم على التراجع عن اتباع محمد (ص) باستخدام أساليب قاسية. كل من لحق بالإسلام في حياة الرسول(ص) أطلق عليهم اسم «الصحابة» لأنهم كانوا «يصحبون» النبي محمد (ص)، بمعنى أنهم كانوا يرافقونه ويعيشون معه ويشاركونه في الدعوة الإسلامية.

يعتبر الصحابة من أهم الشخصيات في التاريخ الإسلامي وقد لعبوا دورًا كبيرًا في نشر الإسلام وتعاليمه. يتفاوتون في قدراتهم وأعمالهم في تاريخ الإسلام ونشره. كانوا من أوائل المؤمنين برسالة النبي محمد (ص). ساهموا بنشر الإسلام في الجزيرة العربية وخارجها، وكانوا من المؤسسين للمجتمعات الإسلامية الأولى. شارك الصحابة في الغزوات والمعارك التي قادها النبي محمد (ص) وساهموا في نقل تعاليمه وحفظ أقواله وافعاله وفي جمع القرآن الكريم. بعد وفاة النبي محمد (ص)، تولى الصحابة إدارة الشؤون الإسلامية، وتأسيس نظام الخلافة. فقاموا بمهام إدارة الدولة الإسلامية وتوسيع حدودها. كان لبعض للصحابة تأثير شخصي على المجتمعات التي عاشوا فيها، فكانت قيمهم وأخلاقهم مصدر إلهام للمسلمين على مر العصور.

ما هي اسباب هجرة الصحابة

التعذيب والقهر الذي واجهه اتباع الدين الجديد من قبل قريش، والقبائل المتحالفة معها بعد السنة الثالثة من الرسالة. قبلها كان الاتصال مقصوراً على من يثق به الرسول(ص) وهم القلة المؤمنة والمصدقة له. هذه الفترة شكلت الجانب السري من الدعوة الى ان امر الله رسوله (ص) ان يعلن الناس بأمر الدين الجديد، وان يدعو الى إظهاره. قال تعالى (فاصدع بما تؤمر واعرض عن المشركين)[٦]. بعد إعلان الرسالة على الملاء والجهر بها تكاثر أصحابه، وزاد امعان قريش في مقاومة الدعوة بشتى السبل، فكان الازدراء والتعذيب والتنكيل بمن تبع محمد (ص). لم ينج الرسول نفسه مما واجهه أصحابه. فحادثة في الطائف والتعرض اليه، دليل واضح على انتشار الرسالة، وكذلك انتشار أساليب المقاومة والمعارضة لها.

الفرق بين الهجرة والوفد

الهجرة تعني خروج او لجوء الأفراد او الجماعات، والابتعاد عن موطنهم او مكانهم الى مكان اخر. وتتعدد أسبابها، ويكون مكان اللجوء اما مقصودا محددا للاستقرار المستديم، او للعبور. والهجرة او اللجوء، غير محددين بزمن معين. كما تعني الهجرة الانتقال من مكان إلى آخر، سواء داخل البلد أو خارجه، بهدف الاستقرار لفترة طويلة أو دائمة. او التنقل خارج المكان المعهود. غالباً ما يكون الدافع وراء الهجرة او اللجوء البحث عن حياة أفضل، في مجالات مختلفة مثل العمل، التعليم، الأمن، أو الحرية. قد تكون الهجرة اختيارية، كما في حالة البحث عن فرص أفضل، أو قسرية، كما في حالة الهروب من الحروب أو الاضطهاد. مثال: هجرة المسلمين من مكة الى ارض الحبشة، او من مكة الى المدينة.

الوفد هو مجموعة من الأشخاص، يتم إرسالهم إلى مكان محدد لتمثيل جهة معينة، سواء كانت قبيلة، دولة، أو مؤسسة. يكون الغرض من الوفد انجاز مهمة مؤقتة ومحددة. يعود أعضاء الوفد عادة إلى مكانهم الأصلي بعد انتهاء مهمتهم.. الفرق بين الهجرة والوفد يكمن في المعنى والغرض من كل منهما. لذلك يطلق على من ارسلتهم قريش للنجاشي (عمرو بن العاص ومن معه وفد. بينما يطلق على الصحابة مهاجرون.

لماذا الهجرة ارض الحبشة؟

* كانت الأوضاع في الجزيرة العربية وما جاورها من كتل اقتصادية وسياسية كبيرة، متمثلة في الإمبراطوريتين الفارسية والبيزنطية، وما دار في فلكهما من ممالك، كقوة داعمة وموالية لهما. وكانت الجزيرة ميدانا لحرب طويلة بين الامبراطوريتين. اضافة الى الحروب التي كانت سجالاً في الوسط بين القبائل وبطونها. وكان عرب قريش يركزون على ضمان مصالحهم التي كانت مع الامبراطوريتين.

* ارتباط أطراف الجزيرة العربية بالتبادلات التجارية، شمالا، جنوبا، شرقاً وغرباً، جعل الاخبر [illegible] عن الدين الجديد واساليب معاداته. شكلت قريش تحالفات داخلية مع كافة القبائل، فجعلت بعضها تعيش في مكة، ويعتبرون

أنفسهم من القبيلة القرشية، عن طريق الولاء على اسلوب العرف القبلي، وقد فسحت قريش صدرها لهؤلاء الموالين واحلتهم منزلة رفيعة فيها، جرياً على سياستها العامة في ربط القبائل بها، فضمنت بذلك ولاء القبائل لها، بل ضمنت تفوقها عليها واعترافها بسيادتها. كما تحالفت قريش مع قبائل البادية الموجودة في الحجاز، ومن اهما الطائف. لذلك عمد سكان بعض المناطق الى المغالاة في مواجهة الدين الجديد (كما حدث للرسول (ص) في الطائف[٧])

* بالولاءات تلك ارتفعت مكانة قريش بين العرب. وبزعامتها ضمنت تنفيذ مخططاتها ضد الدين الجديد، عليه، فان أي قبيلة تقترب من النبي صلى الله عليه وسلم، في مرحلة الصراع بينه وبين قريش، كان يعتبر ذلك عداء لقريش، فاذا امنت به وآزرته فإنها الحرب لا محالة. ولم يكن أحد من العرب راغباً في ان يدخل مع قريش في نزاع[٨]. مما يعني ان كل قبائل الجزيرة العربية كانت متضامنة، او شبه متضامنة مع قريش ضد الدين الجديد. وتحت سيطرة وسطوة قريش والموالين لها المتحكمين على الاوضاع الاقتصادية، والاجتماعية، والروحانية في جزيرة العرب، كل ذلك جعل من المستحيل لجوء الصحابة الى اي بقعة بأرض العرب.

* معرفة الاعراب بالحبشة، حيث كانت هناك تبادلات تجارية مختلفة وتواصل بين ساحلي البحر الأحمر، عمقت دراية الاعراب بطباع الاحباش. يعلمون انهم أصحاب ديانة سماوية، ولهم نظام حكم متوارث. لذلك وصف الرسول ملك الحبشة بانه عادل لا يظلم عنده أحد.

* وجود تعاطف من جانب المسلمين الصحابة تجاه الروم المسيحيين، كما ورد في القران الكريم، (قصة اصحاب الاخدود). تقع الأخدود على ضفاف وادي نجران في منطقة نجران في المملكة العربية السعودية. اشتهرت المدينة بالمذبحة التاريخية التي أوقعها ذو نواس الملك الحميري في سنة ٥٢٠ بحق مسيحيي المنطقة (أصحاب الأخدود) وهو الأمر الذي سجل في القرآن الكريم[٩]. وكانت المذبحة سببا في احتلال عيزانا الملك الحبشي لجنوب جزيرة العرب. إثر تحريضه من الإمبراطورية البيزنطية على غزو اليمن، فأقدم الأحباش على غزو اليمن سنة ٣٣٥، وسيطروا عليها، ذلك يعني ان العرب كانوا يعلمون عن الاحباش علما ينفي الجهالة بطباع الاحباش وامورهم الاجتماعية والدينية.

من هم الاحباش؟

الاحباش هم سكان ارض الحبشة، في منطقة القرن الافريقي، ويعرفون بالحُمر، او الابسينيين، او أصحاب الوجوه المحروقة. الاحباش تواجدوا في الساحل العربي قبل الميلاد بزمن غير معروف. وهم عنصر مستقل غير مخلوط، او ممزوج بدماء أخرى، او بمهاجرين من أجزاء أخرى من العالم، كما يزعم كثير من المؤرخين العرب، وبعض المستشرقين. وحتى يتوصل العلم غير المنحاز الى نتيجة حاسمة بشأن التصنيفات البشرية عن طرق ال DNA او الكروموسومات، لتأكيد انتماءات الشعوب والاجناس الى فئات محددة، فإنني اعتقد ان الجنس الأحمر جنس مستقل، مثل الابيض والأسود والأصفر. اما ما يطلق من مصطلح السامية للدلالة على اجناس البشر فهو تقسيم بني على اللغات وليس الاجناس. وأول من استعمل هذا الاصطلاح مطبوعاً، هو المؤرخ الألماني أوغست لودفيغ شلويستر (August Ludwig Schloester) في مقال له عن «الكلدانيين» عام ١٨٧١ الذي يقول فيه «من البحر المتوسط إلى الفرات، ومن أرض الرافدين حتى بلاد العرب جنوباً سادت كما هو معروف لغة واحدة، ولهذا كان السوريون والبابليون والعبريون والعرب، شعباً واحداً، وكان الفينيقيون (الحاميون) أيضاً يتكلمون هذه اللغة، التي أود أن أسميها اللغة السامية. وقد تولى المحلل النفسي النمساوي أيشهورن بعد ذلك هذا الاصطلاح، والدفاع عنه وإن ادعاه لنفسه[١٠].

كان للأحباش نشاط ملاحي، في ذلك الوقت الرواية هي تلك التي تقول، ان من لجاء الى نجاشي الحبشة من المسلمين، ركبوا سفناً حبشية. اقلتهم من مرسى الشعيبة الى الحبشة. وكذا ما ورد من اعتداءات الاحباش المتكررة على مواني الحجاز فيما بعد. ويرمي حسن صالح شهاب من هذه المقولة، الى انه بعد ظهور الاسلام بقيت الملاحة في البحر الاحمر بيد الاحباش غالباً سواء من كان منهم على ساحل ارتريا او على ساحل تهامة من اليمن او الحجاز[١١].

الاحباش وتقبل الروحانيات والأديان

الانسان في ارض الحبشة محور كل الاحداث التي دارت بها، لذلك نجد ان ارض الحبشة وعلى مختلف العصور منذ الاف السنين قبل الميلاد، كانت بقعة تتقبل، وتحتضن، وتتفاعل مع الاديان والروحانيات المختلفة، الوافدة اليها من مختلف الحضارات القديمة. اذكر على سبيل الأمثلة، تفاعل الاحباش وتعاملهم مع الالهة التالية:

7 الامام ابي الفدا إسماعيل ابن كثر - السيرة النبوية – الجزء الثاني -دار المعرفة -ص ٩٤١

8 عبد الشافي محمد عبد اللطيف – دراسات تاريخ الجزيرة العربية. الجزء الاول – ص ٦٤ –

9 https://ar.wikipedia.org/wiki غزو الاحباش لليمن

10 https://www.aljazeera.net/culture/2024/2/26

11 لمحات من تاريخ الملاحة القديم في البحر الاحمر – حسن صالح شهاب – ص ٩٥

١ - وجود بعض الرموز الفرعونية مثل تمثال أبو الهول بالمتحف الوطني الارتري. واسطورة زيارة الاله الفرعوني (ازوريس) للحبشة التي ذكرها تيودور الصقلي.[12]
٢- الاله (بعل) من بلاد الشام واسيا الصغرى. ويستعمل في اللغات الحالية حتى الان.
٣- الاله (عشتار) عستر من حضارة الرافدين. ويستعمل حتى الان كاسم للسماء، واسم يطلق على الفتيات.
٤- الاله ازقا بحير او (إله البحر ابسيدون) في الحضارة اليونانية القديمة. يستعمل حتى الان للقسم اثناء الحديث.
٥- الاله ازقا مدر وهي (ديميتر)[13] من الحضارة اليونانية القديمة. لا يستعمل كثيرا في اللغات في المنطقة.
٦- اتباع الديانة اليهودية.
٧- اتباع الديانة المسيحية.
٨- اتباع الدين الاسلامي.

الهجرة الأولى للصحابة الى ارض الحبشة

ان الهجرة الى ارض الحبشة تعتبر اشد وأكثر مشقة من الهجرة الى المدينة المنورة، او هجرات الى جهات اخري داخل الجزيرة العربية، لأنها تمثل الانتقال الى العيش في جماعة صغيرة منعزلة في وسط اجتماعي معين، وبيئة وارض تختلف في طبيعتها وتضاريسها. لكن كان الابتعاد عن الأماكن التي عانى فيها أصحاب الرسول من العذاب والقهر، هو الاختيار الصعب والأمثل في ذلك الوقت.

وكانت الهجرة الأولى لأصحاب النبي محمد (ص) في السنة الخامسة للبعثة، وفي الفترة بين رجب (الشهر السابع) واقاموا شعبان (الشهر الثامن) ورمضان (الشهر التاسع) حسب التقويم المعمول به عند العرب ذلك الحين الموافق عام (٦١٥ م) ويوافق هذا التاريخ الفصل الماطر في ساحل البحر الأحمر المرتبط بالرياح الشمالية الشرقية.

وعدد المهاجرين مختلف عليه بين عشرة رجال واربعة نساء وبين أحد عشر رجلا وأربع نساء.

التوجيه بالهجرة الاولى

* وجهه النبي(ص) اصحابه بالكلام والاشارة بالهجرة لأرض الحبشة تحديدا كما هو متكرر ومثبت في كل كتب السيرة النبوية والكتب المؤرخة للهجرتين.
(لو خرجتم الى ارض الحبشة فان بها ملك لا يظلم عنده أحد وهي ارض صدق حتى يجعل الله فرجاً مما أنتم فيه (فلما كثر المسلمون وكثر عليهم العذاب والبلاء اذن الله تعالى لهم بالهجرة الى ارض الحبشة)[14]
*حديث ام سلمة

عن أم سلمة رضي الله عنها قالت لما ضُيِّقَ على أصحاب النبي (ص)، وأُوذوا، وفُتنوا، وأُكرهوا على دينهم، وفتن بعضهم في دينهم، قال لهم الرسول (ص): «إن بأرض الحبشة ملكًا لا يُظلم أحد عنده، فالحقوا ببلاده حتى يجعل الله لكم فرجًا ومخرجًا.» فخرجنا إليها، ونحن نُخاف على أنفسنا من الفتن
وعدد المهاجرين في الهجرة الأولى مختلف عليه في كتب السيرة والتاريخ الإسلامي
الرجال:
١-٢-عثمان بن عفان (رضي الله عنه) - وكان معه زوجته رقية بنت رسول الله صلى الله عليه وسلم.
٣- عبد الله بن مسعود (رضي الله عنه).
٤- أبو حذيفة بن عتبة (رضي الله عنه).
٥- أبو سلمة بن عبد الأسد (رضي الله عنه).
٦- الزبير بن العوام (رضي الله عنه).
٧- عبد الرحمن بن عوف (رضي الله عنه).
٨- مقداد بن الأسود (رضي الله عنه).
٩- سعيد بن زيد (رضي الله عنه).
٠١- الحر بن أبي حرام (رضي الله عنه).
١١- النعمان بن أبي عبد الله (رضي الله عنه).
٢١- أبو الأسود بن عبد الله (رضي الله عنه).
٣١- طليب بن عمير (رضي الله عنه).
النساء:

12 hgl;jfm Diodorus siculus – bibiltheca historica المكتبة التاريخية - من الذكاء الاصطناعي - chatGPT
13 https://ar.wikipedia.org
14 ابن هشام المتوفي ٢١٨ هـ - السيرة النبوية -علق عليها الدكتور عبد السلام تدمري – دار الكتاب العربي - الجزء الأول-٣٤٩

١٤- أم سلمة (رضي الله عنها) - زوجة أبي سلمة.
١٥- زينب بنت رسول الله صلى الله عليه وسلم.
١٦- أم حبيبة (رضي الله عنها) - زوجة النبي صلى الله عليه وسلم فيما بعد.
١٧- رملة بنت أبي سفيان (رضي الله عنها) - وهي زوجة عبد الله بن جحش.

كان هذا الانتقال بداية سلسلة من الهجرات التي تهدف إلى حماية المسلمين من أذى قريش وللعيش في بيئة آمنة.

تجمع كتب السير على ان الصحابة خرجوا متسللين من مكة الى ميناء الشعيبة ومنها الى الحبشة. غير ان خروج بعضهم كان معلن مثل خروج ابوبكر وزوجته الى برك الغماد كما تروي حكاية ابن الدغنة[15] الذي قال عندما خرج الصحابي الجليل أبو بكر الصديق -رضي الله عنه- مهاجراً إلى الحبشة قبل هجرته إلى المدينة، فلما بلغ برك الغماد لقيه ابن الدغنة (زعيم قبيلة القارة) فقال: أين تريد يا أبا بكر؟ فقال أبو بكر: أخرجني قومي، فأريد أن أسيح في الأرض فأعبد ربي.

وكان مع أبي بكر الحارث بن خالد، فقال أبو بكر: فإن معي رجلا من عشيرتي. فقال له ابن الدغنة: دعه فليمض لوجهه وارجع أنت إلى عيالك. فقال له أبو بكر: فأين حق المرافقة؟ فقال الحارث: أنت في حل فامض فإني سأمضي لوجهي مع أصحابي. فمضى حتى صار إلى الحبشة.[16]

تشير هذه الرواية الى انه كان هنالك أكثر من طريق غير طريق الشعيبة (برك الغماد) وكان يمر بها طريق الحجاج على الساحل الغربي بتهامة عسير جنوب غربي السعودية، وعلى مسافة تقدر بنحو ١٣٠ كم جنوب القنفذة، وتبعد عن مكة المكرمة نحو ٥٦٠ كلم. وأيضا كان يمكن الوصول الى ارض الحبشة بالسفر بحرا من عدة أماكن ورحلة الاشعريين الى الحبشة من اليمن دليل على ذلك.

الهجرة الثانية الى ارض الحبشة

عند عودة المهاجرين من الهجرة الأولى، وقبل الهجرة الثانية، تسلطت قبائل وعشائر قريش على الصحابة العائدين فعذبوهم، فأذن لهم الرسول (ص) في الخروج مرة أخرى. وخرج معهم صحابة اخرون.

تم ذكره الهجرة الثانية في روايات متعددة في كتب السيرة لكن لم يتم التوثيق لبداياتها كما حدث في الهجرة الأولى. غير ذكر عددهم البالغ ٨٠ - ٨٣ رجل و١١او ١٠ او ٢٠ امرأة يزيد العدد قليلا او يقل. ثم هجرة الاشعريين من اليمن وإدخالها ضمن عدد المهاجرين فيها.

قائمة بأسماء وعدد الصحابة الكرام -رضي الله عنهم أجمعين- الذين هاجروا الهجرة الثانية:

1- جعفر بن أبي طالب. ٢- أسماء بنت عميس زوجة جعفر ٣- عمرو بن سعيد بن العاص. ٤- فاطمة بنت صفوان زوجة عمرو ٥- خالد بن سعيد. ٦- أمينة بنت خلف زوجة خالد. ٧- عثمان بن مظعون. ٨- عبد الله وعبيد الله ابنا جحش. ١٠- أم حبيبة بنت أبي سفيان زوجة عبيد الله. ١١- قيس بن عبد الله. ١٢- بركة بنت يسار زوجة قيس. ١٣- معيقيب بن أبي فاطمة. ١٤- أبو موسى الأشعري. ١٥- عتبة بن غزوان. ١٦- يزيد بن زمعة. عمرو بن أمية. ١٧- طليب بن عمير. ١٨- سويبط بن سعد. ١٩ جهم بن قيس. ٢٠- أم حرملة بنت عبد الأسود زوجة جهم ٢١- عمرو ٢٢ -وخزيمة ابنا جهم. ٢٣- أبو الروم بن عمير. ٢٤- فراس بن النضر. ٢٥- عامر بن أبي وقاص. ٢٦- المطلب بن أزهر. ٢٧- رملة بنت أبي عوف زوجة المطلب. ٢٨- عبد الله و٢٩ -عتبة ابنا مسعود. ٣٠- المقداد بن الأسود. ٣١ الحارث بن خالد[17]. ٣٢- ريطة بنت الحارث زوجة الحارث. ٣٣- عمرو ٣٤ -وشماس ابنا عثمان. ٣٥- هبار بن سفيان. ٣٦- عبد الله بن سفيان. ٣٧- هشام بن أبي حذيفة. ٣٨- سلمة بن هشام. ٣٩- عياش بن أبي ربيعة. ٤٠- معتب بن عوف. ٤١- قدامة و٤٢ -عبد الله ابنا مظعون. ٤٣- السائب بن عثمان. ٤٤- حاطب بن الحارث. ٤٥- فاطمة بنت المجلل زوجة حاطب. ٤٦- محمد و٤٧ -الحارث ابنا حاطب. ٤٨- خطاب بن الحارث. ٤٩ -فكيهة بنت يسار زوجة خطاب. ٥٠- سفيان بن معمر. ٥١- حسنة زوجة سفيان. ٥٢- جابر ٥٣ -وجنادة ابني سفيان. ٥٤- شرحبيل بن حسنة. ٥٥- عثمان بن ربيعة. ٥٦- خنيس بن حذافة. ٥٧- عبد الله بن الحارث. ٥٨- هشام بن العاص. ٥٩- قيس و٦٠ -عبد الله ابنا حذافة. ٦١- أبو قيس ٦٢ -والحارث ٦٣- ومعمر ٦٤- والسائب ٦٥- وبشر ٦٦- وسعيد أبناء الحارث. ٦٧- سعيد بن قيس. ٦٨- عمير بن رئاب. ٦٩- معمر بن عبد الله. ٧٠ -عروة بن عبد العزى. ٧١- عدي بن نضلة. ٧٢- النعمان بن عدي. ٧٣ -عبد الله بن مخرمة. ٧٤- عبد الله بن سهيل. ٧٥- سليط و٧٦ -السكران ابنا عمرو. ٧٧- سودة بنت زمعة زوجة السكران. ٧٨- مالك بن زمعة. ٧٩- عمرة بنت السعدي زوجة مالك. ٨٠- أبو حاطب بن عمرو. ٨١- أبو عبيدة بن الجراح. ٨٢- سهيل بن بيضاء. ٨٣- عمرو بن أبي سرح. ٨٤- عياض بن زهير. ٨٥- عمرو بن الحارث. ٨٦- عثمان بن عبد غنم. ٨٧

15 (https: llSurahguran .com

16 ابن الملقن - عمر بن على بن احمد الانصاري – التوضيح لشرح الجامع الصحيح – المكتبة الإسلامية - باب جوار ابي بكر ص ١٤٦.

17 الحارث بن خالد ذكره ابن الدغنة انه كان مع ابي بكر الصديق في برك الغماد.

سعيد والحارث ابنا عبد قيس.[18]

من هو النجاشي

معنى كلمة نجاشي من الاصل جئز (نقوس). ومعناها الحاكم ومنها جاءت كلمة منقستي، او الحكومة والتي لا زالت مستعملة. فالحاكم في اكسوم يسمى ملك ملوك اكسوم (نجاشي النجاشين NEGUSI NEGEST). وكذلك يطلق لقب نجاشي (Negus)على ملكي حمير وعلوة الذين كانا خاضعين له. وكان اللقب الاكسومي (نجاشي) عاما وكان لكل شعب ومملكة وامارة ومدينة وقبيلة نجاشيها الخاص بها، وقد ورد ذكر نجاشين خاصين بالجيش. بالإضافة الى قادة الجيوش في الحرب وكان هناك نجاشيون يقومون على عمليات البناء[19].و لفظ نجاشي بالجيم بدلا عن القاف و هو لفظ مستحدث في نطق الاسم.

فالنجاشي او ملك الملوك في اكسوم كان ذلك الحين وحسب تسلسل ملوك اكسوم اسمه ارماح (ARMAH). وهو من سلالة العاميدا (ALLA AMIDAS)، ويقال انه ابنه او أحد افراد اسرته. لم يكن الدين الاسلامي قد انتشر وقتها وكانت ارض الحبشة تدين رسميا بالمسيحية التي دخلتها عن طريق عدوليس.

من هو النجاشي الذي استقبل الصحابة المهاجرين

من واقع البحث الميداني الذي قمت به، تأكدت من معلومات اساسها استقراء جغرافية اليوم، وارتباطها بتاريخ الماضي، توصلت الى معلومات واشارات تدل على ان النجاشي المقصود، الذي استقبل الصحابة هو أحد النجاشين التابعين لملك ملوك اكسوم، وكان موقعه على الساحل الارتري (عدوليس او مصوع). وهو المقصود ببحري نقاش، او ملك البحر. وتذكر كل الروايات ان الصحابة نزلوا عند النجاشي، وان الاشعريين ذكروا ان المركب القى بهم عند النجاشي. مما يعزز ان النجاشي لم يكن على مسافة بعيدة عنهم عند وصولهم.

الصحابة لم يبتعدوا عن الساحل الى اكسوم. وبالرجوع الى تاريخ الهجرة، (رجب في السنة الخامسة للبعثة يوافق شهر يناير، وهو فصل الامطار الشتوية التي تجلبها الرياح الشمالية الشرقية التي من المفترض انها حملت الصحابة الى الساحل الارتري. وفصل الامطار في ساحل البحر الأحمر الغربي تنزل فيه السيول، اضافة الى الخطورة التي تسببه الحيوانات البرية حينها. مع العلم ان الفصول تتغير باستمرار الا ان هذا التاريخ نتيجة لمعادلة التاريخ الفترة المعنية بواسطة البحث الالكتروني.

تذكر الروايات المسجلة عن الهجرة في المؤلفات المنشورة، ان الصحابة خرجوا متسللين. ربما يكون بعضهم خرج متسللا ولكن خروج ابوبكر الصديق الى برك الغماد مع زوجته وصحابي اخر ظهر ضمن المهاجرين في الهجرة الثانية، يعني ان خروج الصحابة من مكة كان معلوما من قريش. ولكنه مموها حيث سلكوا طريقا لا يقود الى البحر وانما الى مناطق قبائل موالية لقريش. مع بسط قريش سيطرتها كما سبق ذكره ،لا يوجد مكان يلجا اليه الصحابة داخل الجزيرة العربية. لذلك قضّوا الطرف عن خروجهم من مكة. ولكن عندما علموا ان الخدعة انطلت عليهم وان الصحابة انحرفوا الى البحر، سارعت قريش للحاق بهم، ثم أرسلوا الوفد المعلوم الى النجاشي لاستعادتهم.

إذا قارنا بين الزمن المستغرق في الرحلة الاولى من مكة الى الشعيبة، او الى أي مكان اخر، ثم حسبنا الزمن المستغرق في السفر بمراكب شراعية صغيرة، ثم نزولهم الى ارض الحبشة حتى اتي وفد قريش ليستدعيهم النجاشي اليه، نجد انه من المستحيل ان يكون الصحابة قد وصلوا الى اكسوم ثم عادوا الي مكة. لم تذكر الروايات يوم الخروج او العودة. هذا من ناحية اما ما تستغرقه الرحلة المزعومة من الساحل الى اكسوم، عبر طرق وعرة وصعود الهضبة مع وجود نساء ضمن المهاجرين، لان الزمن الذي قد تستغرقه هذه الرحلة في الأوقات العادية أطول من الفترة التي خرج فيها الصحابة من مكة ثم عادوا اليها وهي قد تقل عن الثلاثة أشهر أي بين ثمانين وتسعين يوما.

الدلائل على ان الهجرة كانت في سواحل ارتريا

١-من الاستشهادات التي تم ذكرها في رواية السيدة ام سلمة وتناولتها كل كتب السيرة سباحة الزبير ابن العوام على القربة (وهي جلد الشاة او الخروف بنفخ بالهواء ويستعمل للسباحة.) عند مهاجمة النجاشي ممن ينازعونه الحكم والهجوم على ارضه.

٢-تعد هذه الرواية سندا قويا ان الصحابة كانوا على الساحل، لأنه من الحكايات الشعبية المتداولة عند سكان الساحل الارتري انهم كانوا عندما يتعرضون لهجمات من سكان المرتفعات، كانوا يأخذون اسرهم وممتلكاتهم الى جزر البحر الأحمر القريبة. تجدر الإشارة هنا الى ان تاريخ ملوك اكسوم يخلو من حادثة انقلاب، او نزاع على الحكم حدثت في تلك الفترة. لكن قد تحدث في مناطق النجاشين الذين يحكمون مناطق أخرى تتبع لملك الملوك في اكسوم.

18 . علي الشيخ احمد ابوبكر – معالم الهجرتين الى ارض الحبشة -ص ١٦٢-٥٦٢ عن سيرة ابن هشام _ وجوامع السير لابن حزم

19 جين افريك – اشراف جمال مختار- تاريخ افريقيا العام- المجلد الثاني- حضارات افريقيا القديمة - اليونسكو ١٩٨٥ – ص٧٨٣- ٨٨٣

٣-هناك رواية أخرى تدعم وجود النجاشي والصحابة على الساحل ففي ذات حدث منازعة الحكم وهي حديث ابن اسحق، عن جعفر بن محمد (ان النجاشي أرسل الى جعفر واصحابه، فهيا لهم سفنا وقال اركبوا فيها كونوا كما أنتم ...الخ الحديث[20]) فلو كانوا في اكسوم او أي مناطق بعيدة عن البحر لهيا لهم وسائل نقل أخرى غير السفن.

٤-اما عن السباحة على القربة فان مواطني الساحل اللذين كانوا يعملون في ميناء مصوع الارتري كانوا يسبحون على القرب (جمع قربة) من جزيرة قرار الى جزيرة مصوع. وهناك حادثة مشهورة في التاريخ الارتري ان ثلاثة رجال سبحوا (بالقرب) من قرية حرقيقو او دكنو الى مصوع قطعوا فيها عدة كيلومترات تقريبا وقاموا بقتل الحاكم في قصر الوالي الحالي.

٥-لا تذكر وثائق التاريخ الاكسومي اسم (الاصحمة ابن أبحر) وهو ما يذكره كتاب السيرة المسلمين انه استقبل المهاجرين، وانه كان مقره اكسوم. لكن يمكن ان يكون هو (بحر نقاش او ملك البحر). فالنجاشي في ذلك الوقت، او ملك الملوك في اكسوم حسب التاريخ المسجل، وتسلسل ملوك اكسوم كان ارماح (ARMAH) وهو من سلالة العاميدا (ALLA AMI-DAS) ويقال انه ابنه او أحد افراد اسرته.

٦-اما عن الكنائس والقساوسة والاديرة فمن المعلوم ان القرن الافريقي قبل الإسلام كانت تسوده الديانة المسيحية. وان عدوليس كانت البوابة التي دخلت منها المعتقدات الى ارض الحبشة. ومنها المسيحية وقد كشفت الحفريات والدراسات عن مباني وأعمدة ومذابح في عدوليس تفوق في مساحتها ما يوجد في اكسوم. ومن المرجح انها كانت مقر النجاشي الرسمي، بالإضافة الى مصوع. اما المهاجرين فكانوا في مصوع والدلالات على ذلك كثيرة، منها كثرة المساجد ودور العبادة والمقامات فيها كما اورده مفتي الديار الارترية التالي: -

٧- تقرير مفتي الديار الارترية (بتاريخ ٧٢ رمضان سنة ٦٧٣١ هجرية الموافق ٧٢ – ٤ – ٧٥٩١ م ذكر فيه ان زلزال عام ٩٣٣١ هجرية الموافق ٠٢٩١ م الذي ضرب مصوع قد دمر تسعة مساجد (١- مسجد حسن باشا ٢- مسجد اوده باش ٣- مسجد عبدو حامد ٤- مسجد بن علوان ٥- مسجد عبد الله اسعد ٦- مسجد صباح ٧- مسجد الشيخ إبراهيم ٨- مسجد الشيخ إسماعيل ٩- مسجد خواجة صفر الرومي)

كما دمر ومسح الزلزال عدد من المقامات منها ١- مقام جعفر الطيار ٢- ومقام النجاشي ٣ – مقام شيخ ابوبكر ٤- مقام عكاشة ٥- مقام شيخ حامد) وكلها بعد الزلزال أصبحت اثرا بعد عين.[21]

٨-جعفر الطيار هو جعفر ابن ابي طالب ابن عم النبي (ص)، وكان من الصحابة المهاجرين في الهجرة الثانية، وكان المتحدث باسم المهاجرين، ومكث في ارض الحبشة أكثر من عشرة سنين. بالرغم من المكانة السامية لجعفر الطيار الا ان مقامه في مصوع لم يتجدد ولم يعاد الى مكانه بعد الزلزال.

٩- ما كان يعرف بضريح او مقام احمد النجاشي، فقد تمت اعادته، وكان موجودا في رأس مدر عند المدخل الجنوبي للميناء. وكان يقصده الناس من اماكن بعيدة من اثيوبيا للاستشفاء. وقد شاهدتهم في العام ٠٩٩١ ابان تحرير مصوع. ترى هل هو قبرا وليس مقاما؟ ربما ستكشف الدراسات والحفريات عن ذلك. انظر الصورة في الملاحق.

٠١-وجود مسجد الصحابة في راس مدر او راس الأرض في الناحية الشرقية من جزيرة مصوع، وربما هو اول مسجد تمت الصلاة فيه خارج ارض العرب.

كان في مدينة مصوع أكثر من ٢١ مسجداً، بما فيها الذي اندثر وقامت محله بنايات أخرى. ويوجد بها وهو الاهم مزار او ضريح لشيخ نجاش او احمد النجاشي، ويوجد ومسجد الصحابة. هذا الكم من المساجد والمقامات في مساحة جزيرة مصوع التي مساحتها اقل من ثلاثة كلم مربع انما يدل على انها كانت مقر للصحابة ومركزا مبكرا للدين الإسلامي.

تناقضات السيرة عن هجرة الصحابة.

التسجيل لهجرة الصحابة في كتب السيرة اكتنفه كثير من التضارب في المعلومات، مع تشبعه بالعاطفة والرسائل المبهمة التي ترمي الى تهميش دور الهجرتين في التاريخ الإسلامي. ربما لأسباب تتمثل في إحساس غير مرحب به من بعض الجهات بان الإسلام والمسلمين الذين رفضتهم وطاردهم جهات عربية في ارض العرب، تمت حمايتهم من جهة ليست عربية بل ومن جهة دينها المسيحية. اعتقد ان تاريخ الهجرتين الى الحبشة ستتم ازالته من كتب السير وتاريخ الإسلام لأنه تاريخ تم الخلط فيه بين الخيال والواقع، وبين الوقائع وسردها غير الدقيق. اذكر على سبيل المثال لا الحصر المغالطات التالية في معظم كتب السيرة والكتب المؤرخة للهجرتين

١-لا توجد معلومة حاسمة ودقيقة عن عدد وأسماء المهاجرين. سواء في الهجرة الأولى ام الثانية

٢-وفد قريش من كان مع عمرو بن العاص هو عبد الله ابن ابي ربيعة، ام عمارة ابن الوليد، وهل لحق الوفد بالمهاجرين في الهجرة الأولى ام الثانية؟ وهل بعثت قريش عمرو ابن العاض الى النجاشي مرتين ام مرة واحدة؟

20 الدكتور سليمان بن علي السعود -أحاديث الهجرة – مركز الدراسات الإسلامية - ط أولى ٠٩٩١م ص ٥١

21 Https//mukhtar.ca

٣-الخلط بين مملكة اكسوم - الأرض- وبين الملك العادل الذي حدده النبي (ص) هل قابل الصحابة النجاشي في اكسوم ام على الساحل؟
٤-هل التقى الصحابة بالنجاشي في الهجرة الأولى ام الثانية؟ لان جعفر ابن ابي طالب الذي التقى النجاشي وحاوره لم يكن في الهجرة الأولى.
٥-السيدة ام سلمة سند الروايات عن هجرة الصحابة الى ارض الحبشة كانت في الهجرة الأولى ولم تكن في الهجرة الثانية
٦-الزبير ابن العوام كان في الهجرة الأولى ولم يكن في الهجرة الثانية وهو الذي سبح على القربة لمعرفة نتيجة النزاع بين النجاشي والمهاجمين وهل كانت حادثة النزاع في الهجرة الأولى ام الثانية؟
٧هل غادر الصحابة في الهجرة الأولى دون لقاء النجاشي؟ **لان جعفر وام سلمة والزبير ابن العوام لم يجتمعوا في هجرة واحدة.**

النجاشي والامام احمد ابن إبراهيم الغازي (غراي)

خلال ندوة أقيمت في ديسمبر ٧٩٩١ حول مسجد أحمد النجاشي في مكيلي (مقلي) (تغيراي)، وحضرها سفراء إيران وتركيا واليمن وليبيا، وافتتحها رسمياً رئيس ولاية تغيراي الإقليمية، طُرحت فرضية مثيرة للجدل بأن «أحمد النجاشي» كان «أول ملك مسلم إثيوبي» (Addis Z).

هذا الحدث تنطوي تحته أسئلة أولها هل أسلم النجاشي؟ هل القبر المسمى في تغراي هو للنجاشي؟ اين قبر النجاشي الذي استقبل الصحابة؟

شخصيتان لعبتا دورا كبيرا في تاريخ الاسلام في القرن الافريقي، هما النجاشي الملك الاثيوبي، والامام احمد بن ابراهيم الغازي المعروف بأحمد قري او الاعسر.

أولا: النجاشي هو من منح الاسلام والمسلمين الأمان، وشكل بقبوله لهم في ارضه اول قبول للدين الجديد خارج الجزيرة العربية. وهو الذي أسس لانطلاق الدين الجديد من المحلية للعالمية. توفي النجاشي في اكتوبر عام ٠٣٦ م عن ٠٧ عاما تقريبا.

النجاشي المقصود في اكسوم في تلك الفترة هو (ارمها) نجاشي النجاشين. إذا أسلم وكتم اسلامه، او اخفاه، او إذا لم يسلم ومات. فانه في نظر الكهنة مسيحيا، ويدفن على انه ملك الملوك في اكسوم. اما إذا أسلم وتم نفيه الى خارج اكسوم، فارض الحبشة في ذلك العهد لم تعرف الإسلام فموضوع ذفن النجاشي الذي أسلم والمغضوب علية من الكنيسة في ارض مسيحية وعلى مقربة من اكسوم فذلك يدعو لعدم الاستيعاب. ولم يذكر في تاريخ اكسوم ان أحد النجاشين تم طرده، او ابعاده لأنه اعتنق الدين الإسلامي. او أي دين اخر. اعتقد ان القصة تم مزجها مع قصة الامبراطور ايأسوا في العهد القريب ابان حكم الامبراطور هيلاسلاسي.
اما قبر النجاشي الذي استقبل الصحابة، فهو في جزيرة مصوع. انظر الخارطة في الملاحق.

ثانيا: الامام احمد ابن إبراهيم المعروف بالغازي او احمد قراي وتعني الاعسر او الاشول، فهو الذي أطلق لفظ الجهاد وحاول تمكين الدين الاسلامي بالمنطقة. احمد قراني كإمام للمسلمين، ومجاهد لنشر الاسلام وحماية المسلمين في المنطقة حسب قوله، من المفترض كشأن كل المسلمين وتعظيمهم للنجاشي، الا يمر ويعبر او يتجاهل قبره. اذ افترضنا ان القبر الحالي في تقرأي هو قبر النجاشي الذي حمى المسلمين والاسلام، والذي وعندما مات صلى عليه الرسول محمد (ص) صلاة الغائب.

الامام احمد ابن إبراهيم المعروف بالغازي واحمد قراني الذي قاتل، وأطلق الجهاد ضد الامبراطور الاثيوبي، انطلق مع جيوشه من منطقة هرر شرق اثيوبيا من مملكة عدل وأوفات او جبرت، والذي يذكر التاريخ انه اصيب ومات عام ٣٤٥١ وعمره ٧٣ عاما في معركة مع البرتقاليين شرق بحيرة تأنا، وحمله اتباعه الى منطقة قريبة من المكان الذي يزعم انه قبر النجاشي. ثم تبخر بعد ذلك تاريخ احمد قراني، وتوقفت سيرته ولا أحد يعرف مكان قبره سوى اتباعه الذين جاهدوا معه والذين قبروه ويحرسون قبره الى الان تحت اسم احمد النجاشي. لذلك اعتقد ان: -

القبر الموجود باسم احمد النجاشي في إقليم تغراي الاثيوبي، هو قبر الامام احمد ابن إبراهيم قراني. حيث تم الخلط لسبب ما بين الاسمين احمد والنجاشي فصار (احمد النجاشي). ففي كل كتب السيرة لم يطلق اسم احمد النجاشي على الملك الذي استقبل الصحابة، وان النجاشي الذي استقبل الصحابة وعاش بينهم، يوجد قبره في مصوع الميناء الارتري مهما كان اسمه. كتب السير و التراث الإسلامي تذك اسم الاصحم او الاصبحة ابن ابحر، و هو اسم ليس له وجود في تاريخ ملوك اكسوم، ولا في الروايات الشفاهية المؤرخة لتلك الفترة. لكن ربما يكون هو اسم ملك البحر الذي استقبل الصحابة على الساحل وصلي علية النبي (ص) صلاة الغائب. كما انه لا توجد اي علاقة بين جبرت وأوفات، وكل ممالك الطراز الاسلامي بالنجاشي الذي استقبل الصحابة. لأنه باختصار عند قدوم الصحابة الى ارض الحبشة لم يكن فيها اسلام ولا مسلمين سوى الصحابة، ولان الفرق الزمني بين النجاشي الذي استقبل الصحابة والامام احمد قراني ٣١٩ عاما.

الخاتمة

ان الأرض في كل مكان من العالم تعرف او توصف او تسمى او تتميز بثقافة وتاريخ الانسان الذي يسكنها، او سكنها. ارتريا شانها شان كل الأراضي يميزها كذلك تاريخها البعيد والقريب، المرتبط بسبل تكيف انسانها مع الطبيعة.

ربما بسبب الاحداث التي سكنت الساحل الارتري ضمن القرن الافريقي، والتي أسهمت في عدم استقرار المنطقة، الذي أسهم بدوره في إبقاء معظم تاريخ ارتريا محفوظا. كما أسهم المواطنون بغيرتهم على ارضهم، وتراثهم وتاريخهم، في إخفاء او عدم توضيح المعالم الاثرية للغرباء او المستعمرين.

الخلاصة:

• ان هجرة الصحابة الى ارض الحبشة، والى السواحل الارترية تحديدا، شرف للمسلمين وللمنطقة المسماة بأرض الحبشة. والتي كانت تشمل حضارتها معظم القرن الافريقي الحالي الى ما بعد النيل غربا. لذلك علينا دراسة كل ما يتعلق بالأحداث التي واكبت الإسلام في مراحله الأولى من منظورنا الإقليمي المحلي.

1- ان هجرة الصحابة كانت الى السواحل الارترية ولم يغادرونها الا الى بلادهم،

2- النجاشي المقصود والذي كتب عنه المؤرخون الإسلاميون باسم أصبحه ابن أبحر، او (عطية) كما ذكره بعضهم، ربما كان هو بحر نقاش. ومقره في الساحل ربما عدوليس، او مصوع او حرقيقو، المعروفة ب (دكنو) والدكن هو الفيل، او بالقرب من المناطق المذكورة.

3- التناقضات في التوثيق للهجرتين ربما يبدو مقصود، لعدم إعطاء فضل حماية المسلمين والإسلام في مراحله الأولى الى الاحباش، او الى ارض ليست ضمن جزيرة العرب.

4- كانت الهجرتان انفتاحا للدين الإسلامي الى الانتشار العالمي، بعد الاعتراف به من النجاشي المسيحي، الذي يمثل اعترافه بالدين الجديد، اعتراف الامبراطورية البيزنطية كذلك.

5- النجاشي الموجود في اكسوم المعروف بملك الملوك، لم يسلم وربما جنح المؤرخون في كتب التراث الإسلامي الى ذلك لتجميل القول وحماية الإسلام والمسلمين تمت على يد ملك مسلم.. اما إذا كان المقصود هو نجاشي البحر على الساحل، فلا توجد وثائق تثبت اسلامه سوى التاريخ الشفاهي الذي احتوته كتب التراث الاسلامي.

6- ان القبر الموجود في إثيوبيا في إقليم تغراي هو قبر الامام احمد ابن إبراهيم المعروف بأحمد قراني. لان التاريخ لا يعرف ملكا باسم احمد النجاشي في الحبشة، وانما تم الخلط بين احمد قراني والنجاشي، فتم تركيب الاسم الحالي للمزار المذكور.

بما ان الاستقرار الذي تشهده ارتريا الان يدعوا الى البحوث والدراسات العلمية، يتوجب السعي الى إقامة مركز للبحوث الميدانية للتوثيق الثقافي والاثار والتاريخ والحكايات الشعبية والاحداث الطبيعية.. الخ ليكون مركزا مرجعيا للباحثين، وإقامة ندوات ومحاضرات عنه في الداخل والخارج.

التركيز على أهمية دراسة تاريخ المنطقة نقلا عن البحوث الميدانية لتصحيح المعلومات الخاطئة.

ربط تاريخ ارتريا بالمناهج التعليمية، وتنسيق زيارات للطلاب من مختلف المراحل لزيارة مناطق الاثار والمتاحف في ارتريا.

عمل قانون للآثار والمتاحف، لحسم التغول على المناطق الاثرية من كل الجهات بالدولة. ويجرم الاعتداء على الاثار والنماذج الاثرية ومواقعها او تكوين شرطة ونيابة الاثار بالدولة.

الملاحق

صور جوية لجزيرة مصوع الارترية وميناءها الرئيس توضح مواقع قبر النجاشي ومسجد الصحابة[٢٢]

Archives and History

Challenges in Writing History: The Case of the Eritrean Armed Struggle

Zemihret Yohannes
People's Front for Democracy and Justice,
Research and Documentation Center (RDC)

Introduction

Writing the history of the thirty-year-long Eritrean armed struggle for independence exemplifies the inherent complexities and challenges of historical writing. These challenges arise not only from the general intricacies of reconstructing historical narratives but also from specific characteristics related to the Eritrean armed struggle. Among these are its distinct historical background shaped by an aborted decolonization and subsequent annexation by the Ethiopian state; the nascent political and organizational development of the national movement prior to the armed struggle and the spontaneous inception of the armed struggle without the basic and articulated political and organizational orientation and strategy; the emergence of contending organizations within the armed movement and internal conflicts; the scarcity of written records from its early phases of the 1960s; and the passing of key participants.

This paper explores the features and perspectives that make documenting the history of the armed struggle difficult. In addition, it proposes preliminary ideas for producing an objective and judicious historiography of the armed struggle. Three central questions are addressed: What are some of the challenges encountered in writing the history of the Eritrean armed struggle? What perspectives must be considered in the study, analysis, narration, and understanding of the struggle? What possible solutions exist for the historiographical challenges?

Social and Political Characteristics of the Armed Struggle

The Eritrean armed struggle was preceded by two decades of political and social ferment. During this period, Eritrea's political fate was determined not by its people, but by the geopolitical and strategic interests of the major powers, especially the United States. The nascent political movements of the 1940s, formed by small groups of urban Eritreans, were socially and politically inchoate, unable to counter the heavy-handed intervention of the US, Britain and Imperial Ethiopia.

The concept of a distinct Eritrean identity, politically emerging during this period, was contested by various groups with divergent views on what it meant to be Eritrean and the political future of Eritrea. This contestation, rooted in social and cultural divisions, and exacerbated by foreign interventions, had profound political ramifications that ultimately influenced the genesis of the armed struggle two decades later. Because of the absence of a national leadership the early stages of the armed movement were thus marked, like in many national liberation movements in Africa and Asia, by political fragmentation reflecting religious and cultural cleavages.

Historians must carefully study and analyse, the uneven economic and social impact of Italian colonial administration and subsequent foreign rule and interventions; the cultural and religious diversity of Eritrean society; the genesis and limitations of the national movement; and how these factors influenced the development of Eritrean identity and nationalism, as well as their impact on the early phases of the armed struggle.

The intricate nature of social and cultural diversities, particularly religious diversity, and their impact on identity and political affiliations, must be studied deeply within the context of the different periods of the armed struggle. Only then can we comprehend how these factors shaped the movement's internal cohesion, and its long-term vision for a unified national identity. In this regard, historians must navigate two conflicting conjectures: first, viewing existing diversities as immutable aspects of social life and portraying the national movement as a coalition of discrete, bifurcated groups representing distinct interests; and second, overlooking social and cultural diversities as existential realities, thereby failing to engage with them critically.

The armed struggle signalled both continuity and discontinuity in the evolution of the national movement. Initially, continuity trends were prominent, but later, significant social, political and ideological transformations enabled the movement to go beyond the cultural and religious divisions. This new phase fostered identities and solidarities based on shared national ideals, struggles and sacrifices, beyond narrow loyalties. This process of internal transformation was driven and defined by two intertwined dynamics: first, the democratic struggles of nationalist and progressive groups against the dominance of the retrogressive leadership – the Supreme Council – as expressed in its backward policies and divisive structures, and second, the struggle within the opposition groups striving to rectify the course of the struggle and to develop unity of purpose and action based on a national democratic program that transcended narrow loyalties to advance national unity.

The Transformative Period of the Armed Struggle

In the 1970s, the armed struggle underwent significant political, social, cultural and military transformations, culminating in a new and advanced stage of development. This period was marked by a political realignment within the liberation movement favouring revolutionary nationalist and progressive forces; a surge in popular support to the armed struggle, notably in the major urban centres and the countryside of the highlands in the early 1970s; and the crisis of the ancient regime in Ethiopia. These developments transitioned the Eritrean liberation movement into a revolutionary nationwide People's War.

Studying and explaining these far-reaching changes is essential. The extent of public consent and active support, wholehearted participation and mass mobilization had few parallels among twentieth-century national liberation movements. The sheer magnitude and depth of these social and political transformations which led to a high-level of participation of the population and the highly organized support from the diaspora, defies simple and dispassionate historical explanation.

The nineteenth century Scottish historian Thomas Carlyle explored the relationship between faith and historical meaning. One writer paraphrased his view: "Seldom do we find that a whole people can be said to have any faith at all except in things which it can eat and handle. When so ever it gets any faith its history becomes spirit-stirring, noteworthy". These words resonate strongly with Eritrea's experience during the armed struggle, where history and faith converged in extraordinary ways that challenge objective narration.

The Dialectic of Continuity and Discontinuity

The political history of the Eritrean national movement unfolded through three major periods of transition: The first transition was from the proto-parties of the 1940s and their dissipation during the Federation to the social movements of workers and students. These movements became the social base of the Eritrean Liberation Movement (ELM), the first clandestine nationalist organization to build an organized political base in primarily Eritrean towns and among diaspora communities in neighbouring countries.

The second transition was from the clandestine political struggle of the ELM, characterized by its vague and unrealized hybrid strategy of coup d'état and insurrection, to the armed struggle launched under the banner of the Eritrean Liberation Front (ELF), without adequate political orientation and organizational preparation.

The third transition stemmed from the profound political crisis that engulfed the ELF in the late 1960s, threatening the survival of the armed struggle. This extended period of transition was defined by fragmentation, shifting alliances, infighting and armed conflict. Revolutionary nationalist groups attempted to overcome the crisis and reorient the movement toward a progressive nationalist course by constructing a broad-based, ideologically coherent, and united national liberation front. In this transition of the late sixtieth and early seventieth the broad movement for reform in the ELF was met with unrelenting and stubborn resistance from the entrenched leadership. Under this complex political situation marred with political fragmentation unity proved a step too far and the split in the ELF became formalized with the establishment of the Eritrean People's Liberation Forces, a process that started in 1970 and was consummated in 1973-74 with the merger of its three constituent groups. In this defining moment for the Eritrean struggle for freedom and the subsequent trajectory the role played by two clandestine Marxist groups that were set up in the late 1960s and remained active for nearly two decades merits serious study and analysis.

These three transitions are inherently complex and pose significant challenges for historians. The first transition in the 1950s unfolded within a relatively nascent political context, which was relatively uncomplicated. In contrast, the latter two transitions were more complex which generates divergent interpretations. The interpretations surrounding these transitions, coupled with scarcity of documentation, are challenging to historical scholarship.

In essence, the transition from the ELM to the ELF, and the subsequent emergence of two liberation fronts from the ELF, especially the EPLF, embody a dialectic of continuity and discontinuity. Studying how this dialectic unfolded in practice, analysing the internal and external socio-political dynamics, and developing a balanced and judicious understanding of the processes are critical to a deeper appreciating of the historical contours of the armed struggle.

The transition from the ELM to the ELF is often portrayed as a straightforward shift from clandestine urban political activity to open rural armed resistance. However, this interpretation is arguably superficial. While one dimension of this transition did involve a shift toward armed struggle, it also encompassed complex social and political dynamics with long-lasting implications. The leadership and social base of the ELM was urban, while the ELF in the field was predominantly rural. The ELM promoted a nationalist program that emphasised unity superseding the religious divides, while the old ELF lacked a clear political program and sustained and exacerbated ethnic and religious divisions through its practices. As such, the ELF represented a political regression from the progress achieved by the nationalist movement in the late 1950s, a reversal that created the main internal contradiction driving the crisis of the 1960s.

Despite these disparities, the ELF, bolstered by the widespread appeal of armed struggle, became dominant and won over most ELM members. Though weakened, the ELM continued its political activities in Sudan and attempted to establish an armed wing in 1965, which was swiftly and militarily crushed by the ELF. The failed dialogue between the two movements in the early 1960s, followed by the ELF's armed suppression of the ELM, proved historically consequential as a precursor of the wider internal armed conflicts of the 1970s by introducing violent means of resolving internal political differences which had enduring negative ramifications in the political evolution of the national movement. For these reasons, the transition from the ELM to the ELF warrants thorough and nuanced examination.

In its early phase, the ELF faced daunting material and political challenges. The Ethiopian state deployed special police units and military forces in an effort to eradicate the movement in its infancy. The perseverance and dedication of the first generation of ELF veterans in sustaining the fledgling struggle under such adverse conditions represent a truly historic achievement. Therefore, while the ELF in the 1960s was marked by fundamental shortcomings – primarily its lack of strategic direction and political orientation -- it must not be judged solely in hindsight by these limitations. A comprehensive and thoughtful account of this formative period, by recognizing its historical significance, must embrace its full complexity and contradictions.

The Third Transition: Divergence and Dialectics within the ELF and EPLF

The third transition reveals a dialectic of continuity and discontinuity in leadership, political programs, ideology, political culture, organization, foreign policy, and military strategy and tactics in both the ELF and the EPLF. The differences and conflicts between these two fronts were decisive in shaping the internal political developments throughout the 1970s. Understanding the nature of their divergences and the causes of their recurrent confrontations -- particularly the first and the second "civil wars" – remains a critical task for objective historical analysis.

Arguably, the fundamental difference between the ELF and the EPLF lies in how they conceptualized their respective political projects. The post 1971 ELF, whose socio-cultural composition initially mirrored that of the earlier ELF, regarded itself as a democratic extension and custodian of the ELF's legacy. In contrast, the EPLF, with a socio-cultural composition relatively distinct from the old ELF, envisioned itself as its antithesis: the democratic alternative and the revolutionary vanguard previously lacking. Although both fronts espoused similar ideologies, their perspective and policies on basic internal and external issues as well as their political and organizational cultures and praxis, were markedly different. The extent to which these differences constituted an ideological divide remains a vital subject for historical reflection.

Equally significant is the comparative assessment of leadership cohesion, political clarity, and organizational discipline between the two fronts. Ultimately, the ELF and the EPLF, alongside their leading parties as political and ideological core groups, differed notably in the quality of leadership, political coherence, organizational unity and discipline, autonomy from internal societal pressures, and political independence from external influences. These disparities were epitomized in the congresses – serving as the highest political and legal platforms -- held by both fronts. While both the First and Second Congresses of the ELF – 1971 and 1975 respectively -- adopted essentially national and democratic programmes and resolutions, they were marred by internal power struggles and political compromises, and conspicuous foreign interference. Rather than serving to resolve the internal problems to advance the movement, they exacerbated the very political problems that had necessitated their convening.

In contrast, the two EPLF congresses of 1977 and 1987 -- unlike those of the ELF which were overshadowed by internal discord – served as platforms of unity and consolidation, strategic direction, ideological clarity, and developing new policies, all vital for achieving the goals of national liberation. These fundamental differences contributed to the EPLF's rise and the eventual disintegration of the ELF. Crucially, this outcome stemmed from internal developments within both organizations, their respective political and military achievements and failures, and their contest for leadership. This contest by the late 1970st produced a new balance of forces that consolidated the leading role of the EPLF. The fact that this internal struggle led to divergence rather than convergence – despite public declarations of unity and the agreement of October 1977 and subsequent steps which had widespread support of the people -- remains an important issue for historical inquiry.

Within these dynamics, the interplay of continuity and discontinuity between the shared historical legacies and the divergent trajectories of the ELF and the EPLF must be examined with the outmost care. In historical development, even movements that make the most radical and revolutionary claims, are rooted in history. As Marx (1852) famously noted, "Men make their own history, but they do not make it as they please… but under circumstances existing already, given and transmitted from the past." Accordingly, continuity is primary in history, while discontinuity -- at least initially -- is secondary. However, when discontinuity persists and eventually comes to define an era, it can attain a hegemonic character and establish a new continuity. Thus, although elements of both continuity and discontinuity mark the histories of the ELF and the EPLF, their dynamic interplay within and between both fronts warrants study.

In the case of the ELF, while change signalling a departure from the old ELF was acknowledged and celebrated, regressive elements of continuity persisted within its leadership, internal politics and external relations. Traditional practices and attitudes remained entrenched and were further reinforced by a dominant clique led by Abdalla Idris that sustained its power by perpetuating divisive politics. Theses internal political realities defined the ELF more than its declared national and democratic programmes adopted in its congresses. The leadership elected by the congresses was a disparate group that lacked political unity and cohesion which gave the dominant clique of Abdalla disproportionate influence and made it decisive. These dynamics arrested the front's development and eventually culminated in a state of political decay. As a result, the ELF struggled to integrate the new socio-cultural forces that entered the movement en masse after the mid-1970s. This failure of integration precipitated a renewed crisis and gave rise to the 'Falul' opposition within the ELF in 1977 that the leadership attempted to crush militarily which forced it to join the EPLF.

This political upheaval that markedly weakened the ELF was followed by another opposition from a group of conservative veterans. This opposition which was called the 'Yemin' (Rightist) by the leadership was composed by disillusioned conservative veterans who were actively supported by and coordinated with Sabbe's faction, the ELF-PLF. After the 'Yemin' was crushed, the ELF leadership by the end of 1978 till the first half of 1979 waged an all-out military campaign on Sabbe's ELF-PLF in lower Barka and pushed it to Sudanese territory which accelerated its fragmentation.

The ELF's ambiguous and contradictory relationship with its past, compounded by the burden of its regressive experiences; the paralyzing role of its dominant clique; the deepening of the internal political crisis especially after the disorganized retreat of mid-1978; the lack of strategic clarity and consistency on the issue of Soviet intervention and the nature of the Ethiopian military regime which cast doubt on the very legitimacy of the armed struggle; the failure of the agreements with the EPLF and the subsequent civil war; aggravated the political decay and infighting that ultimately culminated in its fragmentation.

Nevertheless, it would be a disservice to the history of the armed struggle to reduce the ELF's legacy to the failures culminating in its political fragmentation in 1981, or to recount its history exclusively from that standpoint. Nor should its later phases be conflated only with the positions and actions of its dominant retrogressive clique. Such reductionist perspectives risk oversimplifying the complex and multifaceted two-decade evolution of the ELF's, which was important to Eritrea's broader national movement.

This history demands a comprehensive and rigorous approach. Its various phases should be explored through meticulous analysis of reliable sources, thoughtful contextualization of events and developments, and a conscious avoidance of biased and judgmental perspectives. Reconstructing the past requires engaging with history as it unfolded, shaped by evolving social and political realities and enacted by groups and individuals with distinct socio-cultural backgrounds, lived experiences, political outlooks, and access to specific cultural and political resources.

Accordingly, the history of the ELF should be studied objectively and judiciously, with due attention not only to contest among political factions and leadership groupings, but also to the role and experiences of the grass-roots of the front -- both within the army and mass organizations -- and more broadly, to its contributions to the overall history of the armed struggle.

In contrast to the ELF, the EPLF emerged as an opposition force, developing under pressure from ELF military campaigns. It consciously rejected the retrogressive experiences of its predecessor and defined itself as a radical alternative, ultimately forging a united and resilient front. This process entailed internal political struggles against influential elements of the old leadership represented by Sabbe's Foreign Mission and its followers in the field and other divisive tendencies obstructing unity, notably the 'Menkae' and 'Yemin' phenomena. The challenges and crisis the EPLF overcame in its formative period (1970-1975) enabled it to transcend divisive legacies by cultivating a new political culture grounded in a strong unity of purpose and unwavering discipline under a leadership forged through struggle. The leadership led by Isaias Afewerki was a central and integral factor in this process. Though the dialectic of building unity in diversity was inherently complex and contradictory, the movement cultivated a sustainable and robust sense of national identity and unity of purpose based on mutual respect and equality.

As a result, the EPLF was able to integrate sustainably the large influx of Eritreans from different socio-cultural backgrounds -- peasants, workers, students, women and professionals -- continuously mobilized after the mid-1970s. It ensured sustainable continuity by securing the necessary political and military advances -- particularly following the strategic retreat of 1978, the fragmentation of the ELF and the subsequent escalation of political interference by some neighbouring countries and their divisive political ploys and the appearance of a fifth column led by Abdalla Idriss – which ultimately led the armed struggle to victory.

In the most challenging post-1978 period in the history of the national struggle, the EPLF confronted and overcame formidable military campaigns launched by the expanded and reorganized Ethiopian army under the Derg, which saw huge expansion with the influx of massive Soviet weaponry and technical assistance. In response, the EPLF developed innovative political, military and organizational strategies that dramatically enhanced its limited human and material resources. These efforts enabled the movement to sustain itself self-reliantly and eventually prevail in a protracted and devastating war.

A strong and effective army was built – politically capable of continuously replenishing its manpower and maintaining military efficacy – able to fight a conventional war for more than a decade along hundreds of kilometres of fortified frontlines. Within the revolutionary base areas that supported the liberation army, various political, economic, social and cultural institutions were developed and self-reliant programs implemented. The ideological and political quality of the movement, particularly its political independence and the unity forged through internal struggles and processes out of diversity, was the decisive factor behind the EPLF's ascent and its consolidation as the leading force in the national liberation struggle.

Moreover, the EPLF took the lead to establish strategic relationships with Ethiopian armed opposition forces and provided them with comprehensive support up to directly participating in their fight against the military dictatorship based on its strategic vision. The alliance cultivated with the Ethiopian movements was part of a broader strategic vision to transition the Horn of Africa to peace, stability and cooperation based on equality and a shared future.

Nevertheless, the EPLF's historical trajectory demands holistic study that consciously rejects approaches focused solely on its outstanding historic achievements. To reach a deeper and more nuanced understanding, its remarkable successes must be examined alongside its inevitable setbacks, errors, limitations and the contingencies inherent to any historical process. Ultimately, the EPLF had to reconcile with its broader historical past and embrace it -- including the aspects it had previously rejected – by developing judgments on the past that were "reasoned, contextualized and judicious."

As Fouad Makki (2023) insightfully noted, "Recurrent narratives of a sharp break from one phase to another, or from one movement to another (are) typical of new movements who naturally wish to establish their distinctiveness…But historians need not uncritically follow them on that path. Nevertheless, it is also just as crucial to register moments of discontinuity in history that need to be analyzed. History is not a simple evolutionary process. One other point concerns judgments we make on the past, which are unavoidable, but they need to be reasoned, contextualized, and judicious."

The political, social, cultural and economic experiences and outcomes of the revolutionary period of the armed struggle warrant in-depth exploration. Following the transition in the mid-1970s from the guerrilla phase to the revolutionary phase of nation-wide People's War -- catalyzed by political and ideological transformation -- the liberated countryside experienced profound changes. Although uneven in scope and depth these developments had enduring ramifications. They included: a rise in political consciousness and organization capacity among both rural and urban populations as well as diaspora communities; new structures of local administration based on popular participation; transformation of women's social status and political participation; land reform and emergence of cooperative movements in certain regions; the introduction of new national curricula, literacy campaigns and expanded access to education and the arts; and the provision of medical services to remote areas. All these factors contributed to the broader revolutionary transformation.

Given the multifaceted nature of these developments and their complex interrelations, it can be argued that the third transition was the most consequential of all. Accordingly, it stands as a fundamental subject for rigorous historical study.

The Problem of Perspectives

In the historiography of the national movement in general, and the armed struggle in particular, it is essential to recognize the constraining presence and influence of the following three key perspectives:

The Colonial Perspective: The Euro-centric colonial perspective was pioneering approach in modern African historiography in general and Eritrean historiography in particular. Although its influence has been significantly diminished, its insidious legacy continues to linger. For this reason, its underlying presence and subtle influence must be consciously recognized and critically examined.

The Ethiopianist Perspective: The Ethiopianist perspective rests upon the denial of Eritrea's independent historical existence. From this viewpoint, Eritrea is regarded as merely an appendage to Ethiopia, devoid of autonomous historical agency. Over the years, this perspective has gained academic traction through Western Ethiopianist scholarship, Ethiopian state sponsorship, and the work of Ethiopian and Western Ethiophile academics.

The Eritrean Perspective: This perspective began asserting itself alongside the development of the national movement, particularly during the advanced stages of the armed struggle. The Eritrean perspective rejects both the Eurocentric and Ethiopianist views by foregrounding Eritrea and its people as autonomous historical agents. It has been articulated by the Eritrean liberation fronts, Eritrean scholars and some Western academics through a body of work that included books, academic articles and essays.

For the purposes of this paper, the discussion of these perspectives will be limited to four salient historical issues: Historical identity of Eritrea, the nature of the Eritrean armed struggle; Ethiopian policy on Eritrea, and differing narratives of the liberation fronts.

1. Historical Identity of Eritrea

The identity of Eritrea has long been contested by foreign powers with vested interests, as well as by colonial and Ethiopianist writers. These entities often portrayed Eritrea as "an Italian creation with no history of unity, statehood or name", implying it possessed no distinct identity and could be readily absorbed. Consequently, history became a political instrument used to legitimize Ethiopia's forceful annexation of Eritrea. As historian Bairu Tafla observed, "history became inevitably an integral part of the struggle" (1993, 506) to determine Eritrea's identity and future.

Ultimately, Eritrea's historical identity was affirmed and consolidated through a thirty-year war of resistance for self-determination. The idea of a distinct Eritrean identity resonated with the majority of Eritreans, motivating them to endure and fight a protracted and devastating war for self-determination. Fouad Makki explained this phenomenon by noting that, "the Eritrean nation was not simply a territorially bounded entity inherited from the era of colonialism. Independence was achieved through a prolonged and protracted war. Something, therefore, had engaged the imagination of a significant portion of the population, making it willing, as Benedict Anderson famously put it, "not so much to kill, as to die for this imagined community."

Thus, unlike most former European colonies in Africa, Eritrea's future was determined through one of the longest wars on the continent. As Alemseged succinctly put it, "Arguably, in no other African independence movement was the issue of a people's identity fought to consummation as it was in Eritrea" (2025, xiv).

2. The Nature of the Eritrean Armed Struggle

The Eritrean armed struggle for independence was depicted deliberately by both the imperial and military Ethiopian regimes as an Arab-instigated secessionist movement, with obvious political and cultural undertones. For much of the thirty-year conflict, this narrative influenced the stance of most African, Western, and other states, due in part to the comprehensive support provided by the United States and the Soviet Union to successive Ethiopian regimes.

In stark contrast, Eritrean nationalists understood their struggle as fundamentally anti-colonial, rooted in the unfinished process of decolonization. By the 1970s, with evolving political and ideological developments within the liberation fronts, the armed struggle came to be conceptualized in Marxist terms as a national and democratic revolution, with interwoven national and social dimensions. This conceptualization of the armed struggle not only as National but a Social Revolution corresponded with the transition of the struggle in the mid-1970s from its earlier guerrilla phase to the revolutionary phase of nation-wide People's War.

3. Ethiopian Policy on Eritrea

In the 1940's, the Ethiopian imperial regime officially adopted an expansionist policy toward Eritrea, which also extended to Somalia. It claimed legitimacy over Eritrea based on historical, economic and security grounds, as well as the strategic imperative of access to the sea. With the full backing of the United States, Ethiopia was able to gain substantial diplomatic support for its claim. Moreover, by deploying its political and financial resources, as well as appeals to historical and cultural connections, Ethiopia was able to secure the support of the hierarchy of the Eritrean Orthodox Church and a significant segment of the Christian elites and notables. Support to Ethiopian claims was not however confined to the highlands, there were many notables and chiefs in the lowlands who also supported it in line with their group interests.

Ethiopia's strategy of exploiting religious divisions, closely aligned with British efforts to partition Eritrea, fomented sectarian politics and fractured the emerging Eritrean elite mostly along religious lines. Yet, as Alemseged observed, when the Italians first came to Eritrea they encountered "a population with no significant homegrown conflict or deeply engrained mutual resentment, especially of an ethnic or religious nature arising from past relationships of dominance" (2025, 7). Therefore, while the division persisted for a few decades, the British and Ethiopian efforts to incite sectarian conflict were eventually absorbed and overcome by the national movement.

Ethiopian policy on Eritrea, during both the imperial and military regimes, remained essentially unchanged despite variations in leadership and ideology. Both regimes attempted to vanquish the resistance by military might and simultaneously pursued a policy of terror by systematically carrying out wide-spread atrocities on the civilian population. But repression proved to be counter-productive, as it increasingly alienated sections of the population that were not actively involved and engendered more resistance and support to the struggle. The military regime, rooted in chauvinist ideology and effective organization, represented a more extreme embodiment of this long-standing policy of repression and divide and rule which was manifested in its futile attempt to divide Eritreans along religious and regional lines in its final years.

4. Differing Narratives of the Liberation Fronts

Within the overall Eritrean perspective of the history of the national liberation movement two main narratives can be identified represented by the ELF and the EPLF. These narratives were articulated during the armed struggle by both fronts as central parts of political education, and played an important role in their political struggles against each other. They represented primarily the conflicting political views of both fronts as represented by their political cores, and served as political justification of their opposing positions.

Today, these narratives continue to shape how former participants interpret and recount their experiences of the armed struggle. As such, historians must remain critically aware of their enduring influence and the limitations they may impose on objective historical inquiry.

These competing narratives offer different accounts and interpretations of major events and developments, including: the initiation of the armed struggle, the nature of the 1960s crisis, the split and civil war of the early 1970s, internal political struggles and developments within each front, the failure of the efforts to unify the fronts, the second civil war, and the role of leading personalities in the different stages of the armed struggle. These are some of the main issues of contention and conflicting interpretations.

The perspectives underlying these narratives – grounded in opposing political positions, distinct lived experiences and differing perceptions – constitute key subjective dimensions of historical formation. They must be critically evaluated during the documentation of the history of the armed struggle. Historians, therefore, must remain conscious of the dual challenge posed by the scarcity of primary sources -- especially from the early phases of the armed struggle – and the influence of partisan viewpoints. To develop an objective and judicious understanding and reconstruction of Eritrean revolutionary history, it is imperative to critically synthesis the partial perspectives into a broader and deeper national narrative.

The Historiographical Gap

A more fundamental challenge lies in the underdevelopment of historical scholarship on the Eritrean national liberation struggle. The Eritrean Revolution, which mobilized unprecedented level of popular participation and forged indigenous organizational structures, strategies and tactics, stands as an extraordinary example of self-reliant resistance against overwhelming odds. The remarkable feats of human solidarity, selfless sacrifice and resilience combined with the ingenuity and initiative of thousands of committed men and women combatants from different social backgrounds, were truly epical. Particularly notable was the steadfast support and involvement of women in all facets of the struggle, a phenomenon with historic implications for the future of Eritrean women. In the process, the Eritrean Revolution became deeply organic to Eritrean society by winning nation-wide support and organizing broad participation of the people all over the villages and hamlets as well as among the urban population and diaspora communities.

It was through innovative organization, high level of mobilization of limited resources, perseverance, and sheer grit that the Eritrean people ultimately prevailed, at great human cost, over successive Ethiopian regimes backed by global superpowers in an unequal contest of arms. The victory achieved after a protracted and herculean struggle against overwhelming odds and long deemed impossible by conventional wisdom, stands today as a major subject of historical inquiry.

Yet, this extraordinary episode in twentieth-century liberation history remains insufficiently studied and poorly documented, particularly by Eritrean historians. That such a history remains largely unwritten constitutes a glaring lacuna, with far-reaching consequences for Eritrea's national historiography and the broader narratives of twentieth-century national liberation movements globally.

Some Historical Writings on the Armed Struggle

Commenting on Eritrean historical writing, Bairu Tafla observed, "The contribution of Eritreans to the writing of their own history is as recent as their decisive successes." He termed this literature as ''liberation history'' which developed "parallel to the armed struggle" (1993, 504 & 506). Thus, historical writing by Eritreans is not only a recent phenomenon, but also still in its formative stages.

Over the past few decades, a small number of Eritrean participants in the armed struggle and scholars have authored works focused on the history of the liberation movement. On the history of the ELM and its relations with the ELF, one of the founders and chairman of the ELM, Mohamed Said Nawud has written 'The ELM: Truth and History'. Tahir Fedab, a member of the ELM leadership, authored 'The ELM and its Historical Journey, (1958-1967)', while Mohamed Brhan Hassen, a local ELM leader in Asmara, penned 'The ELM: a Memoir'. These texts attempt to narrate the history of the ELM from the viewpoints of leadership and cadres, relying primarily on personal memory, except for Tahir Fedab, who has included a number of ELM documents.

Several leaders of the ELF and the EPLF have authored historical and memoir-based texts. Osman Saleh Sabbe published two works: 'The History of Eritrea,' and 'The roots of the Eritrean disagreements', which reflect his bifurcated and binary view of Eritrean society, culture and history. Alamin Mohamed Said, an EPLF leader, wrote 'The Eritrean Revolution: Decline and Progress', focusing on the internal dynamics of the struggle from the EPLF perspective. Ibrahim Mohamed Ali, an ELF leader authored 'Journey of the ELF' from the ELF's vantage point. These books articulate, to varying degrees, differing narratives of the struggle's principal protagonists.

In contrast, Osman Denden's 'The Eritrean War (1961-1970)' is a compilation of important documents from that period. Moreover, the Research and Documentation Centre (RDC) houses an extensive archive of ELF, EPLF, ELF-PLF, and Ethiopian documents related to the armed struggle. It also preserves hundreds of in-depth interviews with veterans from both fronts.

A few Eritrean scholars have produced works of uneven quality focusing on either the general history of the struggle or its specific aspects of it. Ruth Iyob's 'Eritrean Struggle for Independence: Domination, Resistance, Nationalism' covers the national movement from the 1940's through the armed struggle period. Redie Bereketeab's 'Eritrea: The Making of a Nation (1890-1991)' offers extensive coverage of the armed struggle. Additional works by Mahmud Osman Elos, Gaim Kebreab and Saleh Hamed Turki examine internal conflicts and unity-building efforts, social capital and associational life, and the political divisions of the early 1970's, respectively.

Over the past five decades, foreign scholars have also attempted to study and document the Eritrean armed struggle. Several Western scholars have published partial studies supportive of the Eritrean cause and critical of the dominant Ethiopianist perspective that sought to justify Eritrea's annexation. Notable among these are 'Behind the War in Eritrea' edited by Basil Davidson et al, John Markakis's 'National and Class Conflict in the Horn of Africa', David Pool's 'Africa's Longest War', Richard Sherman's 'The Unfinished Revolution', Dan Connell's 'Against All Odds', Roy Pateman's 'Even the Stones are Burning'. Additionally, among the many books written in Arabic, Said Ahmed Kelifa's ''Shameful position on Eritrea" and Mohamed Abdul Gassim Haj Hemed's ''International Dimensions of the Eritrean War" warrant mention.

In contrast, Ethiopian and Western scholars such as Shumet Sisagne, Gebru Tareke, Christopher Clapham, Haggai Erlich and a few others have approached the Eritrean armed struggle from an essentially Ethiopianist perspective. Lately, many former Ethiopian generals who fought in Eritrea have written books of varying quality on their wartime experiences.

While it is beyond the scope of this paper to assess these works in detail, it is important to note that the overall quality of scholarship on the armed struggle is uneven and underdeveloped. Collectively, these writings reflect the diversity of perspectives and underscore the challenges inherent in documenting such a complex historical period.

Developing the Historiography of the Armed Struggle: Tentative Propositions

1. Structures, Politics and Repression

Fouad Makki has argued that any adequate history of a national movement must synthesize three common approaches that are commonly separated from each other: (1) the objective role of socio-economic and cultural structures, (2) the ideological and political character of the nationalist movement and leadership, and (3) the repressive policies of the occupying state. Taken independently, these analytical lenses are insufficient. However, when synthesized, they provide a comprehensive framework capable of illuminating the intricate unity of the national movement. In the case of the Eritrean armed struggle, these three approaches can be employed to study its origins and uneven trajectory of development across time and cultural context.

2. The Challenges of Teleology

Given the formidable organization, commitment, leadership and widespread popular support and participation underpinning the armed struggle -- and its eventual triumph against overwhelming odds -- there exists a strong temptation to idealize it and view its history through a teleological lens. This may reduce its complex historical narrative into a simplistic, deterministic schema, revising the actual sequence of events to fit a preconceived ideological or political framework.

Such approaches are intellectually sterile, as they fail to capture the nuanced, contingent and often contradictory story of real Eritreans who fought to liberate themselves from foreign domination and establish a sovereign national state. From its inception to its victory – including its organizational evolution, internal conflicts, setbacks, successes, and achievements – the Eritrean national struggle must be studied within its own context.

3. Weak and Strong Narratives

Reductionist interpretations of the armed struggle result in weak and superficial historical narrative. These oversimplified accounts often distort certain aspects of the past and obscure the deeper, multidimensional realities of the movement, thereby undermining genuine historical understanding.

By contrast, a critical and methodologically rigorous study based on credible sources can produce strong narratives – objective and judicious accounts that present history in all its complexity. Such narratives acknowledge contradictions, contingencies, limitations, setbacks, resilience and achievements, ultimately contributing to a more profound and comprehensive understanding of the struggle's legacy.

4. Transcending Differing Narratives

The limitations inherent in partial narratives are evident. Accounts that disregard different aspects and trends in the history of the national liberation movement, the perspectives and experiences of the various political and popular organizations, and the role of the people, diminish the depth and complexity of the history.

Thus, to arrive at a deeper and more holistic understanding, we must critically historicize and transcend one-sided perspective and consciously reject the political instrumentalization of history. As one Eritrean scholar aptly noted, "To free history from political instrumentalization does not of course mean that you can have an apolitical history." Historiography is intrinsically bound to politics and culture; there is no truly neutral history, and history without perspectives is an impossibility.

Therefore, transcending partial and narrow narratives leads to a national narrative -- a dialectical synthesis that more faithfully reflects the shared, collective history of the Eritrean people's armed struggle.

5. A National and Inclusive History

Developing a national and inclusive narrative from the dialectical synthesis of the different perspectives is not the endpoint of historiography – it is a new beginning that opens fruitful avenues for further study. This new approach must critically engage with issues arising from the formative period; the crisis and split in the early ELF; the early challenges and crisis in the EPLF; the character and role of the clandestine Marxist parties and their leaderships; the contributions of historical personalities; the unprecedented level of popular support and participation; the equality and participation of women; the disintegration of the ELF; the historic achievements and final victory of the EPLF. Additionally, it should address the need for a revised periodization that reflects the major political and military developments of the struggle.

Thus, dialectical synthesis must not be seen as a conclusion, but as the starting point in the ongoing process of historical interpretation. On this point, Fouad Makki insightfully remarked:

I don't think divergent interpretations can ever be completely overcome through a higher dialectical unity. The same facts are amenable to different interpretations, but these interpretations do not cancel out the autonomous nature of facts as such, which serve as a control mechanism on interpretations, including one-sided and self-serving accounts. The writing of history is in this sense never closed, and subsequent generations will return to the same subject from different analytic perspectives or through engagement with new sources that become available at any given time.

The challenges in developing the historiography of the armed struggle are therefore to establish it on a scientific foundation. As the Eritrean historian Tewelde Beyene emphasized, if the study of history is "truly scientific -- that is critical in its selection of trustworthy sources and objective in its judgment -- it (will) allow us to achieve or, at least to come closer to the truth about people and events of the past" (1992, 2).

By employing rigorous methodology and critically examining primary and secondary sources, it is possible to initiate a project aimed at producing a national and inclusive historiography of the armed struggle. It is my belief that the EPLF/PFDJ, in line with its own progressive and inclusive national vision, will give full and unreserved support to all scholarly efforts directed toward this goal.

Nurturing a national historical memory is a vital dimension in the process of building a united and resilient Eritrean nation -- one that can contribute meaningfully to sustainable regional stability, peace, cooperation, and ultimately integration. A critical and self-reflective historical consciousness will educate confident Eritrean citizens, who are simultaneously citizens of the Horn of Africa and the continent, and of the global community – anchored in equality, mutual respect, and genuine internationalist solidarity.

I hope a new generation of primarily Eritrean historians will take up the challenge.

Reference

Alemseged Tesfai, An African People`s Quest for Freedom and Justice: A political History of Eritrea, 1941-1962 (London: Hurst & Company, 2025)

Bairu Tafla, 'Interdependence through Independence: The Challenges of Eritrean Historiography', New Trends in Ethiopian Studies, 1993, pp. 502, 509-19

Fouad Makki, Personal Correspondence with author, (July 2021)

Marx, Karl, The Eighteenth Brumaire of Louis Bonaparte, trans. Joseph Weydemeyer, 1852.

Tewelde Beyene (fr.), Introduction to the History of Eritrea: Lecture Outlines, (Univeristy of Asmara, 1992)

Reflections on History Writing in Eritrea

Alemseged Tesfai,

Project History, Research and Documentation Center

I am a lawyer by training and profession. By now, I should have been a retired or retiring judge; or, with some luck, a wealthy solicitor. As it has turned out, I am a writer – a profession that, unlike the road I did not take, somehow repels rather than attracts wealth and its attributes. Yet, it was my abandoned legal studies that set me off on a journey into a crucial period of Eritrean history.

In 1970-71, when I was a graduate law student at the University of Illinois in Champaign-Urbana, the topic that I chose for my master's thesis focused on the legal basis for the handling of the Eritrean case by the Four Great Powers and the General Assembly of the UN. In my research at the seven-story university library of over seven million items, I remember being dismayed by the paucity of written material representing the Eritrean side of the story, especially, the absence of anysubstantive material written by Eritreans on their own history. I had to depend Johannes Kolmodin's, "Zanta Hazzega and Tseazzega," to fill in the gap. I may have done well to prove the illegality of the federal decision, but I don't think that I won the history debate.Why have Eritreans been making history and yet, not recording it?

In their book, "Discovering Eritrea's Past," Bairu Tafla and Eva Schmidt state that, the research launched by pioneers like Werner Munzinger, Leo Reinisch and others, which continued with intensity until late 19th Century,

> Eventually gave way to the politics of grandeur. Italian colonial rule had to extend to Ethiopia, and for that very purpose, Eritrean historiography had to be sacrificed. Eritrean studies thus subsided into more or less insignificance, not figuring even as a regional area of focus within Africa Orientale Italiana and still less under the Ethiopian Empire, which equally aspired to centralism and grandeur. In the 1960s and 70s, in fact, historicizing Eritrea became a taboo the breach of which was dangerous for Ethiopians as well as Eritreans and unamenable to foreigners (Bairu and Schmidt, 2016, 17).

Not only Eritrean, but Ethiopian scholarship as well, have suffered major setbacks as a result of this official policy of denial and suppression. Significantly, Ethiopian and expatriate exponents of the Ethiopian version of history, who write on Eritrea, generally do so without referring to or researching Eritrean sources. The Imperial Ethiopian narrative, which views Eritreaas an integral part of an uninterrupted, three-thousand-year-old Ethiopian polity, has generally been deemed adequate enough not to warrant any further inquiry. The hold of this narrative has been so strong and so ardently advocated by an array of international scholars, that attempts at presenting an alternative history are often met with alarming hostility and derision. Many of these, who see Eritrean nationalism as an "aberration," a deviation from an established norm, have been attempting to find specific dates or turns of event that gave rise to it.

Alemseged Abbay and Shumet Sishagne, both Ethiopian writers, argue that the Eritrean identity manifested itself only in the 1950s, when the 'ethnocidal policy of Amharization' and Ethiopia's move to abrogate the UN-sponsored federation between Eritrea and Ethiopia in the 1950s aroused Eritrean sensitivities (Alemseged, Identity Gilted, p. 222-3). The Eritrean armed struggle, instigated by the Italians and the British, was launched in reaction to Ethiopian suppression of Eritrean rights and 'so-called Amhara domination' in the period between 1956 and 1959, which led to the disappointment of expectations (Shumet 2008, 7-8).

For Haggai Erlich, unionist victory in the 1952 elections to the Eritrean Assembly, is proof that Eritrean nationalism had not yet made its appearance. It emerged, he argues, in the second half of the 1950s when the rise of pan-Arabism in neighbouring countries encouraged Eritrean Muslims to initiate separatist activities; while, at the same time, 'many Eritrean Christians, frustrated by Ethiopian methods of doing away with their autonomy, slowly started identifying with Eritreanism'. To Erlich, 'Eritreanism' was 'a negation of Ethiopianism rather than a historically rooted supra-tribal, supra-linguistic and supra-religious sense of Eritrean affiliation' (Erlich 1986, 215).

In 1989, Christopher Clapham attributed Eritrean nationalism to two factors— the 'incapacity of the imperial system of government' of Ethiopia; and the appointment of Andargatchew Mesai, a 'nobleman with a hectoring manner', to govern Eritrea (Clapham 1988, 206). In 2001, during the so-called border war between Eritrea and Ethiopia, Clapham, still arguing that Eritrean nationalism was an imposition from above, wrote the following:

> As a small, poor, artificial and ethnically fragmented territory, all of whose major population groups spill over its frontiers, Eritrea has always been potentially subject to centrifugal forces, from which it has been protected since independence by an awareness of its military strength. The abrupt collapse of the myth of Eritrean invincibility leaves it in a very vulnerable position' (Clapham 2002, 10, 13).

Patrick Gilkes finds the early beginnings of Eritrean nationalism with the start of an armed strugglea gainst Amhara–Tigrayan privileges, not as a 'real expression of anticolonialism'. But he identifies as a landmark the events of January 1975, when government atrocities in response to an attack on Asmara by Eritrean liberation movements induced thousands of youths to join the armed struggle. 'In fact,' he concludes, 'there is a strong case for believing that the attack and the response to it marked the first real appearance of Eritrean nationalism' (Gilkes 2003, 2).

As recently as 2015, twenty-four years after Eritrea's independence, Alex de Waal wrote that the survival of Eritrean nationalism is attributable to three factors – 'a militarized political culture', 'diasporic and online political "imaginaries"'; and 'obsession with the leader, Isaias Afewerki.' after whose departure De Waal expects Eritrea to disintegrate (De Waal 2015, 154). Take away these three, according to De Waal and Eritrean nationalism goes down the drain. Shallow and audacious though they may be, to the extent that they influence opinion, they need to be confronted.

Such attempts to attach specific dates or events to the rise of Eritrean nationalism raise serious questions about the nature of nationalism itself. Is nationalism a sudden happening? Can an election result (Erlich), frustration over bad governance (Clapham, Alemseged Abbay, Shumet Sishagne) or Ethiopian atrocities in response to guerrilla attacks on Asmara in 1975 (Gilkes) independently create a new nationalism or convert a people from a previous identity? Or is it a historical process whose beginning and end one can neither determine nor predict?

The official suppression of their history and sense of identity did not stop Eritreans from forging them on the ground. Eritreans are, obviously, great makers of history. However, the recurrent failure by generations of Eritreans to record and establish it as a force to reckon with has denied it the place that it deserves among the various established narratives in Africa. It has also exposed it to the kind of arbitrary and, often, disparaging assessments that I cited above. Weldeab Weldemariam, great patriot, teacher, exponent of Eritrean nationalism and master of the Tigrinya language was once asked why neither he nor his peers ventured to record their experiences to posterity. In his usual lyrical words, he replied, "When we could write, we did not want to; and when we wanted to, we could not write."

In the 1970's, as the Eritrean struggle for independence started to reach a level of maturity, Eritreans within and outside the liberation movements and mainly western scholars sympathetic to the cause finally started to write the Eritrean story. But the beginning was not without its shortcomings. In a leading article on the topic, Bairu Tafla has this to say,

> Most of the early papers were characterized by inconsistencies and a great deal of defensive expositions. The reason for this was, of course, understandable. As the Eritreans demanded freedom, the colonial and the Ethiopian interpretations of Eritrean history coincided at the point of questioning the identity of Eritrean society. History, therefore, became inevitably an integral part of the struggle ... [However], it is unnecessary and unbeneficial for Eritrea [and Ethiopia for that matter], to produce inflated and unresearched history. Such an attempt would only perpetrate the persistence of ignorance. It is not the quantity of substance or the length of time-depth that makes history worthwhile; but rather the critical evaluation of the source materials, the quality of the analysis, the objectivity of the approach and the genuineness of the reconstruction (Bairu 1993, 502).

Bairu refers here to a tendency in some Eritrean historical analyses that, in reaction to Ethiopian pre-colonial claims, tends to totally reject the varying degrees of historical relationship that existed in those times and that seeks to transport Eritrean nationalism to an era when it had not been born. The concern originates, in my opinion, from the feeling that recognition of those relationships somehow compromises the strength and legitimacy of Eritrean nationalism. We keep in mind that such relationships were not confined to Ethiopia alone, as we have similar ethnic, cultural and historical relationships with Sudan, Djibouti and even Yemen across the Red Sea. The real issue, therefore, is not whether primordial relationships of any origin existed or not. It is, rather, why they did not develop into a common national identity. Something happened in this specific area that we now call Eritrea, something that "engaged the imagination of a significant portion of its population', to borrow Benedict Anderson's famous phrase, and moved it to make untold sacrifices to establish its separate identity.

In 1987-1989, four colleagues within the Department of National Guidance of the EPLF, Semere Solomon, Kidane Solomon, Zemhret Yohannes and myself, took on the task of writing up the ancient, medieval, Italian colonial and British-Federal periods of Eritrean history, respectively. Although the sources at the Front's documentation centre were limited, we were, nevertheless, able to produce locally researched works of modest size and value. I believe that those publications gave a hint of what a broader and more comprehensive treatment of Eritrean history would look like.

In 1997, I was offered the opportunity to lead a project attached to the Research and Documentation Department of the PFDJ, to document and write up the history of the Eritrean People's Liberation Army. Two factors stood on the way to accomplishing that task. First, the 1998 border war made it impossible to work with veterans of the revolutionary army, as they were all engaged in the conflict. But, more substantively, it became difficult for me and my colleagues to separate the military from the political history of the EPLF. The EPLF was a holistic organization, all of whose components merged into a common stream. The EPLA was itself a highly politicized integral part of the Front and attempts on our part to make the cross over from the military to the political did not succeed. Most of the Front's documents of significance were unavailable and still not declassified. More significantly, veterans who could shed light on certain aspects of the armed struggle, were, regretfully, unwilling to share their experiences. Sadly, many of them have now left us without recording their stories and sharing leaving legacy. Except in a few cases of moderate historical importance, mainly based on individual memory, memoirs, biographies and autobiographiesby major figures are still unthinkable. So, with much regret, I went back to my old research on the history of Eritrea, 1941-1962. The saying that history is written by the victors does not seem to be working in Eritrea.

The project has culminated with the production of three Tigrinya volumes of a total of over 2,000 pages. The English version, which is both an abridgement and an expansion of the Tigrinya volumes, but one-fourth the size, is due to be published at the end of this month by Hurst Publishers in London. The American publication by Oxford University Press, in collaboration with Hurst, is due in April of this year.

It is, overall, a highly rewarding and exciting experience. It is about a narrative that attempts to tell the Eritrean story on its own terms, through the long absent Eritrean voices. The extensive use of Eritrean newspapers, government and Assembly documents; British, US, UN, Ethiopian and some Italian references ... have given context and the advantage of comparison to the story. Fortunately, several major protagonists from the period were still alive and willing to share their experiences.

From their crucial recollections, it is easy to discern that, in the 1940's, two mutually exclusive processes were proceeding side by side. One was what Eritreans, unionists and nationalists alike, thought and expected would happen. The second was the Ethiopia-centred, US and UK-sponsored superpower strategic design that was being woven around them. Even the unionists, always operating under the illusion that they were leading the drive to union with Ethiopia, as Eritreans and as equals, were in for a rude awakening when they finally realized that neither their history nor their aspirations and expectations had been taken into consideration. In many ways, theirs was a case of misplaced nationalism.

On the other side, the passionate warning by Ibrahim Sultan, leader of the Independence Bloc, to the UN's Political Committee, that forcing Eritreans into an association with Ethiopia against their wishes would dip, not only Eritrea and Ethiopia, but also the whole region into an indefinite cycle of conflict, was totally ignored by the major powers and the General Assembly of the UN. Ibrahim's prediction may be buried and forgotten in the massive archives of the UN; but as the years pass on, its poignant accuracy persists. Following and during the thirty-year Eritrean war of independence, which I believe was the first major conflict in the region, our Horn of Africa has never seen peace and stability—today, more than ever before. It is a case of a forcefully suppressed people's sense of identity coming back to haunt the suppressor.

Eritreans had every right to resent and militantly reject the way their right to self-determination was usurped and derailed during the African decolonization process; the way their European colonizer was replaced by one who looked like them, claimed a common heritage with them; thus, complicating their case. They had cause to resent the international alienation that they suffered throughout their political and armed quest and struggle for freedom and justice from the 1940's to independence.

In this book, much care has been taken not to let this resentment, legitimate though it was and still is, to take centre-stage. In other words, the suppression and disparagement of the Eritrean people's sense of their own worth by an imposed identity, and the obvious reaction that it provokes, has not been allowed to lead the historical debate. I believe that the Eritrean narrative needs a dispassionate and objective treatment if it is to hold its own place in the history of the Horn, Africa and the world. It needs to be liberated from the polemical diatribe and invective bred of a bitter experience. When narratives clash, let our facts, based on sound, documented evidence expose the falsehood.

Additionally, there is a tendency, a trend within current Eritrean historical and political literature, to reject or negate all or aspects of the past. It actually started in the 1940's—unionists rejecting the nationalists; the liberation movements negating each other; and a tendency today to negate the very essence of the Eritrean nation and nationalism. The causes of such negation, rational or baseless, should be seen within the context of the specific period in which they manifest themselves. But when it comes to history, the paradigms should change. For history is not an arena where present generations sit in judgment over their predecessors. History is a continuum, where the past and the present merge to produce the future. History is the sum total experience of a people's past and present and it consists everything that is positive and all that is negative.

History may contain what the generations do not choose, but they will have to own it. History is not just the good things that one wants to possess. It comes in a bundle—the good, the bad and the ugly. It is incumbent on new generations to embellish, improvise and carry the good side forward; correct the bad; and rectify the ugly. That, I believe, is how nations survive and nationalisms thrive.

We need to take our history more seriously than we are. We need to make the history that we teach our children, exciting and palatable, tune it to their respective ages. We need to encourage our aspiring young historians and students at the concerned colleges to break the reserve and the barriers that have been limiting its scope and depth. We have a wonderful history waiting to be told.

No history book can ever claim finality and perfection, as the writing of history greatly depends upon the availability of source material. My book is no exception. I will expect and welcome assessments and criticism based on a careful reading of its contents. I also look forward to a healthy and constructive debate with Ethiopian writers and scholars aimed at bridging the gap in our respective understanding of history, for the betterment of our future relationships.

Reference

Alemseged Abbay, Identity Jilted or Re-Imagining Identity? The divergent paths of the Eritrean Tigrayan Nationalist Struggles (Lawernsvill: The Red Sea Press, 1998),

Bairu Tafla and Eva Schmidt, Discovering Eritrea`s Past: Selected Documents from the Works of the Poineers (Germany: Verlag J.H. Roll Gmbh, 2016)

Bairu Tafla, 'Interdependence through Independence: The Challenges of Eritrean Historiography', New Trends in Ethiopian Studies, 1993, pp. 502, 509-19

Clapham, C., "War and state formation in Ethiopia and Eritrea" Geopolitics and International Boundaries 1(2): 115-29. 2002

Clapham, C., Transformation and Continuity in Revolutionary Ethiopia, (New York, NY: Cambridge University Press, 1988)

Erlich, Haggai, Ethiopia and the Challenges of Independence (Boulder, CO: Lynne Rienner Publish, 1986)

Gilkes, Patrick, "National Identity and Historical Mythology in Eritrea and Somaliland", Northeast African Studies, 10 (3), 2003, p.2)

Shumet Sishagne, Unionists and Separatists: The Vagaries of Ethio-Eritrean Relations 1941-1991 (Hollywood: Tsehai Publishers, 2007

Waal, Alex de, The Real Politics of the Horn of Africa: Money, War and the Business of Power (Uk: Polity Press, 2015)

The Writer as a Historical Subject

Tej n Dhar

Professor of English,
Shoolini University, India
Email: tejnathdhar@yahoo.com

Abstract

Alemseged Tesfai's writings are distinguishable from a host of other postcolonial writers from Asia and Africa who wrote to expose the falsities the colonials perpetrated to justify their presence in the colonies and pressurized the colonial subjects to internalize them. A brief mention of such writings from Asia and Africa provides an appropriate context for understanding and assessing the contribution of Tesfai in the anticolonial struggle of Eritrea, which is the main theme of the essay. It argues that he is one of the very few writers who, almost like Frantz Fanon, not only wrote to the empire, long before such writings were made visible by *The Empire Writes Back*, but, like a true historical subject, also fought the empire, and inspired his countrymen and countrywomen to do that because of a belief rooted in the cultural ethos of Africa/Eritrea. The essay provides a short exposition of the true historical subject, discusses the multiple roles of Tesfai in the liberation struggle of his country, and examines critically the nature and texture of his writings in different genres written in his native language Tigrinya and translated by him into English in a volume entitled *Two Weeks in the Trenches*.

Keywords: postcolonialism; anticolonial struggle; historical subject; revolution; Eritrean writings

Introduction

Even before literary response to colonialism got international visibility, with the publication of *The Empire Writes Back*, writers from the colonized world had already written back to the empire in different genres, but mostly in the fictional mode. When they found that their people had internalized a falsehood that the colonizers had spread to provide moral legitimacy to their presence in the colonies, they wrote to expose the exploitative nature of the colonial enterprise and inspire their people to fight it. For assessing the contribution of Alemseged Tesfai in the anticolonial struggle of his country in its proper perspective, the essay is structured around the following points. It starts with a brief mention of the literary response of writers from India and Africa to the colonial situation of their day to show that his contribution is different from theirs because he not only wrote to the empire but also fought it like a true historical subject. Then the essay expounds briefly the idea of a historical subject with reference to the work of Franz Fanon, provides details about Tesfai's multiple roles in the liberation struggle of his people, discusses his writings about the function of literature in a colonial setting, and critically analyses the stories, essays, and plays in his *Two Weeks in the Trenches.*

Literary Response of Indian Writers, Mulk Raj Anand and Raja Rao, to the Colonial Situation

Anand and Raja Rao wrote novels in English in response to the colonial situation in their country in 1930s and 40s. Anand's understanding of the exploitative presence of the British in India, in *Two leaves and a Bud*, is revealed in a microcosmic picture of how the British established and legitimized their hold on the country. By dramatizing a significant contest between the role of divine and human agency in the historical process, as, for example, in *The Sword and the Sickle,* Anand lets its protagonist Lalu realize that the omnipotence of the British is a falsity for "The Sahibs were no gods…and they [Hindustanis] had the right to be a separate nation and, what was more, they had the right to their own land" (82-3). Through his writings, Anand also laid the foundations of a vigorous social realism by changing his inherited model of prose fiction that worked through the mode of recital, which had a strong mythical element.

In *Kanthapura*, Rao deals with the British domination of the Indians by using the traditional, myth-dominated oral form. Accepting the British presence as a historical reality, his novel uses the cultural memory of the people, almost in the way Fanon visualizes it in *The Wretched of the Earth* (193), to dramatize "a concrete representation of a significant aspect of Indian freedom struggle, in which the accent is on demonstrating Gandhi's main achievement of arousing the consciousness of the Indian masses against the British. [But he] mythicizes Gandhi and his mission, for purely strategic reasons," (Dhar 144) to make him and his local version in Moorthy acceptable in the popular imagination. That is how Moorthy succeeds in motivating the villagers of Kanthapura to give up the security and comfort of their homes and suffer the brutality of the British forces with calm and dignified restraint.

Thus, confronted with the historical situation of the colonial occupation of their country, Anand and Rao used their novels to create what Fanon calls the "literature of combat" that moulds the "national consciousness, giving it form and contours and flinging open before it new and boundless horizons" (193).

Literary Response of African Writers, Chinua Achebe and Ngugi wa Thiong'o, to the Colonial Situation

Achebe and Ngugi emphasize that they are tradition-bound to work for their community. Achebe says that the writer must educate his people to repair their damaged psyche, lift their spirit, and strengthen their belief in themselves, by showing them that their past was "not one long night of savagery from which the first Europeans acting on God's behalf delivered them" (45). In *Home and Exile*, he says that because their stories had been colonized by the British, the African writers must retell them in their own style, which he calls "re-storying." In his fiction he shows how his society, during the pre-British days, was well-organized and governed by deep-rooted traditions and conventions, and did not need the kind of bureaucratic structure that the British imposed upon it.

Ngugi confirms Achebe's stance and stresses that the duty of the African novelists is to restore "African character to history," and "give back to the African the will to act and change the scheme of things" (43). His early novels not only rewrite the history of his country from the African perspective but also take a firm stand on how to fight the colonizers.

Thus, the two African writers are a little different from the Indian novelists in that they envision a special role for the African writers which they articulate with a high degree of clarity, intensity, and self-consciousness, but fall short of being what I call a proper historical subject.

A Proper Historical Subject and Fanon

Elsewhere, I have stated that "A proper historical subject implies a person whose thoughts and actions are in tune with the requirements of the situation in which he or she functions; in the colonial setting, for example, a person is expected to work for the liberation of his people and not for realizing his private dreams" (Dhar 2007:174). Achebe and Ngugi work towards liberating the minds of their people, which shows that their thoughts are in tune with the requirements of their situation, but their actions are not, because they are not physically involved in the liberation struggle of their people.

In my view, Fanon exemplifies a proper historical subject, for both his thinking and actions addressed the requirements of his time. After his education in France, he returned to Africa to set up his practice but gave it up to join the Algerian Liberation Front. He wrote influential books on the colonial encounter—*The Wretched of the Earth* and *Black Skin White Masks*—took part in the liberation efforts of the Africans, and "played a leading role in the setting up and training of commando units and taught the *moujahedin* how to lay booby-traps, how to control their reflexes when making bomb attempts, and which mental or physical attitude to adopt in order to endure tortures" (Zahar xi).

Fanon's serious illness made him go the U.S. for treatment. From there he wrote to his friend Roger Tayeb, four weeks before his death:

> We are nothing on earth if we are not in the first place the slaves of a cause, the cause of the peoples, the cause of justice and liberty. I want you to know that even when the doctors had given me up, in the gathering dusk I was still thinking of the Algerian people… (quoted in Zahar 20).

Fanon also wrote to his friend in a regretful tone that if he had known that he was so close to death, he would have stayed back to die fighting for the Algerian people.

Tesfai as a Historical Subject

Fanon is a thinker, who was also a fighter, a true historical subject. A writer in the same mould of a thinker and fighter, though not of the same international stature, is Alemseged Tesfai, who holds a place of his own in Eritrea. He is different from the Indian and African writers already mentioned because he left his doctoral research project in the U.S. to join the Eritrean People's Liberation Front (EPLF) in November 1974 and stayed on with it till May 1991, when it won a decisive war against the Ethiopians.

Tesfai lived with the fighters in a huge space within Eritrea called Sahel and moved along with them to fight the Ethiopians. He used his wide reading in world literature to work as a teacher, journalist, writer, and an intellectual. Ghirmai Negash reckons him "the key all-round intellectual and literary figure of this group" of Sahel writers (177).

Unlike the writers already mentioned, Tesfai wrote in his native language, Tigrinya, to ensure a close bonding with his people and to exert a positive influence over their thinking and actions. His literary output during the years he lived in Sahel is not much, but he wrote in varied genres: novel, short story, reportage, essay, play, and history. Much later, he also wrote children's novels. In all, he has written ten works in Tigrinia; one novel, two children's books, three volumes on the history of Eritrea, from 1941 to 1962, two analytical studies on literature—"Literature, Its Development, and Its Role in Revolution" and "Drama" and a collection of Memories, War Diaries and two Plays, which were translated by him into English *Two Weeks in Trenches*: *Reminiscences of Childhood and War in Eritrea*, and provide the basis for this essay. He also translated into Tigrinya Ngugi wa Thiong'o's play "I Will Marry When I Want." Tesfai gave the best years of his life to the cause of the liberation of his country and influenced his generation of Eritrean writers through his writings.

Tesfai's Views on the Role of Literature in a Colonial Setting

Since Tesfai conceived a special role for literature in promoting awareness among his people, he wrote "Literature, Its Development, and Its Role in Revolution," which became a kind of manifesto for other writers to follow. The long essay traces the history of literature in the West from the earliest times to the beginnings of the twentieth century, gives his assessment of literature in the context of colonialism in Africa, and discusses the revolutionary literature of countries like Russia, China, Korea, and Vietnam. This leads him to the history of Tigrinya literature and the role of literature in the Eritrean struggle, which reflects his acute consciousness of what was expected of him in his specific historical location, and the role of literature in fighting colonial oppression.

Tesfai examines critically the emergence of various political, cultural, and literary movements in various parts of Africa in response to colonialism and expresses his special liking for the work of Achebe and Ngugi. His preference for the work of Ngugi is because of his Marxian leanings and Fanonian mode of fighting the colonials by violent means. That is why he considers the soldiers revolutionaries, and fight against the colonials a revolution, and stresses that the fighters should know the revolutionary literature of other parts of the world through its translation, an activity that flourished in Sahel during his time. Apart from this, he pleads for producing revolutionary literature in Tigrinya as well, literature that avoids tendencies associated with the Western literature of his times, which put premium on individual experience, and valorised artistic movements like surrealism. In one of his interviews given after the liberation of his country, Tesfai has confirmed how "socialist realism and the arts served best the needs of the independence struggle [because] … art had to serve revolutionary objectives. Few doubt the significant role that the resulting cultural activities played in the outcome of the struggle for independence" (Dhar 15/1: 3).

Tesfai is for vigorous social realism as a fit vehicle for the writers' engagement with political and social issues of the day. The revolutionary literature, in his view, does not deal only with issues related to fighting the political oppressor but also with problems within society, like inequality among classes. Perhaps sensing that these two concerns might lead to the production of a stereotypical, ideology-driven mechanical kind of realism, Tesfai states that writers need to ensure that their writings combine social commitment with artistic integrity. He insists that literature should not read like reportage. Writers must be proficient in their use of language and skilled in the use of imagery and other literary devices, to make their writings aesthetically pleasing. In this context, it is interesting to note that the essay praises Beyene Haile for the language and style of his first novel *Abidu do tibluwo* (ዓቢዱ ዶ ትብልዎ) but criticizes him for his indifference to the societal concerns of his time. Tesfai is for vigorous social realism that is socially relevant and artistically pleasing.

Since Tesfai is in the field of battle against the colonizing enemy, his writings are different from those of the Indian and African writers. They dwell mostly on the fighters, their psyche and thinking and their qualities. Because of this, heroism forms a powerful theme in them, which is intricately linked with character building. Since such writings are meant to foster a culture suited to the time of resistance, there is stress on idealism, and Tesfai does not hesitate to be explicit in his messaging. Though largely celebratory, they are occasionally punctuated with sombre melancholy. In his plays Tesfai deals with the themes that are the staple of colonial literature, like the insidious varieties of colonial oppression and modes of fighting it.

On Tesfai's *Two Weeks in the Trenches*

Two Weeks in the Trenches has three sections: the first one is of childhood memories, which contains three short stories; the second has writings that are directly related to the field of battle: two weeks in the trenches, the account of a battle, and an essay on the heart of a fighter; the third has two plays that deal with the sinister exploitative ways of the Ethiopian oppressors.

The three stories are like memory recalls in which the narrator remembers childhood experiences, which he considers significant, because they have proved valuable for him. Because of this, they have the shades of an initiation story. Exposure to them has enriched the range of the narrator's experiences, sharpened his understanding of human nature, and helped him learn about vital human values and their importance in life.

The Stories in the Volume

All the stories are in the naturalistic mode and located in a familiar terrain, with recognizable place names, peopled by well-drawn life-like characters, who come alive during their action. They are plotted in a way that each one of them leads to the emergence of a significant value. Since the author-narrator is eager that the readers should share that with him, the value is stated explicitly in the form of a message, although one can easily make out that it would have registered with the reader on its own.

The first story "Shobree" is about a group of children in a school where they must reckon with an authoritarian teacher in their classroom, who lashes them hard and expects them not even to cry. When one of them becomes almost unconscious after getting a good thrashing, they revolt against him by going on a strike, without caring for the consequences. The narrator is happy that all the children respond to their condition with courage and determination. The valuable lesson from this is stated in the form of a message that is full of verve and rooted in a firm belief: "I suppose that there is an internal glow, an overflow of emotion that only defiance provokes. There is something about revolting against despots, something that inflames one's soul" (10).

Another story is named after its protagonist Grazmatch Tsegu, who has fallen into bad days and takes up the menial job of carrying bones in a wheelbarrow. When he talks to one of the children while passing through a lane, he becomes the subject of discussion among the inmates of the houses where the children live, and where his real identity gets known to the readers. When Tsegu realizes that people in the locality have found out who he is, he leaves the place for good. But he makes an impression on the narrator because of his calm dignity: that even in adversity, he embodies "respect for himself and within himself," and greatness, which is "Not the external greatness of birth and appointment, not even that of age, but the inner greatness of spirit and wisdom" (22-23).

The narrator also remembers Hansu, a woman who is at the centre of the story named after her. Her life is a journey of trial and suffering, in which she moves from her village to Asmara and then to Ethiopia, with a small daughter, born to her after a person had forced himself upon her. Pushed out from Asmara by her uncle and pushed out also from their homes in Adis Ababa by her sister and her rich cousins, she settles in a part of the city where she earns her livelihood by braiding the hair of women. When, after many years, people declare her dead mistakenly in an accident, her sister and cousins rush to her home to mourn her loss and apologize for not having helped her when she was alive. When Hansu appears on the scene, she behaves with commendable restraint and does not give up on people who had helped her in her hard times. From Hansu, the narrator learns the virtues of "grace and dignity" (40).

All the three stories enshrine the values that the narrator has learnt to cherish. Interestingly, the fighters, who dominate the second part of the book, cherish the same values.

Writings About War in the Volume

Because Tesfai is right amidst a fight with the enemy, his writings are related to war and warriors, in the same way as some novels of Ernest Hemingway deal with the impact of war on human lives. The key difference between the two however is that though characters in the work of both writers learn the virtue of what Hemingway calls grace under pressure, Tesfai's writing is about historical beings, people he knew and worked with, while Hemingway's writings are about fictional beings. More importantly, the warriors in Tesfai are part of the larger community of warriors, where individual considerations are a taboo. In Hemingway, the main characters sometimes see a divide between the individual and the collective, and, like Frederick Henry in *A Farewell to Arms*, opt for a private peace, which is alien to the thinking of Tesfai's warriors.

The second section of the *Two Weeks* begins with Tesfai's account of "Two Weeks in the Trenches," which is in the form of a diary, with proper dates, describing the moves of the fighters across places and trenches. The account is a feat of acute visualization of scenes of confrontation with the enemy, of the moves of the warriors to different places, and graphic details of the difficult terrain in which the warriors operate.

The diary has also a strong reflective character because the attitude of the warriors to their situation stirs varied kinds of emotions in Tesfai. The major response, though, is a gradual expansion in his understanding of the warriors, of their actions in and off the field and their thinking. The male warrior is *tegadalai*, the female warrior is *tegadalit*, and all the warriors are *tegadelti*.

Right in the beginning, Tesfai makes an interesting observation about the spirit of *tegadelti*, which he finds "difficult to describe, much less to explain," because of its unusualness. It is because for them

> Death and adversity are … accepted as ordinary happenings in the normal course of things. So, any call to active duty, regardless of its nature, always creates such excitement that reluctance is rarely encountered. The sick try to hide their ailment and the lazy are suddenly motivated. This is the norm. (44).

This clearly brings out the mental readiness of the fighters. They are in high spirits, ready to take on any job that is given to them and feel excited about it. Situations that frighten ordinary mortals, like adversity or the possibility of death, are the normal for them. Because of their "love and loyalty" for their country, all the fighters are ready for martyrdom, not just because it is a part of their inherited tradition but because it has already evolved into a culture.

The love of the fighters for their country and their readiness to lay their lives for its freedom is so strong that the sentiment often finds expression in the manner they often dance, which Tesfai does not consider "an ordinary dance. It is an expression of something deep—of youthful desires unfulfilled, of suppressed joys and sorrows, of dead comrades not openly mourned and of loves not consummated. It is a dance of young men and women with a burning love of country and people, about to pay for that love on the altar of self-sacrifice" (49). Dance is a means of erasing sorrow and getting ready to move towards the desired goal.

Interestingly, death for fighters does not signify an "end to a person's journey" (50). Although they do not believe in reincarnation, "yet they have the unflinching conviction that they will live on in their surviving comrades, in the realization of the dream they are dying for" (50). That is why the fighters do not talk about their individual selves. Anybody who does that "would be alienated immediately. Nothing is despised and more condemned than self-aggrandizement and boasting" (50). That makes the warriors fit historical subjects: ready to fight the enemy and prepared to die for their country.

Tesfai records that warriors must rest content with food that is drab and lacks variety, but that hardly makes a difference to their spirit. "To complain about being tired, hungry or thirsty is taboo. Perseverance, suppressing one's personal feelings and desires, resisting all personal problems—even illness—is the prevailing morality and the right way to behave" (53). A warrior "has no time for sorrow, sentimentality and sympathy." He accepts things as they come. This does not denote toughness, but "a calm and serene acceptance of the possibility of death. It is, as if, deep inside them, they have come to terms with martyrdom as the best way to show one's love to one's country" (78). Apart from people's defiance of death, there is emphasis on the individual being a part of the collective. That is how warriors visualize their continuation in the mass that will outlive them. "The wealth of *tegadalai* is his perseverance and his spiritual purity" (57). Because of this, Tesfai raises martyrdom to the level of a mystique, which has a strong spiritual dimension that he finds difficult to understand, believe, and write about. The desire to die for the country is so strong in them that those who survive their peers experience sadness.

When not fighting, the fighters are busy digging new trenches, repairing old ones, and taking "intensive academic lessons and political education" (56).

There are moments in Tesfai's account when overwhelmed by the death of young men and women he has watched grow in his company, he feels that "our revolution is too costly. This struggle is too bitter" (71). He picks and drops his pen for "hours and feeling very inadequate as a writer that it is frustrating" (72). At one moment, struck by the futility of war, he writes:

> That night, I made some final conclusions about the stupidity, cruelty and savagery of war. I was in such a state of agitation that had I maintained it at that level and stayed with those conclusions for much longer, I would probably have gone to the monastery or done some harm to myself ... my dazed mind hated humanity (88).

This reaction reinforces the enormity of the price that the Eritreans paid for their liberation. Hundreds and thousands of people sacrificed their lives willingly. That is why Tesfai feels agitated, but he also feels sad that though each one of them richly deserves to be remembered for their supreme sacrifice as individuals, they are "lumped together into a generic mass called 'patriots.' Brave men and women with names and faces whose individual and collective feats of courage and perseverance have made the difference between oppression and freedom are just one mass of martyrs" (94).

"At the Battle of Afabet" is Tesfai's account of how this decisive battle was fought in a highly professional manner—planned in a time-bound frame according to a well-conceived strategy, in which the enemy was attacked from three different flanks. The battle lasted for three days but the gains started coming in right after day one, leading to the eventual annihilation and rout of the Ethiopians. The statistics about the gains made in terms of captured soldiers and armaments, and the areas captured "could never describe the phenomenon that was the *tegadelai*" (117). As if this were not enough, Tesfai also says that there is "more to a *tegadelai* than what he does or is done to him on the war front" (117). These remarks signify his inability to convey adequately the heights that the warriors touched during the battle.

While walking on the hills of Ad Sherum with some warriors in the joyous mood of victory and ruminating on the "amazing speed and precision" with which the warriors had captured the land under the control of the Ethiopians and the enemy soldiers, he found a patch of blood and right in it a piece of flesh, which stimulates him to write a short lyrical piece titled "Heart of *Tegadalai.*" : "It was a human heart … an oily, bright and red human heart" (130). When his military commander Ali Ibrihim tells him "It is the heart of a comrade," (131) his initial reaction is a cold current going down his spine, but soon after he understands its true significance as a symbol of the entire war effort.

Despite his best efforts, Tesfai cannot "give it flesh and bones and associate it with a face" (132). He soon realizes that that is not even required, because "It was simply the heart of a patriot, the heart of *tegadalai*" (132). That is why his revealing comment that the heart is steeped in the history of the land:

> Our fallen heart has its origins in history, in the ancestral legends and traditions that have come to us through generations. This land of ours have never seen, nor has it enjoyed for ages, the benefits of normal peace and ordinary life. Centuries of foreign rule, invasions and internal strife [have not allowed our people] the air of peace and tranquillity. Consequently, the love of country that the *tegadala*i has inherited is not merely one of ordinary love and compassion. Hate and defiance claim an equal share in his or her passions and emotions. History has pushed the *tegadalai* to be quick and ferocious in retaliation to any threat, any provocation that endangers Eritrea's rights (133).

Tesfai sees the *tegadalai* part of the chequered history of the country. Although the heart is full of love and compassion, it must have a bit of hate too, to defy the enemies that have always beset the country. In fact, beneath the "battlefield cruelty lay an inherent softness, a sea of compassion," (133) which Tesfai sees in evidence in the handling of Ethiopian prisoners of war and their Soviet officers by the *tegadalai.* Because of this, he believes that the heart is "not a heart of grudges and evil intentions. Afabet convinced me that this is a heart of mercy and forgiveness…. It does not want to see repeated, even on its enemies, the pain and ordeal it had to suffer. Nor would it covet from others what is not rightfully its due" (134). That is why it craves for recognition in the polity of nations: "The world has yet to recognize this heart that so willingly sheds all that it possesses for peace among men and women" (134). This heart has fought the power of the enemy for more than a quarter of a century and it has remained steadfast.

Tesfai is happy that "adversity, fortitude and the flames of war have fomented it into maturity, it has reached a new height." So, he also provides a cautionary note: "…it will need careful handling especially in this, its moments of victory, lest it get intoxicated with glory, forget what it has gone through, swell out of proportion and just blow up" (134). It must be guided by "modesty and humbleness" (135) so that it lives up to its responsibility towards its people and reaches new heights in the days to come.

Two Plays in the Volume

Part III of *Two Weeks* consists of two plays. The first, a one-Act play *Le'ul*, has five scenes: two located in a textile factory, where Le'ul and Kidiste must face the disgraceful behaviour of Letegirgish and the malicious gaze of Cadre, whose only job is to announce instant cuts in their wages for flimsy reasons, like talking even after finishing their work. Le'ul and Kidiste must put up with suffocation and oppression. When Le'ul loses her child, she is not only denied leave to mourn its death but also given a shocking piece of wisdom: "...you must swallow your sorrow, keep it within you" (145).

The offensive atmosphere in the workplace disrupts the domestic peace of Le'ul's home, which is dramatized in two scenes of the play. Her marriage with Weldu develops huge cracks because he too is frustrated by happenings in his workplace. They have heated verbal exchanges that border on violence, and end in Weldu's trips to the bar, where he drinks heavily to drown his sorrow. In the fifth scene, located in the bar, Weldu's friend Mussie tries to make him understand that he is needlessly disrupting the peace of his home. To this, he says: "At work, they are deducting our pay even for faults we never committed. Our children die in our arms for lack of care" (153).

The play ends on a note of hope. Weldu realizes his mistake and shows promise of re-building his relationship with Le'ul. There are also hints of a big operation against the oppressors.

The second play "The Other War" has the structure of a conventional five-Act play. Its opening Act functions like the exposition, for it introduces all the major characters: Letiyesus, an Eritrean grandmother and her friend Hiwot; her daughter Astier and her Amharic husband Assefa; Astier's daughter Solomie from her previous marriage with an Eritrean and her son Kitaw from her second and present husband. Through their exchanges, it also builds its main theme of the ongoing war between the Ethiopians and the Eritreans. The war that is being fought away from Asmara looms in the background, but its rumblings are heard in the conversations of the characters. The other war, which is a more complex one, is suggested here but revealed in the action that follows.

The play opens in the home of Letiyesus, which has been opened by Hiwot to let Astier and her family stay there. Letiyesus does not like that her home should provide shelter to the enemy. Her son Miki-el is in Sahel getting ready to take on the enemy. The other war is embodied in the son born to Astier and Assefa, who has been named by his father Kitaw, which means to "punish them."

Letiyesus is shocked by her daughter's marriage to Assefa and calls her a "wicked daughter." Though the daughter thinks that her husband is kind and gentlemanly, Letiyesus is convinced that she has not yet seen the worst in him.

Act II develops this theme by bringing together Letiyesus and Solomie, who experiences isolation in her school because of her mother. She also tells her grandmother how she has been neglected by her parents and how they speak ill of Miki-el. Astier has taken over as the chairperson of the Kebele and is hard on its members, including Hiwot. She blames her mother for having forced her to marry the Eritrean Zecherias and instigating her daughter

Solomie against her. She is also convinced that Assefa loves her enormously. Letiyesus tries to make her understand that just because of one bad Eritrean she has no right to wreak vengeance against all her people. Since the mother in Letiyesus sees how hollow her daughter's life has become, and how blind she is to the abyss that she has fallen into, she tells her:

> When I see you, my own daughter, choosing the way to destruction … wouldn't my womb revolt? Wouldn't my motherhood scream? No, don't be under any illusions. It is not your 'truth' that's making me weep, Astier. You have no truth (184).

Act III intensifies the action in which both the wars figure prominently. The opening of the "Red Star Campaign" of the Ethiopians leads them to partying in the house, in which Letiyesus must slave around. She admits that as a parent she made her daughter look after her family of her first husband but fails to understand its extreme reaction from her, that "she would betray her own country and people" (189). She is also conscious that the daughter is driving her against her, "But my womb, my motherhood fears this confrontation" (189). Through Solomie, she gets to know that the Sahel campaign of the Ethiopians has failed.

This leads to another ugly confrontation between Letiyesus and Astier and Assefa. They tell her that she should help them in bringing back Mike-el and to make him surrender, so that his life can be spared. Her indifference to the proposal makes Assefa angry and he tells her "It looks as if you yourself are a bandit, just like your son" (193).

In Act IV the rumblings of the war on the front are heard louder than before, for Solomie tells her grandmother that the *tagadelti* have won in Sahel. But Assefa is not willing to accept it, though he surely is frustrated. That is brought out in his long conversation with Letiyesus. He tells her that Astier has been arrested by the police because someone from among her friends had complained against her and accuses her of not feeling sufficiently disturbed by this. He also complains to her that she does not love Kitaw as much as she loves Solomie.

This conversation finally leads to the unravelling of the true nature of the other war when Assefa tells her that the new Eritrean generation is Ethiopian at heart:

> Your daughter, Astier, comes from that Eritrean womb you are so proud of. I, Assefa, planted my seed in your own daughter's womb. Kitaw was born. Kitaw walks on this earth.... My roots are firmly planted in Eritrea and no power can ever pull them up. So, get this straight. Eritrea no longer belongs to the Mike-els, it belongs to us, the Kitawas (201).

That is further clarified by Letiyesus, when she tells Assefa: "So another war is going on in our daughters' wombs? I thought you loved and embraced Kitaw as your son. I had no idea he was just bullets and bombs to you" (202).

The last Act sees Assefa and Astier locked in a war of accusations and counteraccusations, with Astier standing alone to cry her heart out.

In his Afterword to the play, Tesfai pleads for seeing the mixed marriage of Astier and Assefa in the context of the colonial setting, where the female body becomes a site of cruelty and domination. And this, he writes, he had come to know during his stay in Ethiopia, when he heard one of the officers of the government say that if they cannot win people politically, then the policy should be "to mix these people with those of the rest of Ethiopia" (213).

We have sufficient evidence now to believe that colonizing the womb is a known colonial practice. It was used in Australia in a massive way. The "Indigenous" women were converted into child-bearing machines and their children made part of the white society that controlled the reins of power. So many commissions in Australia have documented this practice across generations. Such children are a part of what is called the stolen generation. Kim Scott is a living example of this generation and his novel *Benang: From the Heart* is a narrative of this terrible experiment, which changed the colour and mind of the oppressed to make them into different beings.

This essay has not commented on Tesfai's novel *The Son of Hadera*, which was closely related to the needs of the times and became in the words of Negash "the prototype after which [other writers] could mould their own works" (182). He is also "generally credited with being the chief exponent of a lucid and lyrical expository prose style" (178). These are important because Tesfai contributed to the growth and development of Tigrinya language as a creative medium, fit for writings in different genres, when it had just emerged from the shadows of the oral tradition.

Conclusion

The essay clearly establishes that Tesfai is like and unlike some known Asian and African postcolonial writers, and very much like Fanon, a true historical subject, who not only wrote but also fought the empire. This is amply revealed in the multiple roles that he played in the liberation struggle of his people and reflected in his writings in various genres in *Two Weeks in the Trenches.*

Reference

Anand, Mulk Raj. *Two Leaves and a Bud.* 1937 (New Delhi: Arnold Publishers, 1981).

---------------------. *The Sword and the Sickle*. 1942 (London: Asia Publishing House, 1986).

Achebe, Chinua. *Hopes and Impediments: Selected Essays* (New York: Anchor Books, 1990).

--------------------. *Home and Exile* (New York: Anchor Books, 2000).

Ashcroft, Bill, et al, eds. *The Empire Writes Back: Theory and Practice in Post-Colonial Literatures* (London: Routledge, 1989).

Dhar, Tej N. "An Interview with Alemseged Tesfai." *The Commonwealth Review*, 15(1) 1-12.

Dhar, T N. "Historiographic Contest and Postcolonial Theory," *Literature and Ideology: Essays in Interpretation.* Ed. Veena Singh (Jaipur: Rawat Publications, 1998): 27-43.

-------------. "History, Myth and the Post-Colonial: The Indian Context." *Interrogating Post-Colonialism: Theory, Text and Context*. Shimla: IIAS, 1996: 141-152.

Dhar, Tej N. "Ngugi's Retrospective Gaze: The Shape of History in *A Grain of Wheat*." *Kunapipi: Journal of Postcolonial Writing & Culture*. 29(1), 2007, 173-183.

Fanon, Frantz. *Black Skin White Masks*. Tr. Charles Lamm Markmann (London: Paladin, 1970).

----------------. *The Wretched of the Earth*. Tr. Constance Farrington (Harmondsworth, Middlesex: Penguin Books, 1967).

Hemingway, Ernest. *A Farewell to Arms* (New York: Scribner's, 1961).

Negash, Ghirmai. *A History of Tigrinya Literature in Eritrea: The Oral and the Written*. 2nd Edition, (Trenton, New Jersey: Africa World Press, 2010

Ngugi wa Thiongo'. *Homecoming: Essays on African and Caribbean Literature, Culture and Politics* (Heinemann, 1972).

Scott, Kim. *Benang: from the Heart* (North Freemantle: Freemantle Arts Centre Press, 1999).

Tesfai, Alemseged. *Two Weeks in the Trenches: Reminiscences of Childhood and War in Eritrea*. (Trenton, New Jersey: Africa World Press, 2002).

Zahar, Renate. *Frantz Fanon: Colonialism and Alienation*. Trans. Willfried F Feuser (New York: Monthly Review Press, 1974).

The Migration Business

Daniel Sillas

Rome, Italy
Email: danielsillasg@gmail.com

Abstract

Lampedusa Tragedy of October 3, 2013 when 368 people, mostly Eritreans, lost their lives. Since then, 30,000 African migrants have died in the same sea. But what did really happen in Lampedusa that day? Was it possible to save all the victims? Who is responsible for those deaths? And who is behind this "exodus" from Africa? Who are these thousands of people making profits—as if it were an industry—from the arrival of migrants? Who is making fortune from low-cost workforce's exploitation? Who are those abettors rewarded with money, recognition, humanitarian awards, or political support and who is behind all these grants and prizes?

In order to give an answer to such questions, I embarked on an in-depth investigation on the "Migration phenomenon" for almost nine years, focusing my research primarily on the Mediterranean Sea and Italy, the main gateway to Europe. In 2017, I went straight to Italy's Reception System, working as an interpreter for migrants in Italian, Tigrigna, and Amharic. This role granted me access to asylum seekers camps, which is off-limits to journalists, as well as government and private facilities where they were interviewed. I was able to garner insider information from operators and especially "refugees" while consulting confidential documents.

I compiled my findings into a 404-page book titled *Inferno Immigrazione (The Migration Inferno),* where I used a pyramid structure divided into nine levels of responsibility (nine chapters), exposing the key players involved in this affair: human traffickers, human rights activists, NGOs rescue ships, journalists, Italian politicians, the Vatican, UNHCR, Amnesty International & Human Rights Watch, Washington political agendas, and the Open Society Foundation.

"*Inferno Immigrazione*" was published in 2022, an investigative book that exposes uncomfortable truths and, for obvious reasons, faces censorship and boycotts to prevent facts from reaching public opinion. I've become the most censored migrant in Italy related to Migration issues. In this paper I am going to present my book and the results of my studies/ findings.

Keywords: Lampedusa, tragedy, shipwreck, Mediterranean Sea, Europe, Eritrea

Introduction

Migration is a natural phenomenon that began in the times of Neanderthal man and continued with the movements of Homo Sapiens, who left Africa to spread across the globe. However, modern migration is far from natural – it has become a system of human trafficking. More explicitly, it is a return to the era of slavery, with one key difference: there are no longer violent raids to capture and deport people in chains. Instead, invisible shackles drive them toward Libya's shores, where they board vessels unaware that they have, in effect, purchased their own tickets to death. And this is exactly what happened to those who perished in Lampedusa Tragedy of October 3, 2013. My questions are: who is responsible for those deaths? Who are these thousands of people making profits—as if it were an industry—from the arrival of migrants? Who is making fortune from the exploitation of low-cost workforce? Who are those abettors rewarded with money, recognition, humanitarian awards, or political support and who is behind all these grants and prizes?

I think that this paper would help to understand the ins and outs of this *infernal machine* built on the Mediterranean Sea where traffickers, journalists, priests, human rights activists, NGOs and politicians, in the name of human rights, all play their part to facilitate this modern *Slave Trade* from Africa deporting "fifty million migrants in twenty years".

My final aim is to unmask the hypocrisy of those who advocate for increased migration and leave the ports "open" for migrants while portraying themselves as champions of human rights. This paper has 14 chapters and a conclusion, it starts from Lampedusa Tragedy and proceed with what happened after in that island, in the Mediterranian Sea, in Italy, and Europe.

The Mother of All Tragedies

In Lampedusa Tragedy of October 3, 2013, called "*The mother of all tragedies*", a boat leaving Libya with 540 people on board, after two hours of engine stoppage, sank causing the deaths of 368 people, mostly Eritreans. But what did really happen? "It was 6:30 a.m. when I gave the order to call the Coast Guard, but they arrived at 7:30,"[1] reported Vito Fiorino, owner of the fishing boat *Gamar*, which was the first vessel to arrive at the scene of the Tragedy and, with his crew[2], rescued 47 people. Inexplicably, the official rescue operation was delayed by a full hour despite the fact that the sinking occurred just 800 meters away.

Before the shipwreck, many survivors said they sighted two vessels. One remained stationary at a distance, while the other circled their boat before both of them headed back toward the port, ignoring their desperate cries for help. In an attempt to signal for rescue, a Somali smuggler set fire to a bedsheet. But the flames quickly spread out of control, forcing everyone to one side of the already unstable boat. Within minutes, it capsized and was swallowed by the sea. "Something went wrong, the Coast Guard guys have always been very good, but evidently that night they were either waiting for a command or a phone call..."[3] These are the words of Grazia Migliosini, who was also on board of the *Gamar* that night.

Did the Coast Guard really receive instructions to delay the rescue operation? Two years later, in July 2015, the Somali smuggler, the man who had set the fire, was sentenced to 30 years in prison. In December 2020, the court of Agrigento sentenced the Commander of the fishing vessel *Aristeus* to 6 years in prison for failing to provide assistance, while six African crew members received four-year sentences.[4]

However, no action was taken regarding the second vessel—equipped with onboard radar, as reported by survivors—that had circled the doomed boat before returning to port. It was later identified as a patrol boat belonging to *Guardia di Finanza*,[5] but no further investigation followed.

Mare Nostrum Operation

"I hope that divine providence intended for this tragedy to open Europe's eyes," declared the

5 Il naufragio della verità
https://askavusa.files.wordpress.com/2018/09/lampedusa-3-ottobre-2013-il-naufragio-della-verit-c3a0-askavusa-2018.pdf_

Interior Minister Angelino Alfano as he stood before the victims' bodies. In fact, Lampedusa Tragedy had cracked open the wall of European indifference on the migration crisis, prompting José Barroso, President of the European Commission,[6] to visit the island. Standing in a hangar filled with rows of coffins, he pledged to keep "Europe's attention high." Thus, € 30 million were immediately allocated by the EU to help Italy manage the migration emergency. The then-Prime Minister Enrico Letta used that money, not to compensate the families of Eritreans, but to set up *Mare Nostrum Operation.* On October 14, 2013, Italy deployed a large-scale military search-and-rescue operation involving the Navy, Air Force, and Police, supported by 1,000 soldiers, an amphibious ship, two corvettes, two patrol boats, and three helicopters. The operation cost €10 million per month.[7]

The result? A surge in migrant arrivals, skyrocketing from 43,000 to 170,000 in just a few months.

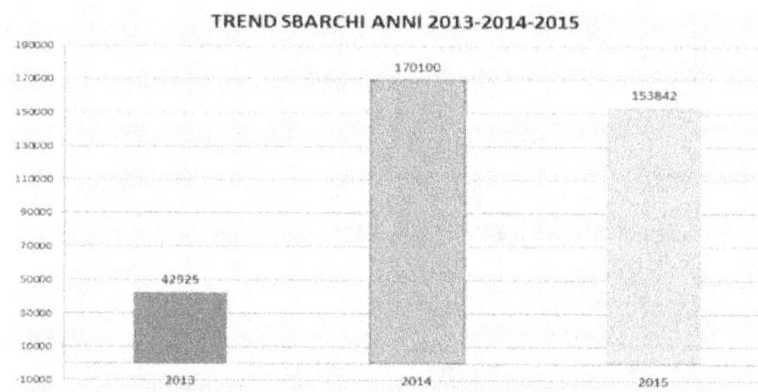

Source: Ministry of the Interior

The Exploiters of the Tragedy

Alongside with the politicians eager for visibility, mainstream media descended on the island, quickly silencing the crew of *Gamar* to prevent the inconvenient truth about the incident from spreading. A *Rai2* journalist named Valerio Cataldi hastily formed the *Three October Committee*[8] and, to lend credibility to the initiative, appointed a "refugee" as president. The committee's main goal was to establish the "National Day of Remembrance and Welcoming". On the first anniversary of the Tragedy, the committee organized the Sabir Festival (October 1-5, 2014), a musical festival featuring artists, politicians and institutions wearing black T-shirts with the motto: *Protect people, not borders.*

In March 2016, the Italian Senate officially approved the bill to establish the *National Day of Remembrance and Welcoming.*

The Humanitarian Awards

In 2014, the journalist Valerio Cataldi won several awards for his work, and by 2017, he was appointed president of *Carta di Roma*[9], an organization tasked with shaping journalistic ethics on migration issues. This Association soon began targeting and silencing the Eritrean community in Italy, labeling it as pro-Eritrean-government.

Two Eritreans who managed the trafficking of migrants, both equipped with satellite phones, were also honored. Abba Mussie Zerai, known as "the angel of refugees," received several humanitarian awards across Italy and Europe and was even nominated for the Nobel Peace Prize in 2017. In October 2015, Dr. Alganesh Fessaha was awarded Italy's highest national recognition by the President of the Republic.[10]

In those dramatic days, both Mussie and Alganesh were very active in Lampedusa, accusing members of the Eritrean community of being "spies". They actively worked against the Eritrean Embassy's efforts to repatriate the victims' remains. As a result, to this day, those who perished in the tragedy remain buried in various Sicilian cemeteries, identified only by numbered tombstones.

The Arrival of NGO Rescue Ships in the Mediterranean Sea

The International Convention[11] establishes that anyone who is at sea in dangerous conditions must be rescued. But who decided that Africans should only reach Europe by sea?

Within a year, the *Operation Mare Nostrum* paved the way for private rescue ships that suddenly poured themselves into the Mediterranean Sea. The first to come was the American vessel *Phoenix,* which arrived in the summer of 2014. Between May 2015 and August 2016, several European NGOs have been engaged in the search-and-rescue work. (*See the figure below*)[12]

NGOs	*Nationality*	*Rescue vessels*	*First mission*
Doctors Without Borders	International	Bourbon Argos	May 2015
Doctors Without Borders	International	Dignity 1	June 2015
Sea-Watch (2015)	German	Sea-Watch 1	June 2015
Proactiva Open Arms	Spanish	Astral	October 2015
MOAS	American	Topaz Responder	December 2015
Sea-Eye	German	Sea-Eye	February 2016
Sos Mediterranée	French	Aquarius	March 2016
Jugend Rettet	German	Juventa	July 2016
Proactiva Open Arms	Spanish	Golfo Azzurro	August 2016

In July 2020, an Italian NGO started its operation with a ship called the *ResQ-People*. One of its founding members is Alganesh Fessaha. To understand the origins of *ResQ-People*, it is necessary to revisit the sad story of Alan Kurdi.

In September 2015, Abdullah Kurdi, his wife, and their two children fled the war in Syria, attempting to reach Greece by boat. Tragically, the vessel sank, claiming the lives of the entire family except for Abdullah. Later, a journalist discovered Alan, his young son, washed up on the shore, his small body clad in a red T-shirt, and blue trousers, face down in the sand – a heartbreaking image that soon became a symbol of the migration crisis. On February 10, 2019, in the port of Palma de Mallorca, the German NGO *Sea-Eye* named its rescue ship *Alan Kurdi* in the child's memory. "I am happy that *Sea-Eye* ship will bear my baby's name. Alan must never be forgotten!"

Unfortunately for the very grateful Abdullah Kurdi, just two years later, the name was erased in exchange for €400,000, and replaced with the more seductive words *ResQ-People*.[13] Then Alganesh Fessaha turned to the Eritrean diaspora, launching a GoFundMe campaign to transform *ResQ-People* into "Eritrean ResQ Boat".[14] Boat & Money seems synonymous. The reality is that these so-called humanitarian ships operate as lucrative enterprises. Just to remain operational, they cost over €10,000 per day.[15] Despite the volunteer label, crew members earn salaries ranging from €2,000 to €10,000 per month.[16]

The number of NGO ships continued to increase. By the end of 2022, 16 NGO ships were carrying out rescue operations at sea.[17] That number has grown since to 24, with over

20 ships and aircraft engaged in these missions.[18] Despite the increase in vessels, 30,000 people have died in the Mediterranean Sea since October 3, 2013. More than two thousands of them were children.[19]

Rescue or Manslaughter?

Some vessel owners come either from the military intelligence or are senior army officers, like Martin Xuereb[20], former Commander of the Armed Forces of Malta, who was the director of the NGO *MOAS*.[21] *MOAS* itself was founded by American businessman Christopher Catrambone, an operational contractor with experience in Iraq and Afghanistan.[22]

On May 24th 2017 a slow-moving boat full of Eritreans was spotted by the *MOAS'* rescue ship *Phoenix*. The news was tweeted[23] by Catrambone at 08:42 a.m. Almost two hours later, at 10:25 a.m., he followed up with grim news: "A great tragedy, many bodies have already been recovered."[24]

What happened during those critical two hours? Few life jackets were thrown from the Rhib.[25] In the ensuing panic, desperate passengers rushed toward them, destabilizing the already fragile boat. Within moments, 400 people were plunged into the sea – most of them without life jackets. Coincidentally, that same day on the *Phoenix* there was Chris McGrath, photographer for *Getty Images*, who took dramatic photographs[26] that would eventually go around the world (*See photos below*) while Catrambone spent the whole day tweeting the rescue operations to update his followers.

The final toll was devastating: 32 bodies were recovered, including four children, while hundreds remained missing. Despite these circumstances, the Italian media misleadingly attributed the disaster to "rough seas." Thirteen days later, *MOAS* was awarded the 2017 *Atlantic Council's Freedom Award*.[27] Nowadays *MOAS* has shifted its focus and is engaged actively in Ukraine.

Fifty Million Migrants in Twenty Years

Today in Italy migrants have become the preferred weapon for obtaining political consensus or dissent. The Mediterranean Sea has transformed into the arena to fight an ideological war. The right-wing would like to close the ports and put-up naval blockades to prevent the "invasion". The left, on the other hand, leaves the "open ports to migrants", believing this policy will somehow address Africa's systemic issues.

The idea of open ports was first promoted in 2011 by a leader[28] from the *Partito*

27 2017 Freedom Award presentation to MOAS accepted by Maria Luisa Catrambone https://www.youtube.com/watch?v=-u52RAFHzQI

28 Massimo D'Alema https://old.partitodemocratico.it/archivio/dalema-sullimmigrazione-litalia-sta-perdendo-una-grande-opportunita-video/

Democratico who stated: "In the next 15 years there will be a need for at least 30 million migrants". A former Foreign Minister increases the figure: "Europe is experiencing a very significant demographic decline; by 2050, it will need 50 million migrants to support the welfare and pension system".[29]

These ideas originated from a United Nations' study titled *Replacement Migration.*[30] But these thirty or fifty million people from Africa will not be brought in with a regular entry visa or through bilateral agreements with the countries of origin, offering them guarantees of employment rights and providing them with airplane tickets to make their travel safe. So far, out of 1.1 million arrivals[31], 2.75% have died at sea. If this model were applied to 50 million migrants, it would mean an estimated 1,350,000 deaths. And for those who survive the journey, what awaits them?

The Jobs That Italians no Longer Want to Do

Without migrants, tomatoes cannot be harvested in southern Italy and agricultural products risk rotting in the fields, so: "Be careful when you say ports closed!"[32] repeated Presidents of regions and Ministers. Even Italy's most important leftist trade union *Cgil* has acknowledged: "Our fields need (migrant) workers". Yet the reality of migrant labor is far from the utopian vision of economic integration. A UN report[33] has condemned Italy for exploiting agricultural laborers, subjecting them to grueling work under harsh conditions for wages too low to cover their basic needs.

But it's not just landowners and gang masters profiting from migration. The business extends to those who manage the reception camps where managers, operators, catering workers, psychologists, lawyers, teachers and police offer their services.[34] And for some criminals, the management of reception camps has become even more profitable than drug trafficking: "You make a lot more money with migrants than with drugs!" they claim.[35] It's true, for each migrant the State spends €45 a day, but the migrants only receive 2 euros and 50 cents a day.

---

29 https://www.ilfattoquotidiano.it/2015/09/09/profughi-bonino-leuropa-in-calo-demografico-ha-bisogno-di-immigrati-per-questo-la-merkel-ha-cambiato-rotta/412355/

30 Replacement Migration - Italy (f) Scenario V does not allow the potential support ratio to decrease below the value of 3.0. In order to achieve this, no immigrants would be needed until 2010, and 34.9 million immigrants would be needed between 2010 and 2040, an average of 1.2 million per year during that period. By 2050, out of a total population of 87.3 million, 46.6 million, or 53 percent, would be post-1995 immigrants or their descendants. https://www.un.org/en/development/desa/population/publications/pdf/ageing/replacement-chap4-it.pdf

31 Gli arrivi di migranti in Italia dal 2013 al 2023 https://www.openpolis.it/numeri/gli-arrivi-di-migranti-in-italia-dal-2013-al-2023/

32 Emiliano replica alla Lega: "Senza i migranti i pomodori a Foggia non li raccogliamo" https://www.pugliareporter.com/2019/02/09/emiliano-senza-i-migranti-i-pomodori-a-foggia-non-li-raccogliamo-video/

Teresa Bellanova: «Porti aperti con politiche inclusive gestite a livello europeo» https://www.la7.it/otto-e-mezzo/video/immigrazione-teresa-bellanova-porti-aperti-con-politiche-inclusive-gestite-a-livello-europeo-12-09-2019-281850

33 L'Onu accusa: in Italia l'industria alimentare sfrutta i braccianti https://www.repubblica.it/cronaca/2020/01/31/news/l_onu_accusa_in_italia_l_industria_alimentare_sfrutta_i_braccianti-247280663/

34 Le 15 coop dalle "uova d'oro": 100 milioni lucrati sui profughi https://www.ilgiornale.it/news/cronache/ecco-i-conti-segreti-delle-coop-ricche-grazie-ai-migranti-1418973.html I soldi per l'accoglienza usati per fini privati. Scoperti danni per 5 milioni di euro https://www.ilgiornale.it/news/nazionale/migranti-corte-dei-conti-contesta-danno-erariale-5-milioni-2321732.html Corruzione su appalti Pnrr per l'accoglienza ai migranti: in manette anche il sindaco di Ceccano, Caligiore (FdI) https://roma.corriere.it/notizie/cronaca/24_ottobre_24/corruzione-negli-appalti-pnrr-e-accoglienza-dei-migranti-imprenditori-e-funzionari-arrestati-a-frosinone-e-napoli-e1963706-109c-4fa7-9daa-513dcca4dxlk.shtml

35 Mafia Capitale, Buzzi: "Con immigrati si fanno molti più soldi che con la droga" https://www.ilfattoquotidiano.it/2014/12/02/mafia-capitale-buzzi-immigrati-si-fanno-soldi-droga/1245847/

In the *Cara*[36] of Castelnuovo di Porto near Rome, considered one of the best camps, the migrants stayed on the first floor, living with mice and mattress bugs. The rooms, divided with plasterboard, had no doors, not even the bathrooms. Therefore, the migrants themselves solved the problem by sticking some plastic sheets to the doorposts to protect themselves from the cold and have some privacy. But on the ground floor the representative offices were bright and fragrant. The Cara could accommodate a maximum capacity of 600 people but before its closure (January 2019) it was holding nearly 1,000.[37] According to the Chamber of Commerce, *Auxilium* (the Vatican's social cooperative that manages it)[38] earned over €56 million in 2015, and more than €61 million the following year.[39] Minister Matteo Salvini's decision to close it sparked several protests from the left, Catholics and the trade union—not because of the migrants' welfare, but because 120 staff members were at risk of losing their jobs.[40]

In the last decade, reception camps such as *Cara*, *Cda*, *Cie*, *Cas*, *Hub*, *Sprar* and *Hotspot*[41] have increased fivefold across Italy. But how many of the millions of migrants who have landed on Italian shores have truly been integrated into society? How many have managed to secure stable housing with a legal rental contract, find honest employment, and build a life beyond these camps?

In major cities, it is not uncommon to see young African migrants begging outside supermarkets, engaging in prostitution, or becoming entangled in drug trafficking. Many sleep on train station platforms or in abandoned buildings, illegally occupied and sustained-by stolen electricity.

The Propaganda Specialists

For years, mainstream media has actively promoted mass migration under the guise of hospitality and integration. For over a decade, in fact, newspapers, talk shows, parliamentary debates, and street protests have consistently framed migration as a humanitarian necessity. To shape public opinion, Western governments and media have pointed fingers at African leadership, portraying them in the worst possible light.

Yet, they conveniently overlook a crucial reality: many Africans are fleeing not just their governments but also the very Western powers that exploit their homelands. These powers continue to plunder Africa's vast mineral resources, transporting them by cargo planes to Brussels, Tel Aviv, London, and Paris. They provoke proxy wars, fuel religious and ethnic conflicts, orchestrate coups, and finance terrorism—all to maintain control over the continent's wealth.

---

36 CARA – Accommodation Centers for Asylum Seekers

37 Migranti nel limbo del Cara, aspettando la relocation. https://www.redattoresociale.it/article/notiziario/migranti_nel_limbo_del_cara_aspettando_la_relocation

38 Castelnuovo di Porto "Anomalie nella gestione dei servizi di Auxilium e inadempienze contrattuali" https://www.ilfattoquotidiano.it/2019/01/24/castelnuovo-di-porto-lanac-sul-cara-nel-2016-anomalie-nella-gestione-dei-servizi-di-auxilium-e-inadempienze-contrattuali/4919436/

39 Il degrado fa business sulla pelle degli asylanten https://ilmanifesto.it/il-business-del-degrado-sulla-pelle-degli-asylanten/

40 Decreto sicurezza. Il Viminale chiude il CARA di Castelnuovo, 120 lavoratori a casa https://www.avvenire.it/attualita/pagine/decreto-sicurezza-il-viminale-chiude-il-cara-di-castelnuovo-visitato-dal-papa

41 CDA – First Reception Centres, CIE – Identification & Expulsion Centres, CAS – Emergency Reception Centres, HUB – Interregional Reception Centres, SPRAR – Protection System for Asylum and Refuge Seekers, HOTSPOT – Centres for Registration Identification Fingerprinting and Debrifieing of asylum seekers

In the case of Eritrea, among other things, the tactic employed by the West was to use "political refugees" as a weapon to push for regime change. For years, the Eritrean government has been vilified in Italian media, and the 2013 Lampedusa tragedy was exploited to reinforce this narrative. The sight of 368 coffins lined up in a hangar was broadcast from every possible angle to maximize public outrage, dismay and international indignation.[42] The President of the Chamber, Laura Boldrini, rushed to Lampedusa that morning—not to question her own government's failure to rescue the migrants in time, but to shift blame onto the Eritrean government.

Among the most vocal critics of Eritrea has been *Avvenire*, the Vatican newspaper, with journalist Paolo Lambruschi producing hundreds of reports over the past decade. No other nation has faced such a relentless media assault. Even on the day of Nelson Mandela's death – when the world's major newspapers honored his legacy—*Avvenire* instead ran a front-page story about a woman allegedly kidnapped in central Asmara and trafficked for organ harvesting.

An Alternative Voice to Italian Mainstream Media

My research is the result of nine years of study and personal experience. I infiltrated Italy's reception system by working as an interpreter in Italian, Tigrinya, and Amharic, giving me a unique perspective on its inner workings. This is why *Inferno Immigrazione* was born.

Despite being a migrant myself and having written a book on migration in Italian, I have rarely been invited to programs discussing the topic. The few migrants given a platform are those who parrot the mainstream narrative. Left-wing circles dismiss my views as supportive of far-right politicians like Salvini or Meloni, while right-wing factions reject my critiques of Western involvement, as they feel directly implicated.

On March 22, 2024, I was invited by an association in Catania as a speaker at the "*Africa and the West*" debate which was to take place in the conference room of the *Cgil*, which is a left-wing trade union that has been committed for years promoting political initiatives in favor of NGOs rescue ships, fighting to open the ports through petitions.[43] The event was canceled at the last moment due to the collapse of the roof of the conference room. Of course, no such collapse had occurred. The real reason was my past public accusations—published in a newspaper five years earlier—condemning these NGOs for perpetuating the modern-day enslavement of young Africans.

UNHCR and the "*Prima Facie*" Approach

Many of those who drowned in the Mediterranean had UNHCR documents in their pockets—just as all the victims of the Lampedusa Tragedy did. This indicates that before attempting the dangerous crossing, these "refugees" had already fled from UNHCR's so-called "safe camps."[44]

For several decades, the UNHCR has been transforming the African continent into a refugee camp, fencing off land and setting up white tents to house them. Some of its guests will grow old in those tent cities and, most likely, will not make it out alive or return to their

42 Il ruolo dei media nella diffusione di una narrativa della tragedia… https://www.UNHCR.org/it/risorse/carta-di-roma/fact-checking/ruolo-dei-media-nella-diffusione-narrativa-della-tragedia-titoli-strategie/

43 Migranti: Cgil a Toninelli, porti aperti a chi salva vite https://www.cgil.it/ci-occupiamo-di/immigrazione/migranti-cgil-a-toninelli-porti-aperti-a-chi-salva-vite-j44u5mzg

44 Assenna: Eritrean Refugees in Ethiopia Comment on Human Trafficking and Sinai Tragedy https://www.youtube.com/watch?v=_5WAx3cj8QM

village. The luckiest few will enter its *Resettlement Program*, which in reality hides the objective of eradicating displaced people from their countries of origin. And transfer them to America, Canada, Australia and Northern Europe so that they can't go back because if they did they would lose their refugee status. Others, however, will try their luck across the desert and the sea helped by human traffickers who live among them in the camps.

In the case of Eritreans, UNHCR welcomes them in camps in Ethiopia (Tigray), Sudan and Libya and then awaits them upon disembarkation in Italy. The UNHCR camps are strategically located along the migration route from the Horn of Africa to Europe. Furthermore, within Italy's Territorial Commission where the asylum seekers are interviewed, one of the four members that make up the commission is a UNHCR official. Although they are supposed to act as neutral observers, in practice, they often take over the questioning themselves.

UNHCR has long been aware that thousands of Tigrayans, originally from Ethiopia, have falsely declared themselves as "Eritreans" to gain access to the *Resettlement Program*. They usually use Eritrean documents confiscated at the first reception camp in Endabaguna. Yet, despite knowing this, UNHCR never exposed the fraud conducted by Ethiopia's *Administration for Refugee and Returnee Affairs* (ARRA), as it was bound to follow Washington's[45] diktats and reinforce the narrative that "as many as 5,000 people flee Eritrea every month."

The journalist Fausto Biloslavo in his investigation on *Panorama* entitled "The Scam of the Fake Eritrean Refugees"[46] who successfully benefited from the *Resettlement Program* through identity theft. However, such revelations have been largely ignored in mainstream discourse, as they challenge the dominant portrayal of Eritrea and its migration patterns.

To prevent in-depth investigations into the true identity of asylum seekers, in its *Eligibility Guidelines 2009*, UNHCR encouraged the adoption of a "*prima facie*"[47] approach. In reality, classifying Eritrean migrants as "prima facie refugees" was conceived as a tool for regime change. The objective was to depopulate Eritrea of its young people, weakening the country from within. To facilitate this, UNHCR permitted *ARRA* in 2008 to establish the Shimelba and Hitsats camps, built just 20 km away from the Eritrean border despite UNHCR's own guidelines requiring a minimum distance of 50 kilometers from national borders for such camps.[48]

It was this collaboration between UNHCR and *ARRA* that President Obama was referring to in 2012 at the *Clinton Foundation Global Initiative* when he stated:

I recently renewed sanctions on some of the worst abusers, including North Korea

45 The U.S are the major UNHCRs donor.

46 L'imbroglio dei finti profughi eritrei https://www.panorama.it/news/limbroglio-dei-finti-profughi-eritrei Even the former Austrian Ambassador in Ethiopia Andreas Melan in an interview with APA said: "We believe that among the thousands of Eritrean migrants in Europe, 30% to 40% are Ethiopians". https://tesfanews.com/40-percent-of-eritrean-migrants-in-europe-are-ethiopians/

47 A *prima facie* approach is the recognition of refugee status for the entire ethnic group, most often used in situations of large-scale displacement in which individual status determination is impractical, impossible or unnecessary. Prima facie approach to recognition of refugee status https://emergency.unhcr.org/protection/legal-framework/prima-facie-approach-recognition-refugee-status

48 Principles & Standards for Settlement Planning https://emergency.unhcr.org/emergency-assistance/shelter-camp-and-settlement/settlements/princples-standards-settlement-planning#:~:text=The%20site%20should%20be%20located,(such%20as%20military%20installations)

See also: Dr. Abiy Ahmed 23 March 2021 (video in amharic, minute 42:07) https//www.youtube.com/watch?v=cAOBqgTtvXM

> and Eritrea. We're partnering with groups that help women and children escape from the grip of their abusers. We're helping other countries step up their own efforts. And we're seeing results.[49]

Amnesty International and Human Rights Watch (HRW)

Human rights multinationals such as Amnesty International and HRW also play an important role in this trafficking of African migrants to Europe. "*Ipse dixit*"[50] is the phrase that is repeated like a mantra by international institutions, governments, associations and mainstream media when referring to them. So, whatever these organizations claim is automatically accepted as credible in the eyes of the Western world, for eg. one of its recommendation: "*...permit UNHCR access to Eritrean asylum seekers in order to screen them for refugee status*."[51]

> Amnesty International and Human Rights Watch, as a combined force of Human rights Defenders, we have received a reasonable grant from the US State Department to try and work with the oppressed Eritrean people to assist and act on behalf of those unable to make their voices heard. (...) The main aim therefore of this Mission to Eritrea is to provide funding and to help in setting up websites and computer centres where people can freely go and chat and exchange ideas and be able to be heard. (...) Our intended goal is that by December this year (2011), the regime of Issayas Afewerky should be shaking and ready to fall..." *Catherine Price, Africa special programmes, AI UK Section.*[52]

Relocation: The European Response to the Refugee crisis

As President Barroso promised, on September 22, 2015, the EU established temporary measures[53] in the sector of international protection for the benefit of Italy and Greece, countries of first arrival for migrants. Thus, was created the *Relocation* program of 120,000 asylum seekers to be transferred "in a safe and legal way" to the participating European countries. However, the program was selective. It did not include all African migrants arriving in Italy and Greece but was limited to Eritreans and Syrians—two nationalities particularly disfavored by Washington. All other immigrants had to remain in Italy, according to the *Dublin Regulation*.[54]

Thus, on October 9, 2015, Interior Minister Alfano went to the airport to personally greet the first 19 Eritreans who left for Sweden.[55] "We broke the ice!" he commented proudly. From 2015 to 2017, at the Rome Police Headquarters, in the *EASO*[56] office, the Applicants (asylum seekers) had their fingerprints taken, photographed[57] and then were subjected to an

---

49 Remarks by the President to the Clinton Global Initiative (September 25, 2012) https://obamawhitehouse.archives.gov/the-press-office/2012/09/25/remarks-president-clinton-global-initiative)

50 Ipse dixit is a Latin phrase which means "He, himself, said it".

51 Service for Life – April 16, 2009 State Repression and Indefinite Conscription in Eritrea https://www.hrw.org/report/2009/04/16/service-life/state-repression-and-indefinite-conscription-eritrea

52 AI's Clandestine and Nefarious Mission to Eritrea http://shabait.com/2019/10/10/from-recent-archives/

See also: Bombshell Linking Amnesty, HRW to US State Dept Regime Change in Africa by Thomas C. Mountain https://www.veteranstoday.com/2015/01/07/ai-eritrea/

53 Decision (EU) 2015/1601

54 The Dublin Regulation dictates that migrants wishing to apply for asylum in the EU must do so in the first country to which they arrive. Therefore, anyone who landed in Italy was obliged to stay there.

55 Alfano: «I ricollocamenti successo senza precedenti» https://www.interno.gov.it/it/notizie/alfano-i-ricollocamenti-successo-senza-precedenti

56 EASO, European Asylum Support Office

57 Then every information is archived in the EURODAC system (European Dactyloscopie)

interview. Those who carried out the interviews were European Experts, with a daily allowance of €500, who had thoroughly memorized the *Country of Origin Information Report*[58] document on Eritrea where Amnesty International and HRW were the main sources of information. *Ipse dixit*, indeed!

It must be said that before the interview the Experts from the *Mobile Team* went to the applicant's reception camps to indoctrinate them on the *Relocation* program and on the interview that awaited them. Their message was clear, as one briefing reportedly stated:

> You came from Eritrea, and for us, it means that you are in clear need of international protection. That's why we are offering you this program. During the interview, our colleagues will ask you many questions about your journey and about what you have experienced before, to assess your vulnerability. For example, if during the journey or in your country of origin you have been tortured or sexually abused, or you have any other problem affecting your mental health, you should mention it. If you are deemed vulnerable, your case will be prioritized.[59]

Knowing that many applicants in the camps had been waiting for over a year without relocation, they quickly understood what was expected of them. Many fabricated stories accordingly. However, there was no actual priority list—the entire notion was a lie.

One day, when asked: "What do you think could happen to you if you returned to your country?" an applicant answered truthfully: "Nothing!" and then the Dutch Expert replied: "Well, if nothing is going to happen to you, then we will send you back." At that point the applicant corrected himself: "No, I meant that I would end up in prison because I ran away from military service and left my country illegally." "Alright, that is the right answer to give," said the Expert.[60] One Expert later confided that, despite serious doubts about the true identities of some applicants, an internal directive within *EASO* prohibited them from questioning these inconsistencies. As a result, thousands of Ethiopians were registered as Eritrean refugees—even when they failed to answer basic questions such as: "How many ethnic groups are there in Eritrea? What are the colors of the Eritrean flag? How does the Eritrean national anthem begin?[61]"

In a Rome refugee camp, I met an Ethiopian woman with two young children. Each time she saw the Experts arrive, she would ask me to help her communicate with them. Her case had been stuck for over a year because, in her initial interview, she had truthfully stated that she was born in Metemma, Ethiopia. This blocked her relocation process. After much persistence, she was granted a second interview, but for her case to move forward, her birthplace had to be relocated—from Metemma, Ethiopia, to Areza, Eritrea, nearly 700 km away. With this fabricated detail, she was no longer an Ethiopian citizen but an Eritrean refugee.

The *Relocation* program ended in 2017. Of the original 120,000 asylum seekers eligible for relocation, fewer than a quarter were actually resettled by the EU member states that had pledged to participate. However, rather than being dissolved, the costly *EASO* agency was repurposed—transformed into the *European Union Agency for Asylum* (EUAA). Now armed with a stronger mandate, the EUAA continues to oversee asylum applications across Europe, with an increased budget of €172 million in 2022 alone. By 2024, its budget had

58 Easo issues Country of Origin Information report on Eritrea: Country Focus https://euaa.europa.eu/news-events/easo-issues-country-origin-information-report-eritrea-country-focus

59 Inferno Immigrazione – L.A.D. Gruppo Editoriale (October 2022), pag. 148

60 Ibidem, (pag. 153)

61 "Eritrea, Eritrea, Eritrea…"

grown to €181 million, with €52 million[62] allocated solely for staff expenditures, which now include 2,000 employees and 500 Experts.[63]

George Soros in Lampedusa

All the key players involved in the mass movement of migrants share a common link: the *Open Society Foundation* (*OSF*), founded by American billionaire George Soros.[64] A few days after the Lampedusa Tragedy, he finances the birth of the *Three October Committee,*[65] and begins to declare, "The EU must accept at least a million asylum seekers annually for the foreseeable future."[66] His financial commitment was substantial—$500 million allocated to migration-related policies.[67]

"*With the support of the OSF*" soon became a trademark in Italy that appeared in hundreds of social, cultural and journalistic initiatives. It makes both small and large contributions for those who carry out pro-migration activities.[68] The *OSF* has penetrated Italian institutions starting with targeted grants, apparently free and without any political purpose, and has made proselytes among left-wing politicians, journalists, artists, associations and NGOs. In the most recent years alone, it has financed 70 projects for a total of $10 million.[69] Also, *Carta di Roma*, the journalistic association that targeted the Eritrean community, until some time ago displayed the *OSF* logo at the top right on its homepage, has received funding of around $50,000 since 2016.[70]

62 EUAA Statement of Revenues and Expenditure 2024 https://euaa.europa.eu/publications/budget-amendment-12024

63 New EU Agency for Asylum starts work with reinforced mandate https://www.euaa.europa.eu/news-events/new-eu-agency-asylum-starts-work-reinforced-mandate

64 Soros was accused of plotting color revolutions over the last 20 years. Live on CNN, he admitted funding the Maidan Square revolution in Ukraine in 2014 https://transcripts.cnn.com/show/fzgps/date/2014-05-25/segment/01

65 October 3 Could Become an Official Day to Remember Fallen Migrants… https://www.opensocietyfoundations.org/voices/october-3-could-become-official-day-remember-fallen-migrants October 3 Committee is a grantee of the Open Society Foundations.

66 6 ways Europe must respond to the refugee crisis – by George Soros https://www.weforum.org/agenda/2015/09/6-ways-europe-must-respond-to-the-refugee-crisis/

67 Why I'm Investing $500 Million in Migrants – by George Soros https://www.wsj.com/articles/why-im-investing-500-million-in-migrants-1474344001)

68 Ecco i finanziamenti di Soros a movimenti e ong in Italia www.adnkronos.com/ecco-i-finanziamenti-di-soros-a-movimenti-e-ong-italiane_1P9qvhw48xrsUs4BhbePoA

69 The Open Society Foundations' Support for Italy https://www.opensocietyfoundations.org/newsroom/open-society-foundations-support-italy

70 www.agi.it/fact-checking/la_carta_di_roma_non_unassociazione_buonista_come_dice_salvini-1519300/news/2017-02-21/

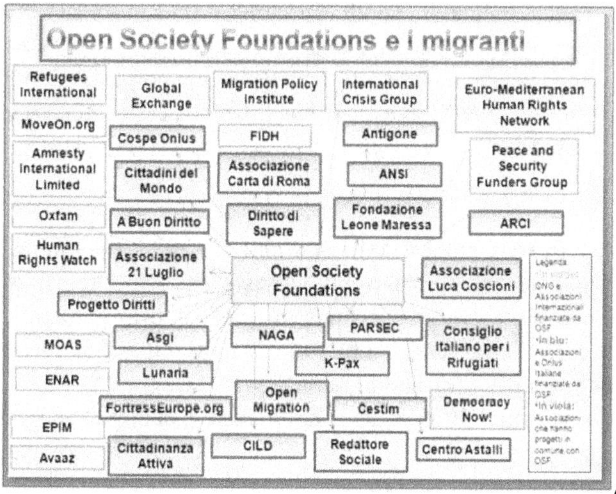

([71])

OSF also finances, albeit indirectly, some of the rescue ships operating in the Mediterranean.[72] To advance its *immigrationist* propaganda and promote its *Open Society* philosophy, the *OSF* has funneled $100 million to HRW[73] over a period of ten year. Aryeh Neier, co-founder of Human Rights Watch, was president of the *Open Society Institute* from 1993 to 2012. Amnesty International and its branches in the USA, Belgium, Ireland, Hungary and Greece have also collected funding provided by the *OSF* for several years, totaling $3.3 million between 2019 and 2023.[74] Nonetheless UNHCR, in the *2021 Global Report* of the High Commissioner, the *Open Society Policy Center* is among the global donors of amounts exceeding $100,000.[75]

In Africa, Soros has focused his interests on the African Union by donating $1,000,000 as part of the "1 million by 2021" initiative. In 2023 he planned to donate $125.2 million into various expenditures across Africa.[76] The idea of exporting its *open society* ideology to the African continent would undermine the millennia-old cultural and historical diversity of the African people.[77] To achieve this, he is financing sixteen African universities in six countries

71 Onlus e migranti in Italia https://www.lucadonadel.it/onlus-e-migranti-in-italia-open-society/amp/

72 The non-profit organization *COSPE*, already beneficiaries of *OSF*, is among the founding members of *Sos Meditèrranee Italia* and among the supporters of the rescue ship *Aquarius*. *ResQ-People* has among its board members the non-profit organization *ASGI* already financed by OSF. In 2023 alone ASGI received from OSF $550,000.https://www.opensocietyfoundations.org/grants/past?filter_keyword=c&page=22&grant_id=OR2023-89644 Among the major financiers of the American NGO *MOAS* was *Avaaz*, financed by *MoveOn.org*, already financed by *OSF*. Thomas Perriello, co-founder of *Avaaz*, was executive director of *Open Society-U.S.* until July 2023. https://www.sfgate.com/politics/joegarofoli/article/MoveOn-a-political-force-online-receives-5-2526436.php In the Mediterranean, Standing Back Is Not an Option https://www.opensocietyfoundations.org/voices/coast-libya-standing-back-not-option

73 Soros and Open Society Foundations Give $100 Million to Human Rights Watch https://www.opensocietyfoundations.org/newsroom/soros-and-open-society-foundations-give-100-million-human-rights-watch

74 Amnesty International Limited https://www.opensocietyfoundations.org/grants/past?page=13

75 Global Report 2021 https://reporting.unhcr.org/globalreport2021/pdf

76 $125.2M expenditures for Africa by 2023 https://www.opensocietyfoundations.org/what-we-do/rgions/africa

77 Open Society Grant: The Africa Institute Receives Major Multi-Year Funding for African Cultural Heritage Fellowship https://www.theafricainstitute.org/open-society-grant/

(Ethiopia, Ghana, Kenya, Malawi, South Africa and Tanzania) to which he has allocated $4.5 million.[78] George Soros is free to spend his fortune as he pleases, one undeniable fact remains: of the millions he has poured into migration-related projects, not a cent has reached the pockets of the migrants themselves, nor has it been used to secure safe air travel to prevent perilous sea crossings.

Conclusions

From 2014 till today, every October 3rd, on the anniversary of the Lampedusa Tragedy, politicians, NGOs, journalists, and human rights activists—all exploiters of the Tragedy—gather on the island for their ritualistic expressions of grief, as they solemnly declare "No more tragedies!" But the tragedies at sea promptly repeat themselves[79] and it seems it will not stop until the arrival of those fifty million migrants from Africa.

Many Italians ask me: "What's the solution to stop deaths at sea?" My answer remains the same, neither Italy nor Europe will solve this crisis; they are too entangled in the migration business. As a Pan-Africanist, I firmly believe that the only real solution rests in our own hands as Africans. We must awaken our youth to the dangers of this false dream and prevent them from risking their lives, only to become modern slaves in a system that makes profits from their suffering. There is no other way!

In the meantime, despite the increase of rescue vessels, 30,000 people have died in the Mediterranean Sea since October 3, 2013, several thousands of whom were children. The *Operation Mare Nostrum* became like a trailblazer for those private ships which arrived from far away—not only to save people, but also to "open ports". Instead, the survivors are destined in the field of slavery, the "fortunate" Eritrean migrants soon became protagonists of the EU *"Relocation program"*, a political project that served as pull-factor, like the "*Prima Facie*" approach of UNHCRs, as well as the "*Ipse dixit*" of Amnesty International and HRWs. At last, this political agenda in emptying Eritrea from its youth have certainly failed.

78 Funding To Universities By George Soros And The Foundation To Promote Open Society https://www.universityphilanthropy.com/funding-by-george-soros-and-open-society

79 Un anno fa il naufragio di Cutro, 94 morti di cui 35 minori https://www.rainews.it/articoli/2024/02/un-anno-fa-il-naufragio-di-cutro-94-morti-di-cui-35-minori-oggi-la-commemorazione-63da637f-3a1b-4742-abea-c7e3142c915b.html

Reference

Archipelago ONG – 2018 – Giuseppe De Lorenzo (Edizioni La Vela)
Inferno Spa – 2019 – Francesca Totolo (Altaforte Edizioni)
IpocriSea – 2022 – Francesca Ronchin (Compagnia editoriale Aliberti)
L'Urlo – 2022 – Michelangelo Severgnini (L.A.D. Gruppo Editoriale)
Profugopoli – 2016 – Mario Giordano (Mondadori)

Reflection Of The Cultural Identity In *The Conscript*: A Postcolonial Perspective

Desbele Tekle Tesfamariam

College of Business and Social Sciences –Adikieh;
e-mail: desbele-teckle@gmail.com

Abstract

This article delves into the intricate exploration of cultural identity within the lesser-known Eritrean postcolonial work, *The Conscript*, penned by Hailu. Rooted in the complexities of postcolonial literature, the study examines how the characters' experiences reflect the multifaceted nature of cultural identity, both shaped by historical context and challenged by external influences. Through an analysis of characters' emotional responses, artistic expressions, and negotiation of loyalty, the article uncovers the nuanced interplay between Eritrea's colonial past, tussle for freedom, and the characters' perceptions of self. The study contributes to the broader discourse on postcolonial literature by shedding light on the voices of the marginalized and highlighting the significance of lesser-known narratives in portraying the intricate tapestry of cultural identity. Ultimately, *The Conscript* serves as a compelling lens through which to explore the enduring power of literature in dissecting the complexities of cultural identity in postcolonial settings.

Keywords: cultural identity, artistic expression, marginalized voices, loss and resilience, identity negotiation

Introduction

The concept of cultural identity is a central theme in postcolonial literature, representing the intricate interplay between individual and collective identities shaped by historical, social, and cultural contexts. "In analyses of culture it is a (generally implicit) assumption that individuals share an essential cultural identity, and it has been a topic of vigorous debate with in post-colonial theory" (Ashcroft *et al* 73). It encompasses how individuals and communities define themselves in relation to their cultural heritage, traditions, language, and history, often in response to the impact of colonialism and its aftermath. As *Post- Colonial Studies* probe:

> Any cultural experience is itself an accretion of many layers, and the term is valuable because it illustrates the ways in which pre-colonial culture as well as the experience of colonization are continuing aspects of a post-colonial society's developing cultural identity" (Ashcroft *et al* 158).

In postcolonial literature, cultural identity takes on a profound significance as it provides a platform for marginalized voices to articulate their experiences, challenge dominant narratives, and assert their agency in shaping their own stories. The exploration of cultural identity in postcolonial literature enables a deeper understanding of the complexities of colonial legacies, cultural hybridity, resistance, and the quest for self-expression and autonomy. As such, the study of cultural identity in postcolonial literature offers insights into the multifaceted nature of human existence and the ongoing negotiations between the self and the larger sociopolitical environment.

Hailu's *The Conscript* stands as a lesser-known yet impactful work within the realm of Eritrean postcolonial literature. Alemseghed Tesfay, the prolific historian and author of *Two Weeks in the Trenches,* describes the novel as "Eritrean literary jewel of global and timeless relevance." While the global literary canon often gravitates towards more recognized postcolonial narratives, Hailu's contribution offers a distinctive perspective rooted in the Eritrean experience. Ali Jimale Ahmed, praises it as "it shows sophistication and maturity in the depiction of the inner turmoil and real-life characteristics of its characters." Set against the backdrop of Italy's colonial presence in Eritrea, *The Conscript* delves into the exploration of cultural identity, shedding light on the often overlooked complexities that emerge at the intersection of personal, collective, and colonial histories. Despite its comparative obscurity, *The Conscript* holds immense value as a literary gem that portrays, according to Laila, "Eritrean man who, after being recruited into the Italian army, is forced to fight in its war to subjugate Libyans," and reveals the struggles, aspirations, and resilience of Eritrean people during a transformative period. Through its narrative, the work unveils the layers of cultural identity in the postcolonial context, inviting readers to embark on a journey of introspection and understanding that goes beyond the more prominent narratives of postcolonial literature.

This article examines how Hailu's *The Conscript* reflects the complexities of cultural identity within the context of postcolonial readings. Through an in-depth analysis of the characters and themes in the novel, this study illuminates how Hailu skillfully portrays the intricate interplay between personal, cultural, and colonial influences on the characters' identities. By delving into the characters' struggles, aspirations, and interactions, this analysis reveals how *The Conscript* serves as a lens through which the multifaceted nature of Eritrean cultural identity in the aftermath of colonialism is both explored and illuminated.

Cultural Identity and Its Complecities

Homi K. Bhabha defines, "identity is never a priori, nor a finished product; it is only ever the problematic process of access to an image of totality" (Bhabha 2012: 73). From this vantage, cultural identity refers to the collective sense of belonging and shared characteristics that individuals or groups derive from their cultural background, heritage, traditions, and values. It encompasses the various elements that shape one's identity, such as language, religion, customs, historical experiences, and social norms. Cultural identity plays a crucial role in shaping an individual's sense of self and their connection to a larger community.

In postcolonial settings, cultural identity takes on added significance due to the historical context of colonization and its aftermath. In his book, Bhabha explores forms of it which says, "This is a lesson of history to be learnt from those peoples whose histories of marginality have been most profoundly enmeshed in the antinomies of law and order – the colonized" (Bhabha 2012: 217). Though "there is no master narrative" (Bhabha 2012: 61), colonialism often imposed foreign values, languages, and systems of governance on indigenous populations, leading to a rupture in their cultural continuity. As a result, postcolonial societies grapple with the task of reclaiming, reconstructing, and reaffirming their cultural identities in the wake of colonial influence. Because, as in Fanon's book, "the Negro enslaved by his inferiority, the white man enslaved by his superiority alike behave in accordance with a neurotic orientation." This process involves navigating complexities, tensions, and negotiations between traditional cultural practices and the modern forces introduced during colonial rule.

In this view of postcolonial literature, *The Conscript* penned by Hailu, "is a harrowing journey into the experiences of an Eritrean man who, after being recruited into the Italian army;" it serves as a platform for exploring and portraying these intricate dynamics of cultural identity within such contexts. The novel describes the conscripts' experiences. The colonialists conscripted them, taking them from their homes, their parents, and their homeland, and moved them to unknown lands. As the text reflects, "Going to a distant land / Not for the honor of my homeland / Leaving my family behind" (56). It illustrates the lasting scars of colonization, where individuals are uprooted from their cultural roots and subjected to foreign authority. "He put down his gun beside him, knelt down before his parents, and asked (3) to bless him "for I do not know what my fate will be in Tripoli" (3). According to Hughes Wadlick, "The events in *The Conscript* occur in Tripolitania, during the Italian government and its armed forces" (Wadlick 2021:31). Furthermore, the characters' interactions with the Italian colonialists reveal the power dynamics and cultural hierarchies that have persisted even after colonial rule. The text portrays the Italians as viewing the *Habesha* people as inferior, emphasizing the deep-seated prejudices engrained through years of colonization; "for the Italian, the *Habesha* was like a weak donkey" (47) underscoring how historical factors have contributed to a distorted perception of cultural identity, where the colonized are dehumanized and marginalized.

The legacy of colonization also reverberates through the characters' language and communication. The Italian language is imposed on the conscripts, creating a linguistic barrier that further alienates them from their cultural heritage. The text portrays the struggle of language preservation, the translator of this text assures, it is "a tribute to celebrate the vitality of African-language" (vii) and the novel "revisits ritualized oral versions of the history of conscription in Eritrea (viii); on the other hand the colonizers forced them to speak the Italian language, threatened with a whip, they would say, 'Speak Italian. The narrative reflects the imposition clearly in its pages, "told them to shout, "*Viva l'Italia,*" (28-9) and "It was heartrending to watch the *Habesha* agents in authority there, who were proudly shouting "Pronto" (53). This imposition of language highlights the multifaceted ways in which historical factors, specifically colonization, shape cultural identity by erasing native languages and imposing external linguistic norms. In essence, Hailu's portrayal in *The Conscript* provides an illustration of how historical factors, particularly the legacy of colonization, intricately shape cultural identity. "It showcases the challenges encountered by individuals like Tuquabo, torn between their own cultural heritage and the influences of the colonizers" (Tekle 2023: 152). The characters' experiences and struggles mirror the broader challenges faced by postcolonial societies as they strive to redefine and reclaim their cultural heritage in the face of enduring historical influences.

Colonizers often imposed their languages- like what Walter Benjamin describes as "foreignness of languages" (Bhabha 2012: 234) institutions, and cultural norms on indigenous populations, disrupting centuries-old traditions and eroding a sense of self rooted in these practices. As postcolonial societies strive to keep safe the "unbroken continuity" (xi) for decolonization, the process of reclaiming and revitalizing indigenous languages and art forms. For example readers comprehend, "a century or so ago, many families in Eritrea tell stories of fathers or grandfathers or other relatives"(x) and belief systems becomes an assertion of cultural identity and a rejection of past oppressors.

Through these instances of cultural hybridity, *The Conscript* exemplifies how postcolonial societies navigate the intricate terrain of cultural identity. The characters' experiences highlight the organic and dynamic nature of identity, which evolves through interactions with external forces. Hailu's narrative invites readers to contemplate the ways in which cultural hybridity can be a source of enrichment, allowing for the preservation of heritage while embracing the complexities of a changing world.

Histories marked by colonization, "the evils of European colonialism on the African continent" (xii), often leave behind a trail of collective historical trauma. The violence, exploitation, and loss experienced during colonial rule shape the psyche of communities, impacting how cultural identity is perceived and transmitted. Narratives of trauma, whether expressed through literature, art, or oral traditions, become integral to cultural memory, informing individuals' understanding of themselves and their place in the world. *The Conscript* captures the intricate interplay between globalization and cultural identity, portraying the characters' struggles as they navigate the modernizing forces while clinging to their traditional roots for globalization and modernization introduce new dimensions to the construction of cultural identity. In the narrative, the conscripts are exposed to the Italians' advanced weaponry and technology, a stark contrast to their own agrarian lifestyle. The text describes the Italians' weaponry: "guns, bullets, binoculars, mortars, rifles, pistols, cannons, bombs..." (27). This exposure not only awes the characters but also underscores the power imbalance between the colonizers and the colonized, highlighting the disruptive impact of modernization on their traditional way of life.

Moreover, the characters' interactions with foreign cultures, particularly the Italians, exemplify the clash between globalization and cultural preservation. The conscripts are not only subjected to forced labor but also exposed to the Italians' way of life. In the novel, the commander describes himself as demigod "We alone are the brave whites; we, Italians, your masters" (27) revealing the clash. In the colony, they lived differently, their culture was different, their clothes, their food, even the look in their eyes was different. Eritreans know that from the early times of colonization. 'When the white snake has bitten you, you will search in vain for a remedy' (Wrong 2005: 23). The book doesn't hide the true experience of the conscripts and their situation, "Not a trace of water. If you asked where the commanding officer was, he would be in his tent safekeeping his water. He had guards around his tent and stayed silent inside" (45). This encounter with foreign culture instills a sense of otherness, "They were, after all, like dogs, if you compared them with the Italians", (46) and prompts reflection on their own cultural practices. Hailu's depiction in *The Conscript* serves as a microcosm of the larger global debate on the impact of globalization and modernization on cultural identity.

As a vessel for cultural expression, language "is a perfect music" (Tesema 2006: 13), and it often becomes a contested terrain. The dominance of colonial languages marginalizes native tongues, compelling individuals to either relinquish their linguistic heritage or engage in the uphill battle of language preservation. It is intricately woven into the fabric of the narrative. The characters' yearning for their homeland is evident in their emotional responses. As Tuquabo returns home, the text reads, "They saw the great mountain chains, which were the topic of their daily conversations and night dreams, and they couldn't resist their tears" (52). Here, the mountains symbolize a connection to their cultural identity, and the language used to describe them emphasizes the deep emotional ties that language can evoke. The loss of language carries profound consequences, as it erodes cultural nuances, historical narratives, and collective memory. The characters in *The Conscript* grapple with this linguistic conflict, reflecting the broader struggle of postcolonial communities to maintain their linguistic authenticity in the face of external pressures.

Moreover, economic and social disparities introduce power dynamics that influence individuals' choices regarding cultural allegiance. The pursuit of economic opportunities or social mobility can compel individuals to conform to the expectations of dominant cultures, often at the expense of their own heritage. **"**The Italian commander-in-chief disappeared on his mule, leaving them behind. He was afraid that they might kill him, but in fact nobody dared... But for the Italian, the Habesha was like a weak donkey, which you couldn't kill for meat or hide and therefore would leave behind to die." (47) The text underscores the power dynamics at play and the exploitation of the conscripts, highlighting the economic disparities between the colonizers and the colonized. This internal conflict is exacerbated by the fear of marginalization or discrimination, forcing individuals to negotiate between their cultural roots and the desire for social acceptance.

The negotiation between cultural roots and external pressures also involves the realm of belief systems and spirituality. The introduction of foreign religions during colonization disrupts indigenous spiritual practices, prompting individuals to choose between embracing the imposed faith and preserving their traditional beliefs. In *The Conscript*, Hailu captures the inner turmoil faced by the characters as they grapple with the clash between their cultural roots and the external pressures of colonial rule. When describing the conscripts' departure for war, the text notes, "Going to a distant land/ Not for the honor of my homeland/ Leaving my family behind/ In agony and tears, for two years" (56) encapsulating the challenges of tradition versus modernity as the conscripts are torn between their duty to their homeland and the foreign authority. They are compelled to serve. This dilemma reflects an internal conflict, as personal convictions are pitted against societal expectations and the desire for a sense of belonging. *The Conscript* portrays characters grappling with this tension, illustrating how external pressures can shape and challenge one's spiritual identity. Spiritual beliefs are interwoven with the characters' struggles. As they navigate the challenges of war and displacement, the text portrays their prayers and pleas for divine intervention. In a moment of desperation, the conscripts turn to their faith, saying**,** " All-Seeing God, we are in distress" (46) illustrating the clash between indigenous spirituality and the religious impositions of the colonizers, showcasing the tension between spiritual beliefs and external pressures.

Loss and Resilience: Artistic Expressions of Identity

In *The Conscript*, the significance of poetry emerges as a means of preserving and expressing cultural identity, particularly in the face of loss and adversity. The characters' emotional landscapes, shaped by grief and regret, find a powerful outlet through poetic expressions that capture the complexities of their experiences. Tuquabo's poignant dirge, mourning the loss of his mother and reflecting on his choices, exemplifies how poetry becomes a vehicle for channeling intense emotions and grappling with cultural identity. Tuquabo's wailing dirge, filled with self-blame and regret, serves as an example of how poetry becomes a medium through which characters navigate their cultural identities. Through the lines, "I am done with Italy and its tribulations / That robbed me off my land and parents," (57) Tuquabo encapsulates his struggle to reconcile his choices with his heritage. The act of crafting a dirge becomes an act of reclaiming agency and asserting his identity in the face of external pressures. Poetry becomes a therapeutic outlet for Tuquabo, enabling him to confront his grief, regret, and the inherent tension between his personal experiences and cultural obligations.

Furthermore, the communal mourning and poetic lamentations of the conscripts for their fallen comrades illustrate how poetry becomes a collective expression of cultural identity and resilience. As they sing dirges like, "One by one they got fewer," (48) they not only memorialize the losses endured but also affirm their shared heritage and unity as Eritreans. These poetic expressions serve as a testament to the conscripts' ability to endure hardship while holding onto their cultural roots. In *The Conscript*, poetry emerges as a form of artistic expression that both preserves and shapes cultural identity. It provides characters like Tuquabo with a means to process complex emotions, confront personal choices, and navigate the challenges of a postcolonial landscape. Through poetry, the characters are able to bridge the gap between their individual experiences and the collective identity of their homeland, underscoring the resilience and enduring strength of Eritrean cultural identity in the face of adversity. The role of poetry in *The Conscript* extends beyond individual expressions of grief and regret to serve as a bridge between generations, preserving cultural memory and transmitting values. The dirges and lamentations that permeate the narrative not only convey the characters' emotional turmoil but also function as oral traditions that connect the past, present, and future of Eritrea. These poetic expressions become a conduit through which cultural heritage is passed down, ensuring that the struggles, sacrifices, and triumphs of the past are not forgotten.

Tuquabo's dirge within *The Conscript* serves as a reflection of his internal conflict and deep sense of regret, "Lacking nothing, I had plenty to eat, drink and clothing to cover/ But left my homeland, oh, such rashness/ Here I return to show my unworthiness" (56) encapsulating the complexities of his cultural identity and the emotional turmoil he experiences. The dirge not only expresses his personal grief but also encapsulates the broader themes of loss, displacement, and the tension between individual aspirations and communal responsibilities. In his dirge, Tuquabo grapples with the consequences of his decision to leave his homeland and participate in the Italian Army. He laments the fact that he chose to abandon his family and cultural roots for the allure of foreign lands, highlighting the internal struggle between his own desires and his responsibilities to his people. The lines "Going to a distant land, / Not for the honor of my homeland / Leaving my family behind, / In agony and tears, for two years" (56) convey the weight of his decision and the resulting emotional toll. Tuquabo's regret is palpable as he recognizes that his pursuit of personal glory came at the cost of his connection to his cultural heritage. Because "the landscape was part of their hearts and souls. It was impossible to forget the land, even when left behind in peace and prosperity" (52).

Furthermore, Tuquabo's dirge underscores the role of external pressures and colonial influence in shaping his actions and regrets. His acknowledgment that he was deceived by "the devil" (56) suggests the manipulation and coercion that colonial powers often employed to recruit and control individuals for their own purposes. The dirge reflects not only Tuquabo's personal remorse but also critiques the larger historical forces that have disrupted Eritrean cultural identity.

Tuquabo's dirge also exemplifies the enduring power of traditional oral forms of expression in conveying complex emotions. The rhythmic and lyrical structure of the dirge carries the weight of Tuquabo's regret, allowing him to articulate his feelings in a way that resonates deeply with both himself and his community. Through his dirge, Tuquabo becomes a poetic voice for the collective experiences of conscripts who grappled with similar conflicts between personal aspirations and cultural allegiance. In a broader sense, Tuquabo's dirge and other forms of artistic expression become acts of cultural preservation. By documenting and sharing their experiences, individuals create a lasting record of their histories, struggles, and triumphs. These artistic expressions become repositories of cultural memory, safeguarding traditions, values, and identities that might otherwise be lost to the currents of time. In this way, art serves as a resilient means of resisting the forces of cultural erasure and displacement, ensuring that the voices and stories of marginalized communities endure.

Negotiating Identity Loyalty and Dislocation

In *The Conscript*, the tension between loyalty to one's homeland and compliance with foreign authority is a central theme that underscores the characters' complex negotiation of their cultural identities. The conscripts, including Tuquabo, find themselves torn between their innate allegiance to their Habesha heritage and the coercive pressure exerted by the Italian colonizers. In *The Conscript*, this internal conflict is portrayed through Tuquabo's inner monologue as he grapples with his dual identities. The text delves into his thoughts. "Go ahead, Leave your family and country behind/For someone else's expanse / That you don't want. /Feel like a stranger / Until you're dead" (14) illustrating the emotional struggle faced by Tuquabo and his fellow conscripts as they navigate the intricate terrain of cultural identity within the context of colonial rule. The juxtaposition of "ancestors" (5) and "Italian officers" (14) underscores the stark contrast between the legacy of their Habesha heritage and the external pressures of colonization. Tuquabo's internal turmoil becomes emblematic of the broader tension between preserving cultural roots for "It was impossible to forget the land" (52) and succumbing to foreign authority, a theme that resonates throughout the narrative.

As the conscripts are coerced into service under the Italian military, their internal conflicts intensify. The text describes the scene where Tuquabo's identity clashes with his new reality. "Tuquabo was dressed in a gray uniform with a colorful belt that embellished his waist, and from his ankle to his knee was bandaged with a thick strip of cloth that looked like a horse blanket" (3). The struggle of the conscripts to reconcile their inherent cultural allegiance with the imposed identity by colonization, further enriching the exploration of the complexities inherent in negotiating cultural roots and external pressures. This tension is evident in the characters' experiences as they navigate the demands of their conscription and the harrowing journey to Tripoli.

The conscripts' forced enlistment under foreign rule exemplifies the clash between loyalty and dislocation. Tuquabo and his companions are uprooted from their homes, "Please leave this one child for us" (4). Many separated from their families, and compelled to serve a foreign power. As they endure the hardships of the desert and the horrors of war, their loyalty to their homeland remains steadfast, but their physical and emotional dislocation fuels internal conflict. The text portrays them as "strangers …in a strange land," (43) where they are denied agency over their own lives and destinies, yet their cultural identity persists as an anchor that keeps them connected to their roots.

The characters' experiences also highlight the internal struggle between survival and maintaining cultural integrity. The conscripts, in their dire circumstances, must make difficult choices to ensure their own survival. These choices often entail compliance with the very authority that seeks to erase their cultural identity. Tuquabo's introspections capture this complex dilemma. "For himself, he carried water on his mule. So did his Italian companions. But who would look out for the conscripts? Even if they dried up from thirst, who would really care? They marched anyway" (45). This passage underscores the excruciating choices that Tuquabo and his comrades face. The juxtaposition of loyalty to "his Italian companions" (45) and meager rations offered by the Italians highlights the harsh reality of their predicament. The text delves into Tuquabo's inner turmoil, revealing the layers of internal conflict: "Go ahead, Leave your family and country behind […] Until you're dead" (15).

By delving into the characters' thought processes and moral dilemmas, the novel provides an exploration of the internal tensions that arise when individuals are forced to negotiate between their cultural identity and the pressures of external authority. The conscripts' plight serves as a microcosm of the broader challenges faced by postcolonial societies, where survival often hinges on navigating the delicate balance between asserting one's cultural roots and adapting to external pressures. For instance, the Italian commander-in-chief's abandonment of the conscripts and their subsequent desperation for water demonstrate how survival instinct can sometimes override loyalty to one's heritage.

However, the characters' resilience in the face of adversity also reflects their unwavering commitment to preserving their cultural identity. "Eritrea is not made of people who cry" (Wrong 2005: 20). This commitment is exemplified through their acts of quiet defiance and their determination to find solace in their cultural heritage. As Tuquabo and his fellow conscripts endure the hardships of military service under Italian rule, their conversations often turn to stories of Eritrean folklore and traditional practices. "One story tells that there was a man whose wife was to deliver a baby…" (33), depicting the conscripts' deliberate effort to strengthen their bond with their heritage.

Tuquabo's inner monologue further underscores commitment. "In your house, you were used to having hot water to wash feet, […] but now, where will you go to after spending all the day toiling"(27)? The excerpt reveals how the characters' resilience is intertwined with their active attempts to preserve their language and cultural traditions, even as they navigate a hostile environment. The characters' resilience, then, becomes a form of resistance against the erasure of their cultural identity. In the midst of adversity, their determination to uphold their heritage becomes a means of asserting agency and autonomy. Their actions echo the broader sentiment of postcolonial societies striving to assert their identities in the face of external pressures. Their camaraderie and mutual support act as a testament to the enduring power of cultural ties, even in the direst circumstances.

The novel captures the oppressive atmosphere of the military camp, where the Italian officers bark orders in their own language, drowning out the conscripts' native tongues. "'Marching,' a new word in Tigrinya, which they had invented as a testimony to their exhausting experience," (27) "Viva" (28) or "pronto" (54) portray the stark contrast between the conscripts' cherished heritage and the foreign elements being forcibly imposed upon them. This cultural erasure is evident in the demeaning treatment of the conscripts, who are treated as expendable mercenaries rather than individuals with their own histories and identities.

The conscripts in *The Conscript* navigate a complex terrain of dual identities as they grapple with the impact of colonial power dynamics on their sense of self. On one hand, they are Eritreans, "I am leaving but I will back, from. ...I assure you" (Tesfay 2022: 33) deeply rooted in their cultural heritage, language, and traditions. As Harold Scheub once noted, reflecting the conclusion of many Africanist scholars, "There is an unbroken continuity in African verbal art, from interacting oral genres to such literary productions as the novel and poetry.'? (ix) On the other hand, they find themselves conscripted into the Italian colonial army, subjected to the authority of a foreign oppressor. "The worst distressing matter of the case is that I am going away" (Tesfay 2022: 32) that seeks to strip them of their Eritrean identity. This dual identity becomes a source of internal conflict, as they are torn between their loyalty to their homeland and their compliance with the foreign colonial power.

The conscripts' experiences illustrate the power dynamics at play. The Italian authorities' treat them as expendable pawns, reinforcing the notion that their lives are of little value. "The Italian officers seated themselves on an upper deck apart from the rest. As for the Habesha, they were stationed in the open, [...] where you put animals" (14). This is evident when Tuquabo and his fellow conscripts are sent to the frontlines, where they confront the horrors of war and the callous disregard for their well-being. "black Eritrean ascari! Those whom you are now going to fight against are but a bunch of shepherds," (27) the text conveys. The conscripts are subjected to perilous conditions, lacking adequate provisions and medical care, further underscoring their disposability in the eyes of their colonial oppressors. The power dynamics are further highlighted in the interactions between the conscripts and the Italian officers. A scene depicts an Italian officer berating a conscript for his exhaustion, accusing him of laziness and incompetence. For them, they were like "a bunch of shepherds" (27), the Massawa slaves" (29) and "confused" (36) the narrative captures. This degrading treatment reflects the racial hierarchies and colonial mindset that reinforce the conscripts' subjugation and erode their sense of self-worth.

As the conscripts endure the brutalities of war and witness the stark contrast between the lives of the colonizers and the colonized, they become acutely aware of their marginalized status. "Pity the conscripts who were on the brink of death from thirst yet were guarding the tent for somebody who carried water.[...] They were like the rich man in hell who longed for a drop of water from Lazarus" (45), the text portrays. The conscripts' experiences serve as a powerful indictment of the systemic oppression and dehumanization perpetuated by colonial forces, offering a visceral portrayal of the power dynamics that shape their lives and impact their cultural identities. The act of conscription itself is emblematic of colonial domination, as their bodies are forcibly taken and exploited for the benefit of the colonizers, "you should fight well for us" (28). This dynamic engenders feelings of powerlessness and dehumanization, causing the conscripts to question their worth and identity in the face of such oppression. As the conscripts endure the hardships of war and dislocation, their dual identities become intertwined with a sense of loss and dislocation. The loss of their homeland, families, and cultural roots is exacerbated by the harsh realities of the battlefield. "'Swept by grief and rocking like a person with a severe stomach pain" (55), they were "moaning and groaning in distress" (55). This exchange underscores the collective grief and shared mourning that the conscripts endure, magnifying the theme of loss that is intricately woven into their dual identities.

The loss of their cultural roots is not solely physical but also emotional, as they grapple with the internal conflict between their innate allegiance to their heritage and the pressures of their wartime circumstances. The dirge composed by Tuquabo eloquently captures this internal struggle, where he mourns not only the loss of his mother but also the erosion of his cultural identity due to his participation in the colonial war machine. However, even within this context of oppression, the conscripts engage in subtle acts of resistance to assert their identity. The griot's songs and Tuquabo's dirge serve as defiant expressions of their cultural heritage, challenging the colonial attempts at erasure. These acts of resistance provide a source of strength and empowerment, allowing the conscripts to reclaim agency over their identities and challenge the dominant colonial narrative.

Thus, *The Conscript* serves as an exploration of the conscripts' negotiation of their cultural identity within the context of postcolonial Eritrea. Through their experiences, Hailu sheds light on the complexities of loyalty, dislocation, and resistance in the face of colonial power dynamics. The characters' emotional landscapes, shaped by grief, regret, and a yearning for their cultural roots, exemplify the multifaceted nature of cultural identity and its profound impact on individuals' lives. As the conscripts engage in acts of artistic expression and resistance, they reclaim agency over their identities and challenge the erasure of their culture. Ultimately, Hailu's work invites readers to reflect on the enduring power of cultural identity and the resilience of individuals who navigate the complexities of postcolonial existence.

Eritrean Literature and the Voices of the Marginalized

Eritrean postcolonial literature, often characterized by its unique blend of cultural heritage and resistance against colonial oppression, provides a rich backdrop for understanding Hailu's *The Conscript*. The narrative portrays the conscripts' marginalized status and their struggles to navigate a world where their identities are manipulated and suppressed. "The sense of shock, sadness, hopelessness, and regret was clearly visible on their faces" (23). When Tuquabo moves to the city, his identity starts to be blended and eroded. He:

> bears the full name "Tuquabo Medhenealem Habte-Mikael Hidru Red'ai," a name that encompasses four generations of his lineage. However, the significant omission of the powerful name "Medhenealem" following "Tuquabo" can be interpreted as a detachment from his true self, his history, and even his creator. (Teckle 2023: 153)

When he encounters an Italian officer who gives him "a gray uniform with a colorful belt" (3), indirectly attempts to erase his identity. The text highlights the power dynamics at play. The imposition of external authority on the conscripts' identities, accentuating their marginalized position and their resistance against erasure as well. The conscripts' experiences and interactions with Italian soldiers provide insights into the hierarchical and exploitative relationships. "A son of Habesha was elected as a privileged orderly to serve a white man"

(26) which emerged from colonialism. A fellow conscript is subjected to mistreatment and regarded as a weak and useless, "cooking his food, or lighting his cigarette" (26). The text portrays the unequal power dynamic. "We alone are the brave whites; we, Italians, your masters" (27) shedding light on the marginalized and dehumanized existence of the conscripts. Hailu's work also delves into the complexities of cultural identity, especially when the conscripts engage in conversations about the bewilderment. A dialogue between Tuquabo and another conscript reveals the intricacies of negotiating identity:

> Dear friend, do you think people live in this dry sand?" Tuquabo was asking, bewildered. "What can I say? I am just like you looking wherever my eyes take me to." "How can I know?" replied the other. Both fell silent. They were all silent (Hailu 2012: 23).

This exchange encapsulates the tension between preserving cultural identity and conforming to external pressures, further substantiating the work's focus on the marginalized grappling with the multifaceted dimensions of their identities. In essence, Hailu's portrayal of marginalized conscripts and their intricate negotiations of cultural identity within the colonial context aligns with the theme of amplifying marginalized voices. In *The Conscript*, Hailu captures the voices of conscripts who are thrust into the Italian colonial war machine, "the Habesha was going to prepare their horses and pack their goods", (26) depicting their struggles "Don't go to a bad wife after spending the day with a bad ox" (27), regrets "you are dying of hunger and wasting away" (27) and resilience. This narrative aligns with the broader themes present in Eritrean postcolonial literature, where the voices of the marginalized take center stage. The conscripts' journey mirrors the collective struggle for liberation, mirroring the Eritrean people's quest to assert their cultural identity and reclaim their agency after years of colonization just as Eritrean literature as a whole serves as a platform for asserting cultural identity and resisting erasure. They "asked to be discharged from the Italian Army" (57). *The Conscript* serves as a microcosm of this larger struggle. The conscripts' acts of resistance through poetry and their expressions of grief reflect the Eritrean people's determination to preserve their cultural heritage and resist the forces, "I was one blessed by his grace and with riches/ Why did I put myself through this" (56)? that seek to marginalize and erase their identity.

Hailu's *The Conscript* exemplifies this importance by shedding light on a specific facet of Eritrean postcolonial history – the experiences of conscripts during the Italian colonial era. As the writer puts in his preface, the story is "about the memory of my fellow-countrymen, the ascari recruits, who were traveling overseas at that time" (xxix). While widely recognized historical events and figures often dominate the narrative, lesser-known stories like that of the conscripts bring to the forefront the lived experiences of ordinary individuals caught in the tide of colonialism. By giving voice to these marginalized characters, the novel humanizes the broader historical struggle and presents a more holistic depiction of the challenges faced by Eritreans during this period.

The characters in *The Conscript*, particularly Tuquabo, embody the multifaceted nature of cultural identity. They grapple with the tension between their inherent cultural roots and the external pressures imposed by colonial rule. As the text reflects:

> The Habesha were fascinated by their encounter with the Egyptians and the activities around them, because knowing only their own people and dress custom, they

had never before imagined the expansiveness of the human race and culture inhabiting the world (Hailu 2012: 19).

It portrays how historical and social contexts intersect to influence the characters' perceptions of self and belonging. Their experiences, shaped by the tumultuous events of colonialism and the struggle for independence, exemplify the complex negotiation between their cultural heritage and the forces that seek to erase or marginalize it. It highlights the emotional landscapes of the characters as they navigate their dual identities. "Thoughts of his mother and father dominated his mind, and Tuquabo saw Massawa running away from him" (14). The yearning for their homeland, the profound grief of loss, and the expression of these emotions through poetic forms all contribute to an understanding of the intricate relationship between cultural identity and emotions. Through Tuquabo's dirge, readers witness the depth of his internal conflict and regret, underscoring the painful consequences of straying from his cultural roots.

Thus, *The Conscript* contributes an understanding of cultural identity by weaving together historical, social, emotional, and artistic dimensions. Through its portrayal of characters' struggles, emotions, and expressions, the novel illustrates the complexities of cultural identity and underscores the resilience of individuals in preserving their heritage in the face of external pressures.

Conclusion

In conclusion, the analysis of Hailu's *The Conscript* through a postcolonial lens offers a rich exploration of the intricate dynamics of cultural identity within the context of Eritrea's postcolonial landscape. The novel illuminates the multifaceted nature of cultural identity, shaped by historical and social factors, and underscores the challenges and conflicts individuals face when negotiating between their cultural roots and external pressures. The characters' emotional responses, particularly their expressions of grief and regret, reveal the impact of colonial experiences on their understanding of self and belonging.

Through the characters' experiences, the narrative examines the tension between loyalty to one's homeland and compliance with foreign authority, shedding light on the complexities of navigating dual identities. The significance of poetic expression emerges as a potent form of resistance against cultural erasure and displacement, serving as a means of preserving and articulating cultural identity. The novel's portrayal of how artistic expressions empower marginalized voices underscores the importance of amplifying lesser-known narratives within the broader postcolonial discourse.

The Conscript" stands as a compelling work within Eritrean postcolonial literature, offering unique insights into the struggles, resilience, and agency of individuals who grapple with the complexities of their cultural identity. By delving into the characters' experiences, emotions, and artistic expressions, the novel contributes to an understanding of the enduring impact of colonialism on identity formation and highlights the ongoing quest to reclaim, express, and preserve cultural heritage in the face of external challenges.

Hailu's *The Conscript* emerges as a powerful literary work that intricately illuminates the multifaceted dimensions of cultural identity within the postcolonial context of Eritrea. Through its exploration of characters' experiences, emotions, and artistic expressions, the novel serves as a reminder of the enduring impact of colonialism on identity formation. It not only delves into the complexities of negotiating between cultural roots and external pressures but also underscores the influence of historical and social factors on shaping individual and collective identities.

The novel's portrayal of characters like Tuquabo captures the internal conflicts, yearnings, and connections that individuals maintain with their cultural heritage. Tuquabo's emotional journey, epitomized by his heartfelt dirge, provides a window into the intricate interplay between personal regret and the broader narrative of cultural resilience. Moreover, the characters' responses to returning to their homeland and their use of artistic expressions as acts of resistance exemplify the ways in which cultural identity serves as a site of negotiation and agency.

References

Bhabha, Homi K. *The location of culture*. Rutledge, 2012.

Hailu, Gebreyesus. *The Conscript: A Novel of Libya"s Anticolonial War*. Ohio University Press, 2012.

Hughes Waldick, Zoë. "Self-Representation in the Contact Zone: An Autoethnographic Reading of *The Conscript*." (2021).

Negash, Ghirmai. "Native intellectuals in the contact zone: African responses to Italian colonialism in Tigrinya literature." Biography (2009): 74-88.

Negash, Ghirmai. *A history of Tigrinya literature in Eritrea: The oral and the written, 1890- 1991*. Research School of Asian, African, and Amerindian Studies, Univ., 1999.

Tesema, Asres. *Teamot. Kab Tarik Medrekawi Sne-Tbeb Ertrea*. Franceskana, Asmara. 2006.

Tesfamariam, Desbele Teckle. "Navigating the Labyrinth of Otherness: A Postcolonial Intervention and Identity Formation in Hailu's Novel the Conscript". International Journal for Multidisciplinary Research (IJFMR) E-ISSN: 2582-2160. IJFMR23056701 Volume 5, Issue 5, September-October 2023.

Tesfamariam, Desbele Teckle. "*The Conscript* as a Hybrid Novel: Exploring Hybridity, Migration, and Identity Negotiation in Hailu's Masterpiece": Praxis International Journal of Social Science and Literature [Volume 6, Issue 9, September 2023].

Tesfay, Mokonen. *North Star: Biography of Dr. Futsum Gebreslasie*, Digy POD Printing and Publishing, Tecumesh, Michigan, USA, 2022.

Wrong, Michela. *I Didn't Do It for You: How the World Betrayed A Small African Nation*. Harper Collins publisher. Australia, 2005.

Young, Robert. *Colonial Desire: Hybridity in Theory, Culture, and Race*. London ; New York : Rutledge, 1995.

Zere, Abraham T. "Narration in Gebreyesus Hailu's *The Conscript*". Diss. Ohio University, 2014

The First Local Periodicals and The Dissemination of Literacy in Eritrea (1891-1919)

Natnael Aron Ghebrehiwet[1] and Massimo Zaccaria[2]

[1] Research and Documentation Center, Asmara, Eritrea
[2] Università degli Studi di Pavia, Italy;
e-mail: massimo.zaccaria@unipv.it

Abstract

The last half of the 19th century saw the arrival of the press in many African countries. In Eritrea, the Catholic Mission and the Evangeliska Fosterlands Stiftelsen, also known as the Swedish Evangelical Mission (SEM), were at the forefront of the print revolution. In 1863, Bishop Lorenzo Biancheri brought the first printing press with Amharic types to Massawa, while the Swedish Evangelical Mission in Ǝmkullu printed its first book in 1885. This paper presents the first periodicals published in Eritrea, and then focuses in particular on two of them: *Mäl'əkti Sälam* [*Peace Message*, 1909-1915] and *Nay Həzbə Itiopya Unätäyna Fätawi* [*The True Friend of the Ethiopian People*, 1916-1919]. Published by the Swedish Mission and Catholic Mission respectively, these are the first two periodicals in Tigrinya language. The complete series of these two very rare periodicals has only recently been recovered and digitized, and are used for the first time in this paper. After the analysis of the history of the two periodicals, the paper will consider the cultural and political context that led to the appearance of the two magazines and discuss their historical relevance.
Keywords: Print culture; local periodicals; Tigrinya language; education; colonialism.

Introduction

The printing presses appeared in Africa between the 19th and 20th centuries. During this period, books and periodicals began to be printed in the major African cities, and many languages were codified for the first time in written and printed form. It was a very rapid, and at times tumultuous, expansion that left a deep mark on the various societies of the continent.[1] Perhaps one reason historians have not fully exploited this production is that, being so rare and difficult to access, most scholars have not been able to work directly on these texts. Even today, few African countries have managed to map their print production. In Eritrea's case, one of the few exceptions is the recent article by Temesgen Tesfamariam, who analysed Italian educational policy in Eritrea from 1890 to 1935 using eleven locally printed teaching textbooks (Tesfamariam, 2023). It is surprising how few other scholars have so far resorted to local printed sources, especially given the wealth of information they could access. What have been the factors that have prevented the use of these sources by scholars for so long? Concerning the Horn of Africa, and Eritrea in particular, attempts have been made in the

1 For an introduction to the history of the book in Africa see: Vlies, 2010; Le Roux, 2012; Davis, 2015; Cagé, 2016; Peterson, Hunter, and Newell, 2016.

past to map this production, with very interesting results. However, some areas have always remained unmapped, and, paradoxically, periodicals have been excluded from this exercise (Zaccaria, 2021: 24). A limiting factor in exploiting the documentary potential of this type of source has always been the rarity of the documents. Most of them were printed in very small numbers. The first two books printed in Massawa in 1867 were a catechism and a syllabary, with the catechism being printed in 500 copies and the syllabary in 1500 copies (Lazzarini, 1989: 20). Todate, only three copies of the catechism seem to have survived. Other titles were even less fortunate, and no copies have survived, or there is only a single copy, a kind of "printed manuscript" as if to remind us that in this region the new technology of printing arrived in a context with a rich tradition of manuscripts.[2] A further factor that contributed to the low level of use of these sources was the extreme simplicity of their typographical layout. They were mostly very modest publications, poorly laid out and poorly bound, falling into the category of what Ami Ayalon has termed "pious print" (Ayalon, 2016: 87-89) and never finding their way into institutional repositories. They therefore struggled to find a place in public libraries. In Italy, apart from the IsIAO library, which has at least occasionally attempted to collect materials printed in the Italian colonies, few public libraries hold anything worth mentioning. The situation is different in the private libraries of some academics who, aware of the potential of these works, have been more careful in collecting them. The Guidi's collection at La Sapienza University (Rome), the Conti Rossini's collection at the Accademia dei Lincei (Rome), and the Faitlovitch's collection at Tel Aviv University all hold numerous copies of books printed in Eritrea.

The potential of these print documents will only be fully realized when the texts, and not just the titles, of these works are available to scholars. In the case of Eritrea, this moment is approaching. An EU-funded pilot project has enabled the recovery and digitization of a large part of Eritrea's printed heritage published between 1867 and 1962, which will soon be available online.[3]

This article will focus on the first example of periodicals printed in Eritrea. The analysis will cover both early periodical publications in Italian and Tigrinya. For the latter category, the existence of *Mäl'əkti Sälam* has been reported on several occasions,[4] but few scholars seem to have had access to copies of this periodical. Even less well known is the history of *Nay Həzbə Ityopya Unätäyna Fätawi*, a periodical that has been almost entirely ignored by scholars.

As we know, the relevance of a historical source depends on its ability to reveal scenarios difficult to reconstruct from other sources: can we use books and magazines to discov-

2 On the relationship between the printed production and the manuscript tradition in East Africa see: Reese, 2022.

3 EU Project CSO-LA/2018|403-679, Digital Technologies and Cultural Heritage Preservation in Eritrea: a Framework for an Improved Action. The activities, which will culminate in making all the digitised materials available online, can be followed on Facebook: Towards the National Digital Library of Eritrea: www.facebook.com/people/Toward-the-National-Digital-Library-of-Eritrea/100092867552115/ (last access 3 June 2025). One of the aims of this experience was to inspire other African countries to do the same.

4 The book by Solomon Ghebrekristos, *Ḥarägatä Tägərəyəna* dedicates a few lines to *Mäl'əkti Sälam* and *Nay Həzbə Itiopya Unätäyna Fätawi*. The *Encyclopaedia Aethiopica* contains an entry by Ezra Gebremedhin (ad vocem); while in 1993, F. Natnael presented a paper dedicated to *Mäl'əkti Sälam* at The First International Symposium on the Tigrinya Language (Asmara, Feb. 20-23, 1993). In most cases, the information provided are inaccurate or wrong: Meseret Chekol Reta (2013: 7) reports that *Mäl'əkti Sälam* appeared in 1912 and lasted for three years.

er local changes that would otherwise be difficult to grasp? This brief note aims to demonstrate the potential of printed sources in reconstructing the past of many African countries, using the case of Eritrea as an example. By their very existence, these newspapers raise questions about an aspect of Eritrea's history that has so far been accepted by the majority of scholars. It is generally agreed that, due to the deliberate neglect of education by the Italian colonial government, Eritrea arrived in the 1940s with very low levels of literacy and schooling. This state of affairs had an inevitable impact on the process of decolonization in Eritrea. In his seminal study on Italian colonialism in Eritrea, Tekeste Negash stated that "In the case of Eritrea, the factors which would have most probably given rise to an Eritrean nationalism, i.e. western educated elites and a local press,[5]did not exist by the end of the colonial period"[5] (Negash, 1987: 164). Doubts about this reconstruction have already been expressed by some scholars, who have indicated how difficult it was to reconcile this view with the vibrancy of public debate of the period 1940-1960s.[6] For example, Uoldelul Chelati Dirar has pointed out that this reconstruction "fails to account for the otherwise mysterious flourishing of written literature in Eritrea during the 1940s and 1950s" (Chelati Dirar, 2007: 265). If we consider the publications which appeared as soon as Italian rule ended, we cannot but be amazed. During the British Military Administration (1941-1949) and British Civil Administration (1949-1952), many newspapers and books written in Tigrinya and Arabic were published: one of the most famous examples was *Nay 'Erətra Sämunawi Gazeṭa* (*Eritrean Weekly News*, 1942-1952), which was edited by Wäldä Ab Wäldä Maryam, who is often called the Father of the modern Tigrinya language. In addition, there were other periodicals like *Ḥanti 'Erətra* (*One Eritrea*, 1950-1952), and the bilingual (Tigrinya and Arabic) *Dähay 'Erətra* (*Voice of Eritrea*, 1952-1954), and *Zämän* (Times, 1953-1962), and *Ityopya* (*Ethiopia*, 1947-1962) (Voigt 2012: 1173). As evidence of the vitality of the political and cultural scene in the years following the end of Italian colonial rule, we can also mention the formation of political, religious, and intellectual groups and circles, such as the *Maḥbär Fəqri Hagär* (Society for love of the country, 1941), *Sänbäte Maḥbär* (Sunday Association, 1943, then *Maḥbär Erətra məs Ityopya*, Association Eritrea with Ethiopia, 1946), and the religious association *Maḥbärä ḥawaryat* (Community of the Apostles, 1949-50). Needless to say, we are not concerned here with denying the repressive nature of Italian colonial rule in Eritrea; rather, we need to aim for a more nuanced understanding of it (Chelati Dirar, 2007: 265). Those were the years in which personalities of great intellectual depth – Ibrāhīm Sulṭān 'Alī, Wäldä Ab Wäldä Maryam, 'Abdälqadər Käbirä, etc. – were at work, stimulating an intense public debate on the nation's destiny, much of which was conveyed by a lively and stimulating press. How, and when, did this generation of intellectuals and great political leaders come into being? Were they the sole result of the educational reforms introduced by the British Military Administration? Or should we consider a longer path dating back to the early 20th century when certain sectors of Eritrean society began using the printed word to debate and circulate new ideas and perspectives on the world?

What we do know, and what we will not dispute, because this is a point on which most scholars agree, that is the Italian government tried to limit education for Eritreans as much as possible to prevent the formation of an educated class. But, the political will of the Italians

5 This idea is also common in several writings. For example, some publications were produced during the British Military Administration in Eritrea. These publications introduced some Tigrinya periodicals of the 1940s. They introduced these periodicals as pioneers. This has led to the obscurity of the pioneering Tigrinya periodicals.

6 On the political and cultural vibrancy of the 1940-1960s see: Negash, 1997; Venosa, 2014; and the trilogy by Alemseged Tesfai, 2001, 2005, 2016; Guazzini, 2024.

had to reckon with the determination of the Eritreans who wanted to educate themselves, perhaps through channels not fully controlled by the colonial government. The printed sources published in Eritrea can facilitate a more profound comprehension of the cultural milieu in the nation before the 1940s, thereby rendering the emergence of this generation of eminent intellectuals and patriots less precipitous. Local printed sources allow us to observe facts and situations that would otherwise be impossible to reconstruct.

In the first part of this article, we will present the first periodical magazines to appear in Eritrea. These magazines were published in Italian by both the government and private individuals. The second part of the article will focus on the first two periodical journals in Tigrinya that were published between 1909 and 1919. These journals provide fascinating insights into the intellectual and social life of early 20th century Eritrea, and should be recognised as valuable historical sources.

The Beginning of the Print Revolution in Eritrea

The arrival of printing in most parts of Africa dates back to the second half of the 19th century. This chronology was shaped by advances in printing technology, which led to the proliferation of new presses and the introduction of lithography in a relatively short time. This was a watershed in the history of printing in many parts of the world (Stark, 2022), and Nile Green has aptly called it the "second printing revolution" or the "Stanhope Revolution" (Green, 2010: 414). This was when, in the early 1800s, a new generation of iron hand presses appeared, helping to industrialize printing. One of these presses was brought to Eritrea at the end of 1863 by the Lazarist bishop Lorenzo Bianchieri alongside some Ahmaric type made in Paris for the French scholar Antoine d'Abbadie (Lazzarini, 1989: 7). The first book printed was a catechism in Tigrinya, published in 1867. In 1879, the printing press was moved from Massawa to Keren, where at least thirty books had been printed by 1895 (Lazzarini 1989: 1-4). In 1885, a new printing press was established in Ǝmkullu by the Swedish Evangelical Mission. In the same year, the *Tipografia Militare* (Military Tipography) was established in Massawa by the Italian military administration. This was soon followed by the *Tipografia del Corriere Eritreo* (1891), the *Tipografia Italiana* (1891), the *Tipografia dell'Africa Italiana* (1893), and the *Tipografia Nazionale* (1894). This proliferation of presses could be because the same printing house changed its name, leading to some confusion. The number of printing presses in Massawa remains surprising considering that in 1893 the Italian presence in the country consisted of 617 civilians and 2.495 Italian military personnel.[7]

The "Settkers" Press

As far as Italian-speaking periodicals are concerned, the first to get published was the *Bollettino degli Annunzi Legali e Commerciali della Colonia Eritrea* [The Legal and Commercial Bulletin of the Eritrean Colony, 1890-1891]. A government-issued journal that published the texts of legislative measures and administrative and commercial news. Starting in 1892, the periodical took the name *Bullettino Ufficiale della Colonia Eritrea*, a name it retained until 1929.

On 12 June 1891, the *Corriere Eritreo. Giornale Settimanale, Politico-Commerciale di Massaua* [The Eritrean Courier. Weekly Political and Commercial Newspaper from Massawa] the first private periodical of the Eritrean Colony, appeared in Massawa.[8] It was an eight-page weekly printed by the *Tipografia del Corriere Eritreo*. A second private newspaper, *L'Eritreo. Giornale Settimanale Amministrativo Commerciale della Colonia Eritrea* [The Eritrean: The Eritrean Commercial Administrative Weekly Newspaper], appeared on 4

7 Censimento della popolazione, *L'Africa Italiana*, 4, 214, 11 Feb. 1894, 2.

8 *Corriere Eritreo. Giornale settimanale, politico-commerciale di Massawa.*

November of the same year. Born in contrast and controversy to the *Corriere Eritreo*, *L'Eritreo*[9] was aimed at the same tiny readership of Italians mainly from Massawa. We do not know how it happened, but eventually the two newspapers overcame their differences and merged on 1 January 1893 to form *L'Africa Italiana. Gazzetta di Massaua. Settimanale politico commerciale* [Italian Africa. Massawa Gazette. Commercial political weekly], a weekly printed by the Tipografia dell'Africa Italiana that lasted until 1897.[10]

Studies have long emphasized how the issue of migration provided Italy with a powerful political and moral justification for its expansion (Labanca, 2001, 2002; Bruner, 2009; Ertola, 2022). The colonies were intended to accommodate the large numbers of Italian emigrants and address what was considered a national emergency. Much of the available research has concentrated on the rhetoric used to legitimize this expansionist project. The studies that have appeared so far have approached the world of the Italians in Eritrea from two main perspectives. On the one hand, some studies treat this presence as a very homogeneous and anonymous group, and as a mere instrument of government policy, of which they were merely the implementers. On the other hand, some works have focused on the micro dimension of the settler experience, using diaries and memoirs to focus on settler individuality and life in Eritrea. The three weekly newspapers were the expression of the Italian community in Massawa, which was small in number but surprisingly divided among itself. The *Corriere Eritreo* and *L'Eritreo*, before merging to form *L'Africa Italiana*, used to be rival newspapers in constant conflict with each other. They represented different sections of the small Italian community but, as is clear from the pages of these journals (see below), their assessments of the situation were often at odds with those of the government. Even then, alongside and often in opposition to the colonial administration, an Italian community of small businessmen, traders and shopkeepers operated. On several occasions, these two souls of the Italian community disagreed on the political choices to be made. The small Italian community found these newspapers effective tools to make their voice heard and express their views.

Anno I. MASSAUA, 12 GIUGNO 1891 Num. 1.

CORRIERE ERITREO

Giornale Settimanale, Politico-Commerciale di Massaua

Ufficiale per i Decreti del Governo e per gli Atti Legali e Giudiziari della Colonia

L'ERITREO

GIORNALE SETTIMANALE AMMINISTRATIVO COMMERCIALE DELLA COLONIA ERITREA

(Conto corrente Postale) MASSAUA, 1 GENNAIO 1893 Anno III — Num. 157

L'Africa Italiana

Gazzetta di Massaua

SETTIMANALE, POLITICO-COMMERCIALE

9 *L'Eritreo. Giornale settimanale amministrativo commerciale della Colonia Eritrea.*

10 *L'Africa Italiana. Gazzetta di Massua. Settimanale, politico-commerciale.*

What emerges from the reading of its pages is an Italian community that regularly expressed its judgments and opinions on the most important issues facing the colony. A community that wanted to have a say on the decisions taken by the government and, in doing so, tried to have a voice at the political level. After all, it was only natural that the Italians in Eritrea should want to express their opinion on the decisions that were bound to affect them. And the more important these decisions were, the more the Italians in Eritrea tried to make sure their voice was heard. A voice that did not always coincide with that of the colonial administration. For example, on one of the most hotly debated issues of the time, Baron Leopoldo Franchetti's colonization project, which aimed to bring Italian peasants to Eritrea to settle on small plots of land on the plateau, the assessment expressed in these newspapers was very critical. From its very first issue, the *Corriere Eritreo* described Franchetti's agricultural experiments as lacking "common sense".[11] From the point of view of the Italian colonists, the priority in the economic field was the exploitation of the commercial potential of the Horn of Africa and the Red Sea region--the settlement of farmers was a waste of time and money. A rather predictable view for newspapers in whose subtitle there was an unequivocal reference to politics and trade, and whose ownership and readership came largely from the commercial sector. These three weeklies offer the opportunity to observe a new dimension of the Italian presence in Eritrea, that of trying to get involved in politics. This is an important novelty. The political dimension had never been considered before due to a lack of adequate sources. Now we have documents that can convincingly show us this aspect.

Between 1903 and 1905, what can be regarded as the first scientific journal in Eritrea was published: the *Bollettino Agricolo e Commerciale della Colonia Eritrea* [Agricultural and Commercial Bulletin of the Eritrean Colony]. Edited by Prof. Isaia Baldrati, the head of the *Ufficio di Colonizzazione della Colonia Eritrea* [Colonisation Office of the Eritrean Colony]. The Bulletin aimed to inform farmers about the main experiments being carried out by the Eritrean Colonisation Office and others in the field of tropical agriculture. The list of the first Eritrean newspapers must also include the weekly periodical *L'Annunzio Eritreo*, which ran from 1916 to 1917. At present, the only existing collection appears to be in the possession of the National Central Library of Florence. As far as we are concerned, we have only been able to consult one issue of this very rare periodical. A cursory analysis of this issue reveals that the *L'Annunzio Eritreo* was often in conflict with the colonial administration. In this sense, it is very reminiscent of the spirit of the *Corriere Eritreo*, *L'Eritreo,* and *L'Africa Italiana*, but it is impossible to draw any kind of conclusion until more issues can be accessed.

"The Natives Press"

By the end of the century, the three Italian newspapers had ceased to exist. The SEM had always regarded the printing press as one of its core activities. The Catholics found it more difficult to continue their work in the field of printing. As we have seen, the Catholics were the first to bring a printing press to the country, but they soon encountered increasing difficulties in maintaining it: lack of materials and staff slowed down and then paralyzed the press that Bishop Lorenzo Biancheri had brought to Massawa in 1863 and moved to Keren sixteen years later (Metodio da Nembro, 1953). While the Catholic Church was forced to cease printing, the SEM strengthened its dominance by increasing the number of publications it printed and distributed.

It has been printed that the "Analysis of journalistic endeavours in Eritrean languages during the colonial period would further contribute to a more comprehensive understanding of the development of linguistic and national identities in Eritrea" (Chelati Dirar, 2007: 267).

11 Dalle zone, *Corriere Eritreo*, 1, 1, 12 Jun. 1891, 3.

From this perspective, the analysis of the first Tigrinya-language newspapers takes on a special significance. In March 1909, the SEM's Printing Press in Asmara published the first issue of *Mälʾək̠ti Sälam* [Peace Message, 1909-1915].[12] In June 1916, the Francescana Printing Press published *Nay Həzbə Ityopya Unätäyna Fätawi* [The True Friend of the Ethiopian People, 1916-1919, hereafter cited as *Nay Həzbə*]. These are the first two journals published in Tigrinya language. Therefore, they jointly represent and reflect a period of eleven years of Eritrean history, which is sometimes difficult to fully understand from other sources and by neglecting these periodicals.

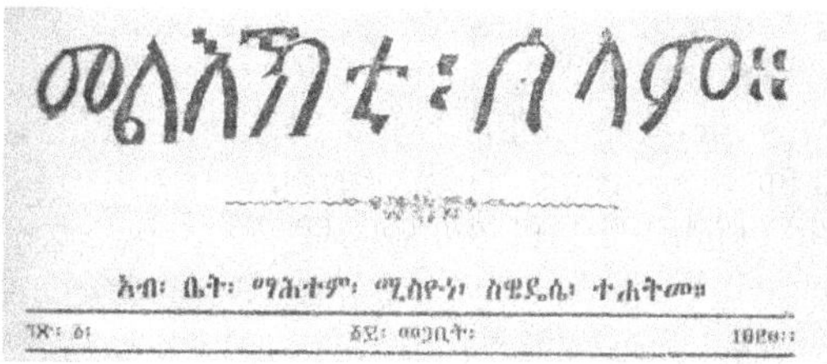

Mälʾək̠ti Sälam [Peace Message, 1909-1915].

Mälʾək̠ti Sälam (1909-1915)

The first issue of *Mälʾək̠ti Sälam* appeared in March 1909 and was edited by the SEM's missionary Olof Eriksson.[13] The journal's stated aim was religious, as was made abundantly clear in the first issue as follows:

> Although this newspaper, Mel'ekhti Selam, is small in relation to the land of Habesha, it aims to spread the word of Christ. However, it cannot reach different villages by itself and we, the believers, should help it to do so. This newspaper will provide useful teachings and advice to all believers and nationals … in this paper, you will find sermons, admonitions, and good narratives beneficial for the faithful as well as for the entire population ... If kept carefully, the issues can later be compiled into a book. From now on, the newspaper will be printed four times a year. If demand increases, we can increase the frequency to five or six times a year. As the price is affordable, we should each buy our own copy to ensure its continuity, rather than sharing a single copy between many of us. We should also contribute our ideas to help develop the newspaper. May the Almighty bless this newspaper."[14]

Based on this, the newspaper's contents consist mostly explanations of biblical verses, motivational Bible stories, and reports on the church's activities.

From time to time, information that helps us to understand how the magazine circulated within the Eritrean evangelical community is provided. On 1 October 1911, the editors complained of poor sales and announced that only three hundred and forty of the one thousand copies printed had been sold, as follows:

> It has been three years since the publication of the first issue of this newspaper.

12 To date, the years of publication of the periodical is often found to be mistaken in many writings. After ceasing for many years, the periodical was able to restart during 1940s. However, the correct starting and ceasing dates of the periodical during its first phase should be understood as mentioned in this article.

13 For a history of the SEC see: Jwarson and Tron, 1918; Aren, 1978 and Gebremedhin, 2011.

14 *Mälʾək̠ti Sälam*, 1, 1st Mar. 1909, 2.

> However, there is a small number of buyers. Only three hundred and forty copies were bought from the thousand copies printed. This might be because of a lack of appetite in members our Association members and students for spiritual lessons. Based on this assumption, to increase its demand, we are thinking of introducing some changes in the contents of the newspaper starting from the coming New Year”.[15]

Numbers on printed copies of a magazine are very difficult to obtain, which makes accurate measurement of readership very challenging. Luckily, in this case we have some idea about the size of the readership community.

Reading *Mäl'əḵti Sälam*, the impression is that sales and public response were disappointing, at least for the magazine's creators. Some of the changes to the format likely helped make the magazine more appealing, with the gradual opening up to non-religious topics probably contributing to an increase in its circulation. In 1915, *Mäl'əḵti Sälam* started including some articles in Italian. The relationship between the SEM and the Italian administration had always been tense, characterised by deep misunderstandings and distrust. The Italian administration accused the SEM of pursuing policies hostile to Italy. A recurring source of friction was the limited use of Italian in SEM activities, as well as the apparent lack of attachment to Italy among some SEM members. In an attempt to reduce these tensions, some Swedish missionaries learnt Italian and took in some Italian members of the Evangelical Church. Although Italy is predominantly Catholic, there are some religious minorities in the country, including an Evangelical church presence in Piedmont. From the early 1900s onwards, members of an Italian evangelical mission began operating in Eritrea. The publication of some articles of *Mäl'əḵti Sälam* in Italian was part of this appeasement strategy with the Italian authorities. It was intended to reassure the authorities of the Evangelical Church's loyalty to the colonial government.

Finally, in 1916, SEM decided to halt the publication of *Mäl'əḵti Sälam*. It should be noted that publishing a periodical is always a complex undertaking. First, the texts have to be written. After being checked carefully and approved by the government, they would be sent to the printer. The last step was distribution, which brought the magazine into readers' homes. The whole process involved numerous people and job profiles, which contributed to the rising cost of production.

FIG. 105. INTERNO DELLA TIPOGRAFIA DI ASMARA

Interno della tipografia di Asmara [Interior of the Asmara print shop] J. Jwarson, A. Tron, Notizie storiche e varie sulla Missione Evangelica Svedese dell'Eritrea, 1866-1916, Asmara: Missione Evangelica Svedese, 1918, fig. 105.

15 *Mäl'əḵti Sälam*, 1st Oct. 1911, 193.

One should not be too harsh with the experience of this first attempt at a periodical magazine in the Tigrinya language. *Mäl'əḵti Sälam* was a pioneering and, in its way, very bold project. There had been no previous experience of this kind, so there was a lack of reference points to guide the creators of this project. Despite this, the Swedish Church decided to invest in this idea and committed itself to its implementation for six years, providing the necessary human and financial resources. The minimalist design and concise pages should not mislead: the publication of *Mäl'əḵti Sälam* was undoubtedly a challenging endeavour in every respect.

Mäl'əḵti Sälam proves the presence within Eritrean society of a small group of readers. This group had begun to debate information and ideas through the press. The Tigrinya journal enabled members of the Evangelical Church living in different parts of Eritrea to form a virtual community with like-minded individuals who wanted to educate themselves, discuss issues, and develop intellectually. Although the sales figure of 340 copies for *Mäl'əḵti Sälam* was considered disappointing, it provides an important indication of the size of this audience. These readers occupy an intermediate position between ordinary people and those with a formal education, namely, individuals who underwent training within the colonial education system. Most likely, the Evangelical Mission trained these readers in the schools that it set up in many Eritrean villages; schools that were often small and modest, but which enabled the formation of groups of believers capable of using the written word to their advantage. Until now, the prevailing tendency has been to play down the phenomenon by dismissing these newspapers as incapable of making an impact, as Reta has argued: "It is hard to say all these newspapers were read by any of the ordinary people, given the massive illiteracy in the society. Most likely, it would have been the few elite within the missionary and colonial camps that read the newspapers, and perhaps surmmarized them for the ordinary people" (Reta, 2013: 7). An assessment that we feel should be at least partly revised, given the commitment and costs involved in these albeit modest periodicals. Indeed, it is not clear why SEM would have invested for six years in a useless magazine.

Nay Həzbə Ityopya Unätäyna Fätawi (1916-1919)

The publication of *Mäl'əḵti Sälam* had marked a new point in favor of the SEM and its primacy in the publishing world. With great timing, when Asmara became the new capital of Eritrea, the Evangelical Church immediately moved its modern printing press and bookbindery there.

The Catholic Church was not as responsive. Its printing press remained in Keren, where its production languished amid technical problems and a lack of personnel. A significant shift in pace occurred when the leadership of the Eritrean Catholic Church passed from the Capuchins of the Roman province to those of the Lombard province (Chelati Dirar, 2002). On 7 February 1911, Pope Pius X named Bishop Camillo Carrara Apostolic Vicar of Eritrea. Mons. Carrara believed that the press was one of the most effective tools in the fight against secularism and that strengthening the Church in the field of education was a necessity. Towards the end of the century, the Church identified the press as a priority area for action and a field of confrontation with secularising tendencies. Among Catholics, this confrontation became known as "the battle for the good press".[16] Committees for the diffusion of the "good press" were set up in almost every place. The congregation to which Mons. Carrara belonged, the Third Order of Franciscans, had supported the "Good Press Society" which,

16 La propaganda della stampa e il Terz'Ordine, *Annali Francescani*, 42, 4, 16 Feb. 1911, 100-103.

in 1913, claimed to have distributed in one year 339,050 daily newspapers, pamphlets, and loose sheets and 31,000 copies of leaflets and special issues.[17] The struggle for the "good press" had a world dimension: embraced by Pope Leone XIII, this movement spread rapidly throughout the Christian world, and in 1873 Father Vincent de Paul Bailly founded the "Maison de la bonne presse" in France (Godfrin, 1965; Véronneau, 2003; Raison du Cleuziou, 2024). Similar activities were recorded throughout the Catholic world. Its main features were outlined in internal discussions and pastoral letters, while bulletins and associations were responsible for putting the theoretical debate into practice.[18] New Catholic newspapers were launched, with attention being paid not only to the content but also to the visual and graphic features. The 'enemy' was anti-clericalism, the "books and newspapers of the ungodly", "the liberal, the masonic and the subversive newspapers".[19] In the early twentieth century, the Catholic press found itself fighting a war it didn't intend to lose. The missionary congregations did not give in and took part in the fight for the "Good Press".

Before he arrived in Eritrea, Bishop Camillo Carrara had been following with great interest the debate on the "Good Press".[20] Once in Eritrea,[21] Mons. Camillo Carrara found that the book market was almost entirely dominated by the Evangelical Church. In his eyes, the Protestant press had "invaded" Eritrea and its printing house was described as "a flourishing printing house supplied in profusion with Latin and Ethiopian characters, from which works continually come out printed in the following languages: Italian, Swedish, Tigrigna, Amharic, Ghez, Cunama, Galla, and Swahili".[22]

Writing back to Italy in May 1912, Mons. Carrara wrote: "I am convinced that printing is one of the most effective factors in spreading education and civilisation in our colony. I have devoted and will devote the most careful attention to the printing press".[23] He then decided to move the small printing press and bookbinding laboratory from Keren to Asmara (Tesfamichael, 2024: 81), he repaired the old machine and then had a new printing press purchased,[24] set it up in special premises, and provided it with an adequate set of Amharic characters. Secondly, a skilled typographer was sent from Italy to take care of the printing activities of the Catholic Mission. It was in this way that the *Tipografia Francescana* [Franciscan Typography] was inaugurated in 1912. Following these investments, the *Tipografia Francescana* was a cutting-edge typography that could compete with that of the SEM in every aspect. The Italian population was small in terms of numbers and not particularly keen on reading, so most of the activity was directed towards the Tigrinya-speaking public. In a short time, religious and scholastic texts, as well as small introductions to reading and writing, were prepared and printed. During this period, the Catholic Church claimed more than 20,000 converts (Tesfamichael, 2024: 88).[25] In 1917, the Catholic Church printed its first periodical: *Parole Buone*, a monthly bulletin which continued to be published until June 1940. It was addressed mainly to Italians in Eritrea. For Tigrinya-speaking audiences, *Nay*

17 L'adunanza della Società 'Buona Stampa', *Annali Francescani*, 44, 11, 1 Jun. 1912, 297.

18 La propaganda della stampa e il Terz'Ordine, *Annali Francescani*, 42, 4, 16 Feb. 1911, 100-103.

19 La stampa, *Annali Francescani*, 42, 12, 16 June 1911, 357.

20 "I, who have always been a firm believer in the Good Press", Mons. C. Carrara to Card. Girolamo Maria Gotti, Asmara 6 Jun. 1912, in Tesfamichael, 2024: 115.

21 On 7 Feb. 1911, Carrara was appointed Apostolic Vicar, and arrived in Asmara on 3 Apr. 1911.

22 Mons. C. Carrara to Card. Girolamo Maria Gotti, Asmara 6 Jun. 1912, in Tesfamichael, 2024: 116.

23 Mons. C. Carrara to Giacomo Agnesa, Asmara May 6, 1912, in Tesfamichael, 2024: 89-90.

24 Il Governatore dell'Eritrea e la Missione Cattolica, *Annali Francescani*, 44, 23, 1 Dec. 1913, 637.

25 *Per una completa libertà*, Asmara, Tipografia Francescana, (1921: 14) estimates at 25,000 the number of Catholics in Eritrea.

Həzbə Itiopya Unätäyna Fätawi made its appearance in June 1916. Originally a sixteen-page monthly publication, it was later reduced to eight pages. Compared to *Mäl'əkti Sälam*, in *Nay Həzbə* the interaction with the readers was much smaller. Occasionally, the journal published list of donors and contributors. This was on the last page of the issue. For the rest, the interaction with the readers was minimal. The last issue of the journal that we have recovered is dated 1919, but we are not entirely sure that the periodical ceased exactly in 1919. Its contents were purely spiritual and religious in nature.

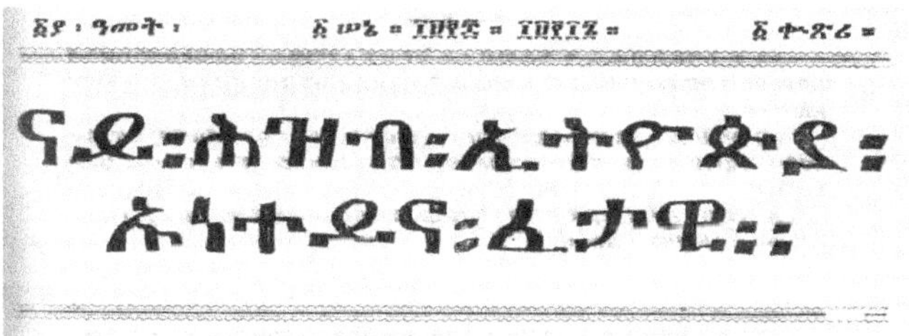

Nay Həzbə Ityopya Unätäyna Fätawi [The True Friend of the Ethiopian People, 1916-1919].

From the very first issue, *Nay Həzbə* polemicised with those who did not venerate St. Mary, Angels, and Saints. *Nay Həzbə* criticized and condemned "some people" who had tried to break the bond between believers and their attachment to saints and icons. In denouncing this situation, the Catholics alluded to "some bad teachers", "some evil people", and "some people who do not want the good of the Ethiopian people", etc. In confidential communications with the authorities, the Catholic Church provided more detailed denunciations and openly referred to the activities of the SEM. The hope was to create an axis with the Orthodox Church, as is evident in some archive documents, to counter the propaganda of the SEM. As with every publication printed in Eritrea, prior authorisation from the authorities was necessary. At the end of May, Archbishop Carrara contacted the public prosecutor's office in Asmara to point out that the purpose of the new Tigrinya magazine was "the religious and moral instruction of the natives".[26] However, only a few weeks earlier, Bishop Carrara had presented his project in more detail. This time, he made it clear that the sixteen pages of the new magazine were intended primarily to respond to the "prejudices and slanders that Swedish Protestants have spread in several of their books and pamphlets against sound Christian doctrine". The new magazine was intended for both Catholics and Orthodox Christians. The controversy surrounding the SEM was central, as can be seen from articles such as "First Impressions of a New Missionary" (Issue 1), in which the preaching of the "enemies of Mary" (Protestants) was denounced. They were defined as enemies to be distrusted.[27]

From Amharic to Tigrinya

The local press can highlight many changes within Eritrean society. Some of this information is linguistic. The appearance of *Mäl'əkti Sälam* and of *Nay Həzbə* marks an important stage in the process of transition towards the adoption of Tigrinya as the official language of Er-

26 Mons. C. Carrara to Public prosecutor's office, Asmara 30 May 1916, Archivio Storico Diplomatico del Ministero Affari Esteri, Rome [herafter ASDMAE], Archivio Eritrea [hereafter AE] 431, f. "Missioni religiose".

27 Mons. C. Carrara to [Alessandro Allori], Asmara 14 Apr. 1916, in ASDMAE, AE 431, f. "Missioni religiose".

itrea. As Gianfrancesco Lusini reminds us, this was a long and complex process, not always linear: "Paradoxically, Tigrinya always struggled to become a written language" (Lusini, 2021: 165). The spread of Tigrinya as a written language was hampered by the spread of Amharic, which over time became the official heir of Gəʼəz. Amharic found its environment of diffusion at the imperial court and in the army (Lusini, 2021: 166-167). In the early stages of their rule, the Italians maintained Amharic as the language of administration. In the few government-run schools, Amharic was the language of instruction. In 1893, male and female students enrolled in the colony's schools were awarded "for diligence in study and the Amharic language".[28]

When it came to navigating the complex multilingual landscape of Eritrea, the Italians were at a loss as to what to do. Initially, some thought of using Amharic because it was already used for official communications. Some people thought that Italian could serve as a lingua franca. In 1893, the magazine *L'Africa Italiana* presented the linguistic landscape of Eritrea as follows:

> How can we understand each other in this Babylon? Not by introducing Arabic, which nobody knows except in Massawa, and which is a language imported from the Turkish and Egyptian domination; not by spreading Tigre, a semi-wild language with no writing, no history, no way of expressing civil or simply abstract ideas; not by promoting Tigrigna, a minority language, slightly superior to Tigre. So there is nothing left to do but to use Italian: and nobody can complain about that.[29]

In reality, Italian was far from being the ideal solution, being, first of all, a language that no one spoke and that was not easy to learn, given that colonial schools were few and concentrated in a few urban centres. Gradually, the Italian government began to look more favourably on Tigrinya. By 1919, the process of transition from Amharic to Tigrinya seemed almost complete, and Tigrinya was used as the written language in the Administration. The Italian authorities had begun to issue laws, decrees, and all official documents in Tigrinya, but less and less in Amharic because the population did not speak it (Cerulli, 1920: 186-187). Politically, the adoption of Tigrinya also had the great advantage of helping to loosen the bonds with Ethiopia, as the Italians themselves admitted: "To make the separation of our subject populations from those of the Ethiopian kingdom more and more great, this is of no small advantage for our policy" (Nallino, 1919: 127). For this reason, the colonial authorities did not raise any objections to the fact that the two magazines were in Tigrinya and did not hinder their distribution.

Conclusion

Printing was introduced to Eritrea in the second half of the 19th century when the Catholic Church and the SEM each established a printing house in the country. Around 1887, the "Military Printing House" began operating on behalf of the Government. The first private printing house, the "Italian Printing House and Library", was inaugurated in 1891. Initially, production was almost exclusively focused on books, pamphlets, and administrative materials. The first periodical in Italian appeared in 1890.

These early prints are usually very rare and difficult to find. However, when we are fortunate enough to be able to consult the complete collections, the amount of information

28 Cronaca, *L'Africa Italiana*, 3, 203, 26 Nov. 1893, 3.
29 La lingua itaaliana, *L'Africa Italiana*, 4, 226, 5 May 1895, 2.

we can recover is truly astonishing and can help us to revise some commonly accepted judgments. In the case of the first Italian journals, these help us understand how the small Italian community used the press to give voice to their ideas, which were often in conflict with those of the colonial government. The same goes for the first journals in Tigrinya *Mäl'əḵti Sälam* (1909-1915), and *Nay Həzbə Ityopya Unätäyna Fätawi* (1916-1919). By their existence, these two journals offer a window into a society where certain groups were already using the press to communicate news and ideas in the early 1900s. The very existence of these two journals reveals the existence of unexpected pockets of literacy and readers within Eritrean society. Despite their simplicity and unpretentious appearance, these magazines managed to create new communities of readers. The regularity of these publications created a sense of a community of readers and contributed to the regularity of these reading practices.

The two Tigrinya periodicals thus reveal an environment already permeated by writing and animated by groups of people with some education. These are still small groups, but the trend is confirmed by the marked increase in the number of titles printed by Catholic and Protestant presses in the years around the First World War. It was a clear sign that more and more people were turning to the written and printed word as a means of communication. Education, reading and writing were fundamental aspects of the work of the Evangelical Church as clearly expressed in this passage from a 1918 book illustrating the history and work of the SEM in Eritrea: "Evangelical missions, whatever their field of activity, have as their principle that every people must be enabled to read and understand the Holy Scriptures in their language" (Jwarson and Tron, 1918: 36). This policy was also embraced by the Catholic Church, albeit in a less pronounced way, in order not to leave the field free to the Evangelical Church alone.

Although we have not yet found any documents to support this view, it is easy to imagine that the Orthodox Church did not simply stand by and watch these innovations take place. The activism of the SEM and the Catholic Church aimed to enlarge their respective flocks, mainly by drawing from the Orthodox Church. Not reacting and remaining passive in the face of the rediscovered centrality of the written word in colonial society could have turned out to be a fatal error. Perhaps the time has come to address the issue of the almost total illiteracy of the Eritrean Society before the 1940s in a more nuanced manner. Most likely, the cultural effervescence of the 1940s and 1950s is therefore part of a longer process of transformation that began to manifest itself almost forty years before. A process that encouraged the population to make greater use of the written word, often bypassing the colonial schools. These are fascinating avenues of research that could be developed in the years to come.

References

Alemseged, Tesfai (2005). *Fadarasen Ertra mes 'Ityopya. Kab Matiyanso kesab Tadla, 1951-1955*, Asmara, Hdri.

Alemseged, Tesfai (2001). *Aynfelale. Ertra 1941-1950.* Asmara: Hdri.

Alemseged, Tesfai (2016). *Ertra kab Federation nab GobeTan Sewran 1956-1962*, Asmara, Hdri.

Aren, Gustav (1978). *Evangelical Pioneers in Ethiopia. Origin of the Evangelical Church Mekane Jesus.* Uppsala: EFS.

Arribas, Delia Guijarro, Mercier, Charles, and Raison du Cleuziou, Yann (eds.) (2024), *De la Bonne Presse à Bayard. 150 ans d'histoire d'un groupe de presse et d'édition catholique, 1873-2023.* Montrouge et Lyon: Bayard et Lahra.

Ayalon, Ami (2016). *The Arabic Print Revolution.* Cambridge: Cambridge University Press.

Bruner Stephen C. (2009). Leopoldo Franchetti and Italian Settlement in Eritrea. Emigration, Welfare Colonialism and the Southern Question. *European History Quarterly*, 39, 1, 71-94.

Cagé, Julia and Valeria, Rueda (2016). The long-term effects of the printing press in sub-Saharan Africa. *American Economic Journal: Applied Economics*, 8, 3, 69-99.

Cerulli, Enrico (1920). I rapporti italo-abissini nel dopo-guerra. In Istituto coloniale italiano, Sezione studi e propaganda, *Atti del Convegno nazionale coloniale per il dopo guerra delle colonie. Roma, 15-21 gennaio 1919. Relazioni, comunicazioni e resoconti delle sedute.* Roma: Tip. dell'Unione Editrice, 1920, 186-187.

Chelati Dirar, Uoldelul (1999). Fra Cam e Sem. L'immagine dell'"Africa Italiana" nella letteratura missionaria (1857-1895). In Alberto Burgio (ed.), *Nel nome della razza. Il razzismo nella storia d'Italia 1870-1945*, Bologna: Il Mulino, 1999, 183-199.

Chelati Dirar, Uoldelul (2002). Collaborazione e conflitti: Michele da Carbonara e l'organizzazione della Prefettura Apostolica dell'Eritrea (1894-1910). *Quaderni storici*, 1, 149-188.

Chelati Dirar, Uoldelul (2003). Church-State Relations in Colonial Eritrea: Missionaries and the Development of Colonial Strategies. *Journal of Modern Italian Studies*, 8, 3, 391-410.

Chelati Dirar, Uoldelul (2007). Colonialism and the Construction of National Identities: The Case of Eritrea. *Journal of Eastern African Studies*, 1, 2, 256-276.

Davis, Caroline and David Johnson (eds.) (2015). *The Book in Africa: Critical Debates.* London: Palgrave Macmillan.

Ertola, Emanuele (2017). *In terra d'Africa. Gli italiani che colonizzarono l'impero.* Bari-Roma: Laterza.

Ertola, Emanuele (2022). *Il colonialismo degli italiani. Storia di un'ideologia.* Roma: Carocci.

Forno, Mauro (2009). *Tra Africa e Occidente. Il cardinal Massaja e la missione cattolica in Etiopia nella coscienza e nella politica europee.* Bologna: Il Mulino.

Gebremedhin, Ezra (2007). *Mäl'əkti Sälam. Encyclopaedia Aethiopica. Volume 3. He - N.* Wiesbaden: Harrasowitz, 695.

Gebremedhin, Ezra (2011). *Kenisha. The Roots and Development of the Evangelical Church of Eritrea, 1866-1935.* Trenton: Red Sea Press.

Godfrin, Jacqueline and Philippe (1965). *Une centrale de presse catholique. La Maison de la Bonne Presse et ses publications.* Paris: Presses Universitaires de France.

Green, Nile (2010). Persian Print and the Stanhope Revolution: Industrialization, Evangelicalism, and the Birth of Printing in Early Qajar Iran. *Comparative Studies of South Asia, Africa and the Middle East*, 30, 3, 473-490.

Guazzini, Federica (2024). *L'Eritrea nella Seconda guerra mondiale. Dinamiche di controllo e cambiamento.* Soveria Mannelli: Rubbettino.

Jwarson, Jonas and Tron, Alessandro (1918). *Notizie storiche e varie sulla missione evangelica svedese dell'Eritrea, 1866-1916.* Asmara: Missione evangelica svedese.

Labanca, Nicola (ed.) (2001). *Posti al sole. Diari e memorie di vita e di lavoro dalle colonie d'Africa.* Rovereto: Museo storico italiano della guerra.

Labanca, Nicola (2002). *Nelle colonie*, in P. Bevilacqua, A. De Clementi, E. Franzina (a cura di), *Storia dell'emigrazione italiana. 2. Arrivi.* Roma: Donzelli, 2002, 193-204.

Lazzarini, Vincenzo (1989). *Note storiche su gli inizi della stampa in Eritrea.* Typescript held at the Library of the Pavoni Social Center, Asmara.

Le Roux, Elizabeth (2012). Book history in the African world: The state of the discipline. *Book History*, 15, 1, 248-300.

Lundström, Karl Johan and Ezra Gebremedhin (2011). *Kenisha. The Roots and Development of the Evangelical Church of Eritrea, 1866-1935.* Trenton: Red Sea Press.

Lusini, Gianfrancesco (2021). Lingua e storia in Africa. Considerazioni sul caso del tigrino (Eritrea ed Etiopia), in F. Aiello, I. Micheli, A. Pensabene, M. Toscano (eds.). *Language and Identity. Theories and Experiences in Lexicography and Linguistic Policies in a Global World.* Trieste: Edizioni Università di Trieste, 157-173.

Metodio da Nembro (1953). *La missione dei Minori Cappuccini in Eritrea (1894-1952).* Roma: Institutum Historicum Ord. Fr. Min. Cap.

Nallino, Carlo Alfonso (1919). L'istruzione nelle colonie. In Italia, Commissione per il dopoguerra, *Relazione della 7. sezione della Commissione del dopo-guerra. Quistioni coloniali.* Roma: Tip. della Camera dei deputati, 125-137.

Natnael, F. (1993). Presented a paper dedicated to *Mäl'əkti Sälam* at The First International Symposium on the Tigrinya Language (Asmara, February 20-23, 1993).

Negash, Tekeste (1987). *Italian Colonialism in Eritrea, 1882-1941. Policies, Praxis and Impact.* Uppsala: Uppsala Universitet.

Negash, Tekeste (1997). *Eritrea and Ethiopia. The Federal Experience.* Uppsala: Nordiska Afrikainstitutet.

Nosnitsin, Denis (2007). Early Prints in Ethiopia and Eritrea. In Siegbert Uhlig (ed.) *Encyclopaedia Aethiopica. Volume 3. He - N.* Wiesbaden: Harrasowitz, 138-141.

Pankhurst, Richard (2010). Printing. In Siegbert Uhlig (ed.) *Encyclopaedia Aethiopica. Volume 4. O - X.* Wiesbaden: Harrasowitz, 217-218.

Peterson, Derek R., Emma Hunter, and Stephanie Newell (eds.) (2016). *African Print Cultures: Newspapers and Their Publics in the Twentieth Century.* Ann Arbor: University of Michigan Press.

Reese, Scott (2022). "The Ink of Excellence": Print and the Islamic Written Tradition of East Africa. In S. Reese (ed.), *Manuscript Tradition and Print in the Islamic Tradition.* Berlin-Boston: De Gruyter, 217-242.

Reta, Meseret Chekol (2013). *The Quest for Press Freedom. One Hundred Years of History of the Media in Ethiopia.* Lanham: University Press of America.

Stark, Ulrike (2022). Calligraphic Masterpiece, Mass-Produced Scripture: Early Qur'an Printing in Colonial India. In Scott Reese (ed.), *Manuscript Tradition and Print in the Islamic Tradition*, Berlin-Boston: De Gruyter, 141-180.

Tesfamariam, Temesgen (2023). Teaching and Preaching. Missionary Education and Colonial Subjects in Italian Eritrea (1890-1935. *Journal of Religion in Africa*, 3-4, 407-426.

Tesfamichael, Andemariam (2024). *Mons. Camillo Carrara primo vicario apostolico dell'Eritrea. Semplice memoria di un centenario 1924-2024*. Gorle: Velar.

Venosa, Joseph (2014). *Paths toward the Nation. Islam, Community, and Early Nationalist Mobilization in Eritrea, 1941-1961*. Athens: Ohio University Press.

Véronneau, Pierre (2003). Le Fascinateur et la Bonne Presse. Des médias catholiques pour publics francophones, 1895. *Mille huit cent quatre-vingt-quinze*, 40, http://journals.openedition.org/1895/3282.

Vlies, A. (2010). The History of the Book in Sub-Saharan Africa. In *The Oxford Companion to the Book*. Oxford: Oxford University Press, 313-320.

Voigt, Rainer (2012). 69. Tigrinya as National Language of Eritrea and Tigray. In Stefan Weninger (ed.), *the Semitic Languages. An International Handbook*. Berlin-Boston: De Gruyter Mouton, 1153-1169.

Zaccaria, Massimo (2021). From Incunabula to Book History: Ethiopia, Eritrea, and the Search for Their Printed Past. *Aethiopica. International Journal of Ethiopian and Eritrea Studies*, 24, 141-174.

The Impossible Eritrean Code: Mariano D'amelio, Law and Colonial Governance

Marco Fioravanti

University of Rome 'Tor Vergata';
e-mail: marco.fioravanti@uniroma2.it

Introduction

This article aims to address a period and topic that have been overlooked in legal studies concerning Italian colonialism in Eritrea. Overlooked between the interest in the early Italian settlements following the acquisition of Assab Bay in 1869 and the fascist era, characterized by the proclamation of Italian East Africa and the Empire, the period from the nineteenth to the twentieth century has been underestimated by historiography. This historiography has primarily focused on the social, institutional, and, in some cases, constitutional aspects of Italian colonialism in the Horn of Africa, while studies on codification are either absent or largely insufficient. This issue will be examined through the interpretative lens of a jurist like Mariano d'Amelio, who significantly influenced the evolution of legislation, jurisprudence, and the administration of justice in Italy and its colonies from the late nineteenth century to the first half of the twentieth century, always remaining at the top of the Kingdom's magistracies.

The story of this judge, endowed with precocious legal skills and a keen political sense, will be reconstructed through his numerous writings, particularly on colonial law, and through archival sources that legal historiography has often undervalued. By following the codification process in the Eritrean colony, we can identify the stages of colonial rule's establishment and the local populations' reactions. This unfinished process reveals much about the dynamics of domination between the colonial power and Eritrean resistance, as well as between Italian administrators and local elites, within the system of "legal apartheid" that Italy imposed upon its colonial subjects.

Through the prism of the "failed" Eritrean codification, this article attempts to reconstruct a broader world of compromises, perspectives, ideas, conflicts, and resistance between two peoples and two differing, though not always incompatible, conceptions of state and society.[1]

Mariano D'amelio, Overseas Magistrate

Residing in Eritrea from 1899 to 1905, where he served first as judge and then as president of the court in Massawa, d'Amelio also oversaw the reorganization of the colonial judicial system and the establishment of the Court of Appeal in Asmara, of which he became the first president.[2]

1 Locatelli, 2004; Tewolde, 2018; Deplano, 2019.

2 Saraceno, 1986; Martone, 2003; Rosoni, 2006; Giorgi, 2010; Auletta, 2013.

D'Amelio's judicial career – certainly not confined to his African experience – culminated in his appointment as First President of the Supreme Court of Cassation in 1923, at just over 50 years of age. His time in the colony not only entailed a considerable salary increase but also allowed the time spent in overseas service to count double for promotion purposes.[3] His political career was further solidified in the following year with his appointment as Senator of the Kingdom.

Thanks to the judicial roles he held in the colony, alongside other administrative positions (such as Director of Civil Affairs for the colony's government),[4] d'Amelio, leveraging his collaborative relationship with Governor Ferdinando Martini, sought to establish a more stable legal structure for the colony from juridical, political, economic, and religious perspectives. The governor, whose seat was in Asmara – effectively the colony's capital – was appointed by royal decree upon the proposal of the Minister of Foreign Affairs and after consultation with the Council of Ministers. He was accountable exclusively to the Minister of Foreign Affairs, directing the colony's policy, governing its population, and managing its budget according to received instructions, while the civil and military administrations operated under his oversight.[5]

As president of the commission tasked with drafting the Eritrean codes (civil, criminal, civil procedure, commercial),[6] d'Amelio aimed to establish a legal system for the colony based on sound principles, drafting the "Draft Civil Code for Publication in the Eritrean Colony" in 1905.[7]

He lamented the excessive emphasis placed on the juridical nature of the system, recalling how the regulation of settlements (Assab) and the colony was initially addressed through straightforward government interventions (via emergency decrees) and later by Parliament, beginning with the law of 24 May 1903, no. 205, which established the colonial system.[8] In this regard, he leaned towards viewing the colony as belonging to a distinct internal legal personality separate from that of the state.[9]

Although there was an evident close connection between the exercise of public powers in the mother country and the colony, the latter was recognized as an autonomous legal entity, capable of contracting, litigating, and inheriting. Article 765 of the Eritrean Civil Code stipulated that in the absence of designated heirs, the inheritance would devolve to the State's assets, while Article 7 of the Eritrean Commercial Code allowed the colonial public administration to engage in commercial activities without assuming the status of a merchant. Chapters I and II of Book II of the Eritrean Penal Code specified crimes against the security of the colony as distinct from those against the state's security. While Article 7 of the Italian Commercial Code stated that the State, Provinces, and Municipalities could not acquire the status of merchants but could engage in commerce, the Eritrean Commercial Code included the colonial public administration, with the accompanying Report stating: "it is quite natural that, like the State, the Provinces, and the Municipalities, the colonial administration should perform acts of commerce without acquiring the status of a merchant".[10]

---

3 Saraceno, 1986; Giorgi, 2010, p. 272.

4 Auletta, 2013 p. 636.

5 D'Amelio, 1911, p. 122.

6 Sagù, 1986; Martone, 2002, pp. 3-18; Rosoni, 2006, pp. 215 ss.; Martone, 2008, pp. 45-65.

7 Sagù,1986, pp. 576 ss.; Zaccaria, 2006, pp. 317-395; Mancuso, 2013, pp. 72-75; Giorgi, 2010; Panza, 2015, pp. 124 ss.

8 D'Amelio, 1911, p. 69.

9 Ivi, p. 90.

10 Ivi, p. 90 and n.

Upon his arrival in the colony, d'Amelio encountered a legal system established in 1886 that had remained in force until 1894, when Gennaro Celli, Deputy Attorney General at the Supreme Court in Rome, drafted a text incorporating certain laws relevant to the colony. However, this text was never formally approved by ministerial or governmental decree:

> When, with the establishment of the Court of Appeal in Asmara, the necessity arose to amend the Celli regulation, it became clear that no decree or authoritative measure had ever sanctioned the validity of this regulation. D'Amelio viewed this situation as emblematic of a tacit acceptance by those who found the law agreeable, as it respected local traditions.

The law of 1890 had already mandated the establishment of civil and penal legislation for the colony, yet in the years preceding the publication of the codes, Italian colonial legislation amounted to a few inconsequential interventions, presuming that the laws of the kingdom automatically applied in the colony, a notion repeatedly disproven by case law.[11]

Regarding the judicial organization, d'Amelio noted upon his arrival in the colony that:

> For eight years, until 1894, when the new order was published, Celli's work served as the procedural norm regulating competence and judgment forms. There were instances where judgments were annulled for violating provisions of the Celli regulation, without anyone questioning who had approved those provisions and what authority they held. This stands as the highest praise for that initial judicial order, which prevailed due to its intrinsic merit and was universally accepted as having the force of law.[12]

The Administrative and Judicial Organization of the Colony

Following the purchase of Assab Bay by the Rubattino company in 1869 and the subsequent takeover by the Italian government in 1882, English caution emerged, tolerating Italian commercial interests but not political ones in the Red Sea, which had become one of the world's most strategic locations following the opening of the Suez Canal and served as Britain's communication route to the Indies. Britain strategically encouraged Italian settlements in Massawa to counterbalance French presence in the Horn of Africa, while in West Africa, it supported Italy to mitigate Paris's hegemony over Tangier.

Having overcome Egyptian resistance, which initially claimed territories behind Assab Bay based on agreements between the Sultan of Constantinople and Egypt, Italy recognized Turkish-Egyptian sovereignty over the remaining western Red Sea coast. Jurisdiction over Assab Bay, acquired by the Rubattino company,[13] became effective on 5 July 1882 by Law No. 857. Article 3 addressed the relationship between colonizers and the colonized, invoking the medieval principle of personality of law: for Italians, the codes and laws of the Kingdom applied, while for the natives, their customary laws were to be respected:

> Their religious beliefs and practices will be honoured. Their personal status, family relations, marriages, successions, and all private law relationships will be governed by the customary legislation previously in effect for them, provided that such legislation does not contradict universal morality and public order, nor is overridden by express provisions. Legislation shall be applied towards the natives in these matters, and in judgments occurring between them without the involvement or interest of

11 Rosoni, 2006, pp. 216 ss.

12 D'Amelio, 1911, p. 172.

13 Scovazzi, 1996.

other Italian or foreign individuals, by a magistrate who is a scholar of Muslim law (cadí); this magistrate shall be appointed by the Royal Commissioner and shall administer justice in the name of the King of Italy.[14]

This principle established a differentiated legal regime, where commands and sanctions varied according to one's personal status, allowing various rights to coexist within the same territory based on the model of personality of law, thereby reinstating singular law and the privileges of commune, which had been supplanted by European legal systems for centuries.

The doctrine was divided on this matter, although the thesis of the personality of colonial law prevailed. Umberto Borsi, a foundational figure in Italian colonial law between the liberal and fascist periods[15], argued that the sources of colonial law comprised only general laws, metropolitan laws extended to the colonies, and colonial laws. Indigenous law had no place in this framework[16]. Emanuele Ghersi, also a professor of colonial law at the University of Florence in the 1930s, echoed this sentiment, asserting that indigenous law could not be classified as colonial since it pertained to relations unrelated to the phenomenon of colonization.[17]

Consequently, the dominant approach among colonial powers regarding the administration of justice was to retain, even if partially, indigenous customary institutions, systematically collecting not so much the substantive law of indigenous peoples but rather the judgments issued by indigenous courts, thereby recognizing colonial autonomy more in terms of procedural than substantive law.[18]

From a penal perspective, the authority of the colonizing state was asserted through a legal system characterized by greater flexibility and adaptability, accommodating indigenous peoples who lived in feudal-like societies.[19] Thus, a differentiated judicial system persisted between colonizers and the colonized, irrespective of the civil code model. This represented an intermediate choice between the English model of indirect rule and the French model of assimilationist administration. The Italian doctrine of the time perceived the issue of indigenous justice almost exclusively as a matter of public order.[20]

The first legal act regulating the administration of the colony dates to 1880, when a ministerial order issued on 24 December stipulated that the commissioner was to administer justice according to the laws of the state in effect, while considering the needs inherent in the circumstances, character, religion, and customs of the residents.[21]

Law No. 857 of 5 July 1882 introduced the critical distinction – fundamental for colonial governance – between two subjects of law, each with different legal statuses and political capacities: Italian citizens, who were subject to the laws of the Kingdom and the Italian codes, and the natives, who would be governed by customary law unless it conflicted with principles of public order and universal morality. In response to objections that the norms in force in Italy, beginning with the Statute of the Kingdom, had not been published in the colony, d'Amelio countered that this argument lacked legal merit: "since there is no prece-

14 Volterra, 2008.
15 Borsi, 1917; Borsi, 1932.
16 Volterra, 2008.
17 Ghersi, 1942.
18 Mazza, 2013.
19 Martone, 2002, p. 28.
20 Volterra, 2008, p. 90.
21 Mori, 1914-1915, vol. I, p. 1.

dent in Italy for the formal extension of the statute; and just as in the successive annexations of Italian provinces, our fundamental charter was considered extended by plebiscites, in the colony it must be deemed imported by military occupation".[22]

"As for the natives of Eritrea" d'Amelio elaborated, "they are not considered citizens, but subjects, *cives non optimo jure*, according to the various laws in force, particularly the Civil Code."[23] While Article 1 of the Eritrean Civil Code stated that colonial subjects enjoyed civil rights modified by law, the accompanying Report, largely contributed to by d'Amelio, indicated that these modifications primarily pertained to personal status.

> They do not – d'Amelio wrote – constitute a set of civil rights inferior to those of Europeans. On the contrary, they are inspired by respect for indigenous peoples' religious beliefs, their secular traditions, and aim to facilitate their legal relations. As less evolved individuals, they require simpler, less formal laws, more generous protection from authority, and laws that align with natural law and their current social conditions. Thus, being subject to their special law does not denote inferiority but rather a legal status granting greater protection under the law. In this manner, the equity of the legal system is affirmed before them. If certain provisions necessitate special protection for the interests of European civilization, the thought of facilitating the natives in their legal affairs, including interactions with Europeans, is a recurrent theme throughout the codification process[24].

The natives were deprived of political rights, especially the right to vote, thus lacking political representation. Moreover, following this logic of inclusion/exclusion, colonial subjects residing in Italy would not become voters or enjoy the political rights of the kingdom. D'Amelio bluntly stated that they could not consider themselves Italian, as they were "ethnically and geographically of a different race."[25]

Napoleone Colajanni, one of the few critics of Italian colonial expansion, a former supporter of Giuseppe Garibaldi, a professor of statistics at the University of Palermo, and a ten-term deputy of the Republican Party, conducted a well-known and severe inquiry into Eritrea in 1891. This inquiry culminated in the same year's publication of his fundamental yet poorly received book, *Colonial Policy.*[26] He noted that the "psychic contagion" of the European civilizing mission had also affected Italy, which, driven by "greed for domination, intolerance of command, and military power"[27] embarked on a venture characterized by war and abuse.

Before d'Amelio's arrival in Eritrea, justice, administered by the military government, was aptly described by Colajanni, who characterized it as scandalously poor, marked by financial mismanagement and governmental inexperience: "justice and administration were in the hands of Livraghi and Cagnassi [accused of torturing, robbing, and executing eight hundred Eritrean notables without trial], backed by bandit groups who called themselves indigenous bands, irregulars, etc.; these sinister figures were protagonists in a drama whose final act unfolded in the Court of Assizes."[28]

In 1892, the prevailing enemy for the (few) critics of Abyssinia's occupation was militarism:

22 D'Amelio, 1911, p. 82.
23 Ivi, p. 85
24 *Relazione*, 1905, p. 9.
25 D'Amelio, 1911, p. 86n.
26 Colajanni, 1892.
27 Ivi, p. 52.
28 Ivi, p. 54.

> To civilize the natives by imposing our legal and administrative methods upon them would only create infinite artificial duties for us and for them, which they would be unable to fulfil, resulting in the necessity of inflicting daily indignities upon our subjects, undermining everything they hold dear. Such well-intentioned yet tormenting impositions would certainly be less tolerated than a rapacious and bloodthirsty tyranny, breeding hatred against us and potentially inciting dangerous uprisings more due to the passive complicity of the docile than the strength of the insurgents.[29]

This advice went unheeded, both in the immediate future and in the following decades. The colony would be governed through governmental rather than legislative acts, thus maintaining its state of exception – not geographically Italian but politically so. With the new legal system established in 1902, the military character that had permeated the 1894 regulation was softened in favour of civil administration. To accomplish this, Governor Martini commissioned d'Amelio to develop a new judicial system for the colony that would simplify procedures and provide the natives with justice administered by political authorities (9 February 1902, n. 51).[30] The "d'Amelio" regulation innovated upon the previous system by instituting a single judge as the foundation of the system, establishing a court of appeal in Asmara to hear appeals against single-judge decisions, abolishing military tribunal jurisdiction for specific crimes committed by natives and colonists, and recognizing the governor's authority to review criminal judgments against natives, even ex officio.[31] Overall, the 1902 regulation centralized justice within the judiciary, enhancing its role and placing it at the heart of the colonial experience.

Conversely, after d'Amelio's return to Italy and under a new governor, a new regulation was enacted in the colony, the Legal Order for Eritrea, approved by royal decree no. 325, dated 2 July 1908, under the administration of Governor Giuseppe Maria Salvago Raggi, who succeeded Martini that same year. This new order aimed to simplify the judiciary while granting greater autonomy to the administrative personnel in the colony. D'Amelio was highly critical of this shift, as it reversed his approach, delegating broader powers to the judiciary while allowing more discretion for the colonial administration.

Upon his arrival in the country, Salvago Raggi, with the 1908 judicial order (Ordinamento giuridico per l'Eritrea, 2 July 1908, n. 325), approved a text that sanctioned the supremacy of justice administered by government bodies over that administered by judges or courts, significantly increasing the governor's autonomy».[32] The new judicial order, having effectively abandoned the notion that codes could take effect in the colony, reaffirmed the differential regime more vehemently, sidelining the assimilationist idea. For the native population, justice was predominantly administered by political authorities (the Commissioners and Residents), with the Governor retaining jurisdictional functions. Excessive power was thus granted to administrative officials lacking adequate legal training, undermining the judicial courts' justice.[33]

Article 2 of the new order stated:

> An individual shall be considered a colonial subject if they are not Italian or a citizen of a foreign state but are native to the Colony or belong to tribes or lineages thereof. An individual belonging to a population from Africa or other regions of the Red Sea,

29 Ivi, pp. 55-56.
30 Rosoni, 2006, pp. 204 ss.
31 Martone, 2002, p. 66.
32 Zaccaria, 2006, p. 318.
33 Martone, 2002, pp. 159 ss.

who has worked or has been resident in the Colony for two uninterrupted years, shall also be considered a colonial subject. A foreigner from a population deemed to lack civilization equivalent to that of Europe shall be assimilated.[34]

D'Amelio strongly criticized the reformist choice of 1908 and the mentioned Article 2, arguing that it mistakenly equated Italian citizens with colonial subjects, particularly placing Indian and Egyptian citizens on the same level as "the semi-barbaric peoples of Abyssinia and the Muslim tribes of Eritrea."[35] However, this might have been the least serious limitation, as the removal of jurisdiction from the judicial authority for disputes in civil, commercial, or administrative matters involving private individuals and the colonial government to be decided by the Governor reinstated the infamous Article 48 of the 1894 order, which, in his view, represented a "significant regression in the legal order of Eritrea."[36]

As Luciano Martone observed, on one hand, an autonomous codification solution was sought through commissions for the promulgation of colonial codes, while on the other, the colonial administrative apparatus, bolstered by the law of 1908, continued to maintain a differentiated judicial system between colonized and colonizers, irrespective of the codification model.[37]

Italian jurisdiction was represented by the Conciliator (an Italian appointed by decree of the governor from among the colony's notables, upon the recommendation of the President of the Court of Appeal and the King's Prosecutor), the single regional judge (assisted by two assessors for criminal cases), the Court of Appeal in Asmara, which functioned as an Assize Court, and the Court of Cassation in Rome. Indigenous jurisdiction was constituted at the first instance by the "Cadì" or local chiefs for disputes in civil, criminal, and commercial matters between indigenous peoples; also at the first instance by the Regional Commissioners or Residents, assisted by indigenous elements and interpreters (but without the provision for legal representation) for cases that could not be adjudicated by the Cadì or local Chiefs due to the involvement of individuals from different tribes, religions, or regions. In the appeal stage, the Regional Commissioners or Residents judged with the Governor's right of review. Cases involving indigenous people under the jurisdiction of the Court of Assizes were judged by the Court of Appeal.

Regarding local customs, d'Amelio's stance appeared influenced by Édouard Lambert's theory, which at the time spurred debate within the French doctrine traditionally opposed to any approach other than the sovereignty of law[38]. According to Lambert, only jurisprudence or judicial practice could bestow upon customs the force of law.[39] The concept of living law, cherished by authors like Raymond Saleilles and François Gény, as jurisprudential rather than legislative law – *horribile dictu* in positivist France dominated by Raymond Carré de Malberg's teachings – could find its realization in colonial realities, perceived as "at a lower level of legal civilization", according to the terminology of the time.

This perspective reflects a certain understanding of the relationship between law, rights, and custom, prevalent even among authors sensitive to the significant changes underway, such as Santi Romano. His well-known formula for framing this relationship in the colonial context remains illuminating. According to the future president of the Council

34 Mori, 1914-1915, vol. VI, p. 233.
35 D'Amelio, 1911, p. 186.
36 Ivi, p. 184.
37 Martone, 2002, pp. 30-31.
38 Lambert, 1903.
39 D'Amelio, 1910, pp. 37-45.

of State and a scholar of colonial law – still largely understudied, with notable exceptions[40] – primitive peoples (as he termed them) or semi-barbarians (as d'Amelio described) do not perceive laws in a strict sense but rather customs: "primitive peoples indeed view justice as the highest and most concrete expression of authority. [...] More than in the legislator, the State for them is personified in the judge."[41]

The underlying assumption, present in both Italian and French literature, is that colonized populations are in a state of minority, conceptualized by influential figures in early twentieth-century European colonial sciences, such as Paul Leroy-Beaulieu. Colonized peoples are often likened to children in need of protection. An encyclopaedic economist, a liberal supporter of colonization, and a key interlocutor for Italian jurist and political deputy Attilio Brunialti, Leroy-Beaulieu described Algeria as having "barely emerged from the lengthy difficulties of its childhood."[42]

The "Failed" Eritrean Codification

From the onset, extending metropolitan codes to the colonies was deemed unfeasible; instead, there arose a need to adapt them to local customs, especially concerning the legal status of colonial subjects and their interactions with Italian citizens residing in the colony. Law No. 205 of 24 May 1903 mandated the promulgation of five codes for the colony within 18 months, through a commission composed of magistrates and lawyers residing in the colony, with the objective of adapting the codes (civil, commercial, and criminal) to the needs of colonial subjects[43]. The five codes aimed to represent a true Eritrean corpus juris.[44] Upon completion in 1905, the work was reviewed by the Colonial Council, which, involving additional jurists, concluded the codification process with the approval of the Council of Ministers and the promulgation between 1908 and 1909. However, the implementing decrees, which included provisions for translation into indigenous languages (Amharic and Arabic), never materialized.

The dual legislation for Italian citizens and Eritrean subjects mandated that native's resort to local customs for regulating interpersonal relationships in civil law, while criminal cases would rely on a special criminal law, based on local customs, except for modifications introduced by the governor (Art. 3 of the 1903 regulation). For Italian citizens and assimilated foreigners – a distinction that posed interpretative challenges – the application of national law and codes issued in the Metropolis was prescribed. The failure to publish the codes in local languages likely stemmed from the shared awareness, paradoxically held by both d'Amelio and Vittorio Scialoja, the Dioscuri of colonial codification, that in the colony, it was more prudent to allow local magistrates to navigate the coexistence of different legal regimes. "The path through which the Italian legislator sought to progressively achieve legal assimilation of the colony by the metropolis was not through legislation but through jurisprudence."[45]

According to Gennaro Mondaini, a scholar and interpreter of colonial law, the Eritrean codes did not come into effect in the colony:

40 Bascherini, 2013.

41 Romano, 1918, p. 196.

42 Leroy-Beaulieu, 1887.

43 Copy of Report No. 3975/15 dated 4 June 1905, addressed by the Government of the Eritrean Colony to the Ministry of Foreign Affairs, in «Archivio storico diplomatico del Ministero degli Affari Esteri», of Rome, Eritrea 1894-1909, 12/5, Fasc. 46.

44 Mondaini, 1927, p. 144.

45 Rosoni, 2006, pp. 217-218.

> Due to indigenous political considerations expressed by the governor of the time, Marquis Salvago Raggi, regarding the inappropriateness of applying to indigenous populations the provisions contained in the codes for regulating their legal relationships with Italian citizens and Europeans or assimilated individuals; provisions with which, conversely, all other codes formed an inseparable whole.[46]

In contrast, Scialoja, in his Report to the Colonial Council of 1909, raised numerous concerns regarding the necessity of codification, particularly civil codification:

> To ensure that the relationships of Italians in the colony are governed by the same legal principles that prevail in Italy, it is not essential – and experience over many years demonstrates this – to transport the cumbersome mass of our codes, which presuppose a social condition found in nations with centuries of history and a complex organization of civil, administrative, and judicial offices, which the Colony lacks. Instead, a few general rules sufficed to state that the laws applicable to them in Italy also apply to citizens, as already stipulated in a certain way by Art. 9 of the judicial system, and that for colonial subjects, their customs remain in force unless contrary to civilization, along with some rules to resolve potential conflicts arising from the dual legal regime and to regulate some of the most significant mixed relationships.[47]

Conclusions

According to d'Amelio himself, what he perceived as a mere suspension of the application of the codes (which, in truth, would have continued indefinitely) did not negate the significance of the colonial codification effort, which would remain a corpus juris containing an ideal order likely to influence judicial decisions.[48] However, the Civil Code – which its drafters deemed "indispensable" for a few hundred Italians living in the Colony – despite being approved by Royal Decree No. 589, dated 28 June 1909, and published in both the Official Compendium and the Official Bulletin of the Colony itself, never came into force due in part to the lack of translations into Arabic and Amharic. Except for the penal code, which also never took effect, all colonial codes were unenforced, and in practice, judges continued to supplement Italian legislation with local customs.

This represents a "melancholic and undignified epilogue"[49] to a narrative that began with commercial and financial dealings, colonial governance, and military abuses, ultimately culminating in legal failure.

References

Auletta, F. (2013). *D'Amelio, Mariano*, in *Dizionario biografico dei giuristi italiani*, diretto da I Birocchi, E. Cortese, A. Mattone, M.N. Miletti, Bologna, Il Mulino, vol. I, pp. 635-638.

Bascherini, G. (2013). *Ancora in tema di cultura giuridica e colonizzazione: prime note sul* Corso di diritto coloniale *di Santi Romano*, in "Giornale di Storia costituzionale", pp. 117-135.

Borsi, U. (1917). *Studi di diritto coloniale*, Torino.

Borsi, U. (1932). *Corso di diritto coloniale*, Padova.

Colajanni, N. (1892). *Politica coloniale*, Palermo, Carlo Clausen Editore.

D'Amelio, M. (1910). *La giurisprudenza etnologica e la revisione della teoria della consuetudine*, in "Rivista di diritto pubblico", pp. 37-45.

D'Amelio, M. (1911). *L'ordinamento giuridico della Colonia Eritrea*, estratto dalla *Enciclopedia Giuridica Italiana*, Milano.

Deplano, V. (2019). *Archivi d'Africa. Le carte dell'amministrazione coloniale in Italia e nei territori di nuova indipendenza*, in *Archivi sul confine. Cessioni territoriali e trasferimenti documentari a 70 anni dal Trattato di Parigi del 1947*, a cura di M. Gattullo, Ministero per i beni e le attività culturali e per il turismo, Direzione generale Archivi, pp. 41 ss.

Ghersi, E. (1942). *Diritto coloniale*, Milano, Cetim.

Giorgi, C. (2010). *Magistrati d'oltremare*, in "Studi storici", pp. 855-879

Lambert, É. (1903). *La fonction du droit civil comparé*, Paris, Giard & Briere.

Leroy-Beaulieu, P. (1887). *L'Algérie et la Tunisie*.

Locatelli, F. (2004). *The Archives of the Municipality and the High Court of Asmara, Eritrea: Discovering the Eritrea "Hidden from History"*, in "History in Africa", Vol. 31, pp. 469-478.

Mancuso, S. (2013). *Terra in Africa. Diritto fondiario eritreo*, Trieste, Edizioni Universitarie di Trieste.

Martone, L. (2002). *Giustizia coloniale. Modelli e prassi penale per i sudditi d'Africa dall'età giolittiana al fascismo*, Napoli, Jovene.

Martone, L. (2003). *Magistrati italiani nella colonia Eritrea Immagini d'Africa e riflessioni giuridiche (1886-1941)*, in *Amicitiae Pignus. Studi in ricordo di Adriano Cavanna*, Milano, Giuffrè, 3 voll., vol. II, pp. 1375 ss.

Martone, L. (2008). *Diritto d'oltremare: legge e ordine per le Colonie del Regno d'Italia*, Milano, Giuffrè.

Mazza, M. (2013). *Giustizia indigena e giustizia coloniale. Codificazione e formazione del giudice*, in «Il Politico», pp. 161-178.

Mondaini G. (1927). *Manuale di storia e legislazione coloniale del Regno d'Italia*. Parte I - *Storia coloniale*, Roma, Attilio Sampaolesi Editore.

Mori, A. (1914-1915). *Manuale di legislazione della Colonia Eritrea*, Roma, L'Universelle, vol. I.

Panza, V. (2015). *La colonizzazione giuridica dell'Eritrea: diritto coloniale tra scienza giuridica, antropologia, etnografia giuridica (1880-1912)*, Tesi di Dottorato, Università di Napoli.

Relazione al Codice civile eritreo (1905). Roma, Tip. Unione Coop.

Romano, S. (1918). *Corso di diritto coloniale*, Roma, Athenaeum.

Rosoni, I. (2006). *La Colonia Eritrea. La prima amministrazione coloniale italiana (1880-1912)*, Macerata, Eum.

Sagù, M. L. (1986). *Sui tentativi di codificazione per la Colonia Eritrea*, in "Clio", XXII, n. 4, pp. 567-616.

Saraceno, P. (1986). *La magistratura coloniale italiana*, in "Clio", pp. 275 ss.

Scialoja, V. (1909). *Relazione sul Codice civile per la colonia Eritrea*, Roma, Stamperia Reale D. Ripamonti.

Scovazzi, T. (1996). *Assab, Massaua, Ucciali, Adua. Gli strumenti giuridici del primo colonialismo italiano*, Torino, Giappichelli.

Tewolde, A. (2018). *The Archives of Eritrea as a Primary Source of Information for the Eritrean Cultural Heritage*, in "Rassegna di Studi Etiopici", 3° serie, Vol. 2 (49), pp. 11-22.

Volterra, A. (2008). *La giustizia coloniale in Eritrea tra diritto comune e diritto consuetudinario (con in appendice un repertorio del materiale conservato presso alcune istituzioni eritree con sede ad Asmara)*, in "Africa: Rivista trimestrale di studi e documentazione dell'Istituto per l'Africa e l'Oriente", pp. 82-107.

Zaccaria, M. (2006). *"Tu hai venduto la giustizia in colonia". Avvocati, giudici e coloni nell'Eritrea di Giuseppe Salvago Raggi (1907-1915)*, in "Africa. Rivista trimestrale di studi e documentazione dell'Istituto italiano per l'Africa e l'Oriente", pp. 317-395.

Archieving and Nation Building: RDC`S Role in Salvaging and Documenting

Azieb Tewold

Research and Documentation Center, Asmara, Eritrea;
e-mail: aztew777@gmail.com

Abstract

The production, conservation, and transmission of archival documents across generations serve as a cornerstone for reproducing collective national memory and fostering cultural identity. Eritrea`s history, marked by centuries of resistance against European colonizers and Ethiopian invasions and occupations, forged a unique nationalist ethos rooted in self-reliance and unity. However, international narratives on Eritrea remain dominated by colonial perspectives. This paper argues that establishing robust national archives is critical to preserving Eritrea`s historical authenticity, countering distorted narratives, and transmitting values to future generations. By analyzing the role of archives in nation-building, the paper highlights gaps in current archival practices, underscores the Research and Documentation Center`s (RDC) contribution as a de facto national archive, and proposes actionable steps for institutionalizing state archives.

Keywords: Armed Struggle, Eritrea, Generation, Self-Reliance

Introduction

Politically and technically, the production, preservation, and transmission of archival documents across generations serves as a vital mechanism for reproducing a collective national memory—an essential foundation for cultural identity. Eritrea has endured successive waves of European and non-European colonization, including nearly three decades under the harsh occupation of Ethiopia. In response, Eritrean society initially resisted these forces through spontaneous and disorganized efforts, which eventually evolved into a sophisticated and highly organized political and armed struggle.

Throughout this resistance, the Eritrean people forged a rich, shared history rooted in strong nationalist values and customs, characterized by principles of self-reliance and unity in diversity—an experience unparallel in most African contexts. Yet these historical realities have long lacked international visibility. Much of the content in academic journals, newspapers, and books written by foreign journalists and scholars—often consciously or unconsciously—acted as instruments of colonial interests, distorting the true narrative of Eritrean history. For example, in his book's subsection titled "Eritrea and the Dergue (1974-1978)", Machida highlights the global media's increased coverage of Eritrean events, though many portrayals were misleading. On May 16, 1976, Ethiopia proposed a nine-point plan to resolve the war in Eritrea peacefully. This offer was dismissed by the Eritrean liberation fronts, especially as simultaneous reports (such as one published in The New York Times on

May 11) indicated Ethiopian peasants were promised land grants and framed the conflict as a Christian "holy war" against Muslim Eritreans.[1]

To ensure the transmission of Eritrea`s hard-won national values—and to present its true historical narrative to the global stage—it is crucial to establish a robust national archival institution. While initial efforts toward founding such a system are laudable, the challenge remains significant. A key solution is to integrate the establishment of a national archive and library into Eritrea`s broader nation-building agenda. National archives safeguard historical assts and serve as essential stepping stones in cultivating national identity. The continuity of Eritrean nationalism relies on the intergenerational transfer of authentic historical experiences, including the struggles and resistance of its forebears.

This paper aims to examine the role of archive in the nation-building process and evaluate the contributions of the Research and Documentation Center (RDC) in realizing this vision. It offers a general assessment of the current state of public archives relative to their importance as a resource for nation-building and for Eritrean historians and writers. Moreover, the article discusses the significance of national archives as custodians of society`s collective memory and outlines their role in shaping national identity. It further highlights the urgency of repatriating relevant colonial-era archives, with particular attention to those from the Italian period.

Given that the Italians were among the first to introduce a modern economy, large-scale agroindustry, an extensive road network, and urban development in Eritrea, their records are indispensable reference for understanding national development. However, the lack of a centralized Eritrean archive has left scholars deprived of crucial sources not only from the Italian era, but also earlier periods under Turkish and Egyptian rule. To date, many works on Eritrean history—especially relating to colonial and liberation struggles—have been shaped by foreign or Ethiopian scholars and reflect external biases rooted in colonial mentalities. The genuine Eritrean perspective remains underrepresented.

As a provisional solution, RDC has emerged as a de facto national archive. Its efforts in salvaging abandoned colonial archives and preserving documents from the Armed Struggle are commendable. This paper explores RDC`s contributions in collecting and protecting multi-format documents, and how these efforts have shaped Eritrean Scholarship. This article is structured into six parts: the role of National archives in state-building, tradition of documentation in Eritrean society, RDC`s initiative in establishing National archives and national library, the importance of archives in nation building, current records management practices and their implication for the future, and conclusion.

National Archive

Archival records form the bedrock of a nation`s cultural identity and its public historical memory. The acquisition, organization, preservation, and accessibility of archives—along with support for all types of research and researcher—are core mandates of national archive and library. The absence of such institutions leads to irreparable cultural and historical loss. According to the International Council on Archive (ICA), archives constitute the memory of nations and societies, shape their identity, and are a cornerstone of the information society. By providing evidence of human action and transactions, archives support administration and underline the rights of individuals, organizations, and states.[2]

1 Machida, Robert, *Eritrea: the struggle for Independence*, New Jersey: The Red Sea Press, 1989), P.64

2 International Council on Archives (ICA), 2024, http://w.w.w.ica.org/index.php?plangue=eng

Eritrean society has been shaped by a dynamic internal history and repeated external interventions. Despite having endured various forms of colonialism, few scholarly works have been written or translated into Tigrigna or other Eritrean languages, leaving much of this history underexplored. Colonial-era books and archival records are crucial references for developing history textbooks and reading materials, making them essential to the nation-building process. Researchers must examine the impact of Eritrea`s colonial past—both positive and negative—on the country`s development to accurately document its legacy.

The creation of national archives in Eritrea today offers dual benefits: it supports ongoing state-building efforts and aids in the preservation of historical heritage that must be maintained and passed onto future generations as part of the collective memory.[3]

European success in archival development began notably with the French Revolution in 1789. As French legislation recognized, collective memory is pillar of nationalism. Following this model, national archives were later established in countries such as Holland (1802), Poland (1808), Prussia (1831), Italy (1874), the United States (1889), and Russia (1917).[4] Emphasizing governmental responsibility in archival preservation, Lord Mantagu of Beauliu once said "A nation must be judged on how it conserves its archives. Our future depends on our past, and we must make sure that is always available."[5]

History of Documentation Culture in Eritrea

Eritrea holds abundant valuable relics, manuscripts, graphics, icons, religious artifacts, and related cultural items—primarily safeguarded by churches, monasteries, mosques, and private individuals. Prior to colonialism, literacy was embedded within religious communities; priests wrote manuscripts on goat or sheep skins, and customary laws across Eritrean ethno-linguistic groups were recorded by literate elders.

Modern records management in Eritrea began with the Italian colonial administration (1885-1941), which stablished Eritrea as a distinct entity with defined borders and eight administrative regions. Archives and registries were introduced, modeled after Italian systems, which treated archives as classified repositories for inactive records. This marked a departure from the medieval European tradition, where state papers were royal property—secrete and inaccessible to the public.[6]

Under British rule (1941-1952), Eritrea maintained similar records practices. Even during the Ethiopian Federal period (1952-1962), records management continued systematically, including well-organized filing systems. However, after Ethiopia annexed Eritrea and dismantled the Federal Act, these systems collapsed, replaced by Ethiopian methods—arguably less effective than those used by European colonizers.

During the thirty-years Armed struggle for Independence (1961-1991), both the Eritrean Liberation Front (ELF) and Eritrean People`s Liberation Front (EPLF) placed high

3 Alessandro Triulzi, Istituto Universitario Orientale, Napoli, Istituto Italiano per l'Africa e l'oriente (Roma) on the occasion of conducted work shop on to promote Records and Archives Management Asmara from 3-5 September, 2001

4 Alessandro Triulzi, Istituto Universitario Orientale, Napoli, Istituto Italiano per l'Africa e l'oriente (Roma) on the occasion of conducted work shop on to promote Records and Archives Management Asmara from 3-5 September, 2001

5 Kitching, Christopher, "Archives: the Very Essence of our Heritage." *Chichester: Phillimmore*, 1996, 9.

6 Alessandro Triulzi, Istituto Universitario Orientale, Napoli, Istituto Italiano per l'Africa e l'oriente (Roma) on the occasion of conducted work shop on to promote Records and Archives Management Asmara from 3-5 September, 2001

importance on documentation, despite operating under difficult conditions. While not standardized, nearly all records were carefully preserved, including personal biographies and photographs of members. These archives hold immense value for researchers and are foundations to Eritrean nation-building. Their existence reflects the liberation movement`s consciousness regarding historical preservation, making the Eritrean struggle distinct among Africa`s liberation movements.

The leadership`s commitment to cultural documentation is captured in the National Charter, which states, "A national culture is the foundation of a developed economy and modern political system. No society has become modern without developing its culture."[7] The Charter emphasizes that during the liberation movement, fighters overcame sectarian divisions and cultivated a unified national culture built on the strengths of Eritrea`s diverse communities. This revolutionary culture was marked by self-reliance, confidence, determination, patriotism, ingenuity, heroism, and sacrifice—qualities indispensable to the success of Eritrea`s independence.[8]

Preserving this legacy in a national archive ensures its transmission to younger generations and serves as a guiding principle for the future. The National Charter further asserts that while Eritrea`s modern borders were established under Italian colonialism, the people Eritrea had rich civilizations, legal systems, and cultural identities long before colonization. The charters notes, "it is critical to understand the internal dynamics of Eritrean society, in addition to the legacies of the various colonizers."[9]

Therefore, exploring and documenting the legacy of colonial rule necessitates the diplomatic repatriation of archival materials—either through negotiation or through acquiring digital copies. These records are not only historical assets but also vital instruments of memory, identity, and sovereignty.

Research and Documentation Center (RDC), Eritrea's De Facto National Archives and Library

The Research and Documentation Center (RDC), currently operates as Eritrea`s de facto National Archives and library. Its long-term mission is to formally establish these institutions. Over the past three decades, RDC has dedicated itself to salvaging and preserving abandoned and scattered colonial archives—including records from Italian, British, Ethiopian, and Federal periods—with the goal of transmitting theme to future generations as vital components of Eritrea`s cultural identity and national consciousness.

During the Ethiopian occupation, many archival records were deliberately discarded in basements and in abandoned government buildings in an apparent effort to erase Eritrean history. Since its inception, RDC has worked tirelessly to rescue these materials, arranging and making them accessible to researchers, historians, and writers. "Commitment to recovering and preserving this heritage is increasingly urgent… No library can claim to possess even one complete collection of any periodical printed in Eritrea before 1941." Political turbulence during the colonial and post-colonial eras led to the disappearance of titles from Africa, while surviving copies found their way into European archives and libraries.[10]

7 People`s Front for Democracy and Justice, *A National Charter For Eritrea*: Approved by the Third Congress of the Eritrean People's Liberation Front (EPLF), Asmara, Adulis printing press, 1994, 30.

8 People`s Front for Democracy and Justice, *A National Charter For Eritrea*: Approved by the Third Congress of the Eritrean People's Liberation Front (EPLF), Asmara, Adulis printing press, 1994, 30.

9 People`s Front for Democracy and Justice, *A National Charter For Eritrea*: Approved by the Third Congress of the Eritrean People's Liberation Front (EPLF), Asmara, Adulis printing press, 1994, 3.

10 Recovering the African Printed Past Virtually Re-membering a dispersed collection in Eritrea by Mas

Despite these challenges, RDC has made remarkable progress in preserving written, audiovisual, photographic, cartographic, and other historical records from the colonial era. Materials from the Armed struggle period—including photographs, audio recordings, video footage, and documents—have been digitized through RDC`s own initiatives. Collaboration with others such as the Mari Curie Fellowship[11] and a current partnership with the University of Pavia under the project "Digital Technologies and Cultural Heritage Preservation in Eritrea"[12] have enabled the scanning of published Eritrean materials both domestically and abroad. This work is being carried out in phases; 1867-1941 and 1941-1962.

Beyond archival recovery, RDC actively promotes the importance of robust records management. It conducts awareness seminars for senior managers of public institutions and offers short-term training for employee handling records and archives. RDC also provides consulting services on request. However, due to the lack of standardized procedures, regulations, and guidelines on records disposition, consistent records management across government offices remains unaccomplished—endangering the preservation of the nation`s future memory.

In 2015, the government of Eritrea issued Proclamation No. 177/2015, "The cultural and Natural Heritage proclamation." It defines a library as, "A collection of written, printed or otherwise recorded materials including—but not limited to—books, manuscripts, pamphlets, posters, photographs, motion pictures, videotapes, sound recordings, and computer databases, as well as the facility housing them and the institutions responsible for their maintenance."[13] While the proclamation acknowledges the cultural importance of libraries and encourages their establishment, it does not address the need for state archives. This highlights a limited public understanding of the distinct roles of libraries and archives.

Moreover, the archival records salvaged from various colonial and federal eras pose challenges. Their original arrangements in disrupted, and some records are missing due to the absence of filing systems. As kitching Christopher observes:

> One good reason for using original archival sources in the study of history is that the evidence must be carefully assessed within the context in which the archives were created: Why and by whom the information was recorded, how it was recorded, and whether it represents incontrovertible fact, the writer`s opinion or preconception, or perhaps even a deliberate attempt to deceive."[14]

Notwithstanding these limitations, historian Alemseged Tesfai illustrates the immense value of these salvaged archives. His three seminal books--ኣይንፈላለ (Let us not separate 1941-1950), ፈደርሽን ኤርትራ ምስ ኢትዮጵያ ካብ ተድላ ክሳብ ኣንዞማተዮንሶ (Federation of Eritrea with Ethiopia from Tedla to Anzo Matinzo 1951-1955), and ኤርትራ ካብ ፌደርሰሽን ናይ ጓቦጣን ሰውራን (Eritrea from Federation to Annexation and Revolution 1956-1962)—were entirely based on salvaged archival materials. Tesfai emphasized the uniqueness of the Federal period ar-

simo Zaccaria ....Chapter 7 ....The Written World in Africa p. 151-152 2014

11 The project's title was "Developing New Approaches to the preservation and sharing of Africa's 'Africa's Printed Past: The Eritrean case'. It obtained a Marie Curie International Outgoing Fellowship for the period 2009-11 (proposal number 236699).

12 Joint research activities between the University of Pavia and the Research and Documentation Centre began in 2012, when the two institutions signed a Memorandum of Understanding. Since then, the University of Pavia has hosted Eritrean researchers and participated in two joint projects funded by the EU

13 Government of Eritrea. Proclamation No.177/2015. The cultural and Natural heritage proclamation. Sabur printing service, 2015, 4

14 Kitching, Christopher, "Archives: the very Essence of our Heritage." *Chichester: Phillimmore*, 1996, 18-19.

chives, stating "almost all parliamentary members` records are available. The contents provided factual evidence of the efforts made by political figures and the Eritrean people during that time."[15]

He further argues that historians who previously wrote about the Federation period without consulting these archives could not fully grasp the enthusiasm of the Eritrean people. These once—inaccessible records—discarded in basements and scattered throughout the country—now served as invaluable sources for Eritrean historians seeking to reclaim their narrative. Tesfai urges young scholars to follow in the footsteps and use these archives to write Eritrea`s history from within.[16]

Archives as Heritage in Nation Building

As a young nation, Eritrea must prioritize the establishment of state archives and the repatriation of colonial era records. Nation-building encompasses cultural identity, historical continuity, and societal transformation, all of which rely on access to past and present information. Repatriated colonial archives can provide immediate benefits by offering critical insights that strengthen national identity and inform developmental planning.

The absence of materials from the colonial period has two major consequences. First, it limits scholarly engagement with Eritrea`s historical narrative, hindering a comprehensive assessment of its past and the growth of a unified national identity. Second, the lack of documentation on urban infrastructure—such as sewage networks, water systems, and road planning—affects public health and the delivery of basic services.

According to Palma, materials from Eritrea, Italy`s first and longest-lasting colony, were prominent in the archive—not limited to documentary content. The map library housed Asmara`s master plans, town planning schemes and amendments, and public works maps from Eritrea`s colonial offices, unavailable elsewhere.[17] In her article "colonial Archives, Memory and Political Culture in Italy: The Italo-Eritrean Historic Heritage Workshop," Palma outlines efforts by African nations to decolonize archival memory.

> The Struggle for independence in Africa also entailed a re-appropriation of history. In the 1960s, colonial archives were the object of disputes and claims that in some cases became real `wars for archives`, such as that which occurred in France between 1980 and 1981 for the Algerian colonial records stored at the Aix-en-Provence Overseas Archive—a significant example of the real and symbolic weight that archives can have as sites of collective memory. [18]

She emphasizes the necessity of archival repatriation, the new states needed to reclaim their history: precolonial history erased by Eurocentric narratives of an `Africa without history,`and colonial history—including the dynamics it had triggered.[19]

Archives and Records Management Program (RMP)

Compared to archival studies, records management is a newer and continually evolving dis-

15 Alemseged Tesfay, Individual Interview, March 2024

16 Ibid

17 Silvana Palma "Colonial Archives, Memory and Political Culture in Italy", *The Italo-Eritrea Historic Heritage workshop,* 1987

18 Silvana Palma "Colonial Archives, Memory and Political Culture in Italy", *The Italo-Eritrea Historic Heritage workshop,* 1987

19 Silvana Palma "Colonial Archives, Memory and Political Culture in Italy", *The Italo-Eritrea Historic Heritage workshop,* 1987

cipline—particularly shaped by advances in information Technology (IT). There are three core values that make a record indispensable: (i) operational continuity, when records help organizations take informed actions and manage current operations smoothly; (ii) accountability, whereby records serve as proof of adherence to legal, policy, and best-practice standards. Records are vital for audits, regulatory compliance, and safeguarding rights; (iii) cultural and historical values, that records promote research, cultivate organizational memory, and enhance public awareness of institutional history.[20]

Record management also supports transparency in public service and reinforces the protection of individual rights. For organizations lacking institutional knowledge among their staff, archives become the only reliable sources of corporate memory. It enables strategic decision-making and preserving continuity. Historical records also answer specific factual inquires, which are critical in areas such as architecture, engineering, and legal claims. Maps and plans can have many practical uses—from verifying rights-of-way, identifying dangerous old mines, or revealing structural features of buildings. Architects often request old plans when renovating historic structures.[21]

Emmerson highlights the need for strategic integration of records management by saying, "It is important to devise a strategy for the introduction of records management in to organization."[22] This is especially crucial in an era where digital systems dominate. The power of computers to store, process, and retrieve vast amounts of information makes them indispensable in records management. The widespread adoption of IT is no longer optional. It is essential to global communication and heritage preservation.

Information professionals must proactively ensure that today`s digital records remain accessible and trustworthy, safeguarding tomorrow`s cultural legacy. Establishing a strong records management system is therefore vital not only for operational efficiency but also for preserving institutional and national history.

Conclusion

The paper has emphasized the critical role of salvaging archival records and repatriating colonial archives to meaningfully narrate Eritrea`s history. It underlines the significance of state archives as custodians of national memory, tasked with preserving and transmitting that memory onto future generations. The article explores global precedents in state archives establishment and examines the RDC`s pivotal contribution as Eritrea`s de facto archive. It highlights how RDC salvaged archives from Italian and British periods, assesses their present value, and discusses the pressing need for formal legislation and infrastructure to support archival development.

Furthermore, the paper stresses the importance of implementing proper records management programs (RMPs) across public institutions. Guided by clear policy, strategy, and standards, such programs preserve organizational memory and ensure a smooth transition into future archival efforts. By reinforcing the role of national archives in supporting researchers and challenging distorted historical narratives, the paper calls for increased awareness, reform, and urgent action. Public institutions must prioritize proper records management to protect vital documents, build institutional memory, and contribute meaningfully to the nation-building process.

---

20 Shepherd, Elizabeth and Geoffrey Yeo. *Managing Records: A Handbook of Principles and Practice.* (London: Facet Publishing, 2003) pp. xi-xiii.

21 Kitching, Christopher, "Archives: the very Essence of our Heritage." *Chichester: Phillimmore*, 1996.10

22 Emmerson, Peter. *How to Manage Your Records: A guide to Effective* Practice. Hertfordshire: ICSA Publishing Ltd, 1989. p.11

Reference

Alessandro Triulzi. Istituto Universitario Orientale, Napoli, Istituto Italiano per l'Africa e l'oriente (Roma) on the occasion of conducted work shop on to promote Records and Archives Management Asmara from 3-5 September, 2001.

Christopher, Kitching. "Archives: the very Essence of our Heritage." *Chichester: Phillimmore*, 1996.

Emmerson, Peter. *How to Manage Your Records: A guide to Effective* Practice. Hertfordshire: ICSA Publishing Ltd, 1989. p.11.

Government of Eritrea. Proclamation No.177/2015. The cultural and Natural heritage proclamation. Sabur printing service, 2015. http://w.w.w.ica.org/index.php?plangue=eng

Machida, Robert. *Eritrea: the struggle for Independence*, New Jersey: The Red Sea Press, 1989), P.64.

People`s Front for Democracy and Justice. *A National Charter For Eritrea*: Approved by the Third Congress of the Eritrean People's Liberation Front (EPLF), Asmara, Adulis printing press, 1994.

Zaccaria, Massimo, "Recovering the African Printed Past Virtually Re-membering a dispersed collection in Eritrea" *The Written World in Africa*, 2014.

Shepherd, Elizabeth and Geoffrey Yeo. *Managing Records: A Handbook of Principles and Practice.* (London: Facet Publishing, 2003) pp. xi-xiii.

Silvana Palma. "Colonial Archives, Memory and Political Culture in Italy", *The Italo-Eritrea Historic Heritage workshop,* 1987

Economics And Development

The Economic Impact of Microfinance Institutions on Improving the Welfare of their Clients: The Case of SMCP-Eritrea

Habteab Tekie[1] and Dawit Teclemariam Bahta[2]

[1, 2] College of Business and Social Sciences, Adi Keiyh, Eritrea
E-mail- dawitt7@gmail.com

Abstract

Microfinance institutions supply loans, savings and other financial services to the needy and other low-income clients on a sustainable basis and play a role in the alleviation of poverty. The objective of this article is to examine the impact of Saving and Micro-Credit Program (SMCP) in improving the livelihood of its clients. SMCP is one of the two microfinance institutions active in the provision of micro loans in Eritrea. The target population was all those whom credit was forwarded by the SMCP. The sample has covered beneficiaries from six Zobas and 56 out of 58 Sub-Zobas of Eritrea. Two-stage stratified cluster sampling method was carried out to select the 550 individual respondents who were clients of the program. A sample design was prepared to collect quantitative data from clients and use it in a quasi-experimental data analysis and appraise groups to define a counterfactual outcome. In order to accomplish such an objective and compare the impact of the program on clients', three alternative approaches were used. In the first approach comparison was made by dividing the 550 clients into senior-long time (n= 461) and new clients (n= 89). The second approach took into consideration the amount of loan cycle a client has reached. Here we compare clients who are in the early loan cycles with clients who have reached higher loan cycles. The third and final approach compares Tier one (group organized) clients with tier two (individual) clients. To realize the objective, three different research questions and appropriate hypotheses were employed to be tested.

Through the help of nonparametric independent samples test, namely Mann-Whitney U and Kruskal-Wallis K tests the paper has proved that there is statistically significant difference in monthly mean income between the groups of clients. Therefore, the study has traced that the mean income between the groups is different, proving that the program has a positive economic impact on clients.

Keywords: impact, nonparametric test; mean income

Introduction

SMCP aims to provide the needy access to financial services, regardless of their social-economic background. SMCP focuses on two groups of clients: economically deprived people and women. Via the use of these relatively small loans into their financial management practices, clients are expected to consistently improve their business performance and increase their household welfare. Thus, by providing access to financial resources SMCP aims to contribute to local and rural development in Eritrea. The object of this study is to ascertain whether the activities (products and services) of SMCP have positive economic influence (impact) on its beneficiaries.

The research design is based on analysis of data by using Mann-Whitney *U* and Kruskal-Wallis Test. The Kruskal-Wallis test, an extension of the Mann-Whitney *U* test, is the nonparametric analog of one-way analysis of variance and detects differences in distribution location and does not assume normality or homogeneity of variance. This statistical tool is used as an alternative to the traditional ANOVA which is based on the assumption of equal variances of the group population and the data at hand violates the homogeneity of variance assumption.

The data used in this study is collected by using a survey questionnaire and provides information on the direction and magnitude of impacts on clients. In addition, qualitative data is also collected through the use of focus group discussions and key informant interviews. The finding of this qualitative data is used to augment the questionnaire used in the survey.

The sample design for the survey is prepared to collect quantitative data from clients and use it in a quasi-experimental data analysis and appraise groups to define a counterfactual outcome. In order to accomplish this objective and compare the impact of the program on clients', three alternative approaches are used. In the first approach comparison is made by dividing the 550 clients into senior-long time (n= 461) clients and new clients (n= 89). In this case clients are clustered into two groups. The second approach takes into consideration the amount of loan cycle a client has reached. Here we compare clients who are in the early loan cycles with clients who have reached higher loan cycles. In this approach the clients are clustered into three groups. The third and final approach compares Tier one (group organized) clients with tier two (individual) clients'. In this case clients are clustered into two groups.

Using these variables and with a help of statistical package, attempts have been done to give answers to the three different hypotheses and research questions. The general inquiry that applies to all the three different research questions and hypotheses is whether a statistically significant monthly mean income difference exists between the cluster groups. Explicitly the study investigates whether clients from different groups have the same monthly mean incomes or not. In summary the study endeavors to investigate whether the loans forwarded by SMCP led to income growth of its clients or not.

Background of SMCP

This study was carried out among clients of SMCP. The mission of SMCP is to give the needy access to financial services, independent of their social-economic background. SMCP focuses on two client groups: economically deprived people and women. By providing access to financial resources SMCP aims to contribute to local and rural development in Eritrea.

The SMCP was active since 1996 as part of the Eritrean community development fund (ECDF). During the first five years, it had been working under the auspices of the Ministry of Local Government (MoLG). Since 2002 SMCP become an autonomous unit operating under the Ministry of Local Government (MoLG). Nowadays the sole support of the program comes from the government of Eritrea.

The aim of the SMCP is to provide financial services to the vulnerable groups in both rural and urban areas who have no access to formal banking services. Its long- term objective is to promote the private sector in Eritrea by encouraging the establishment and expansion of micro and small enterprise managed by individuals or groups to increase their income generating ability and make them contribute their part in national development.

The program has three operational strategies to achieve its objectives. The first strategy is to provide access to saving and credit to individuals who are currently unable to get access to credit from formal banks. The second strategy is to strengthen community representative structures, from village to higher levels, and to involve communities in the development and sustainability of the program. It also endeavors to establish a legal, regulatory and judicial framework for the microfinance sector of Eritrea so that the SMCP can become an autonomous financial institution that is organizationally and financially sustainable.

Over the last twenty-eight years SMCP has been providing two loan categories. In the first category, borrowing groups known as Solidarity Groups (SGs) organized in Village Banks (VBs) form the backbone of the program. This category of borrowers belongs to Tier I. These borrowers have to organize themselves into groups of three to seven members in order to acquire loans from the program. Generally, the beneficiary groups that fall into this category will obtain short-term microloans that range in loan size from ERN 6,000 to ERN 20,000. Over 50 percent of the SMCP clients fall into this category.

On the other hand, individuals who graduated from the group-based lending and entrepreneurs whose credit requirements cannot be met through the Tier I facility can have access to a Tier II window. This window should enable individual clients, without recourse to collateral, to borrow starting from ERN 7,000 up to ERN 100,000. The interest rate is set at 16% for all products calculated in a declining balance amortization method.

Eritrea is divided into six administrative zobas (Zones), 58 subzobas with 2,606 village clusters, organized into 701 legally registered administration villages (*Kebabis*). Nowadays SMCP covers all administrative zobas and subzobas. As of the end of 2023, there were around 75,000 active clients of which women clients comprised 52% (which is around 39000). During the same period there were 538 village banks in all six regions of the country spread over 58 subzobas.

Some previous impact assessments studies conducted also attempted to measure the impact of the program on target clients. Generally, these studies agree that SMCP was making inroads in mitigating its clients' economic hardships and improving the welfare of their households.

Literature Review

Microfinance is widely known as a provision of financial services such as credit, saving, deposit, insurance and repayment services to those who are deprived of accessing conventional financial services because they are underprivileged and they cannot offer collateral (Littlefield, Murduch, and Hashemi 2003). Since its inception in Bangladesh, during the mid-1970s, microfinance institutions have begun to establish themselves and expand their operations in both Bangladesh and throughout the world. In 2018, there were around 27 million active borrowers in Bangladesh alone (Global Outreach Financial Benchmark Report, 2018). Among the major microfinance institutions (MFIs) in Bangladesh the Grameen Bank (GB), and the Bangladesh Rural Advancement Committee (BRAC) are the pioneers of microfinance service providers.

In Africa most MFIs are replication of the Bangladesh financial institutions especially that of the Grameen Bank. These institutions which provide financial needs to low-income clients include, non-bank financial institutions, non-governmental organizations, rural banks etc. In Africa MFIs aim at growth and reaching out to potential clients who are ignored by the mainstream financial service providers (Lafourcade et al. 2005). A study conducted by

Lafourcade et al. (2005) across African regions show that MFIs in East Africa are most profitable, and those in West Africa generate positive revenues, whereas MFIs in Central and Southern Africa generate negative returns. Moreover, Rooyen et al. (2012), stated that microfinance play an important role on poverty reduction and socioeconomic development in Sub-Sahara Africa countries

Enormous empirical literatures have been done to evaluate the nature and effects of microfinance on livelihood of its clients. Generally, microfinance is recognized as an effective tool to fight against poverty alleviation and improvement in socioeconomic wellbeing of clients. However, studies also show conflicting effects ranging from positive to no impact and even to negative impacts. However, many studies on the impact of microfinance, revealed positive impact on poverty alleviation and improvement of livelihood. According to Morris and Barnes (2005) the Uganda microfinance has a positive effect on the rural clients "*households*" income diversification and assets accumulation. The Zimbabwe microfinance has positive impact on poverty reduction and the average income of microfinance clients was more than the average income of new clients or non-clients (Morduch and Graduate 2002). Aaron and Williams (2018), using a sample size of 380 clients from various microfinance institutions (MFIs) within the research area in Ghana and a survey with descriptive analysis, found that the clients enjoy opportunities for savings which enable them to increase their income levels by way of using their savings to acquire basic necessities of tools, equipment and materials and are able to create employment and improve their standard of living. Erulkar and Chong (2005) applied a longitudinal dataset covering 300 borrowing households matched with a comparison group from the same neighborhoods in urban Kenya, and reported significant positive impact of microfinance on per capita income. Borrowing households reported statistically significant higher incomes in the order of 20 per cent relative to the control group. Using adjusted risk ratios (ARR) and a cluster randomized control trial (RCT), Kim et al. (2009) in South Africa found significant improvements in the capacity of borrowers to meet their basic consumption needs relative to the control group.

Dadson et al. (2012) conducted study in Ghana covering 300 respondents from five districts to measure the impact of group lending on women's income from business activities. By utilizing average treatment effects on the treated estimates and based on propensity scores matching (ATT-PSM), they found that female borrowers reported statistically significant higher incomes relative to the control group. Similarly, in Mali, Koloma (2013) opted PSM estimators and found that access to various microfinance credit programs produced positive and significant impacts on poverty reduction, with women exhibiting higher effects than men. According to Crepon et al. (2015) in rural Morocco, a microfinance institution named Al Amana generated a significant increase in asset investment, including livestock used for self-employment activities.

Sayed et al. (2015) used multinomial logistic to study impact of microfinance on poverty reduction in Malaysia and found that Amanah Ikhtiar microfinance institution has positive effect on household income of women clients who spent three years in the scheme as compared to new clients who have not received treatment. According to Ghaliba et al. (2014) microfinance in rural Pakistan has positive effect on poverty alleviation which was exhibited in household income and expenditure particularly in clothing and health. Similarly, Khandker (2005) used panel data in Bangladesh and microfinance was found to have positive impact on poverty reduction and household expenditure particularly on food and non-food ingredients. Yunus (2007) stated that the surplus income that microfinance provides empowers people to lift themselves, and their families out of poverty. This is pertinent as it exhibits how microfinance programs can have a compounding effect on a client's entire household welfare and living standards. Mazumder and Lu (2015) conducted a study with the primary focus of examining the impact of microfinance programs in Bangladesh on rural livelihood. By utilizing a series of data collection and regression analysis and propensity score matching methods, they established that microfinance improved the clients quality of life by enabling them access to higher quality nutrition, education, and health facilities. James et al. (2009) evaluating effect of microfinance institutions in Guatemalan using univariate and multivariate analyses and their result showed that MFIs do produce a measure of improvement in the lives of microfinance clients. This improvement is concentrated along the social dimensions of housing, health, and client empowerment.

Recent study in Bangladesh revealed that microfinance, played a dynamic role to decrease poverty in the selected study areas by income generating activities of the poor women clients and by improving their living standard. The study confirmed that microfinance has positive effect on income, expenditure, condition of dwelling house, education, health and decision-making ability of the poor women clients who spent at least five years in microfinance institutions comparing to the non-borrowers (Shakina: 2019). Islam, Hossain and Afrin (2014) investigated the impact of microfinance on poverty reduction from 100 recipients of microfinance institutions using structured questionnaire and found that MFIs play vital role to reduce poverty by generating employment and improving living standard by better education, health facilities and higher health expenditure. Khandker et al. (2015) used panel data of households covering 87 villages of rural Bangladesh and estimated long run effects of microcredit on household per capita income, expenditures, poverty, non-land assets, household net-worth, male and female labor supply and schooling of children. Kondo et al., (2008) conducted study to evaluate the impact of Rural Microenterprise Finance Project (RMFP) in the Philippines by applying the difference-in-difference estimation method. Study result showed that a slightly significant positive effect on per capita income, per capita total expenditure and per capita food expenditure which indicated that microfinance is an effective tool for poverty-alleviation and improving livelihoods of participants. Neponen (2003) used Internal Learning System (ILS) technique to collect data and to keep track of the changes in clients living situations. In the ILS study, he found better results for long term members, that is 64 percent lived in tile roof and concrete houses, compared to only 50 percent of newer members; 80 percent of their sons and their daughters of long-term members attended school compared to 74 percent of sons and 65 percent of daughters of newer members. Cintina and Love (2017) used PSM to evaluate the effectiveness of microfinance. In their study they set up two comparison group i.e. comparing MFI borrowers to those without any other loan and to those with other type of loans and found a positive significant effect on housing improvement and in purchasing durable goods: in both cases while the difference in temptation goods and festivals was slightly lower as compared to MFI borrowers with other type of loans. According to Haque et al. (2017), borrowers who completed at least three loan cycles, revealed that microfinance program has positive and significant impact on their household income, expenditure and saving. These various studies significantly show that enormous literatures agree on the impact of microfinance on the improvement of local economy and livelihood of clients.

Methodology

In this study data is collected at the beneficiaries' level to appropriately address the study objectives and in gathering both primary and secondary information was required. Specifically, three key research methods are used in the collection of primary data, namely survey, focus group discussion and interviewing key informants. Primary quantitative data was collected through survey from beneficiaries in sample sub-zobas. Qualitative data was collected using in-depth key informant interview and focus group discussions. Moreover, the study used secondary data acquired from the archives of SMCP. The study was conducted over a period of six months. While the study primarily depended on the quantitative survey data, the secondary and qualitative data gathered are used to develop and enrich the quantitative data.

The target population was all those people to whom credit was forwarded by the SMCP. The sample has covered beneficiaries from five out of six Zobas. The sample size was set to ensure a statistically significant sub-set of program participants in each analytical category of interest. Analytical categories were the categories used to analyze the survey results. In 2015 SMCP had branches in 56 out of 58 sub-zobas, and about half of the participants were females. As the SMCP documents indicated in 2015 there were about 52,300 beneficiaries. During the same period, there were 538 village banks in all the six zobas of the country spread over 56 sub-zobas.

The analytical categories selected for this survey were zoba and gender. This means the survey has included statistically significant number of program participants for each zoba*gender combination. Thus, the survey has 12 analytical categories (six zobas* two gender categories). Thus, our sample size became 600 i.e., (12 analytical categories*56 participants). However, the study is forced to discard 50 because of missing information.

A two-stage stratified cluster sampling method was carried out to select the 600 individual respondents. Cluster sampling technique was used to select the participants first, by dividing our sampling frame into clusters of sub-zobas. Every sub-zoba in the survey area was assigned a measure of size equal to the total number of beneficiaries in that sub-zoba. Then a sample of sub-zobas with a predetermined sample size is selected independently with probability proportional to this measure of size. Then we selected a sample of beneficiaries from the selected sub-zobas. The sample comprises about 48% male and 52% female beneficiaries.

The appropriate way of drawing out a counterfactual design requires the use of baseline data. However, during the study period the program did not have systematically kept baseline data. It was, therefore, not possible to carry out a "before and after" impact evaluation design. To compensate this shortcoming, our assessment work is based a quasi-experimental design and analysis that requires treatment and control groups to define a counterfactual outcome.

In order to accomplish this objective and compare the impact of the program on clients', three alternative approaches are used. In the first approach comparison takes place by dividing the 550 clients into senior-long time (n= 461) and new clients (n= 89). The second approach takes into consideration the amount of loan cycle a client has reached. Here we compare clients who are in the early loan cycles with clients who have reached higher loan cycles. The third and final approach compares Tier one (group organized) clients with Tier two (individual) clients.

The impact evaluation assessed the difference in outcomes on key impact variables between the treatment (target) group and a control group. The control group was used as a proxy for what would have occurred had there been no intervention. To measure impact, it is necessary to compare the outcomes of target group clients with that of the control group that had identical social and demographic characteristics.

Results and Discussion

As aforementioned, the sample design is prepared to collect quantitative data from clients and use it in a quasi-experimental data analysis and appraise groups to define a counterfactual outcome. The analysis has three different sections. In the first section comparison takes place by dividing the 550 clients into senior-long time (n= 461) and new clients (n= 89). In in order to ensure the robustness of the analysis we decided to use nonparametric independent samples test. Our dependent variable is monthly income of clients and the independent variable is clients of the program divided into two groups 461 (Group 2= 461) and 89 (Group 1 =89). New clients who have been with the program not for more than two years make the first group and group two consists of senior client, clients who have been with program for more than two years. The research question and hypothesis to be tested are the following:

RQ_1: Does a statistically significant difference exist between duration of cliental membership and monthly mean income?

H_0: There is no statistically significant difference between duration of cliental membership and monthly mean income.

H_1: There is a statistically significant difference between duration of cliental membership and monthly mean income.

By means of the available data and use of Mann-Whitney test (nonparametric test) the study attempts to answer the following research question and hypotheses. Mann-Whitney U test which is known as independent-samples t test is used to see if there is statistical evidence that the population of the two means are significantly different.

The logic of this test is based on the simple observation that if there is a real difference between scores in two samples then the scores in one sample should be generally larger than the scores in the other sample. Thus, instead of comparing mean scores, as in the t test, we are comparing mean ranks. Thus, the Mann-Whitney test ranks scores from the two samples into one set of ranks and then tests to determine whether there is a systematic gathering into two groups paralleling the samples.

Table 4.1.1: Descriptive Statistics.

	N	Mean	Std. deviation	minimum	maximum
Amount of monthly income	550	4885.25	9686.176	0	90000
SMCP's membership in Years	550	1.84	.368	2	

Table 4.1.2: Mann-Whitney Test Ranks

	SMCP's membership in years	N	mean rank	sum of ranks
Amount of monthly Income	1-2 Years	89	235.62	20970.50
	3+ Years	461	283.20	130554.50
	Total	550		

Table 4.1.3: Test Statistics [a]

	Amount of monthly Income
Mann-Whitney U	16965.500
Wilcoxon W	20970.500
Z	-2.600
Asymp. Sig. (2-tailed)	.009

a. Grouping Variable: SMCP's Membership in Years

The tables display data comparing new and old groups on the variable rank of seniority. Table 4.1.2 provides the mean rank given to each group on each variable. For example, new members rank mean is 235.50 while the average rank given to senior members is 283.20. This means that the rank of the senior clients tends to be slightly larger than those for new once. Table 4.1.3 displays the Mann-Whitney U value and the significance level and these values are 16965.5 with a two-tailed significance of 0.009 ($U = 16965.5$, $p = 0.009$). This value is therefore significant and we must reject the null hypothesis and accept the alternative hypothesis as the difference in mean rank between the two groups is significant and is not a product of chance. Thus Mann-Whitney U-test showed that there is a significant difference between new and old clients.

In the second section the paper takes into consideration the amount of loan cycle a client has reached. Here we compare clients who are in the early loan cycles with clients who have reached higher loan cycles. Some clients are in the early loan cycles and others are in the middle loan cycles and the senior clients have already reach the 9th loan cycle. Thus, the paper divides clients into groups based on the amount of loan cycles that they have reached. Clients who have reached the first, second and third loan cycles are members of group one (G1) and clients who have reached the fourth, fifth and sixth loan cycles are members of group two (G2) while group three (G3) includes clients who have reached seventh, eighth and ninth loan cycle respectively. The research question and hypotheses to be tested are the following:

$RQ_{2:}$ Does a statistically significant difference exist between the amounts of loan cycle a client has reached and monthly mean income.

Ho: There is no statistically significant difference between the amounts of loan cycle a client has reached and monthly mean income.

H_1: There is statistically significant difference between the amounts of loan cycle a client has reached and monthly mean income.

In this section the appropriate test is the Kruskal-Wallis K independent sample test which is the non-parametric equivalent of one-way ANOVA and an extension of the Mann-Whitney test. It is used when the independent variable has more than two groups and, in this case, our independent variable has three groups.

Table 4 2.1: Descriptive Statistics

	N	Mean	Std. Deviation	Minimum	Maximum
Amount of monthly Income	550	4885.25	9686.176	0	90000
Amount of Loan-cycles	550	1.5051	.64873	1.00	3.00

Table 4.2.2: Kruskal-Wallis Test Ranks

	Loan Cycle	N	Mean Rank
Amount of monthly Income	1.00	320	265.26
	2.00	186	279.48
	3.00	44	333.14
	Total	550	

Table 4.2.3: Test Statistics

	Amount of monthly Income
Chi-Square	7.315
Df	2
Asymp. Sig.	.026

A a.Kruskal Wallis Test

b. b. Grouping Variable: Loan cycle

Table 4.2.1 gives the descriptive statistics and table 4.2.2 lists the mean ranks on mean monthly income for each group while the test statistics table 4.2.3 shows a chi square result. As the results show (table 4.2.2), those who have reached higher loan cycles have a higher mean rank than the other two groups. The chi-square value of the Kruskal-Wallis test is significant ($\chi^2 = 7.315$, $p = .026$). Therefore, one should conclude that the mean ranks of the three groups differ significantly from each other. In order to see which group differs significantly from the other group we need to see pairwise comparison. Table 2.4 shows that there is significant difference between G1 and G3 as well as G2 and G3 but not between G1 and G2.

Table 4.2.4 Pairwise Comparison of Groups

Groups	Test statistic	Std. error	Std. Test Statistic	Significance
G1 Vs G2	-14.200	-14.570	-.975	0.329
G1 vs G3	-67.875	25.408	-2.671	0.008
G2 vs G3	-53.665	26.491	-2.025	0.043

In this third section the paper endeavors to compare whether the there is significant mean income difference between clients of Tier one (group organized) and Tier two (individual) clients. Clients who belong to Tier one need to organize themselves into a group of three to seven people and start to save before they get a loan. On the other hand, clients of Tier two are either former members of Tier one who got promoted after staying with the first tier for many loan cycles or individuals whose credit needs could not be accommodated in Tier one. In this section the relevant research question and hypotheses are the following.

RQ_3: Does a statistically significant difference in monthly mean income exist between clients belonging to Tier one and Tier two.

Ho: There is no statistically significant difference in monthly mean income between Tier one and Tier two client.

H_1: There is statistically significant difference in monthly mean income between Tier one and Tier two client.

By means of Mann-Whitney U test the study gives solution to the above research question and hypotheses. The Mann-Whitney U test is used when testing for differences between two independent groups. The Mann-Whitney U test ranks scores from the two samples into one set of ranks and then tests to determine whether there is a systematic gathering into two groups paralleling the samples. The test is based on the simple observation that if there is a real difference between scores in two samples then the scores in one sample should be generally larger than the scores in the other sample. Thus, instead of comparing mean scores, as in the t test, Mann-Whitney U test compares mean ranks.

Table4.3.1: Descriptive Statistics

	N	Mean	Std. Deviation	Minimum	Maximum
Amount of monthly Income	550	4885.25	9686.176	0	90000
No. of Clients	550	1.6032	.48959	1.00	2.00

Table 4.3.2: Mann-Whitney U Tests Ranks

	Tiers	N	Mean Rank	Sum of Ranks
Amount of monthly income	1.00	212	246.49	52255.00
	2.00	338	293.70	99270.00
	Total	550		

Table 3: Test Statistics[a]

	Amount of monthly income
Mann-Whitney U	29677.000
Wilcoxon W	52255.000
Z	-3.410
Asymp. Sig. (2-tailed)	.001

a. Grouping Variable:

The above table's displays data comparing Tier one and Tier two groups on the variable rank of clusters. Table 4.3.2 provides the mean rank given to each group on each variable. For example, Tier one rank mean is 246.49 while the average rank given to Tier two members is 293.70. This means that the rank of Tier two clients tend to be slightly larger than those of Tier one. Table 4.3.3 displays the Mann-Whitney U value and the significance level ($U = 29677.000$, $p = 0.001$). This value is therefore significant and we must reject the null hypothesis and accept the alternative hypothesis as the difference in mean rank between the two groups is significant different and not product of chance. Thus Mann-Whitney U-test showed that there is a significant difference between Tier one and Tier two clients.

Conclusion

Microfinance institutions supply loans, savings and other financial services to the needy and play a positive role in the alleviation of poverty. SMCP is one of the two microfinance institutions active in the provision of micro loans to the needy section of Eritrean society. The objective of this paper was therefore to examine the economic impact of SMCP in improving the livelihood of its clients. By dividing the analysis into three sections the paper analyzed the performance of SMCP and has established that there is statistically significant mean income difference between the clustered groups in each section. Thus, one can claim that the program has positively influenced the economic livelihood of its clients.

Hence, it is high time that sponsoring government bodies give due consideration and support to the SMCP so that the program could continue its services to the economically deprived rural population and venture to new heights.

References

Aaron Kumah and Williams Kwasi Boachie (2016), An Investigation into the Impact of Microfinance in Poverty Reduction in Less Developed Countries (LDCs): A Case of Ghana. American Scientific Research Journal for Engineering, Technology, and Sciences (ASRJETS), ISSN (Print) 2313-4410, ISSN (Online) 2313-4402

Cintina, I. and I. Love (2017) 'Re-Evaluating Microfinance: Evidence from Propensity Score Matching', *The World Bank Economic Review*: p.lhw069.

Crepon, B., F. Devoto, E. Duflo, and W. Pariente (2015). 'Estimating the Impact of Microcredit on Those Who Take It Up: Evidence from a Randomized Experiment in Morocco'. *American Economic Journal: Applied Economics,* 7(1): 123–50.

Dadson, A.-V., V. Abankwah, and J.K.K. Kwansah (2012). 'Women Participation in Microcredit and Its Impact on Income: A Study of Small-Scale Businesses in the Central Region of Ghana'. *American Journal of Experimental Agriculture,* 2(3): 502–15.

Erulkar, A.S., and E. Chong (2005). *Evaluation of a Savings & Micro-Credit Program for Vulnerable Young Women in Nairobi.* Nairobi: The Population Council.

Ghaliba, Asad K., Issam, Malki, & Katsushi, S. Imai. (2014). Microfinance and household poverty reduction: Empirical evidence from rural Pakistan. *Oxford Development Studies,* 84–104.

Global Outreach Financial Benchmark Report (2018) *Global Outreach Financial Benchmark Report - 2017-2018.* Available at: https://content.centerforfinancialinclusion.org/wp-content/uploads/sites/2/2021/07/MIXMarket_GlobalOutreachFinancialBenchmarkReport_2017-2018.pdf

Haque, A.C., A. Das and A. Rahman (2017) 'The Effectiveness of Micro-Credit Programmes Focusing on Household Income, Expenditure and Savings: Evidence from Bangladesh', *Journal of Competitiveness* 8(2).

Islam, M. N., Hossain, T., & Afrin, S. (2014).*Impact of micro-credit to reduce poverty: A case study of Mymensingh*(Vol.31, pp.227-240). Retrieved from https://www.researchgate.net/publication/327944192

James C. Brau, Shon Hiatt, Warner Woodworth, (2009) "Evaluating impacts of microfinance institutions using Guatemalan data", Managerial Finance, Vol. 35 Issue: 12, pp.953-974, https://doi.org/10.1108/03074350911000025

Khandker, S. R., Khalily, M. A., &Samad, H. A. (2015). Non-borrowing effects of microcredit participation: Evidence using long panel survey data in Bangladesh. *Working Paper No. 41,*Institute of Microfinance (InM), Dhaka.

Khandker, Shahidur. (2005). Micro-finance and poverty: Evidence using panel data from Bangladesh.

Kim, J., G. Ferrari, T. Abramsky, C. Watts, J. Hargreaves, L. Morison, G. Phetla, J. Porter, and P. Pronyk (2009). 'Assessing the Incremental Effects of Combining Economic and Health Interventions: the IMAGE Study in South Africa'. *Bulletin World Health Organization,* 87(11): 824–32.

Koloma, Y. (2013). 'The Effects of Microfinance on Poverty Reduction in Mali: An Evaluation by the Method of Propensity Score'. Available at SSRN: https://ssrn.com/abstract=2467036 or http://dx.doi.org/10.2139/ssrn.2467036

Kondo, T.O., Aniceto Jr., Dingcong, C., &Infantado, C. (2008).Impact of microcredit on rural households in the Philippines.*Discussion Paper Series no. 2008-05*, Philippine Institute for Development Studies. Retrieved from http://www.eaber.org/node/22639

Lafourcade, Anne-Lucie, Jennifer Isern, Patricia Mwangi, and Matthew Brown.(2005) "Overview of the Outreach and Financial Performance." Mix, 2005: 1.

Littlefield, Elizabeth, Jonathan Murduch, & Syed Hashemi. (2003). is microfinance an effective strategy to reach the millennium development goals?

Mazumder, M.S. and Lu, W. (2015) "What impact does microfinance have on rural livelihood? A comparison of governmental and non-governmental microfinance programs in Bangladesh," *World Development*, 68, pp. 336–354. Available at: https://doi.org/10.1016/j.worlddev.2014.12.002.

Morduch, Jonathan, & Robert F. Wagner Graduate. (2002). Analysis of the effects of microfinance on poverty reduction.

Morris, Gayle, & Carolyn Barnes. (2005). An assessment of the impact of microfinance. *Journal of Microfinance, 7(1),* 39–54.

Neponen, H. (2003). ASA-GV microcredit impact report 2003. Trihcirappalli, India.*The Activists for Social Alternatives*, 2003.

Rooyen, C., R. Stewart, & T. de Wet. (2012). The impact of microfinance in Sub-Saharan Africa: A systematic review of the evidence. *World Development, 40,* 2249–62.

Sayed Samer, Izaidin Majid, Syaiful Rizal, M. R. Muhamad, Sarah-Halim, and Nlizwa Rashid (2015). The Impact of Microfinance on Poverty Reduction: Empirical Evidence from Malaysian Perspective. Procedia - Social and Behavioral Sciences 195, 721 – 728

Shakina Sultana Pomi (2019). Impact of Microcredit on Rural Poverty Alleviation in the Context of Bangladesh. International Journal of Economics and Finance; Vol. 11, No. 6; 2019. doi:10.5539/ijef.v11n6p70

Yunus, M. and Weber, K. (2007) *Creating a world without poverty: Social Business and the future of Capitalism.* New York: Public Affairs.

Smallholder Farming Development in Sub-Saharan Africa: Potentials, Constraints and Policy Implications for sustainable Development

Melake Tewolde,[1] Natnael Aregawi,[1] Yemane Ateshim,[1] and Hafiz Suleman[1]

[1]College of Bussiness and Social Sciences Adikeih

Abstract

A well performing agricultural sector in Sub-Saharan Africa (SSA) to a greater extent depends on the sustainable farming system of Smallholder Farmers (SMFs), the dominant producers in SSA. Sustainable farming system is a prerequisite to sustainable human development: sustaining livelihoods of SMFs, eliminating poverty, promoting the advancement of women-headed rural households, and regenerating natural resources and the environment. Thus, empowered SMFs with adequate capital assets have the ability to increase their productivity, reduce food insecurity, poverty and reduced dependency on cereal imports. Using the sustainable livelihood framework, the overall objective of the study is to explore the necessary conditions for the transformation of smallholder farms to contribute to household and national level food security, reduce poverty and dependency on food imports for SSA countries.

The study found that SMFs are subject to several constraints, including: traditional farming techniques, lack of access to formal markets, lack of skills, lack of access to financial services and inappropriate government policies that are obstructing their path toward sustainable farming. Declining agricultural performance is a major driving force behind growing poverty among African smallholder farming populations. Its recovery offers the greatest prospects for rural populations to escape out of poverty. As a result, SSA countries are facing triple challenges: (i) ensuring food security and adequate nutrition for all, (ii) strengthening livelihoods of SMFs and reduce the level of poverty incidence (iii) ensure sustainable development through sustainable use of natural resources and other resources.

Notwithstanding the constraints facing SMFs in SSA, the way out of poverty and ensuring food security is development of the smallholder farms to increase their productivity. To those ends, the promotion and empowering of producers' associations to increase their bargaining power, increase public investments in agricultural research and development that takes into account the needs of SMFs, promotion of agricultural extension services in favour of small holder farmers, increase public funding for natural resources development with a focus on soil and water conservation and water harvesting, promotion of integrated land-use strategies that actively consider the various environmental and resilience benefits of natural vegetation are suggested as key strategies.

keywords: Agriculture, Farming, Poverty, Smallholders

Introduction

Agricultural development is one of the requisites for better quality of life in all societies because until countries had made scientific and institutional advances to reliably meet their food needs through improved production as well as processing and trade, the process of modern economic growth becomes challenging as observed in SSA countries (FAO, 2016, p. 8). All past cases of rapid widespread economic development have been causally associated with the transformation of agricultural systems, from 18th and 19th century Europe and North America to late 20th century East Asia. Rapid increases in agricultural productivity, improvements in food processing and storage, and markedly reduced costs of food distribution improved the quantity, quality, safety and variety of food available at affordable prices.

In SSA countries, 70% of the countries' agricultural holdings are owned by SMFs (smallholder farmers) with farm sizes of two hectares or less while supplying 70% of the countries' total food requirements (UNCTAD, 2015, p.2). They, however, have limited access to modern farm inputs causing low productivity in a rapidly growing population (Poole, 2017, p.17).

In SSA, SMFs are major producers of crops, livestock and fisheries products. However, most of them remain food insecure, confronted by various interlocking challenges including water stress, lack of access to agricultural inputs, ecosystem degradation and a policy bias in favour of large-size farms (UNCTAD, 2015, p.10).

Of the total cultivated land in Sub-Saharan Africa, 95 percent is rain-fed (WB, 2013, p.14). Climate change and rainfall variability have disproportionately affected smallholder productivity reducing nutritional value of their crops and deepening their food insecurity and poverty (AGRA, 2023, p.15; WB, 2013, p.17). The inherent challenges are to manage and adapt to weather variability, and to use water from rainfall more productively (FAO, 2020, p.23). With little application of productivity-enhancing farm technologies in smallholding farms; food insecurity, hunger and malnutrition have become major challenges for most SSA countries, where more than one in five people faced hunger in 2020, and approximately 282 million people were undernourished (AGRA, 2023, p.15).

At present SSA countries face triple challenges: (i) ensuring food security and adequate nutrition for all, (ii) strengthening livelihoods of SMFs and reducing the level of poverty incidence (iii) ensuring sustainable development through sustainable use of natural resources and other resources. The 2030 Agenda for Sustainable Development has set two agriculture related goals: (i) Goal 1: End poverty in all its forms everywhere and (ii) Goal 2: End hunger, achieve food security and improved nutrition and promote sustainable agriculture" (UN, 2015, p.18). Agenda 2030, underlines all countries by 2030, double the agricultural productivity and incomes of small-scale food producers, in particular women, indigenous peoples, family farmers, pastoralists and fishers, through secure and equal access to land, other productive resources and inputs, knowledge, financial services, markets and opportunities for value addition and non-farm employment (UN, 2015, p.19). Doubling productivity and income over a 15-year period implies, for any initial value, an average growth rate of 4.6 percent per year (FAO, 2016, p.6). In the face of the formidable challenges posed by climate change, there is growing concern that food security for all must be pursued in a climate-resilient and environmentally sustainable manner. Persistent and severe droughts, exacerbated by climate change, are causing increasingly serious water shortages in rain fed agriculture, posing a higher risk to the livelihoods of rural people by reducing crop and livestock yields (FAO, 2020, p.20).

Objective of the study

Empirical studies have shown that agriculture based economic growth is up to two times more effective in increasing poor household income (AGRA, 2014, p.16). The success of the Green Revolution in Asia put SMFs firmly on the development agenda. SMFs are rational and efficient in resource allocation and that they remain poor not because they are unresponsive to economic incentives but because they respond to limited technical, and market opportunities available to them (Hayami, 1998, P.303). Productivity growth in smallholder farms contributes towards growth not only by reducing the price of staple food, but also by increasing the demand for labour in rural areas, generating jobs for the poor and raising the unskilled labour wage rate.

Development of smallholder farming also contributes to reduction of food import, improving food security, creation of employment and combating land degradation and water stress (Fadcyi, Ariyawardana and Aziz, 2022, p.14). While focusing on smallholder farming development, the strategy, agricultural development-led industrialization that aims to stimulate farm output and rural incomes would improve the livelihoods of rural populations in SSA. The focus of the strategy is on increasing smallholder production and productivity through complementary interventions such as promotion of improved agricultural technologies, provision of credit services and improvement of health care services, primary education and infrastructure. To these ends, public investments that encourage SMFs for adopting modern farm technologies would be required. This means SSA countries must at least commit 10 percent of total government expenditure on agricultural as per Maputo Declaration (WB, 2013, p. 42).

Without purposeful attention paid to the SMFs in SSA Africa, smallholder farming may not be able to effectively contribute to achievement of agriculture related SDGs: (i) End poverty in all its forms everywhere (ii) Goal 2: End hunger, achieve food security and improved nutrition and promote sustainable agriculture (iii) Goal 15: Protect, restore, and promote sustainable use of terrestrial ecosystems, sustainably manage+ forests, combat desertification, and halt and reverse land degradation and halt biodiversity loss (UN, 2015, p. 18).

The overall objective of the study is to explore the necessary conditions for the transformation of smallholder farms to contribute to household and national level food security, reduce poverty and dependency on food imports for SSA countries. The specific objectives are to: analyze the food related situation of SSA African countries by taking key indicators of food –related status of the countries; analyze the challenges facing SMFs in SSA; suggest policy measures that would contribute to sustainable smallholder farms production so that food security will be improved in SSA.

Taking into account the problem statement and objective of the study, the central research questions this paper will address are: (i) what are the critical policy measures required to enable SMFs increase their productivity and in environmentally sustainable way? (ii) What major constraints do smallholder farms face? (iii) How can SSA successfully reconcile population growth, food production and sustainable management of its natural resources?

The research is an extensive literature review focusing in SSA. Eritrea being part of the region, Eritrean SMFs show similar characteristics with the farmers in broader region and are not immune from the prevalent problems. This indicates that the prospects and challenges of Eritrean SMFs are included in the broader spectrum of this research.

Methodology

This study is conducted based on relevant literature review. Data used in this study are secondary data generated through extensive literature review of relevant sources, including books, relevant publications and reports of international organizations as well as relevant national policy documents of different SSA nations and academic journals. The data selection, representing the region (as a whole) as well as several nations, was based on the relevance of the data to the objective of the research and representation of the nations from different regions of SSA. This helped in identifying patterns of similarity as well as shared challenges across the whole region as well as between nations.

Following sections 1-4, section 5 deals with review of pertinent development theories with the objective to come up with appropriate theoretical framework in the context of socioeconomic and political conditions prevailing in SSA countries. Section 6 is the main part of the paper and focuses on food related development outcomes in SSA countries in the past decades, while section 7 deals with the major constraints of SMFs in SSA countries. Section 8 deals with discussions and conclusions, while section 9 presents suggested interventions that would enhance transformation of smallholder farms that would overcome food related problems in SSA countries.

Conceptual framework: Farming and Sustainable livelihood

In SSA, SMFs are the dominant producers of food. However, for most SMFs market participation is limited because they lack access to capital assets which limit their capabilities to increase their productivity and sustain their livelihood (Poole, 2017, p.28). A livelihood is the activities and assets that jointly determine the living conditions of an individual or a household (Ellis, 2000). A livelihood is sustainable when it can cope with and recover from stresses and shocks, and maintain or enhance its capabilities and assets both now and in the future, without undermining the natural resource base (Chambers and Conway, 1992, p.10).

Sustainable livelihood is a holistic framework that captures the fundamental causes, and manifestations of poverty, food insecurity, allowing for effective prioritization of actions that aim to improve the livelihood of a target population (Scoones, 1998). Without building the capital asset of SMFs, through access to productive resources, a vicious circle towards sustained rural poverty and dependency is likely to perpetuate. The five capital assets that are central to SMFs' livelihood improvement include the following.

Human Capital

Human capital is the most important source of livelihoods and is acquired through education and training (Goodwin, 2003). Human capital also plays a key role to strengthen farmers' ability to adapt farm technologies to a particular situation and to changing needs (UNCTAD, 2015, p.87). SSA countries have underdeveloped human resources which lack the critical skills and knowledge and is still a constraint to productivity increase to those involved in crop and livestock production. Thus, education related to modern farming practices, including bookkeeping, marketing and entrepreneurship would increase SMFs' skill and their capabilities as well as adaptability of modern farming.

Natural Capital

One of the requisites for improved livelihoods of SMFs is sustainable use of the natural resources including land, water and biodiversity (Poole, 2017, p.61). Under the present circumstances, land and water are the main natural resource for agricultural production and income generation for the largest segment of the population in SSA. However, because of climate change and increasing land degradation, SMFs' productivity has been low. Especially, climate change is a cause for the increased frequency and severity of natural shocks that negatively affect smallholder farms.

Financial Capital

Financial capital is essential for investment that enhances farmers' productivity (SIDA, 2001). The saving and capital formation capacity of the SMFs in SSA is still low. Farmers tend to underinvest in farm equipment, land improvement and human capital, limiting their productivity. Access to microfinance is thus an effective tool to support investment in smallholder agriculture and that governments need to improve financial services and the banking system so that they work more effectively for smallholder agriculture.

Physical Capital

Physical capital includes basic infrastructure such as irrigation canal, dams, energy, transport, communications which support the livelihoods of the population in a given area (BGHRC, 2001, p.11). In the context of SSA, physical capital comprises of basic infrastructure such as main and feeder roads connecting various villages, towns and zones of the countries. For smallholders, lack of access to roads constrains access to markets increasing their vulnerability to shocks and hinders economic opportunities that could arise from trade with adequate transportation facilities.

Social Capital

The social capital includes networks, social claims, social relations, affiliations and associations upon which people draw when pursuing different livelihood strategies (SIDA, 2001). In SSA, social capital refers to the networks among members of the same communities within and outside an area, membership of groups, relationship, and access to wider institutions of society upon which people draw resources in their pursuit of livelihoods. SMFs are vulnerable to different shocks such as drought, floods, earthquake, illness, accidents and death of breadwinner which requires strong social network to overcome the effects of the adverse shocks. In most SSA countries, there is a strong system of social capital in the traditional societies, which reflects mutual support during good times such as marriage and bad times such as death of breadwinner. The support takes different forms such as in cash, in kind and in labour. The support system is one of the strongest mechanisms for resilience and coping mechanism during difficult times and need to be nurtured.

Empirical Evidence on Crop Yield, Fertilizer Use and Food Status of SSA Countries

Crop Yield

Increasing cereal yields in SSA allows SMFs to achieve food security, reduce poverty and diversify their diets (AfDB, 2010, p.16). Studies indicate that under conducive policy environment, average productivity growth has been higher in countries where smallholder farming dominates agricultural production, taking average farm size of less than 2 hectares per worker as the threshold (FAO, 2016, p. 6). Table 1 show cereal yield per hectare in SSA countries has increasing trend, although there are variations in the rate of growth in land, labour, and total factor productivities depending on country and region. The overall yield of SSA countries increased from 1129 kg per hectare in 2001 to 1456 kg per hectare in 2014, an encouraging trend. However, the yield per hectare in SSA is far less from the yield in other regions, reflecting the low productivity of SMFs in the region. Cereal yields in SSA are less than half those of other developing regions (OECD, 2006, p.7).

Declining agricultural performance is a major driving force behind growing poverty among African smallholder farming populations, and its recovery offers the greatest prospects for rural populations to escape out of poverty. The causes of low crop yields in most of SSA includes land degradation, inadequate fertilizer use, lack of access to high-yield resilient crop varieties, lack of access to irrigation and climate change (UNCTAD, 2015, p.67).

Table 1: *Cereal Yield in selected SSA countries, 2001-2014, selected years*

Country	Year		
	2001	**2010**	**2014**
SSA(average)	1129	1385	1456
Angola	585	629	888
Benin	1067	1272	1460
Chad	635	763	856
Ethiopia	1198	1833	2325
Kenya	1640	1710	1628
Mozambique	880	1028	703
Rwanda	913	1919	1981
South Africa	2423	4149	4894
Tanzania	2044	1648	1678
Zimbabwe	1160	735	541

Source: AGRA 2017, pp.155-156

Fertilizer Utilization

The decrease in soil nutrients, due to intensive farming, must be replaced by using fertilizers to increase crop yields. The use of fertilizers refers to agricultural use of mineral or chemical fertilizers for the three main plant nutrients, namely nitrogen, phosphorous and potassium (FAO, 2014, p.32).SSA is at an early stage in fertilizer use as shown in Table 2. The proportion of fertilizer use in Africa to the world fertilizer used was fraction of that used by other regions of the world. Low use of inorganic fertilizer is thus one of the reasons for low crop productivity in Africa. Moreover, the fertilizer use was mainly by large-scale farms because of policy bias as the government distribution favoured the larger holdings which have high fertilizer buying capacity.

Table 2: Inorganic fertilizer use in Africa by nutrient, 2021, (thousand tons)

Region	Soil Nutrient		
	Nitrogen (N)	Phosphorus (P2O2)	Potassium (K2O)
World	108,688.2	46,269.7	40,128.2
Africa	4,788.4	1,784.1	1,045.2
% of world	4.4	3.8	2.6
Asia	60,285.5	24,849.8	18,685.9
% of world	55.4	53.7	46.4
Americas	26,909.3	14,272.5	15,355.2
% of world	24.7	30.8	38.2

Source: (FAO, 2023, p.146)

Given that SMFs dominate agricultural production in SSA, the main reasons for low uptake of fertilizer are poor infrastructure that hinder transport of fertilizer to rural areas, inadequate supply of fertilizer, inappropriate fertilizer to local soil qualities, lack of credit at planting, unfavourable fertilizer/crop price ratios and poor soil quality leading to low crop to fertilizer response rates (Gilbert, 2020, p.88). Studies indicate that, at the current levels of use, a rapid agricultural growth program would require a ten percent or higher growth rate of fertilizer use (Mellor, 2017, p.177).

Food Security Status

Food security exists when all people, at all times have physical, social and economic access to sufficient, safe and nutritious food that meets their dietary needs and food preferences for an active and healthy life (FAO, 2003). Food security relies on four main components: availability, access, utilization and stability. "Availability" refers to the obtain-ability of sufficient food quantities of appropriate quality, supplied through domestic production or importation. "Access" refers to the physical and economic access to food. "Utilization" integrates water quality sanitation and nutritional quality as well as food distribution within the household. "Stability" exists when the other three dimensions are verified over time (FAO, 2003). The food security status of SSA countries is a reflection of the crop and livestock production of SMFs, the dominant producers in the countries. Table 3 shows the prevalence of severe food insecurity of Africa, the percentage of people in the total population who live in households classified as severely food insecure (FAO, 2020, p.65).

Table 3: Prevalence of severe food insecurity (percent) in SSA, selected countries, 2014-2022

Country	Year			
	2014-2016	2017-2019	2019-2021	2020-2022
Angola	21	24	30.4	31.2
Burkina Faso	10	12.6	18.5	21.2
Cameroon	22.3	28.1	26.7	26.7
Ethiopia	14.5	14.1	19.6	21.1
Ghana	5.1	6.2	5.6	6.2
Malawi	47.7	48.3	51	52.2

Namibia	28.9	31.3	32.6	33.0
Togo	16.1	16.9	18.8	19.4
Tanzania	20.6	25	25.8	26.3
Zambia	22.4	26.2	32.6	32.1
Africa	17.7	19.8	22	23.4
Asia	6.7	7.6	9.5	9.9
Latin America and the Caribbean	7.9	9.7	12.3	13.0
World	7.7	8.9	10.7	11.3

Source: FAO (2023, pp. 267-271)

The prevalence of severe food insecurity in Africa increased from 17.7% during 2000-2002 to 23.4% during the period 2020-2022 (FAO, 2023, p.267). Severe food insecurity is much higher in Africa than in any other part of the world with an increasing trend. Taking into account the rapidly growing population in SSA countries and the adverse effects of climate change, the countries' food insecurity is likely to increase.

Food Import Dependency

As in most SSA countries domestic cereal production does not meet the demand of the rapidly growing population, the countries have been forced to depend on food imports. Cereal import dependency ratio provides a measure of the dependence of a country or region from cereal imports. Specifically, the cereal imports dependency ratio indicates how much of the available domestic food supply of cereals has been imported and how much comes from the country's own production. It is computed as (cereal imports - cereal exports)/ (cereal production + cereal imports - cereal exports) * 100). Negative values indicate that the country is a net exporter of cereals (FAO, 2022, p.121). Table 5, shows that cereal import dependencies of SSA countries varied among countries and during different period even within a country. Compared to other regions of the world, Africa had the highest cereal import dependency ratio and with upward trend, increasing from 28.2% during 2000-2002 to 30.6% during the 2018-2020 periods. Persistent food import becomes a problem when the high and rising food import bills take foreign exchange away from other important development agendas without resolving food insecurity.

Table 4: *Cereal import dependency of SSA (percent), 2000-2002 to 2018-2020, selected countries*

Country	Year		
	2000-2002	**2013-2015**	**2018-2020**
Senegal	53.1	56.6	43.3
Benin	13	46.2	42.2
Botswana	90.9	76.6	81.6
Cameroon	31.2	31.0	31.4
Ethiopia	10.2	6.0	6.6
Ghana	26.1	31.0	36.2
Kenya	23.9	37	41.1
Madagascar	11.3	14.4	19.6

Uganda	5.9	5.1	8.9
Zimbabwe	11.7	45.6	60.3
Africa	28.2	30.2	30.6
Asia	7.9	6.7	8.5
Latin America and the Caribbean	12.1	-5.6	-7.6
World	-0.5	-1.7	-1.7

FAO (2023, pp.282-286)

Constraints faced by SMFs

Traditional Farming Methods

In most countries, small-sized farms tend to have higher agricultural crop yields per hectare than larger farms because they manage resources and use labour more intensively (FAO, 2014, p.13). However, in SSASMFs use traditional farming techniques such as oxen drawn plough with reduced fallow periods and use of traditional seed varieties, while degrading soil fertility decrease low yields or fail altogether, particularly during shortage of rainfall, in a region where only 4 percent of arable land is irrigated (UNCTAD Secretariat, 2015, p.5).

With traditional farming practices, yields have remained low with little contribution to food security and diet improvement. Notwithstanding traditional methods of farming, smallholder farms are expected to ensure food security, to care for and protect the natural environment and to end poverty, undernourishment and malnutrition. These goals can be achieved if public policies support small farms to become more productive and sustainable. With public support to SMFs, their productivity would increase in turn improving food security and diets of the SSA peoples. Yet public support to SMFs decreased significantly since 1980s, leaving smallholders with few resources to improve their farming practices (FAO, 2014, p.8).

Climate Change and Land Degradation

The impacts of climate change and land degradation disproportionately affect SMFs, increasing the incidence of hunger, and exacerbate poverty (Assefa, et al., 2020. p.209). Climate change has affected the amount of rainfall and evaporation rate creating pressure on available water resources. The increase in temperature and subsequent increase in water evaporation will reduce the yield of staple foods as the convenient environment for production of such foods is altered (AGRA, 2014, p.19). The rise in temperature is also creating new unprecedented phenomena. Plant and animal diseases as well as pests are becoming frequent threat for African agriculture (FAO, 2015, p.22).

In addition, land degradation has become serious problem in SSA. The land degradation has also affected the ability of land to flexibly adapt changes in climate in particular droughts. Continuous farming without fallow period have reduced soil nutrients in most SSA, deepening the adverse effects of climate change. The problem of drought is exacerbated for heavily rainfall reliant SSA agriculture given that the region is the highly drought vulnerable region of the world (OECD/FAO, 2016, p.73). SMFs are unable to fully utilize their land all year round as they cannot produce outside the rainy season.

Inadequate Infrastructures

The World Bank in 2019 stated that infrastructure expenditure in SSA is 3.5% of annual GDP which is much lower than the 7% needed to attain the 2030 sustainable development (Chinzara, Dessus and Dreyhaupt, 2023, p.1). Most SMFs are located in rural areas where lack of physical infrastructure limits their expansions. With poor Infrastructure, markets for agricultural inputs and outputs are often missing and unreliable (Republic of South Africa, 2012, p.3). Moreover, inadequate infrastructure increases transaction cost of SMFs, constraining their investment and growth. Delivery of quality product to customers is also colossal problem due to absence of post-harvest storage, cold-rooms and processing facilities (Balyoi, 2010, p.25; Kamara, Conteh, Rhodes, and Cooke, 2019, p.14058). This creates huge post-harvest loss and further worsens financial condition as they cannot cover their cost and generate profit. The harvest that reachs the market has lower quality and force the farmers to sell at lower price.

Lack of Access to Formal Markets

Raising smallholder productivity and strengthening market linkage are strategic necessity. Inadequate linkage to efficient and inclusive markets is a key reason for low levels of adoption of available productivity-increasing technologies (FAO, 2013, p.7). Lack of access to reliable markets is a major constraint faced by SMFs. Many SMFs receive low prices for their products by selling them at their farm gate or local markets. However, these SMFs could receive much higher prices by selling their goods based on marketing knowledge and by recognizing opportunities for product diversification or the limits between market research and product development.

Not all farmers can take advantage of market developments. SMFs' access to agricultural markets is often constrained, including geographic barriers such as remoteness, or biophysical limits to productivity. Limited productive asset holdings– of land, livestock, labour, critical equipment– may constrain smallholders' capacity to generate a sufficient level and consistency of marketable surplus (FAO, 2013, p.21). Another problem is SMFs participate primarily in domestic markets, hindering creation of new capital as only domestic capital is circulating in the economy (Chikazunga and Goyer, 2013, p.9). The risks facing smallholders associated with market integration are high. This is due to difficulties faced in accessing market information, credit and other inputs, and technical assistance, which together with inefficient and conflicting policies, laws and regulations and weak infrastructure create significant uncertainties in the returns that smallholders can expect from engaging in agricultural markets (FAO, 2013, p.21).

Lack of Skill

Lack of skill is a serious constraint for SMFs. They are often illiterate with poor technological skills, which are serious obstacles in tapping useful technological knowledge. The majority of SMFs are not capacitated with financial and marketing skills and are unable to meet the quality standards set by fresh produce markets and food processors. Lack of production knowledge leads to lower quality in production (Republic of South Africa, 2012, p.4). Furthermore, many smallholders have little formal education, which limits their ability to keep adequate written records or educate themselves about improved agricultural practices. They may have only a vague idea of basic metrics, such as farm size, crop yield, and real costs, on their own farms (IFC, 2013, p.13).

Lack of Access to Financial Services

Majority of SMFs in SSA have limited access to financial services because financial institutions often view smallholders as unattractive clients due to insufficient collateral, lack of written records on their sales and expenditures, and the small size of loans requested by farmers which has high transaction costs for the credit providers. Financial constraints are more pervasive in agriculture because agricultural loans are typically paid after the harvest, which may be several months after the loans taken. This delay creates a further disincentive for financial institutions to provide credits, particularly to SMFs (IFC, 2013, p.13). As the financial markets are imperfect, governments have to provide credits to SMFs at lower cost to increase investment that would increase their productivity. Empirical findings in SSA have indicated that SMFs with access to credit increased their production by 38.5% (Peprah, Afoakwah, and Issac, 2016, p.23).

Inappropriate Government Policies

Government policies are central to the development of smallholder farming. Policy design should incorporate essentiality, innovation and enabling institutional environment in order to be labelled as good (HLPE, 2013, p.14). However, in most SSA countries policies are not conducive to the development of smallholder farming. Most government policies have inbuilt bias in favour of large-scale farming in the provision of farm inputs, including better seed varieties, fertilizers, extension services. The research in agriculture is also often tuned to the needs of large-scale farming (Ajwang, Atela, and Arora, 2019, p.1).

One important factor that has implications on the productivity of farmers in SSA is land tenure system. Secure land ownership creates sense of security since farmers can access credit and make long-term investments. This is because the level of certainty farmers have is that their land rights is recognized and protected, especially against the encroachment or loss of use rights over the land will encourage forms to increase their investment on their farmlands. The importance of land policy is all the more critical at a time when social and economic inequalities are on the rise, the world is becoming increasingly urbanized, new needs stemming from demographic and economic change are threatening to marginalize a whole swath of the farm population, and the urgency of environmental challenges is greater than ever.

With the rapidly increasing population in SSA the size of smallholder farms has been rapidly decreasing. Shrinking of arable land in SSA countries has also been aggravated by land policy of many SSA countries which attract foreign investors seeking large arable land on a lease basis. Foreign investors have become increasingly interested to acquire farm lands in SSA, threatening to marginalize a whole swath of the farm population. This phenomenon of 'land grabbing' has displaced smallholders from their farm lands, with negative effect on their food security. For instance, the Gulf States such as Saudi Arabia, Qatar and the UAE have become the main buyers of agricultural land in SSA. In Nigeria, Namibia, Morocco and Ghana; in Sudan, Emirati investors have bought more than 400,000 hectares of arable land (Garrido, 2023, p.10).

Discussions and Recommendations

Small-scale agriculture is the main source of food in SSA producing up to 70 percent. However, main characteristics of production systems of SMFs in SSA are of simple, outdated technologies, inadequate use of modern farm inputs and low cereal yield per hectare, a major cause for food insecurity and malnutrition in the region, which is far from the yields in other regions. Declining agricultural performance is a major driving force behind growing poverty among African smallholder farming populations, and its recovery offers the greatest prospects for rural populations to escape out of poverty.

Currently, SSA countries face triple challenges: (i) ensuring food security and adequate nutrition for all, (ii) strengthening livelihoods of SMFs and reduce the level of poverty incidence (iii) ensure sustainable development through sustainable use of natural resources and other resources. As a result, SSA countries are far from the track of SDG1 and SDG2. Reflecting the low cereal production by SMFs, SSA is also highly dependent on cereal imports. Persistent food import has also become a major challenge because of high and rising food import bills take big chunk of foreign exchange away from other important development agendas without resolving food insecurity. The food insecurity status of SSA countries is a reflection of the crop and livestock production of SMFs, the dominant producers in the countries have been far below desired levels Low productivity in smallholder farms is a major cause for the prevalence and persistence of undernourishment in SSA. Persistent and severe droughts, exacerbated by climate change have been causing increasingly serious water shortages in rained agriculture, posing a higher risk to sustainable farming that improves the livelihoods of rural people by improving crop and livestock yields.

Sustainable farming is crucial to sustainable human development: sustaining livelihoods of SMFs, eliminating poverty, promoting the advancement of women-headed rural households engaged in farming activities and regenerating the environment. Notwithstanding the various challenges facing SMFs in SSA, there is urgent need to significantly increase the productivity of SMFs' to ensure long term food security. This can be achieved by, among others, encouraging SMFs to pursue sustainable intensification of production through use of improved farm inputs. Thus, smallholder farming development is also central to the achievement of the two SDGs through sustained increase in the level of production and productivity of SMFs over a reasonable length of time and the subsequent improved wellbeing of farmers that would be reflected in their higher per capita income and standard of living. From a broader perspective, rural development in which coordinated development programs of agriculture, education, health and nutrition, rural electrification, promotion of producers associations implemented properly will result into a sustained increase in the level of production and productivity of all rural dwellers, including SMFs, and a sustained improvement in their wellbeing, manifested by increasing per capita income and standard of living. This in turn would improve food security, reduce poverty and improve nutrition of rural communities, including the SMFs, while reducing countries' food dependency on food imports.

Based on theories and empirical findings the paper concludes that SMFs are not in a subsistence-oriented mindset. They can boost their production with conducive and supportive policy environment. This means the food, nutritional, climatic, environmental, social and economic challenges of SSA demand conducive policies for SMFs. Thus governments in SSA should increase public investments which focus on improving SMFs' access to modern farm inputs, credits, skills and building rural infrastructure to connect smallholders to markets. Only then will SMFs be able to respond to market incentives, adopt modern farm technologies and increase their productivities that would have positive impact on food security and nutrition at household and national levels. The future of smallholder agriculture in SSA and their contribution to their respective national economic development to a greater extent will depend on policy environment friendly to smallholder farming rather than evicting them by attracting foreign investment in agriculture that exacerbates food insecurity and poverty.

African agricultural transformation has to embrace and safeguard the future of African SMFs and boost the development of individual countries in a sustainable way and avoid provision of farm land to foreign investors through lease system or selling, as it exacerbates poverty and landlessness. Sustainable smallholder farming development involves ensuring and maintaining productive capacity for the future and increasing productivity of SMFs without damaging the environment, while guaranteeing their land on ownership. Thus taking into account the current status of SMFs in SSA and the constraints they face, the following interventions are suggested to increase their productivity and ensure sustainable farming system that would contribute to food security and poverty reduction.

Recommendations

(I) Promotion of producers association that involve smallholder farms to increase their bargaining power in accessing farm inputs such as seeds and fertilizer; transport and storage and in selling their products to buyers.

(II) Public investments in agricultural research and development (R&D) that takes into account the needs of SMFs and appropriate to local agro-ecological conditions of each country to boost their productivity of SMFs and to contribute to food security, improve nutrition and reduce poverty.

(III) Promotion of agricultural extension services in favor of small holder farmers with a focus on promotion of sustainable agricultural advisory services so as to enhance the adoption of farming technologies and knowledge and the skills necessary for the increase in their productivity.

(IV) Increase public funding for natural resources development with a focus on soil and water conservation and water harvesting so that SMFs would respond to the adverse effects of climate change, land degradation and water stress for ensuring sustainable production system.

(V) Promotion of sustainable agricultural advisory programs capable of providing smallholder agricultural producers with the skills necessary to be more productive, and adaptive to different environmental constraints.

(VI) Promotion of Integrated land-use strategies that actively consider the various environmental and resilience benefits of natural vegetation in order to ensure that cropland expansion does not lead to land degradation, loss of carbon stocks, and loss of biodiversity.

(VII) Strengthen government policieswith focus on SMFs through a provision of public goods to rural areas including roads, health services, clean water, and schools; investing in agricultural research and extension.

(VIII) Provide support by increasing investment in resilient and sustainable agriculture and food systems that benefit and empower poor and vulnerable farmers, their families, and communities.

(IX) Establish a business-friendly environment for smallholders and make small-scale farming attractive to the younger generation to increase youth employment and emigration.

(X) Increase public investment for developing non-farm activities to create employment opportunities in rural areas and to enable SMFs diversify their livelihoods.

(XI) Promotion of integrated land-use strategies that actively consider the various environmental and resilience benefits of natural vegetation in order to ensure that cropland expansion does not lead to land degradation, loss of carbon stocks, and loss of biodiversity.

References

African Development Bank (AfDB). (2010, April). Smallholder Agriculture in East Africa: Trends, Constraints and Opportunities. (Working Paper No. 105).

AGRA. (2014). *Africa Agriculture Status Report 2014: Climate Change and Smallholder Agriculture in Sub-Saharan Africa.* Nairobi: AGRA.

AGRA. (2023). *Africa Agriculture Status Report 2022: Realigning to Transform Africa's Food Sytems.* Nairobi: AGRA. Retrieved from https:// www.agra.org

Ajwang, F., Atela, J., & Arora, S. (2019). *Agricultural Policy making in Kenya: Why must SMFs' agency be made central?* Relationl Pathways: MApping Agency and Poverty Dynamics.

Assefa, T., Jha, M., ., M. R., Worqlul, A., Doro, L., & Tilahun, S. (2020). conservation Agriculture with drip irrigation: Effects on Soil Quality and crop yield in Sub-Saharan Africa. *Journal of Soil and Water Conservation, 75*(2), 209-217. doi:10.2489/jswc.75.2.209

Balyoi, J. K. (2010). *An Analysis of constraints facing SMFs in the Agribusiness value chain: a case study of Limpopo province (Master's Thesis).* Pretoria, Republic of South Africa: University of Pretoria.

Benfield Greig Hazard Research Centre (BGHRC). (2001). Sustainable livelihoods and vulnerability to disasters: Disaster Managment. (Working Paper 2).

Chambers, R., & Conway, G. (1992). Sustainable rural livelihood: Practical concepts for the 21st century. *Institute for Development Studies* (Discussion paper 296).

Chikazunga, D., & Goyer, P. (2013). Smallholder farming: A panacea for employment creation and enterprise development in South Africa? Lessons from the Pro-Poor Value Chain Governance Project in Limpopo Province. (Working Paper 27: PLAAS, UWC).

Chinzara, Z., Dessus, S., &Dreyhaupt, S. (2023, February). INFRASTRUCTURE IN AFRICA: How Institutional Reforms Can Attract More Private Investment (Working Paper).

Ellis, F. (2000). *Rural Livelihoods and Diversity in Developing Countries.* Oxford: Oxford University Press.

Fadeyi, O. A., Ariyawardana, A., & Azziz, A. A. (2022). Factors influencing technology adoption among SMFs: a systematic review in Africa. *Journal of Agriculture and Rural Development, 123*(1), 13-30. Retrieved from https://www.jarts.info

FAO. (2003). *Trade Reforms and Food Security: Conceptualizing the Linkages.* Rome.

FAO. (2013). *Climate-Smart Agriculture Sourcebook.* Rome: FAO.

FAO. (2013). *Smallholder integration in changing food markets.* Rome. Retrieved from https://www.fao.org

FAO. (2014). *Statistical Yearbook 2014: Africa Food and Agriculture.* Accra.

FAO. (2014). *The State of Food and Agriculture: Innovation in family farming.* Rome: FAO.

FAO. (2015). *Regional overview of food insecurity: African Food security prospects brighter than ever.* Accra: FAO.

FAO. (2016). Can Smallholders Double their productivity by 2030? by Orsolya Mikecz and Rob Vos. *ESA working Paper, 16*(04).

FAO. (2020). *2019 Regional Synthesis Report based on country studies in eight countries in Europe and Central and Asia; Empowering smallholders and family farms in Europe and Central asia* . Budapest: FAO.

FAO. (2020). *The State of Food and Agriculture: Overcoming Water Challenges.* Rome. doi:10.4060/cb1447en

FAO. (2020). *World Food and Agriculture- Statistical Yearbook 2020.* Rome. doi:10.4060/cb1329en

FAO. (2022). *Statistical Pocketbook 2022.* Rome.

FAO. (2022). *World Food and Agriculture- Statistical Yearbook 2022.* Rome. doi:10.4060/cc2211en

FAO. (2023). *World Food and Agriculture – Statistical Yearbook 2023.* Rome. doi:10.4060/cc8166en

Garrido, G. O. (2023). Emiratis, Saudis and Qataris cross the Redsea: What are the Gulf monarchies doing in Africa? (IEEE Analysis Paper 45). Retrieved from https://www.ieee.es/Galerias/fichero/docs_analisis/2023/DIEEEA45_2023_OSCGAR_Emiratos_

Gilbert, J. R. (2020). Inorganic Fertilizer Use Among SMFs in Sub-Saharan Africa: Implications for Input Subsidy Policies. In S. G. Paloma, L. Riesgo, & K. Louhichi (Eds.), *The role of smallholder Farms in Food and Nutrition Security.* Switzerland: Springer Nature Switzerland AG Publisher.

Goodwin, N. R. (2003). Five Kinds of Capital: Useful Concepts for Sustainable Development. (Global Development and Environment Institute working Paper No. 03-07).

Hayaami, Y. (1998). *The Peasant in Economic Modernization.* (C. Eicher, & J. Staaz, Eds.) Baltimore: The Johns Hopkins University Press.

HLPE. (2013). *Investing in smallholder agriculture for food security.* Rome: High Level Panel of Experts. Retrieved from https://www.fao.org/cfs/cfs-hlpe

International Finance Corporation (IFC). (2013). *Working with Smallholders: A Handbook for Firms Building Sustainable Supply Chains.* Washington DC: Sustainable Business advisory.

Kamara, A., Conteh, A., Rhodes, E., & Cooke, R. (2019). The Relevance of Smallholder Farming to African Agricultural Growth and Development. *African Journal of Food, Agriculture, Nutrition and Development, 19*(1), 14043-14065. doi:10.18697/ajfand.84.BLFB1010Mellor, J. W. (2017). *Agricultural Development and Economic Transformation: Promoting Growth with Poverty reduction.* Washington DC: Palgrave Macmillan.

OECD. (2006). Institutional Bottlenecks for Agricultural Development: A Stock-taking Exercise Based on Evidence from Sub-Saharan Africa. (OECD Development Centre Working Paper No. 248).

OECD/FAO. (2016). Agriculture in Sub-Saharan Africa: In OECD/FAO, *OECD-FAO Agricultural Outlook 2016-2025.* Paris: OECD Publishing. doi:10.1787/agr_outlook-2016-5-en

Peprah, J. A., Afoakwah, C., & Issac, K. (2016). Crop Yield Volatility among SMFs in Ghana. (UNU-INRA Working Paper No.15).

Poole, N. (2017). *Smallholder Agriculture and Market Participation.* Rugby, UK: Practical Action Publishing. Retrieved from http://dx.doi.org/10.3362/9781780449401

Republic of South Africa. (2012). *A framework for the development of SMFs through cooperative development-.* Pretoria. Retrieved from https://www.daff.gov.za

Scoones, I. (1998). Sustainable Rural Livelihoods: A Framework for Analysis. (IDS Working Paper 72).

Swedish International Development Cooperation Agency (SIDA). (2001). *The Sustainable Livelihood approach to Poverty Reduction: An Introduction.* Uppsala. Retrieved from https://www.sida.se

UN. (2015). *Transforming Our World: The 2030 Agenda for Sustainable Development.* New York: UN.

UNCTAD. (2015). *The Least Developed Countries Report 2015: Transforming Rural Economies.* Geneva.

UNCTAD. (2015). *The role of SMFs in sustainable commodities production and trade.* Geneva.

World Bank (WB). (2013). *Unlocking Africa's Agricultural Potential: An Action Agenda for Transformation.* Washington DC. Retrieved from http://worldbank.org

Introducing Sodium-ion Batteries into the African Market: Eritrea as a Gateway of the New Arcade

Yemane Ateshim,[1] Natnael Issak,[2] and Kibra Hayle[1]

[1]College of Business and Social Sciences, Adi-Keih
[2] College of Engineering and Technology, Mai-Nefhi
Email: yemanenegash18@gmail.com,

Abstract

Energy sustainability is a demanding issue for governments, industries, and researchers worldwide. Over the past few decades, the excessive use of non-renewable fuels has significantly elevated CO_2 levels, harming the environment, climate, and public health. While Lithium-ion battery (LIB) consumption has surged since 1991, the limited availability of lithium and its growing demand has raised economic and ecological concerns. These challenges highlight the urgent need to explore alternative battery technologies and develop sustainable energy solutions. Sodium-ion batteries (SIBs) are emerging as transformative solutions to energy storage challenges, driven by the limitations of LIBs and the global need for sustainable and renewable energy systems.

This study explores the potential of introducing SIB technology in the African market, positioning Eritrea as a strategic hub. Key motivations include the abundance and low cost of sodium, which can be extracted from sea salt, and Eritrea's geographic and economic advantages, including its location along the Red Sea and substantial salt production capacity. Eritrea's salt resources, including the highly saline waters of the Red Sea and the productive capacities of sites like Halibay and Menkaekae, present a unique opportunity to develop a cost-effective supply chain for SIB production. Using a qualitative approach, the paper discusses SIBs' advantages over LIBs, such as higher safety, lower environmental impact, extended cycle life, faster charging, and affordability compared to LIBs. The findings point out the significance of Eritrea's resources and strategic position for renewable energy integration and distribution across the continent. The study highlights collaboration with international stakeholders, particularly China, as a critical enabler for scaling SIB technology. This research emphasizes the potential economic, environmental, and technological benefits of SIB adoption, offering a roadmap for Eritrea to lead Africa in sustainable energy innovation.

Keywords: Eritrea, Sodium-ion batteries, Market, Renewable energy, Sustainable

Introduction

It is well recognized that renewable energy sources, such as solar, wind, hydropower, geothermal, biomass, waves, ocean currents, and tidal energy, will be the energy of the future. On the other hand, solar energy is thought to be the most plentiful and cleanest renewable energy source. However, this energy needs to be effectively stored in a high-quality rechargeable battery. The development of new-generation rechargeable batteries is a determining factor in the future of energy storage, which is key to decarbonization and the energy transition in the face of the challenges of climate change.

Research on LIB was conducted in 1970-1980 and Sony became a successful pioneer in the commercialization in 1991, since their emergence, lithium batteries have dominated the energy storage sector (Lin *et al.,* 2017). Nonetheless, the demand for the mineral has significantly increased as a result of this leadership and does not appear to be declining. The possibility of lithium supply constraints, which could lead to price increases and delays because of the shortage, has so gained attention. Furthermore, lithium-ion has a short lifespan, safety issues, an adverse effect on the environment, and temperature sensitivity (Brunelli *et al.,* 2023).

The study aims to introduce SIB in Eritrea which can be carried across to the greater African markets. This initiative is expected to significantly enhance economic growth, promote renewable energy, address environmental issues, improve energy storage efficiency, and contributes to technological advancements in nation-building. The findings could also be used in other African nations to launch comparable initiatives. This study is conducted in Massawa salt production–Salina Company over three months, using interviews and secondary sources. The study does not include other substitutes for lithium in the production of energy storage devices such as LiFePO4 (lithium iron phosphate), graphene, aluminum-ion batteries, and others.

This paper pursues to provide actionable insights into the potential of SIBs to transform Africa's renewable energy landscape while emphasizing Eritrea's strategic role in sustainable energy innovation. Following the introduction, Sections 2, 3, and 4 outline the problem statement, objectives of the study, and the research methodology consecutively. Section 5 reviews existing literature and elaborates on the central argument, offering a detailed analysis of SIBs. The core of the paper, Section 6, examines why Eritrea is well-positioned to serve as a gateway to new energy markets in Africa. Section 7 presents a thorough analysis and interpretation of qualitative data gathered through interviews with the key stakeholders. Section 8 concludes the study by summarizing the findings, while Section 9 offers targeted recommendations informed by the research, and finally Section 10 shapes the proper citations of the references used in the article.

Problem Statement

Traditional energy storage solutions, primarily dominated by LIBs, face serious limitations including high extraction costs, resource scarcity, and environmental concerns, which exacerbate the global supply chain disruptions (Brunelli *et al.,* 2023). The environmental impact of LIBs, from the energy-intensive production processes to the hazardous disposal of spent batteries, further complicates their widespread adoption (Patel, 2023). Moreover, access to energy continues to be a major issue in sub-Saharan Africa, where about 568 million people were without electricity in 2019, hindering progress toward achieving the United Nations Sustainable Development Goal of providing affordable and clean energy for everyone (Opoku & Song, 2023). The Ministry of Energy and Mining in Eritrea indicated that the current annual energy consumption of 216.76 Gigawatt (GWh) is far below the expected demand of 650.28 GWh, highlighting a significant energy shortfall that hinders economic growth and development. Despite growing interest in SIBs due to their enhanced safety, lower costs, and sustainability, the technology remains underdeveloped in Africa, and Eritrea has yet to capitalize on this opportunity (Desai *et al*., 2022).

Research Objectives and Questions

The primary objective of this research is to explore the potential of introducing SIBs into the African market, with Eritrea positioned as a strategic gateway for this transition. Additionally, the research aims to evaluate the economic, environmental, and technological advantages of SIBs over traditional LIBs, focusing on their cost-effectiveness, safety, and environmental benefits. By leveraging Eritrea's abundant salt resources and its strategic location along the Red Sea, the study explores the feasibility of establishing a local SIB production industry capable of supplying regional and global markets. The specific objectives of the study are:

- To assess the contribution of SIBs to sustainable energy solutions and environmental benefits over LIBs in terms of resource extraction, manufacturing, and disposal, and how SIBs could contribute to reducing Africa's carbon footprint and mitigating the environmental challenges posed by the growing demand for LIBs.
- To analyze the rationale for selecting Eritrea as the primary hub for the introduction of SIB technology, particularly focusing on its vast natural salt reserves, its geographic advantages, and its position as a growing economic center within the Horn of Africa.

In light of the study's objective and problem statement, the main research questions this paper will try to answer are:

- How do SIBs compare to LIBs in terms of cost, performance, safety, and environmental impact, particularly for large-scale energy storage applications in Africa?
- What factors make Eritrea an ideal location for becoming a hub for SIB production in Africa, considering its natural resources, infrastructure, and geographical position?

Methodology

This research employed a combination of primary and secondary data collection methods. Primary data was gathered through non-structured interviews with key stakeholders. Participants were purposively selected based on their expertise and institutional roles. Secondary data was collected by extensively reviewing academic articles, books, reports, and reliable online sources.

Why Sodium-ion batteries?

SIB is a type of rechargeable battery that uses sodium ions (Na^+) as the primary component of its electrolyte for conducting an electrical charge (Luke, 2022). Sodium ions can be found from evaporation of sea salt (brines) or rock salt (Halite). Halite naturally occurs in crystalline form with an extensive deposit of 90–99% sodium chloride, while brine is typically only 23.3% of sodium chloride, meaning it is 76.7% water (City of Coralville, 2016). Desalination batteries also extract sodium and chloride ions directly from seawater. Through the input of electrical energy, they create fresh, drinkable water and batteries. They are an efficient and sustainable technology. They address the dual challenges of water scarcity and energy demands. This approach highlights the potential of renewable energy integration. (Nurohmah *et al.*, 2022).

By 2031, the global market for SIBs is expected to have grown from $528 million in 2021 to $4.37 billion. This growth is fueled by the advantages of sodium-ion technology, the rising use of renewable energy sources like solar and wind, and the increasing adoption of electric vehicles (BIS Research, 2022). Currently, 31 startup companies are working on manufacturing SIBs, indicating a rising interest in this technology (Tracxn, 2024).

So why Sodium? When you take a look at the periodic table, you might notice that sodium and lithium are neighbors sharing similar electrochemical principles and physical features (Mogensen, 2020). SIB is considered one of the candidates for LIB succession in the following 5-10 years because of its abundance. Sodium covers about 2.7% of the Earth's crust, versus lithium's 0.006% (Energy T.S, 2016). Kevin in his article "Rising demand for LIBs' may lead to raw material supply shortages" described that the demand for electric vehicles is expected to be more than 30% after 2030 which only accounted for 4% in 2021. Lithium demand has risen 30 times in the energy transition, per the new models of solar panels, electric vehicles, and other technological devices, which surely every element of lithium on this planet will be used after those demands are met; it is these fears of raw material shortages and price increases that constitute the primary reason for the research devoted into the SIB technology (Jost, 2022).

Sarah (2021) stated that SIBs are a promising solution for grid storage because they are inexpensive as it is abundant, sustainable, and eco-friendly; have no safety issues, and have suitable energy densities. Materials used in LIBs, such as cobalt and copper, are expensive to mine and environmentally damaging, leading to ecological toxicity, resource depletion, global warming, and adverse effects on human health. SIBs' have more charging cycles than LIBs', it is also charged faster and can deliver energy more quickly than LIBs' (Sarah, 2021). Desai, Huang, Foix, Tarascan, & Mariyappan (2022) mentioned that SIB is more stable and safer than LIB when stored at 0 volts, whereas the maximum state of charge in LIB is 30% voltage which leads to a diminution of performance in the battery if the discharge rate drops. The article also added that LIB's have many safety issues, during transportation LIB was labeled as "dangerous goods" by the United Nations Economic Commission for Europe (UNECE) and the International Civil Aviation Organization (ICAO) due to its high thermal runaway (-25/40°C). But still, after the restrictions there have been about 354 incidents in the last 15 years be it a fire, smoke, or even an explosion that happened during a cargo shipment or air load (Desai, 2022).

LIBs currently exhibit an energy density between 200 and 300 Wh/kg (Thundersaid Energy, 2024). In comparison, a new generation of high-density sodium-ion battery (SIB) technology, set to be introduced by 2025, is projected to achieve an energy density between 380 and 700 Wh/kg under room temperature conditions. This technology is also expected to demonstrate full-cycle efficiencies ranging from 6,000 to 12,000 cycles, with a calendar life surpassing 30 years (Itani & De Bernardinis, 2023). In contrast, LIBs typically offer a charge cycle life of only 500 to 1,000 cycles, which is substantially lower than that of SIBs (Mossali *et al.*, 2020).

The price of lithium carbonate, a key material used in battery manufacturing, has experienced significant fluctuations. As of November 2024, it ranges between $10,000 and $23,000 per metric ton, depending on the market and region (ChemAnalyst, 2024). In contrast, sodium, which is more abundant and cheaper to extract, is priced significantly lower. Sodium carbonate, a common sodium compound, costs approximately $150 to $200 per metric ton on average (ChemAnalyst, 2023; Procurement Resource, 2023).

Table 1 - The table below summarizes the difference between SIBs' and LIBs'.

Grounds	Lithium-ion batteries	Sodium-ion batteries
Safety	Low	High
Price	\$10,000-\$20,000 per metric ton	\$150-\$200 per metric ton
Energy density	200-300 Wh/kg	380-700 Wh/kg
Mining process	Threat to environment	Environmentally Friendly
Abundance	Scarce (0.006%)	Earth Abundant (2.7%)
Charge Cycling	500-1000 cycles	6000-12000cycles
Thermal runaway	–25°C to 40°C	–40°C to 60°C
Min. Discharge rate	30%	0% (0 Volt)
Transportation	Restrictions on discharge rate	Easy Transportation (0 volt)
Charging speed	Low	High
Market availability	More	New and progressive

Source: Authors' Construction

SIBs are emerging as a promising alternative to LIBs due to their high energy density, low-cost raw materials, abundant availability, sustainability, and environmental safety. These batteries are particularly suited for large-scale renewable energy storage applications, such as solar power plants, electric vehicles, and data centers. Their potential to drive down the costs of solar power grids, automobiles, and electronic devices makes them an attractive choice for future energy solutions. As research and development continue, sodium-ion technology is poised to play a crucial role in advancing sustainable energy systems and reducing dependence on scarce resources. This innovation could revolutionize energy storage while promoting environmental and economic benefits.

Eritrea as a Gateway: Strategic Opportunities and Potential

Eritrea, located in the Horn of Africa, is a relatively small but strategically significant country that offers considerable potential for economic growth and regional integration. It has a significant geographic position, bordering to the west by Sudan, to the south by Ethiopia, to the southeast by Djibouti, and to the east by the Red Sea (Pateman, 1998). Eritrea is ideally positioned to become a gateway to new markets in Africa because of its wealth of natural resources, geographic location, strategic international alliances, and dedication to regional integration.

Availability of the Raw Materials

Eritrea's natural resources present a significant advantage in its quest to establish itself as an economic hub in Africa. Among these, the country's salt production is one of its most notable assets. The Red Sea, bordering Eritrea to the east, is known for its high salinity levels, making it one of the saltiest bodies of water in the world. The Red Sea contains the saltiest ocean water, with an evaporation rate of almost 40%, according to the National Weather Service, giving Eritrea access to abundant natural resources (Mulroy, 2024).

In 2013, Eritrea's salt production saw a notable increase of 21%, with two major salt companies, Assab and Salina Salt Works, producing 290,000 tons of salt annually. Assab Salt Works alone produced 150,000 tons per year, while Salina Salt Works contributed 80,000 tons (Yager, 2016). Furthermore, about 93% of the salt produced is exported, with only 7% consumed domestically (Woldu, 2013).

Eritrea's potential for expanding salt production is also evident with the discovery of two new salt mining sites: Menkaekae and Halibay. The Menkaekae site near Assab is projected to yield up to 10 million tons of salt, while the Halibay site holds the promise of producing one million tons, valued at billions of dollars (Tesfamichael, 2024). Additionally, the discovery of thick salt formations and salt pans in areas such as Lake Asale and Erta Ale (an active volcano) further emphasizes the potential for large-scale salt extraction (Itani & De Bernardinis, 2023)including electromobility and stationary domains. For e-mobility, batteries are essential components in various types of electric vehicles (EVs.

Beyond salt, Eritrea's raw materials, particularly in the mining sector, position the country as a key player in the global supply chain. The country could leverage these resources, including the salt, to produce value-added products such as SIBs. Eritrea could tap into the growing demand for renewable energy storage solutions by establishing partnerships with global battery manufacturers, especially those in China.

Strategic Location

Eritrea's geographical location is another significant factor that positions it as an emerging gateway to African and international markets. Situated along the Red Sea, Eritrea has historically been a point of interest for regional and global powers due to its access to major international shipping routes (Connel, 2005). The country's strategic location provides access to the Middle East, Asia, Europe, and other parts of Africa, making it a valuable transit hub for trade and commerce. Due to its geographical position, investment, climate, people, and resources, Eritrea has a significant edge in developing a competitive industrial sector (Hailemariam, 2001).

Eritrea's importance has been recognized in various national policies aimed at fostering economic growth through international trade. In 2001, the government enacted the Eritrean Free Zones Proclamation, which aimed to create an environment conducive to both domestic and foreign investment, particularly in export-driven industries (Government of Eritrea, 2001). This initiative was reinforced by the country's macro-policy of 1994, which emphasized the importance of export promotion and industrial development (Government of Eritrea, 1994).

Eritrea's government has also established two export processing zones (EPZ): in its port cities of Massawa and Assab. By providing tax breaks, customs exemptions, and expedited business registration processes, these zones aim to draw in foreign investment (Conference, N., & Trade, O. N, 1985). The strategic locations of Massawa and Assab along key maritime routes provide logistical advantages for international trade, further enhancing Eritrea's potential as a global trade gateway.

Massawa's position along one of the world's busiest sea lanes enhances its significance, as the port can facilitate the easy flow of goods for import and export purposes. This logistical advantage, combined with the Eritrean government's investment policies, positions the country as a prime candidate for foreign direct investment, which is crucial for driving economic development in a growing market (Yohannes, 2014).

Eri-China Diplomacy

One of Eritrea's most promising international partnerships is with China, a country that has long been an advocate for developing nations. Eritrea and China share a rich history of cultural and commercial ties, dating back to as early as the first century AD, as evidenced by the discovery of a Chinese wine vase near the ancient port of Adulis (Zigang, 2017). Over the years, China has remained a key partner in Eritrea's economic development, particularly in the mining, infrastructure, and agriculture sectors. Up until now, Eritrea has mostly engaged in bilateral economic relations with Russia, India, and the People's Republic of China (PRC). All of Eritrea's mines are owned by the People's Republic of China (PRC), which also has significant commercial investments in the country (U.S. Embassy Asmara, 2023).

China's Belt and Road Initiative (BRI), which aims to enhance global connectivity and trade, has also been a key factor in Eritrea's economic engagement with China. Eritrea has responded positively to the BRI, seeing it as an opportunity to strengthen its economic and social development. The cooperation between Eritrea and China is a win-win situation, with China offering infrastructure development and Eritrea providing valuable resources (Zigang, 2017). This collaboration has the potential to drive further industrialization and economic growth in Eritrea, positioning the country as a central player in global trade.

The bilateral relationship between China and Eritrea has expanded significantly in recent years, with China investing heavily in the country's resources and infrastructure. Nearly 20 Chinese companies operate in Eritrea, spanning industries such as mining, electricity, telecommunications, and construction. The partnership between the two nations has been particularly beneficial in terms of trade, in the last 5 years the exports of Eritrea to China have increased to an annualized rate of 21.3% with Eritrea exporting key products like zinc and copper ore to China. In 2022, Eritrea's exports to China amounted to $412 million, driven by high-value products such as zinc ore and copper ore (OEC, 2024).

Eritrea's Integration in to International Organizations

Eritrea's involvement in international organizations has strengthened its position as a key gateway to Africa. During his annual speech in March 2023, the President of Eritrea outlined new policies aimed at fostering economic growth, focusing on sectors such as mining, renewable energy, water conservation, and agriculture. The country then rejoined the Intergovernmental Authority on Development (IGAD) in 2023, signaling a renewed commitment to regional integration and cooperation. Eritrea's return to IGAD is seen as a positive step toward enhancing relations with neighboring countries and promoting regional stability (U.S. Embassy Asmara, 2023).

Eritrea is an active participant in several international organizations, including the African Union (AU) and the United Nations (UN), and it holds observer status in the Arab League. These affiliations enable Eritrea to engage in international trade agreements, foster diplomatic relations, and participate in development initiatives. Notably, in 2022, Eritrea entered into a five-year Sustainable Development Cooperation Framework with the UN, prioritizing areas such as renewable energy, agriculture, and water conservation (United Nations, 2022). Furthermore, Eritrea has collaborated with the African Development Bank (AfDB) to advance solar energy projects, addressing persistent electricity shortages. These initiatives stress Eritrea's commitment to sustainability and economic growth, which are critical to developing a competitive industrial sector (AfDB group, 2023).

Trans-African Highways (TAH) Network

The trans-African highways (TAH) network is one of the most ambitious infrastructure projects in Africa, linking major cities and regions across the continent. Spanning over 60,000 kilometers and passing through 18 countries, the TAH network is seen as a vital route for trade and economic integration in Africa. Eritrea's participation in this network, especially through its connections with the Red Sea, further strengthens its role as a gateway to regional and international markets. One of the key benefits of the TAH network is its potential to enhance economic growth across Africa by facilitating the movement of goods, services, and people. As part of this project, Eritrea could contribute to solving the power supply challenges in Sub-Saharan Africa by producing and distributing SIBs, a technology that could address power surge issues across the region. The TAH project is projected to cost around $30 billion, with a focus on boosting trade, improving infrastructure, and creating economic opportunities across the continent (Pee, 2022).

Table 2 - TAH routes

Route Number	Route	Road Lengths
TAH 1	Cairo – Dakar	8636 kilometer (km)
TAH 2	Algiers – Lagos	4504 km
TAH 3	Tripoli – Windhoek – Cape Town	9610 km
TAH 4	Cairo – Gaborone – Cape Town	8860 km
TAH 5	Dakar – N'Djamena	4500 km
TAH 6	N'Djamena – Djibouti	4220 km
TAH 7	Dakar – Lagos	4760 km
TAH 8	Lagos – Mombasa	6260 km
TAH 9	Beira – Lobito	3520 km
TAH 10	Djibouti – Libreville - Bata	7000 km

Source: African Union, 2023

Key Findings and Discussion of the Study

This section offers a thorough exploration of the key factors influencing Massawa Salt Factory (MSF) production capacity, the opportunities for expansion, and the potential to export. The findings highlight the historical evolution of MSF, its current operations, and prospects, especially at the Halibay site, which has become central to Eritrea's salt production ambitions. With a combination of favorable environmental conditions, high evaporation rates, and vast untapped resources, the country is poised to become a key player in the global salt market.

Historical Development of MSF

MSF's general manager, Mr. Negassi Tesfay, started his interview by narrating the history of MSF. Massawa Salt Factory, founded 117 years ago in 1907 by two Italian investors, originally bore the name "Societa` Italiana Salina per L'Eritrea." In its early stages, the factory faced significant challenges related to labor-intensive production methods. Workers utilized basic tools, including wagons, mattocks, shovels, and baskets, to harvest salt. Despite the difficult working conditions, the factory managed to operate for several decades, though its production gradually declined during the Second World War and the subsequent period of the Ethiopian Federation.

The Ethiopian government's policies, particularly during the period of the Dergue regime in the 1970s, led to the confiscation of the factory in 1975 from its original owners: a family of Caveleri Michealle Gostala. This move marked a significant turning point, as mismanagement, poor coordination with workers, and an inability to meet customer demands resulted in a sharp decline in production. The factory's diminished performance led to the loss of its international market, further hampering its growth potential. However, with the liberation of Massawa in 1990, the Eritrean government undertook efforts to renovate the factory and revitalize its operations. This revival resulted in a steady increase in production capacity, which reached 83,000 tons per year by 1995.

In 1998, a collaboration between the British company Mandorstand PLC and the Eritrean firm JEDEC conducted a comprehensive study on potential areas for salt production. The study focused on three key locations: Aftat, Beremi, and Halibay. The results indicated that Halibay, located approximately 40 kilometers southeast of Massawa, had the highest-quality salt deposits. As a result, the study recommended the establishment of a large facility in Halibay, which was deemed to have the capacity to produce up to 1 million tons of salt per year across 3,800 hectares. More recent studies by Mr. Tesfay have suggested that the Halibay area could be expanded to cover 7,600 hectares, enabling the production of up to 3 million tons annually.

Modernization and Expansion Efforts

MSF's expansion efforts continued into 2016, with the installation of new machinery, including evaporators, concentrators, and crystallizers. Additionally, two generators (330 kVA) and two high-speed pumps (capable of pumping 2,000 cubic meters of water per hour) were added, with support from Sawa Building Construction. These upgrades have contributed to an increase in the factory's production capacity to 100,000 tons of salt annually across 84 hectares of land. Despite the 1998 study's projections, which suggested that 1,200 hectares of land would be needed to produce 200,000 tons and 3,800 hectares to reach 1 million tons, the factory has demonstrated that just 200 hectares are sufficient to meet the 200,000-ton production goal.

One of the most striking findings from Mr. Tesfay's interview was the exceptionally high evaporation rate at Halibay. Both the northern and southern Red Sea coastal areas benefit from an extended period of evaporation, with the prime harvesting months from June to September. However, Halibay enjoys a nearly year-round evaporation period, offering a significant advantage in the salt production process. This high evaporation rate, coupled with the region's salinity, which is saturated at 3.5% brine, is a key factor in Halibay's potential for salt production, making it one of the most fertile sites for salt production in Eritrea.

MSF primarily produces standard sodium chloride (NaCl) with a purity of 97%, which meets the Eritrean standard quality of salt in the global market. According to Mr. Tesfay, 94% of MSF's production is exported for industrial uses, such as textile dyeing, fish preservation, glass manufacturing, leather processing, animal feed, water treatment, soap making, rubber production, and aluminum manufacturing. The remaining 6% of the output is utilized for domestic consumption. MSF exports salt to neighboring countries like Ethiopia, Kenya, Tanzania, Uganda, and Rwanda, as well as to international markets in Germany and Japan. Despite the challenges posed by outdated refinery machinery, the factory has managed to maintain its position as a leading supplier of high-quality salt.

Looking ahead, Mr. Tesfay envisions continued growth for the factory, particularly at the Halibay site. MSF plans to expand the factory's land area to 200 hectares by 2025, to produce 200,000 tons of salt annually. Furthermore, the company aims to extend the production area to 500 hectares, which would increase the factory's output to 1 million tons per year. Based on projections by Mr. Tesfay, the potential production capacity of Halibay could reach 3 million tons of salt, with an estimated market value of $1.2 billion. If the Menkakae site is developed as well, the combined production capacity of both Halibay and Menkakae could reach 13 million tons of salt, potentially yielding $5 billion in revenue.

Technological Concerns

Despite the significant progress made at MSF, several encounters remain. The factory's refinery machinery, which is critical to the production process, is aging and worn out. The lack of modern machinery is a major obstacle to expanding export capacity and meeting the increasing demands of the global market. As a result, MSF struggles with producing high-quality salt at scale, which limits its potential for growth and global market penetration.

Furthermore, the technological landscape for salt production in Eritrea is evolving. The Director-General of the Ministry of Energy and Mining (MOEM), Mr. Tesfay Zekarias, discussed the potential for SIBs as an alternative to the more commonly used LIBs in the country's renewable energy sector. He also added that while LIBs are currently in use, the introduction of SIBs could present significant advantages for Africa's sustainable energy future. According to Mr. Tesfay, the adoption of SIBs could not only benefit Eritrea's energy grid but also have a broader impact on the African and global energy markets.

On November 30, 2024, Eritrean President Isaias Afewerki gave an explicit interview on national television (Eri-Tv), revealing a major initiative for the development of the country's energy infrastructure. He outlined plans to enhance both thermal and renewable energy capacities, to reach a national electricity supply of 2000 MW. This ambitious plan is aimed at strengthening the nation's energy resilience and supporting its future growth. In line with this, Mr. Ataklti Tewelday from the MOEM discussed that the total annual energy consumption in Eritrea is approximately 216.76 GW/h. The country's electrification rate remains underdeveloped, and there is a pressing need for increased energy capacity to meet the demands of the population. Engineer Tewelday notes that Eritrea's energy demand is projected to be three times higher than its current consumption, reaching an estimated 650.28 GW/h. The government has planned for the installation of additional solar power capacity, such as the upgrade of solar panels in Mai'Dma-Areza from 22 megawatts (MW) to 30 MW in Deckemhare, which will help improve the country's renewable energy capacity.

Expanding the Salt Production and SIBs

Eritrea's government, through its various ministries and agencies, is committed to expanding the country's salt production capabilities. Mr. Mussie Tewolde, from the Mapping and Information Center in the President's Office, discussed the government's plans to develop salt production fields across the country. In particular, the area between Bera'sole and Rehhayta, spanning 50,000 to 60,000 hectares, is identified as having a high potential for salt production due to its clay-rich soil. Additionally, the area between Halibay and Menkakae, stretching for approximately 600 kilometers, contains vast salt deposits that remain largely untouched.

Mr. Zeray Leake, country manager of South Boulder Mines, demonstrated that the Colluli project is a large multi-salt resource located 75km from the Red Sea coast and 180km from the port of Massawa: the key import-export facility in Eritrea. The Colluli project has uncovered an extremely massive reserve of Rock salt (Halite), with an estimated thickness of 1 meter in the upper layers that cover the potash deposit and up to 10 meters in the lower layers underneath the potash. This discovery underscores the vast untapped potential of Eritrea's salt resources, which, if fully developed, could place Eritrea among the leading global producers of salt.

Chinese Solar Energy Expansion in Eritrea

Mr. Qiu Yunlong, a Chinese representative at the Chinese Embassy, discussed the ongoing progress in solar energy projects in Eritrea, noting the significant advancements made through China-Eritrea cooperation in the development of renewable energy infrastructure. Completed projects include a 2 MW small-scale photovoltaic (PV) power station and a two-phase solar street lighting project in Asmara.

One of the major ongoing projects in Deckemhare is the 30-megawatt solar PV power plant, funded by the AfDB. The contract for this project has been awarded to China Energy Engineering Group Shanxi Electric Power Construction Co., Ltd., with a total value of 31,370,277.26 ERN, commenced on March 1, 2024, and is expected to be completed within 24 months, including a one-year defect liability period. The Dekemhare project represents a significant milestone in Eritrea's renewable energy strategy, underscoring the country's commitment to sustainable infrastructure development and strengthening its international partnerships.

Furthermore, the Chinese delegates from the Department of Economics and Marketing at the Embassy expressed concerns about the premature introduction of SIBs in Africa, particularly given their current stage of development and production capabilities. They agreed that the scarcity of lithium, projected to occur within the next 8-10 years, could present a challenge for the continued use of LIBs. In this context, SIBs are considered a viable alternative, with Eritrea potentially playing a crucial role in addressing this challenge. The delegates highlighted Eritrea's strategic position as key to securing the future of SIB technology in Africa. By embracing new energy technologies such as SIBs, Eritrea could significantly enhance its standing in the global energy market, positioning itself as an essential player in Africa's renewable energy future. Currently, while China operates two subsidiaries in Africa focusing on the production of electric vehicles in Morocco and Egypt, no battery manufacturing subsidiaries exist on the continent to power these vehicles, which are assembled in Africa and sold to larger international markets.

Conclusion

The research has demonstrated that SIBs present a transformative opportunity for sustainable energy storage, particularly in addressing Africa's renewable energy challenges. Their low cost, environmental friendliness and accessibility make them a viable alternative to LIBs. Moreover, the research validated, that Eritrea's vast salt resources and strategic geographic location position it as an ideal hub for introducing SIBs into the African market. Leveraging its abundant natural resources, such as the high-quality sodium chloride deposits in the Red Sea region, alongside favorable trade partnerships, Eritrea can play a pivotal role in this energy revolution.

In conclusion, SIBs offer a practical, scalable, and eco-friendly solution to Africa's energy storage needs. As Ambassador Yang Zigang of China (Zigang, 2017) in Eritrea stated "Eritrea has an advantage to exploit solar energy since most of the country is highlands in low geographical latitude with very strong sunshine". Eritrea's unique resources and strategic initiatives make it an ideal gateway for this technology, highlighting the country's potential to lead the region in renewable energy innovation is vigorous. Further research and investment are critical to realizing this vision, paving the way for a sustainable and energy-secure future.

Recommendations

Therefore, based on the analysis and conclusions presented in the paper, the researcher offers the following recommendations:

- Eritrea should prioritize modernizing its salt production facilities to increase the quantity and quality of sodium chloride for use in SIB production.
- Establish partnerships with international battery manufacturers and research institutions to accelerate the transfer of SIB production technology.
- Promote Eritrea's strategic advantages to global markets, emphasizing its geographic location and abundant raw materials for renewable energy solutions.
- Engage with regional bodies, such as the African Union (AU) and IGAD, to promote SIB adoption across the continent.
- Establish a sustainable supply chain for raw materials in Africa and battery components to ensure scalability and long-term production.
- Expand research funding to explore further applications of SIBs and improve their energy density and lifespan.

References

African Development Bank Group. (2023). *Eritrea strengthens its partnership with the African Development Bank for growth.* https://www.afdb.org/en/news-and-events/

African Union. (2023). *TAH network annex I.* African Union. https://au.int/sites/default/files/newsevents/workingdocuments/29736-wd-e_-_tah_annex_i_tah_network.pdf

BIS Research. (2022). *Sodium-ion batteries market size forecast - 2031.* BIS Research. https://bisresearch.com

Brunelli, K., Lee, L. Y., & Moerenhout, T. (2023). *Fact Sheet: Lithium Supply in the Energy Transition.* Columbia SIPA, December, 1–2. https://www.energypolicy.columbia.edu/publications/fact-sheet-lithium-supply-in-the-energy-transition/

ChemAnalyst. (2023). *Sodium carbonate price analysis: Market trends, forecasts, and outlook.* ChemAnalyst. https://www.chemanalyst.com

ChemAnalyst. (2024). *Lithium carbonate price analysis: Market trends, forecasts, and outlook.* ChemAnalyst. https://www.chemanalyst.com

City of Coralville. (2016). *Anti-icing brine fact sheet.* Coralville, IA. https://www.coralville.org/DocumentCenter/View/6806/Anti-Icing-Brine-Fact-Sheet?

Conference, N., & Trade, O. N. (1985). *Trade And Development.* https://unctad.org/system/files/official-document/tdr5_en.pdf

Connel, D. (2005). *Eritrea on a slow fuse.* In Battling Terrorism in the Horn of Africa (pp. 62–94). Brookings.

Desai, Parth; Huang, Jiaqiang; Foix, Dominique; Tarascan, Jean-Marie; Mariyappan, S. (2022). *Zero volt storage of Na-ion batteries: Performance dependency on cell chemistry.* Power Sources, 551[15].

Energy, T. S. (2016). *CATL : sodium ion battery breakthrough ?* https://thundersaidenergy.com/downloads/catl-sodium-ion-battery-breakthrough/

Government of Eritrea. (1994). *Macro-policy.* Asmara, Eritrea: The Government of Eritrea.

Government of Eritrea. (2001). *Proc-115/2001 The Eritrea Free Zones Proclamation* (pp. 6–25). http://www.eri.gov

Hailemariam, S. (2001). *Corporate value creation, governance, and privatization.* University of Groningen. https://www.rug.nl

Itani, K., & De Bernardinis, A. (2023). *Review on New-Generation Batteries Technologies: Trends and Future Directions.* Energies, 16(22).

Jost, K. (2022). *Rising demand for lithium-ion batteries amid material supply shortages.* Futurride. https://futurride.com/

Lin, D., Liu, Y., & Cui, Y. (2017). High-energy batteries. *Nature Publishing Group*, 12(3), 194–206.

Luke, J. (2022). *What is a sodium-ion battery? Definition, structure, and more.* Power & Beyond. https://www.power-and-beyond.com

Mogensen, R. (2020). *Realization of Sodium-ion Batteries.* Uppsala University.

Mossali, E., Picone, N., Gentilini, L., Rodrìguez, O., Pérez, J. M. & Manuel, J. (2020). *Lithium-ion batteries towards a circular economy : A literature review of opportunities and issues of recycling treatments.* Environmental Management, 0–21.

Mulroy, C. (2024). *Why is the ocean salty? Ocean salinity explained, plus the world's saltiest ocean.* USA TODAY. https://www.usatoday.com

Nurohmah, A. R., Nisa, S. S., Stulasti, K. N. R., Yudha, C. S., Suci, W. G., Aliwarga, K., Widiyandari, H., & Purwanto, A. (2022). *Sodium-ion battery from sea salt: a review*. In Materials for Renewable and Sustainable Energy (Vol. 11, Issue 1, pp. 71–89).

Observatory of Economic Complexity (OEC). (2024). *Eritrea (ERI) trade partners with China (CHN)*. MIT Media Lab. https://oec.world/en/profile/bilateral-country/eri/partner/chn

Opoku, P., & Song, H. (2023). *Sustainability and affordability of Chinese-funded renewable energy project in sub-Saharan Africa...*, Environmental Science and Pollution Research, 30(33), 80768–80790.

Patel, T. (2023). *A Comparative Study of Lithium-ion and Sodium-ion Batteries : Characteristics, Performance, and Challenges*. FAU

Pateman, R. (1998). *ERITREA Even The Stones Are Burning (2nd ed.)*. The Red Sea Press, Inc.

Pee, O. (2022). *TransAfrica Highway, the next transport project of the century? Transport Africa*. Transportation and Travel Trends in Africa.

Procurement Resource. (2023). *Sodium carbonate price and market analysis*. Procurement https://www.procurementresource.com

Sarah, H. . (2021). *Sodium-ion Batteries for Grid Storage: Investigation of Bulk and Interfacial Properties of Electrode Materials*. University of California San Diego.

Tesfamichael, H. (2024). *Massawa Salt Factory: Revitalizing Eritrea's Salt Industry*. Ministry of Information. http://www.shabait.com/Shabait.com

Thundersaid Energy. (2024). *Lithium-ion batteries energy density*. Thundersaid Energy. https://thundersaidenergy.com/downloads/lithium-ion-batteries-energy-density/

Tracxn. (2024). *Startups in sodium-ion batteries*. Tracxn. https://tracxn.com/d/trendingbusinessmodels/startupsinsodiumionbatteries/

United Nations. (2022). *Eritrea signs five-year Sustainable Development Cooperation Framework with the UN*. https://eritrea.un.org

U.S. Embassy Asmara. (2023). *Integrated Country Strategy: Eritrea for public release*. https://www.state.gov/wpcontent/uploads/2023/11/ICS_AF_Eritrea_Public.pdf

Woldu, B. (2013, May 31). *Eritrea's investment opportunities*. Ministry of Information, 1–4. http://www.shabait.com

Yager, B. T. R. (2016). *2013 Minerals Yearbook*. U.S. Geological Survey. https://www.usgs.gov

Yohannes, M.(2014). *Eritrea as a gateway for investments in Africa : the Export Processing Zone*.

Zigang, Y. (2017). *Eritrea profile: Promoting friendship and cooperation (pp. 1–3)*. Embassy of the People's Republic of China in the State of Eritrea. https://www.chinaembassy.er

Education and Language

Bayesian Ordinal Modelling of Educational Attainment among Four Cohorts of Eritrean Women born Between 1945 and 1989[1]

Gebrenegus Ghilagaber

Stockholm University, Department of Statistics
e-mail: Gebre@stat.sus.se

Abstract

This contribution is both methodological and substantive. It presents a family of modelling approaches for ordered outcome variables (like those in Likert scale) and illustrates them using data on educational attainment among four cohorts of women from rural and urban Eritrea. We demonstrate that the Bayesian ordinal modelling approach adds flexibility in specifying the models and provides goodness-of-fit measures to compare alternative models within the family. Data used for illustration come from about 24000 women extracted from the 1995 Eritrean Demographic and Health Survey (one cohort of women born 1945-59) and the 2010 Eritrean Population and Health Survey (three cohorts of women born 1960-69, 1970-79 and 1980-89, respectively). Results obtained using the *Bayesian Regression Models using 'Stan' (brms) R-Package* show that there has been significant educational progress among the younger cohorts of women (relative to the oldest cohort) and that there were notable urban-rural differences in educational levels within the cohorts of women studied. Further, and more interestingly, goodness-of-fit tests reveal that *an adjacent-category model with category-specific effects provides the 'best' fit for the data than the cumulative logit model that is most common in the literature*. This, in turn, implies that cohort-gradients of educational progress have not been uniform across the four educational levels recorded in the surveys (No Edu- cation, Primary, Secondary and Post-secondary). Ignoring such a lack of uniformity could lead to spurious significance of the cohort variable and erroneous conclusion that educational progress across cohorts differed at all levels of education.

Introduction

In social science research, variables are quantified using measures that fall into four scales: nominal, ordinal, interval and ratio scales. These scales are a hierarchical and possess fewer or higher mathematical properties depending on their hierarchy. For instance, common measures of central tendency (such as the arithmetic mean) and of dispersion (like the standard deviation) do not make sense in variables measured in nominal or ordinal scales.

More importantly, the scales in which variables are measured determine the statistical techniques to be used for analyzing them. Failure to use the appropriate statistical method to analyze a given data set often yields misleading results which, in turn, can mislead policy interventions. For instance, a study by Liddell and Kruschke (2018) shows that using t-tests and analysis of variance (methods that are appropriate for response variables measured in ratio scale) for ordinal responses leads to false alarms (detecting effects that do not exist) while failing to detect effects that exist.

In the present contribution, we demonstrate how educational attainment, which is measured in the lower hierarchy ordinal scale, can be modelled using an advanced statistical model that is appropriate for ordinal response variables. The Bayesian framework on which our modelling is based provides further flexibility in specifying the ordered models and comparing them using goodness-of-fit tests.

We illustrate our methods using data on educational attainment among four cohorts of Eritrean women from the 1995 Eritrean Demographic and Health Survey and the 2010 Eritrean Population and Health Survey. We have fitted four variants of Bayesian ordinal models using the *Bayesian Regression Models using 'Stan' (brms) R-Package*. Our results indicate there has been significant educational progress among the younger cohorts of women and that the urban-rural difference in educational attainment was still notable.

Goodness-of-fit tests reveal that an *adjacent-category model with category- specific effects* provides the 'best' fit for the data than the cumulative logit model that is most commonly used in the literature. We interpret this as unequal progress in the four educational levels recorded in the surveys (No Education, Primary, Secondary and Post-secondary).

Thus, while our contribution is mainly methodological, it also presents evidence on trends and patterns of educational progress among Eritrean women born 1945- 89. Lastly, it adds to the very few works that are based on data from the Eritrean Demographic and Health Surveys.

Our contribution is organized as follows. In Section 2, we introduce the data used for illustration and define our response and explanatory variables. In Section 3, we argue for the choice of our analytical framework and outline the four variants of models that we fit to the data. Our empirical results are presented in Section 4 while we summarize our findings with concluding remarks in Section 5.

The Dataset, Response and Explanatory Variables and Preliminary Examination of Their Relationships

The Dataset

The data used for illustration come from two sources: the 1995 Eritrean Demographic and Health Survey (National Statistics Office, Eritrea, and Macro International Inc. USA, 1997) and the 2010 Eritrean Population and Health Survey (National Statistics Office, Eritrea and Fafo Institute for Applied International Studies, Norway, 2013).

One cohort of women, born between 1945 and 1959, is selected from the former survey while three cohorts of women (born 1960-60, 1970-79 and 1980-89) are selected from that later survey. The oldest cohort, born 1945-59, were 36-49 years old at the time of survey (1995) while the three younger cohorts were 21-49 years at the time of the survey (2010). We believe these cohorts were old enough to achieve the educational levels recorded in the surveys. We didn't include those born 1990-95 and where 15-20 years old in the 2010 survey because they were too young to complete all the educational level recorded.

Response Variable

Our response variable Y is educational attainment and takes one of four possible values: 1 if a woman has reported she had no education at the time of the survey; 2 if she has reported she has primary-level education; 3 if she has reported secondary- level education and 4 if she has reported post-secondary-level education.

Thus, the response variable is ordinal in the sense that $1 < 2 < 3 < 4$. But, it is not metric because the educational levels are not equidistant. For instance, though $4 - 3 = 3 - 2 = 2 - 1 = 1$, this does not imply that the difference between post- secondary and secondary educational levels is the same as the difference between secondary and primary levels or that between primary-level education and no education. Consequently, we have to analyse the data using methods that take due account of the ordinal nature of our response variable. Overlooking this and analyzing the data using methods that treat it as metric will lead to biased estimates and misleading conclusions (Liddell and Kruschke, 2018).

Explanatory Variables

For illustration purposes, we use two explanatory variables, the woman's birth cohort with 4 levels (1 for those born between 1945 and 1959; 2 for those born between 1960 and 1969; 3 for those born between 1970 and 1979 and 4 for those born between 1980 and 1989) and the woman's residence area with two levels (1 for those from urban areas and 2 for those from rural areas). Since the first levels of each variable (those born in 1945-59 and live in urban areas) are used as baseline (reference) levels for the corresponding variable, we will need only 3 dummy variables (x_1, x_2, x_3) for birth-cohort and 1 dummy variable (x_4) for residence.

In other words, $x_1 = 1$ for women born between 1960 and 1969 while $x_1 = 0$ for each of the other three cohorts. Likewise, $x_2 = 1$ for women born in 1970 – 79 while $x_2 = 0$ for the other three cohorts of women. Finally, $x_3 = 1$ for women born in 1980 – 89 while $x_3 = 0$ for the other three cohorts. Thus, for women born in 1945 – 59, $x_1 = x_2 = x_3 = 0$. In other words, this oldest cohort of women is used as a baseline/reference group and no parameter is estimated for this category. Instead, results for the other three cohorts are interpreted relative to this oldest cohort. For the residence variable, women from the urban areas are used as baseline/reference. Hence, $x_4 = 1$ for women from rural areas while $x_4 = 0$ for women from urban areas. No parameter estimates for the urban area but results for women from rural areas are interpreted relative to those from urban areas.

Distribution of our data set across the response and explanatory variables is shown in Table 1.

Table 1: Distribution of the data across cohort, residence & education

		Educational level				
Birth Year	Residence	No Educ	Primary	Secondary	Post-Sec	Total
1940-59	Rural	691	24	4	0	719
	Urban	407	206	73	15	701
	Total	1098	230	77	15	1420
1960-69	Rural	2626	424	15	3	3068
	Urban	811	863	351	97	2122
	Total	3437	1287	366	100	5190
1970-79	Rural	3249	1039	97	16	4401
	Urban	774	1503	1001	146	3424
	Total	4023	2542	1098	162	7825
1980-89	Rural	2708	2299	427	64	5498
	Urban	351	1823	1844	438	4456
	Total	3059	4122	2271	502	9954
Total	Rural	9274	3786	543	83	13686
	Urban	2343	4395	3269	696	10703
	Total	11617	8181	3812	779	24389

Association Between the Response and The Explanatory Variables

In Tables 2 (a) - 2 (c), we attempt to examine association between the response variable and each of the explanatory variables using the standard Chi-squared test of association.

In Table 2 (a) we extract the observed frequencies from Table 1, compute their corresponding expected frequencies in Table 2 (b) and finally the resulting Chi- squared values in Table 2 (c). The resulting total Chi-squared value of 2556.26 for birth cohort is by far larger than the corresponding 95 % and 99 % critical values of the Chi-squared with 9 degrees of freedom (16.92 and 21.67, respectively). Similarly, the total Chi-squared value of 6342.34 for residence is much larger than the corresponding 95 % and 99 % critical values of the Chi-squared with 3 degrees of freedom (7.81 and 11.34, respectively).

The results in Table 2 (c) indicate a strong association between educational attainment and birth cohort as well as between educational attainment and residence. But, they also indicate that not all cells contribute equally to the total value of the Chi-squared. For instance, the cells belonging to the cohort born 1970-79 and those of post-secondary education contribute much less to the total Chi-squared for birth cohort. Similarly, the two cells corresponding to post-secondary education contribute much less than the other cells to the total Chi-squared value for residence.

These results give early indications of the fact that educational progress might not have been uniform across all cohorts and/or across all educational levels. We will return to this after we fit our models to the data in Section 4. Before that, we introduce our analytic framework and the four variants of models we fit in the next Section.

Table 2 (a): Observed freq. of educ. levels by cohort & residence

	Educational level				
Birth Year	No Educ	Primary	Secondary	Post-Sec	Total
1940-59	1098	230	77	15	1420
1960-69	3437	1287	366	100	5190
1970-79	4023	2542	1098	162	7825
1980-89	3059	4122	2271	502	9954
Total	11617	8181	3812	779	24389
Residence	No Educ	Primary	Secondary	Post-Sec	Total
Rural	9274	3786	543	83	13686
Urban	2343	4395	3269	696	10703
Total	11617	8181	3812	779	24389

Table 2 (b): Expected freq. of educ. levels by cohort & residence					
	Educational level				
Birth Year	No Educ	Primary	Secondary	Post-Sec	Total
1940-59	676.38	476.32	221.95	45.36	1420.00
1960-69	2472.11	1740.92	811.20	165.77	5190.00
1970-79	3727.21	2624.80	1223.05	249.94	7825.00
1980-89	4741.30	3338.95	1555.81	317.94	9954.00
Total	11617.00	8181.00	3812.00	779.00	24389.00
Residence	No Educ	Primary	Secondary	Post-Sec	Total
Rural	6518.93	4590.81	2139.12	437.14	13686
Urban	5098.07	3590.19	1672.88	341.86	10703
Total	11617	8181	3812	779	24389

Table 2 (c): Chi-squared values across cohort & residence					
	Educational level				
Birth Year	No Educ	Primary	Secondary	Post-Sec	Total
1940-59	262.82	127.38	94.66	20.32	505.18
1960-69	376.61	118.35	244.33	26.10	765.39
1970-79	23.47	2.61	12.79	30.94	69.81
1980-89	596.91	183.64	328.77	106.56	1215.88
Total	1259.82	431.99	680.54	183.91	**2556.26**
Residence	No Educ	Primary	Secondary	Post-Sec	Total
Rural	1164.36	141.09	1190.96	286.90	2783.31
Urban	1488.88	180.41	1522.89	366.86	3559.03
Total	2653.24	321.50	2713.84	653.76	**6342.34**

Analytical Framework

The analytical framework presented here follows that Bürkner & Vuorre (2019). A conceptual illustration of the framework for three models for ordinal response variable is shown in Figure 1. The common assumption in all models is that our observed ordinal variable Y, the attained educational level reported by a randomly selected woman, originates from the categorization of a latent (not observable) continuous variable Z (say continuous educational progress).

For instance, in the cumulative logit model (upper panel in Figure 1), 3 thresh- olds T_1, T_2 and T_3 partition the latent (continuous) variable, Z, into 4 observable, ordered categories. In our case, the 3 thresholds partition the unobserved continuous educational progress into 4 observed discrete educational levels coded as 1 for

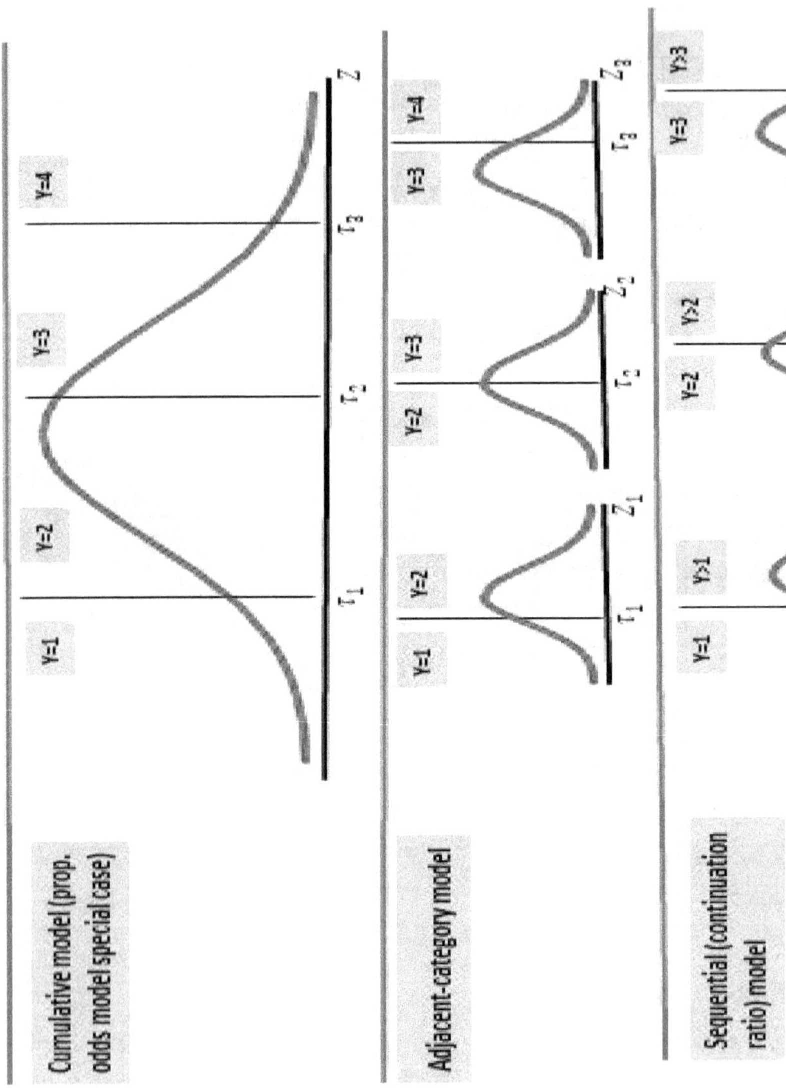

Fig. 1 Relationships between the latent variable (continuous educational progress), Z, and the observed ordinal educational attainment, Y, in three ordinal models (adopted from Bư rkner & Vuorre, 2019).

no education, 2 for primary education, 3 for secondary education and 4 for post- secondary-education. If we assume the continuous latent variable Z has a certain distribution (like a normal distribution), with cumulative distribution function F, we can express the discrete probabilities p_j as

$$p_j = P(Y = j) = F(\tau_j) - F(\tau_{j-1}) \,. \tag{1}$$

Thus, the probability that a randomly selected woman will have no education is given by

$$p_1 = P(Y = 1) = P(Z <= \tau_1) = F(\tau_1) \,. \tag{2}$$

Similarly, the probabilities that a randomly selected woman will have primary education, p_2, secondary education, p_3, and post-secondary education, p_4, are given by

$$p_2 = P(Y = 2) = F(\tau_2) - F(\tau_1) \,. \tag{3}$$

$$p_3 = P(Y = 3) = F(\tau_3) - F(\tau_2) \,. \tag{4}$$

and

$$p_4 = P(Y = 4) = P(Z > \tau_3) = 1 - F(\tau_3) \,. \tag{5}$$

In the adjacent-category model (middle panel in Figure 1), the probabilities of attaining category j and not its adjacent category j + 1 are obtained using a latent variable Z_j, with thresholds r_j and cumulative distribution function F:

$$p_j = P(Y = j \mid Y \epsilon \{j, j+1\}) = F(\tau_j) \,. \tag{6}$$

In the bottom panel (sequential model), the probabilities are obtained as in the adjacent category model; but adjacent-category models compare the probabilities of Y = j and Y = j + 1 whereas sequential models compare the probabilities of Y = j and Y > j:

$$p_j = P(Y = j \mid Y >= j\}) = F(\tau_j) \,. \tag{7}$$

Equations (1), (6) and (7) above establish the relationship between the latent and observed response variable in three ordinal models and how the corresponding probabilities are obtained in each model.

We now shift attention to our main goal: to use these probabilities in modelling the relationship between the observed response variable and the explanatory variables in two variants each of first two models in Figure 1 (cumulative model and adjacent- category model). The third model in Figure 1 (sequential or continuation ratio model) is not used in this study as it is closely related to the second model.

Model 1: Cumulative logit model with equal variances across the levels of the explanatory variables

In the cumulative logit model, the log-odds of having the jth level of education or higher is modelled as a linear function of the explanatory variables x_j:

$$\ln\left(\frac{p_1 + \ldots + p_j}{p_{j+1} + \ldots + p_4}\right) = \beta_{0j} + \beta_{1j}x_1 + \beta_{2j}x_2 + \beta_{3j}x_3 + \beta_{4j}x_4 \text{ for } j = 1, 2, 3 \quad (8)$$

Thus, when j = 1 (no education), the above equation becomes

$$\ln\left(\frac{p_1}{p_2 + p_3 + p_4}\right) = \beta_{01} + \beta_{11}x_1 + \beta_{21}x_2 + \beta_{31}x_3 + \beta_{41}x_4 \quad (9)$$

Similarly, we have the following two equations for j = 2 (primary education) and j = 3 (secondary education) respectively:

$$\ln\left(\frac{p_1 + p_2}{p_3 + p_4}\right) = \beta_{02} + \beta_{12}x_1 + \beta_{22}x_2 + \beta_{32}x_3 + \beta_{42}x_4 \quad (10)$$

$$\ln\left(\frac{p_1 + p_2 + p_3}{p_4}\right) = \beta_{03} + \beta_{13}x_1 + \beta_{23}x_2 + \beta_{33}x_3 + \beta_{43}x_4 \quad (11)$$

We cannot have a similar equation for j = 4 (post-secondary) because that is the highest recorded level of education. Hence, we do not have any higher educational level we can compare with (that can appear in the denominator of the log-odds in the left-hand side of the equation).

If the lines in the three equations (9), (10) and (11) are parallel (the slopes are equal), then we have the **proportional odds model** as a special case:

$$\ln\left(\frac{p_1}{p_2 + p_3 + p_4}\right) = \beta_{01} + \beta_1x_1 + \beta_2x_2 + \beta_3x_3 + \beta_4x_4 \quad (12)$$

$$\ln\left(\frac{p_1 + p_2}{p_3 + p_4}\right) = \beta_{02} + \beta_1x_1 + \beta_2x_2 + \beta_3x_3 + \beta_4x_4 \quad (13)$$

$$\ln\left(\frac{p_1 + p_2 + p_3}{p_4}\right) = \beta_{03} + \beta_1x_1 + \beta_2x_2 + \beta_3x_3 + \beta_4x_4 \quad (14)$$

Thus, while equations (9), (10) and (11) require estimating 15 parameters (one intercept and four slopes in each of the three models), equations (12), (13) and (14) require estimating only 7 parameters, the 3 intercepts for each model and the 4 slopes that are the same in each equation.

Because of this ease of computation and interpretation, the **proportional odds models** in equations (12), (13) and (14) are the most commonly used model in analyzing ordinal response variables. There are methods for testing if the **proportional odds** assumption is violated (for instance by comparing the difference between the respective deviances which will have a Chi-squared distribution).

Model 2: Cumulative logit model with unequal variances across the levels of the explanatory variables

A tacit assumption in the cumulative logit models in equations (9), (10) and (11) or their proportional odds equivalents in equations (12, (13) and (14) is that the latent response variable has equal variances across the levels of the explanatory variables.

In our illustrative example, this implies the variances of the latent educational progress are the same across the four birth cohorts and across the two residence areas

To allow for possible violation of this assumption, the models in equations (9) - (14) can be fitted by specifying unequal variances. There is no change in the equations but, in the R-codes used to fit the models, an additional regression formula is specified for the variance component of the latent variable, Z.

Model 3: Adjacent-category model with constant effects of explanatory variables across the levels of the response variable

In adjacent-category models, the log-odds of having the jth level of education or the next higher level is modelled as a linear function of the explanatory variables x_j.

If we can assume that the effects of the explanatory variables do not vary across the levels of education (an assumption that can be tested formally by comparing the deviance in the two models), then we have:

$$\ln\left(\frac{p_j}{p_{j+1}}\right) = \beta_{0j} + \beta_1 x_1 + \beta_2 x_2 + \beta_3 x_3 + \beta_4 x_4 \text{ for } j = 1, 2, 3 \qquad (15)$$

Thus, for our three possible log-odds, we have the following equations:

$$\ln\left(\frac{p_1}{p_2}\right) = \beta_{01} + \beta_1 x_1 + \beta_2 x_2 + \beta_3 x_3 + \beta_4 x_4 \qquad (16)$$

$$\ln\left(\frac{p_2}{p_3}\right) = \beta_{02} + \beta_1 x_1 + \beta_2 x_2 + \beta_3 x_3 + \beta_4 x_4 \qquad (17)$$

$$\ln\left(\frac{p_3}{p_4}\right) = \beta_{03} + \beta_1 x_1 + \beta_2 x_2 + \beta_3 x_3 + \beta_4 x_4 \qquad (18)$$

Model 4: Adjacent-category model with category-specific effects of explanatory variables across the levels of the response variable

Finally, if we allow even the coefficients (slopes), 0_j, in (16) - (18) to vary between categories, we get the following equations:

$$\ln\left(\frac{p_j}{p_{j+1}}\right) = \beta_{0j} + \beta_{1j}x_1 + \beta_{2j}x_2 + \beta_{3j}x_3 + \beta_{4j}x_4 \text{ for } j = 1,2,3 \qquad (19)$$

$$\ln\left(\frac{p_1}{p_2}\right) = \beta_{01} + \beta_{11}x_1 + \beta_{21}x_2 + \beta_{31}x_3 + \beta_{41}x_4 \qquad (20)$$

$$\ln\left(\frac{p_2}{p_3}\right) = \beta_{02} + \beta_{12}x_1 + \beta_{22}x_2 + \beta_{32}x_3 + \beta_{42}x_4 \qquad (21)$$

$$\ln\left(\frac{p_3}{p_4}\right) = \beta_{03} + \beta_{13}x_1 + \beta_{23}x_2 + \beta_{33}x_3 + \beta_{43}x_4 \qquad (22)$$

We now turn our attention to fitting the above four models to our dataset presented in Section 2.

We began by fitting the four models to our dataset with only birth cohort as explanatory variable. Thus, we don't have estimates of 0_4 in these four models.

We followed this by fitting the four models with only residence as explanatory variable which implies don't have estimates of 0_1, 0_2 and 0_3 in these second set of four models.

Lastly, we fitted six models with both birth-cohort and residence as explanatory variables. The first four models are as described above (model 2 allows for unequal variances for birth cohort only and model 4 allows for category-specific effects in the adjacent-category model for birth cohort only). Model 5 allows unequal variances for residence only while model 6 allows for category-specific effects in the adjacent- category model for residence only.

Results

The full set of estimates from a univariate model with only birth cohort are displayed in Table A1 in the Appendix. Thus, when we fitted equations (12), (13) and 14) without the last term, $\beta_4 x_4$, we got $\widehat{\beta}_{01} = 0.69$, $\widehat{\beta}_{02} = 1.7$, $\widehat{\beta}_{03} = 2.71$, $\widehat{\beta}_1 = 0.31$, $\widehat{\beta}_2 = 0.67$, and $\widehat{\beta}_3 = 1.15$. Standard errors and lower and upper limits for 95 % credible intervals for each estimate are also provided in the next three columns.

The estimates of the intercepts $\widehat{\beta}_{01}$, $\widehat{\beta}_{02}$ and $\widehat{\beta}_{03}$ are related to the threshold of the latent variable τ_{01}, τ_{02} and τ_{03}, respectively, and will not be discussed further.

We then exponentiated the estimates $\widehat{\beta}_1$, $\widehat{\beta}_2$ and $\widehat{\beta}_3$ to obtain the corresponding odds ratios and their 95 % credible intervals in the last three columns. Thus, the odds ratio for the birth cohort born in 1960-69 (relative to the baseline cohort born 1945-59) is given by $exp\left(\widehat{\beta}_1\right) = e^{\widehat{\beta}_1} = e^{0.31} = 1.36$ with 95 % credible interval limits $e^{0.24} = 1.27$ and $e^{0.39} = 1.48$.

Likewise, the odds ratio for the birth cohort born in 1970-79 (relative to the baseline cohort born 1945-59) is given by $e^{0.67} = 1.95$ while that of the youngest birth cohort born in 1980-89 (relative to the baseline cohort born 1945-59) is $e^{1.15} = 3.16$.

We note that the odds ratios for the three younger cohorts are greater than 1 and that their 95 % credible interval limits do not include 1. Thus, we can conclude that the younger cohorts of women attained significantly higher educational levels than the oldest cohort of women. The relative gains were 36 %, 95 % and 216 % higher for the cohorts born in 1960-69, 1970-79 and 1980-89, respectively.

In model 2, we fit equations (12), (13) and 14) again but we allow for unequal variances for the latent variable across the levels of the birth cohort. The R-program, *Bayesian Regression Models using 'Stan' (brms) R-Package* (Bürkner et al., 2024), that is used to fit the models yields additional estimates labeled as *logdisc* (logarithm of discrimination).

These estimates are converted into standard deviations (square roots of variances) through the relationship standard deviations (SE) = 1/exp(*logdisc*). Thus, the standard deviation for the cohort born in 1960-69 (relative to that of 1945-59 cohort) is $SE(1960-79) = 1/\exp(0.11) = 0.90$. The corresponding 95 % credible interval limits need to be reversed since SE is inversely related to *logdisc*. Thus, in computing the lower 95 % credible interval limit for SE, we use the upper 95 % credible interval limit for *logdisc*: Lower $SE(1960-79) = 1/\exp(0.21) = 0.81$ and upper $SE(1960-79) = 1/\exp(0.01) = 0.99$.

What appear under the columns "Odds Ratio", Lower 95 % and Upper 95%? corresponding to the rows labeled by *logdisc* in model 2 are, therefore, not real odds ratios but rather standard deviations obtained using the relationships above.

Relevant parts of the results from these 14 models in three sets (with 4, 4 and 6 models, respectively) are summarized in Tables 3 (a) - 3 (c).

Results from Four Univariate Models for Birth Cohort

We have extracted the most relevant parts of Table A1 in Table 3 (a).

The odds ratios for the younger cohorts in Models 1, 2 and 3 are all greater than 1 with 95 % credible intervals that don't include 1. These imply that the younger cohorts of women (born 1960 - 1989) have attained significantly higher educational levels than their counterparts in the baseline cohort of women born in 1945-59.

Further, the results from Model 2 show that the standard errors are lower in the younger cohorts than in the oldest cohort (whose standard error is set to 1 by design). This is indicated by the estimates of standard errors that are less than 1 and with 95% credible intervals that don't include 1.

Model 4 (last column) in Table 3 (a) displays results from an adjacent-category model with category-specific effects. In other words, we let the effects of birth cohort to vary over the different categories (educational levels) instead of constraining them to be equal across all levels.

Here, we note that the cohort of women born in 1960-69 attained primary level (than no education) at significantly higher odds than the baseline cohort of women born 1945-69. This is indicated by an odds ratio of 1.40 with 95 % credible interval (1.28, 1.54). On the other hand, the odds of attaining secondary level education (than primary) and post-secondary level (than secondary) were not significant because their 95 % credible intervals include 1.

The same is true with the cohort of women born 1970-79. They attained primary level education (rather than no education) at significantly higher odds than the base-line cohort of women born in 1945-59. But, their odds of attaining secondary (rather than primary) and post-secondary (rather than secondary) were not significant.

Lastly, the youngest cohort of women born 1980-89 attained primary and secondary level education at significantly higher odds than their counterparts of women born in 1945-59. However, they didn't differ from the baseline cohort of women in terms of attaining post-secondary level education.

Table 3 (a): Estimated posterior odds ratios (and 95 % CI) by birth-cohort in 4 models				
Birth Year	**Model 1**	**Model 2**	**Model 3**	**Model 4**
Born 1960-69	1.36 (1.27, 1.48)	1.45 (1.30, 1.58)	1.26 (1.19, 1.34)	
Born 1970-79	1.95 (1.82, 2.12)	2.05 (1.90, 2.23)	1.62 (1.54, 1.72)	
Born 1980-89	3.16 (2.94, 3.39)	3.03 (2.83, 3.29)	2.20 (2.08, 2.32)	
SE(1960-69)		0.90 (0.81, 0.99)		
SE(1970-79)		0.79 (0.72, 0.88)		
SE(1980-89)		0.72 (0.66, 0.79)		
Born 1960-69 (1)				1.40 (1.28, 1.54)
Born 1960-69 (2)				0.91 (0.77, 1.07)
Born 1960-69 (3)				1.21 (0.87, 1.73)
Born 1970-79 (1)				1.93 (1.77, 2.10)
Born 1970-79 (2)				1.17 (1.00, 1.38)
Born 1970-79 (3)				0.86 (0.62, 1.21)
Born 1980-89 (1)				3.10 (2.86, 3.35)
Born 1980-89 (2)				1.36 (1.15, 1.58)
Born 1980-89 (3)				1.07 (0.78, 1.49)
Goodness of fit				
ELPD (SE)	-26070.7 (100.2)	-26009.6 (99.6)	-26138.8 (100.1)	-26000.3 (99.6)
Diff. (SE)	-70.3 (13.1)	-9.2 (5.0)	-138.6 (18.2)	0
95 % CI (Diff.)	(-95.98, -44.62)	(-19, 0.60)	(-174.27, -102.93)	-

Results from four univariate models for residence

Table 3 (b) displays results from the four models with only residence as an explanatory variable.

Here, we see in all four models, that women from rural areas had significantly lower odds of attaining higher levels of education compared with the baseline women based in urban areas. This is indicated by the estimated odds ratios which are less than 1 and their 95 % credible interval limits that are also lower than 1. Further, model 2 shows that the standard error of rural woman didn't differ significantly from that of urban women.

Table 3 (b): Estimated posterior odds ratios (and 95 % CI) by residence in 4 models				
Residence	**Model 1**	**Model 2**	**Model 3**	**Model 4**
Rural	0.28 (0.28, 0.29)	0.29 (0.28, 0.30)	0.41 (0.39, 0.41)	
SE(Rural)		0.99 (0.95, 1.02)		
Rural (1)				0.39 (0.38, 0.40)
Rural (2)				0.38 (0.36, 0.40)
Rural (3)				0.84 (0.73, 0.95)
Goodness of fit	Model 1	Model 2	Model 3	Model 4
ELPD (SE)	-23946.4 (104.4)	-23947.1 (104.5)	-23958.0 (105.0)	-23906.7 (103.7)
Diff. (SE)	-39.7 (9.4)	-40.5 (9.5)	-51.3 (11.9)	0
95 % CI (Diff.)	(-58.12, -21.28)	(-59.12, -21.88)	(-74.62, -27.98)	-

Results from Six Multivariate Models for Birth Cohort and Residence

In Table 3 (c), we present results from 6 models with both birth cohort and residence included in the models.

We expected that the two explanatory variables have an interaction effect, not least because we expect women with higher education to reside in urban areas.

But, including both of them in the model didn't change the previous results appreciably. If any, we now see women born 1970-79 have even higher odds of attaining secondary (rather than primary) education - an effect which was not significant in the univariate model. We also note that the standard error of the cohort born 1960-69 didn't differ from that of the baseline cohort (born 1945-59) and that the standard error of rural residents is now significantly lower than that of urban residents.

Else, the urban advantage in educational attainment still persists and that, generally, there was higher odds of attaining higher level of education among the younger cohorts.

Comparison of Models Using Goodness-Of-Fit Tests

Which model fits the data best in each of Tables 3 (a), 3 (b) and 3 (c)

Table ??: Estimated posterior odds ratios (and 95 % CI) by birth-cohort & residence in 6 models

Birth Year and Resid	Model 1	Model 2	Model 3	Model 4	Model 5	Model 6
Born 1960-69	1.68 (1.54, 1.82)	1.68 (1.52, 1.82)	1.45 (1.36, 1.55)		1.67 (1.54, 1.80)	1.48 (1.38, 1.58)
Born 1970-79	2.51 (2.32, 2.72)	2.48 (2.29, 2.69)	1.97 (1.86, 2.10)		2.48 (2.29, 2.69)	1.99 (1.88, 2.14)
Born 1980-89	4.44 (4.14, 4.81)	4.01 (3.71, 4.39)	2.97 (2.80, 3.16)		4.35 (4.01, 4.71)	3.00 (2.83, 3.22)
Rural	0.25 (0.24, 0.26)	0.32 (0.29, 0.35)	0.36 (0.35, 0.37)	0.36 (0.35, 0.36)	0.26 (0.25, 0.28)	
SE(Born 1960-69)		0.91 (0.84, 1.00)				
SE(Born 1970-79)		0.84 (0.76, 0.91)				
SE(Born 1980-89)		0.79 (0.72, 0.85)				
Born 1960-69 (1)				1.68 (1.54, 1.84)		
Born 1960-69 (2)				0.99 (0.84, 1.16)		
Born 1960-69 (3)				1.26 (0.91, 1.75)		
Born 1970-79 (1)				2.44 (2.25, 2.66)		
Born 1970-79 (2)				1.35 (1.15, 1.57)		
Born 1970-79 (3)				0.91 (0.67, 1.27)		
Born 1980-89 (1)				4.39 (4.01, 4.81)		
Born 1980-89 (2)				1.77 (1.52, 2.08)		
Born 1980-89 (3)				1.21 (0.89, 1.67		
SE (Rural)					0.96 (0.92, 0.99)	
Rural (1)						0.34 (0.33, 0.35)
Rural (2)						0.34 (0.32, 0.36)
Rural (3)						0.76 (0.67, 0.87)
Goodness of fit	Model 1	Model 2	Model 3	Model 4	Model 5	Model 6
ELPD (SE)	-22308.6 (110.6)	-22270.7 (110.4)	-22411.1 (111.6)	-22250.5 (111.0)	-22305.8 (110.8)	-22353.9 (110.4)
Diff. (SE)	-58.0 (17.9)	-20.2 (11.5)	-160.5 (19.8)	0	-55.3 (16.2)	-103.4 (25.0)
95 % CI (Diff.)	(-93.08, -22.92)	(-42.74, 2.34)	(-199.31, -121.69)	-	(-87.05, -23.55)	(-152.40, -54.40)

The software we used to fit our models provides goodness-of-fit statistics in terms of Expected Log-Predictive Density (ELPD). These are shown together with their standard errors (SE) for each model at the bottom of each table.

Below the ELPD-row are differences in ELPD between the 'best' model (the model with the highest ELPD) and the other models.

In the last row, we have 95 % credible intervals for the differences. These are obtained as (Diff - 1.96*SE(Diff), Diff + 1.96*SE(Diff)). These 95 % credible intervals are used to test if the corresponding model fits the data significantly worse than the 'best' model.

Thus, we see in Table 3 (a), that model 4 provides the highest ELPD (-26000.3) and, hence, fits the data 'best'. Further, the 95 % credible intervals for models 1 and 3 show that these models are rejected in favour of model 4. In model 2, the 95 % credible intervals include 0, meaning that its fit to the data is not significantly worse than that of model 4.

The results are the same in Table 3 (b). We see even here that model 4 fits the data 'best' and that the other three models fit the data significantly worse than model 4.

The results of goodness-of-fit in Table 3 (c) are more like those in Table 3 (a). Model 4 still fits the data best but that the fit of model 2 is not significantly worse. The other four models (models 1, 3, 5, and 6) provide significantly worse fits than model 4.

Based on the above results, we provide plots of posterior probabilities of attaining the four educational levels by birth cohort and residence in Figures 2 and 3, respectively. These plots are based on model 4 in Table 3 (c).

Figure 2 shows that there was a steady decrease in the proportion of women with no education (orange) as the cohort got younger. On the contrary, there has been increase in the proportion of women attaining primary education (green) & secondary education (blue) - though not as steady as in the decrease of no education. The increase in proportion of women with post-secondary education (pink) is not as marked as in the other levels of education.

In Figure 3, we note that there is a much higher proportion of women with no education (orange) in rural areas than in urban areas. There are more women with primary and secondary level education in urban areas - though the differences are not as marked as in the proportions of women with no education. Lastly, the difference in the proportions of women with post-secondary education looks to be small and may have to do with the low frequency of women in this category, as we saw in Table 2 (a).

Summary and Concluding Remarks

This contribution was intended to be both methodological and substantive. Methodologically, it demonstrates the use of Bayesian framework and its flexibility in modelling response variables measured in ordinal scale — such as responses to opinion surveys that are recorded in Likert scale. Substantively, it sheds light on trends and patterns of educational progress among four cohorts of Eritrean women born between 1945 and 1989.

Our empirical results reveal that there has been significant educational progress among the younger cohorts of women (relative to the oldest cohort). Further, the goodness-of-fit tests revealed that an adjacent-category model with category-specific effects fitted the data 'best' both in the univariate and multivariate models.

We interpret this result as that there were differential gains in educational progress among the cohorts. This was clear from the plots of the posterior probabilities of educational attainment where the difference between the cohorts was mainly in the decrease of women with no education among the younger cohorts. We also found that the urban advantage in educational attainment still persisted among the cohort of women studied.

We hope that this contribution adds on the very few available works based on data from Eritrean Demographic and Health Surveys. The survey has been conducted in three phases (1995, 2002 and 2010) but still remains underutilized in terms of scientific output based on analyses of the data. It is our hope this study inspires readers with access to the data to attempt similar analyses on other outcome variable or on other part of the data.

References

P. -C. Bürkner, & M. Vuorre (2019), *Ordinal regression models in psychology: A tutorial.* Advances in Methods and Practices in Psychological Science 2, pp.77 - 101.

P. C. Bürkner, J. Gabry, S. Weber, A. Johnson, M. Modrak, H. S. Badr, F. Weber, A. Vehtari, M. S. Ben-Shachar, H. Rabel, S. C. Mills, S. Wild and V. Popov (2024) *Package brms: Bayesian Regression Models using 'Stan'*, CRAN: The Comprehensive R Archive Network. Last updated on 23 September 2024; last accessed on February 2025.

G. Ghilagaber (2022), *Diagonal Reference Modelling of the Effects of Educational Differences between Couples on Women's Health-Care Utilization in Eritrea.* Pages 9-20 in Chen, D. -G. (Din) Chen, Manda, S. and Chirwa, T. (eds.), Modern Biostatistical Methods for Evidence-Based Global Health Re- search. Springer Series on Emerging Topics in Statistics and Biostatistics.

T. M. Liddell, & J. K. Kruschke (2018), *Analysing ordinal data with metric models: What could possibly go wrong?* Journal of Experimental Social Psychology 79, pp. 328 - 348.

National Statistics Office, Eritrea, and Macro International Inc., USA (1997), *Eritrea Demographic and Health Survey 1995.* National Statistics Office and Macro International.

National Statistics Office, Eritrea and Fafo Institute for Applied International Studies, Norway (2013), *Eritrea Population and Health Survey 2010.* National Statistics Office and Fafo Institute for Applied International Studies

Appendix: Table A1 and Figures 2 and 3

Table A1: Parameter estimates and corresponding odds ratios (or SE) from four models for cohort

Model 1 (Eq. 12 - 14)	Estimate	Std. error	Lower 95 %	Upper 95 %	Odds Ratio	Lower 95 %	Upper 95 %
Int. 1 (θ_{01})	0.69	0.03	0.63	0.76	1.99	1.88	2.14
Int. 2 (θ_{02})	1.7	0.04	1.63	1.77	5.47	5.10	5.87
Int. 3 (θ_{03})	2.71	0.04	2.64	2.79	15.03	14.01	16.28
Coh60-69 (θ_1)	0.31	0.04	0.24	0.39	1.36	1.27	1.48
Coh70-79 (θ_2)	0.67	0.04	0.6	0.75	1.95	1.82	2.12
Coh80-89 (θ_3)	1.15	0.04	1.08	1.22	3.16	2.94	3.39
Model 2 (Eq. 12 - 14)							
Int. 1 (θ_{01})	0.75	0.04	0.67	0.82	2.12	1.95	2.27
Int. 2 (θ_{02})	1.53	0.05	1.44	1.62	4.62	4.22	5.05
Int. 3 (θ_{03})	2.3	0.08	2.16	2.45	9.97	8.67	11.59
Coh60-69 (θ_1)	0.37	0.05	0.26	0.46	1.45	1.30	1.58
Coh70-79 (θ_2)	0.72	0.04	0.64	0.8	2.05	1.90	2.23
Coh80-89 (θ_3)	1.11	0.04	1.04	1.19	3.03	2.83	3.29
logdisc (60-69)	0.11	0.05	0.01	0.21	0.90	0.81	0.99
logdisc (70-79)	0.23	0.05	0.13	0.33	0.79	0.72	0.88
logdisc (80-89)	0.33	0.05	0.23	0.42	0.72	0.66	0.79
Model 3 (Eq. 16 - 18)							
Int. 1 (θ_{01})	0.73	0.03	0.67	0.78	2.08	1.95	2.18
Int. 2 (θ_{02})	1.07	0.03	1.01	1.12	2.92	2.75	3.06
Int. 3 (θ_{03})	1.61	0.04	1.54	1.68	5.00	4.66	5.37
Coh60-69 (θ_1)	0.23	0.03	0.17	0.29	1.26	1.19	1.34
Coh70-79 (θ_2)	0.48	0.03	0.43	0.54	1.62	1.54	1.72
Coh80-89 (θ_3)	0.79	0.03	0.73	0.84	2.20	2.08	2.32
Model 4 (Eq. 20 - 22)							
Int. 1 (θ_{01})	0.94	0.04	0.86	1.02	2.56	2.36	2.77
Int. 2 (θ_{02})	0.68	0.08	0.52	0.83	1.97	1.68	2.29
Int. 3 (θ_{03})	0.98	0.16	0.67	1.31	2.66	1.95	3.71
Coh60-69 (θ_{11})	0.34	0.05	0.25	0.43	1.40	1.28	1.54
Coh60-69 (θ_{12})	-0.09	0.09	-0.26	0.07	0.91	0.77	1.07
Coh60-69 (θ_{13})	0.19	0.17	-0.14	0.55	1.21	0.87	1.73
Coh70-79 (θ_{21})	0.66	0.04	0.57	0.74	1.93	1.77	2.10
Coh70-79 (θ_{22})	0.16	0.08	0	0.32	1.17	1.00	1.38
Coh70-79 (θ_{23})	-0.15	0.17	-0.48	0.19	0.86	0.62	1.21
Coh80-89 (θ_{31})	1.13	0.04	1.05	1.21	3.10	2.86	3.35
Coh80-89 (θ_{32})	0.31	0.08	0.14	0.46	1.36	1.15	1.58
Coh80-89 (θ_{33})	0.07	0.16	-0.25	0.4	1.07	0.78	1.49

Factors Affecting Female Teachers' Participation in School Leadership: A Case of Mendefera Subzone

Idris Ali Osman,

PFDJ Zoba Debub

Abstract

Females' potential in the holistic development of schools is undeniable. Their less visibility in leadership positions in the Eritrean schools in the Mendefera subzone (sub-region) within the Debub zone (Southern Region), where this study was conducted, is however worrying despite the presence of more than 71% of the teaching force being females. The study aimed to study why such social and educational incongruity prevails in the elementary schools in the subzone. The researcher employed mainly qualitative but also quantitative methods. He had the opportunity to collect adequate data from and around the schools from teachers, school leaders, and community members to explore the extent of participation of females in leadership positions in elementary schools and disclose the contextual challenges using interviews (20 interviewees), focus group discussions (70 discussants in six groups), and questionnaires (89 respondents) through a combination of purposive and random sampling procedures. The researcher used permissible and acceptable research integrities and abided by human as well as honest ways to conduct the research process. The information was organized, compared, abridged, analyzed, and related to literature reviews. The outcomes were thus important for helping the researcher to understand the attitude of the surrounding communities about female leadership, the ability and interest of female teachers to hold leadership ranks, the support and encouragement females get from the school system, the policy values and practices available in and around the schools, and essential obstacles that hinder female teachers from coming to leadership positions. The study suggested the need for more detailed studies in each elementary school community for specifically encouraging female teachers to hold leadership positions considering females' professional merit in the schools.

Keywords: Female teachers; leadership; elementary school; Eritrea

Introduction

In many countries, women actively take part in areas of public leadership where they were not previously visible. In Eritrea, women are more than 50% of the population. They have played crucial role in the struggle for liberation as well as in safeguarding the sovereignty of the country. Their potential is indispensable in the holistic development of the country in general and in the education sector in particular, where a great number of females are employed both as teachers and as learners. Women have practically shown their special talents in caring and helping young learners and hence they can play reliable leadership roles in schools at all levels. However, despite the government's emphasis and encouragement in its education policy towards gender equality, reality reveals that the participation of females in school leadership positions is insignificant. Males still hold school leadership positions overwhelmingly as compared to females. Only two of the 12 subzones in the Debub zone were led by a female, and only 6 among the 78 school supervisors were females. For example, in 2022/23 school year, the Mendefera sub zone, where this research was conducted, had none female elementary school principals among the 276 despite the proportion of female teachers being more than 71%. Besides, among the 778 teachers deployed in all the schools in the sub zone 57% were females but there were no female school principals in the eight middle and two secondary schools during that year.

Concepts of Educational Leadership

Educational leadership is primarily concerned with educational purpose of guiding and directing teaching and learning to improve educational outcomes for all learners (Liyod, 2009). According to Tarnve (2000, as cited in Leliftu 2014, p. 10), school leadership is a continuous work of activating school communities to believe and behave in regard to shared vision that result in high achievement for every child. Sergiovanni has indicated the dependability of school success on effective leadership stating that "tomorrow's school success will depend upon the ability of leaders to harness the capacity of locals to enhance sense and meaning and build a community of responsibility" (as cited in Harris, et al., 2003, p. 1).

Females and School Leadership

Women do possess the capabilities and skills to be excellent school administrators attuned to their feminine modes of leadership behaviors. Female attributes of nurturing, being sensitive, empathetic, intuitive, compromising, caring, cooperative, and accommodative are increasingly associated with effective administration. While these characteristics are innate and valuable, women possessing the qualities of a good leader still face higher attrition and slower career mobility particularly in educational administration (Porat, 1991). Data on equality of opportunity in educational administration reveals that gender, more than age, experience, background, or competence decides the role an individual will be assigned to in education (Lane & Whitaker, 1990). The balance of gender roles in schools changed following World War II as many men returned from the armed services and sought employment in the school settings. Soon many females also steadily lost ground in the pursuit of administrative positions. In the 1940s in the United States for example, about 41% of elementary principals were women but dropped to 20% by 1980s (Paddock, as cited in Grady, 1992). Aburdene and Naisbitt (1992) described 25 behaviors that characterize women's leadership. The behaviors clustered in six central patterns were identified as behaviors that empower, restructure, teach, provide role models, encourage openness, and stimulate questioning. Gillet-Karam (1994) on the other hand, used essential female behaviors that result change, care and respect for individual differences, collaborate, and build trust and openness. No matter how the leadership behaviors of women are delineated, the fact is that women do possess the capabilities and skills to be excellent educational leaders.

Conceptual Models on Women Participation in School Leadership

Over the past two decades, several interrelated conceptual models have been adapted from the social sciences and used to explain the underrepresentation of women in educational leadership. These models included the social model, organizational model, individual model, and international model. The social model looks for explanations neither in women as individuals nor in educational systems, but in society as a whole (Estler, 1975; Schmuck, 1980; Shakeshaft, 1989). It suggests that in order for women to move into positions of administrative leadership, they must first deny societal norms and values. The organizational model (Schmuck, 1980) explains men's and women's differential career aspirations and achievements not as a function of different psychological predispositions but as an effect of the limited opportunities for women that accompany systemic gender bias (Burstyn & Tallerico, 1996). It assumes that people adjust their aspirations to that which is possible and that women in education do not aspire to leadership as a result of limited opportunity (Lyman & Seizer, as cited in Grady, 1992). The individual model assumes people are promoted according to their ability as a basis for promotion and that men occupy the most senior positions in educational administration because they hold the best qualifications (Schmuck, 1980; Ortiz & Marshalls, 1988; Skakeshaft, 1989). It seeks to explain the persistent and continuing gender segregation in the teaching profession from a psychological orientation of women to see themselves as the cause in things like personal traits, abilities, qualities, motivation, self-image, confidence, and aspirations (Burstyn & Tallerico, 1996). This belief is often reflected in statements such as; they are not assertive enough, they don't want power, they lack self-confidence, they don't apply for the jobs. The assumption of the international model emphasizes a number of variables influencing women's aspiration and access to educational leadership encompassing international variables such as the existence of international agreement about gender equality and how national and organizational variables respond to these international standards in employment and access to leadership positions. The above four leadership styles of conceptualizing the under-representation of women in educational leadership reflect long-standing debate over the causes and meanings of inequalities in the education profession. An assumption underlying each model is the need to remedy the inequalities (Burstyn & Tallerico, 1996) that females encounter in educational leadership.

Research Methodology

The research was carried out in Mendefera subzone located in the Debub zone (Southern Region). The study employed qualitative methods of research where the participants' views and descriptions of situations about current status of female roles in elementary school leadership were explored. It also attempted to collect relevant and adequate quantitative data which greatly enriched the quality data in providing adequate picture of the situation of females as school leaders. The quantities were crucial in introducing the complex socio-cultural, organizational, as well as individual barriers that females encounter in the Mendefera subzone. Likert's four-scale was very helpful to fairly qualify and introduce qualitative information in the study. The study has therefore benefited from the mixed approach of research to a very extent.

A combination of both random and purposive sampling procedures was used in the research process. Interview participants were selected purposefully. Purposive sampling was also used in selecting six schools where the focus group discussions (FGDs) were carried out. Simple random sampling from an alphabetical list of names was used to select the female teachers who completed the questionnaires.

The items in all of the three data collection instruments were developed along coherent issues related to the confidence of school communities in female teachers, competence and interest of females to hold elementary school leadership, practices and policies that support females to hold elementary school leadership positions, and main obstacles that hinder female teachers to become school leaders. The organization of the items in all of the three data collection tools were meant to help the research process to collect logically sequenced data both within one tool as well as among each of the tools depending on whether the data is qualitative or quantitative. The issues, scope, and sequence of the items in the research tools has thus greatly facilitated the development of a relevant and sound framework for both organizing the data into further appropriate substantial themes and sub themes for developing the findings of the research. This has also supported the research process to greatly structure the literature search, review, and write up. Moreover, the organization of the items among the research tools was meant to triangulate the coherence of the responses of the participants.

Analyses of the data obtained in the questionnaires were first organized into the sequence of the issues mentioned above but with careful examination of the similarities and differences the respondents provided. The information from interviews and FGDs were separately examined and organized along both the sequence of the items as well as along the emerging information that the participants gave. Then the information in the three tools of the study were organized, compared, abridged, and developed the report of the findings. The findings were then related to the literature outcomes. In the write up of the output of the study the information of the quantitative data was used to introduce the quality information that the interviewees and focus group discussants provided.

Significance of the Study

Female teachers in elementary schools constituted close to 71% but still do not hold any school leadership positions in the subzone. This deserves a more detailed study in each of the schools in the subzone though this research has come up with essential indicators that the subzone community should develop strategies and practices to impartially position the role of female leaders in the school system. Most of all, the study was significant to females in general and female teachers in particular to merit their own professional commitments and to secure what they are qualified for.

Mendefera encompasses the zonal capital city as one of its 15 administrative areas (Kebabis). The study happened to reveal an important gender related concern in school leadership by listening to the voices of more than 30% of the 387 teachers working in the 13 elementary schools located in the subzone.

Findings and Discussion

The findings of the study are organized to the following eight main issues raised by the interviewees and focus group discussants along with the information provided by questionnaire respondents. These issues are discussed afterward.

- Confidence of school communities in female school leaders
- Competence of females to hold elementary school leadership positions
- Interest of female teachers to become elementary school leaders
- Practices that support females to hold elementary school leadership positions
- Modeling practices and issues for encouraging females to become leaders

- Policies to encourage female teachers to become school leaders
- Main obstacles that hinder female teachers from becoming school leaders
- General views of participants to encourage female teachers to become school leaders

School Communities' Confidence in Female Roles as School Leaders

Though the findings indicate a general negative bent of the participants' views about community attitudes to confide in females to become school leaders, a worrying finding was that of the teacher questionnaire respondents (89) who were *all females* to doubt about their own confidence and that of their female colleagues to hold leadership titles in their respective schools. The implication is that there is much work to be done at all levels in the school system to primarily convince females that they can be trusted by their respective communities if they first happen to think about themselves and other females' capability to hold leadership positions. Secondly, if males can do it, there is little barrier for them cross over to becoming leaders. All socio-cultural barriers should not be underestimated but it all depends on their own resilience to break the barriers that also include being highly family planners that can encourage their leadership commitments in their schools.

The lack of confidence of school communities on females to become school leaders as indicated by the FGD discussants and interviewees though generally discouraging, did not seem to be well articulated in the sense that their views included clouded hesitations that implied the need for female teachers to demonstrate leadership capabilities that can merit them to earn the confidence of their respective school communities. Such mixed observations of both the interview and FGD participants underestimated the strong influence of socio-cultural and organizational barriers that can disallow females silently or explicitly to demonstrate their capability to get adequate support from their school communities. For example, one crucial issue mentioned by both the interviewees and FGD discussants was that schools indistinctly try to ignore the importance of bringing female leadership matters and concerns, while actually males are often taken for granted candidates for leadership posts. Schools should thus ask themselves on whether females' caring and motherly qualities could also become essential virtues to allow them to become especially elementary school leaders where very young learners in need of such care are in place. Some of the very good examples brought by some of the interviewees and FGD discussant was the need of school commutes to refer to the country's historical struggles and female teachers' competence for school leadership.

Competence of Female Teachers to Hold Leadership Positions

The information the study has been able to find was generally positive about the competence of female teachers to climb to leadership positions. Besides the informants in the study have provided a very long list of qualities that women teachers possess both as professionals and as members of the female sex that permits them to suitably become school leaders. The professional ones include merits like general seriousness that enhances their effectiveness at work and qualities to generate positive change in their classroom duties and schools. These also included prompt decisions, communication and conflict resolutions skills, participatory mechanisms, close follow up of routines and task-oriented leadership, abiding by rules and regulations etc. The participants have also listed female characteristics like motherly and household experiences that can benefit them to manage schools with calmness and seriousness. Their resilience as character can help them endure challenging school situations.

A few of the participants however disclosed that the professional capability of females should be further developed considering both their very limited pedagogical and managerial knowledge in particular, as well as the very limited professional education that is provided to teachers in elementary schools in general. Female teachers who are well educated, supported, and encouraged may tend to become very good school leaders in managing the young elementary school learners as well as teachers and parents advantaged by the observed characteristics the respondents have provided.

Female Teachers Interest for School Leadership

Responses to the two items in the questionnaire for the female teachers were greatly inconsistent. The items were meant to check dependability of the responses of the participants and inquired the level of interest (affirmative) and disinterest (negative) of female teachers to hold leadership positions in elementary schools. Expectedly, if the female teachers respond that female teachers' *interest is low*, they are also expected to respond that female teachers' *disinterest is high* in the four Likert scale. However, actual responses to such interest-disinterest levels greatly differed by about 28 (61%-33%) percentage points indicating the considerable occurrence of inconsistency in the responses of the same female respondents.

An essential finding from the views of the FGD discussants was their 50-50 interest-disinterest observations. This may mean that the discussants were split into two camps - may be showing the situation as partially promising for schools if encouraging situations are created in schools to manage particularly organizational arrangements in schools such as creating day care facilities. Interviewees also indicated that schools should be very watchful that teaching for females can become a "dead lock" post if schools do not design assuring promotions for females to hold leadership positions, meaning that females will 'prefer' only classroom responsibilities and nothing more. Some among the FGD discussants were of the idea that female teachers disinterest in leadership roles is only a "hidden interest" or forced disinterest due to socio-cultural and organizational barriers that prevail in and around schools. Arrangements like purposeful promotion policy based on solid educational and economic incentives were also suggested by the interviewees and FGD discussants to alleviate females professionally as well as their value of self-concept to accept leadership positions.

Support for Females to Hold School Leadership

High level policy intentions to give opportunities to females to become leaders seem to have succeeded more in other sectors of the country than in the education sector. The probing questions for us are then why has the school system failed to see more females in leadership positions like in elementary schools when the school system has more educated females than all the other sectors in the country. All socio-cultural and organizational factors that may disallow females to reach leadership titles can be raised but again why do such apologies emphasize in the school system and not in the other sectors relatively. The other crucial issue about the low prevalence of female leaders in elementary schools above all is the inability of the school system to strategically and practically make arrangements to ensure females to lead many of the schools at least as equal as males during the last 30 or so years after the liberation of the country.

An upsetting educational account mentioned by the interviewees and FGD discussants was unavailability of effective education and training programs both for teachers in general and school leadership in particular. These worrying educational drawbacks will not only deprive elementary schools from having competent female and male leaders but also deprive the country from having successful learners. It is then not difficult to imagine unsuccessfully managed learners at elementary schools to become successful learners and professionals at higher levels of the formal education system. It is highly worthy for the school system to look at the critical role of school leaders to earn quality education for learners in and above the elementary school system. Considering the trend of women's role in the schools during the liberation struggle, the current school system should have had high level educational programs for both females and males with great focus to KGs and elementary school leaders and teachers. Other practices that do not encourage females in schools to assume leadership like the prevalence of day care facilities would them be out of justification in the school system if both females and males are well prepared to understand and practice effective school leadership qualities.

Modelling and Mentoring Practices that Encourage Females to Become Leaders

The interviewees and FGD discussants largely argued that professional models that inspire female teachers to aim and seek to become school leaders may not depend on the existence of female models in the school system alone. Both the interviewees and FGD discussants for example suggested that female teachers should follow and learn from the presence of exemplary females that are often identified and broadcasted not only in TV programs but also in the widely distributed radio programs in the country that could reach all elementary schools. Since modelling is an indirect learning protocol for female teachers to seek and follow, participants have suggested that schools should intentionally look for parents, school board members, and other community members around and beyond school boundaries to expose female teachers to leadership qualities. This is very important especially for female teachers in remote areas. Besides, there is no reason why female teachers should not choose to learn from the positive sides of the practices of many males to ensure that they pick relevant ones for them to cultivate.

In general, all participants confirmed the absence of mentoring arrangements for female teachers both in schools and the school system.

Policy to Encourage Female Teachers to Become School Leaders

Participants in the study agreed that the school system from the Ministry of Education and to the schools uphold general principles of gender equality. Practically, however, most of the participants informed that they do not know of any clear policy guideline that requires schools to encourage female teachers to hold school leadership positions.

Main Obstacles that Hinder Female Teachers from Becoming School Leaders

The study was able to learn from the views of the informants about obstacles that hinder female teachers from becoming school leaders as listed in previous sections. To avoid repetitive discussions, here it will be adequate to mention that the barriers that females face along their two-fold responsibility at home as mothers and the demanding responsibilities of school leadership require males, females, the school system, and society to accommodate females mainly through a process of developing their competence in multiple, coordinated, and practical measures at school level. The process can then benefit female teachers to hold leadership posts beyond the school level higher to the sub zonal and zonal leadership ranks.

Outcomes of the Literature Review as Related to the Findings in the Study

Leadership is a multifaceted issue in the science and practice of management. School leadership is a continuous work of activating school communities and learners to believe and behave in shared vision for creating an environment of organizational culture for learners' quality learning (Tarnve, 2000 as cited in Leliftu 2014 2, p. 10). The literature reviews as well as the findings of the study showed that females do possess the capabilities to be effective school leaders as equal as males. However, both the literature review and the outcomes of the study indicated that females are much less supported by communities, families, and organizational setups such as schools, compared to the support given to males, and consequently, very few of the females are observed to lead even elementary schools in the Mendefera subzone in Eritrea.

A number of ways to overcome existing barriers have been noted and one of the crucial ways is for female aspirants to have a mentor. Despite the benefits of mentoring, both the literature review and the findings of the study showed that the availability of female or male mentors for women is still lacking. Both in the literature and the findings of the study suggest that even if females' and males' leadership styles are different, we must embrace that difference and open adequate opportunities for females to lead schools especially in elementary schools.

Conclusions

The difference in percentage points (about 14 percentage points) in the four-level Likert scale of the negative versus positive views of the female teachers about how much their respective communities' trust females to hold leadership positions in schools may not be adequate to firmly tell us the confidence level of the communities on female leadership roles. This was also strengthened by the doubts that the FGD discussants expressed saying the communities have inadequate observation about females that hold leadership roles to help them entrust on female school leaders.

However, the discussants recognized females' educational abilities as well as their caring and motherly approaches towards learners and school issues that can qualify them to become good leaders. Besides, leadership roles of females in other sectors of the zone and subzone were enough to accept female teachers' leadership roles in schools who are often more educated and experienced in school practices. This was again confirmed by most of the interviewees who mentioned practical examples that females are effectively heading local and sub zonal positions in the country. Both interviewees and FGD participants have mentioned that what matters is the ability and interest of female teachers to hold leadership ranks in elementary schools. Participants, however, were convinced that male members of communities at various levels have still open and wide reservations to confide in females to hold school leadership positions, and do not feel "comfortable" to be led by females. Some of the interviewees mentioned females can be discouraged by the prevalence of demanding home chore along with family and social engagements to ignore school leadership responsibilities. Moreover, a female promotion in school leadership does not occur as a serious formal agenda in schools and educational meetings.

Competence of Females to Hold Elementary School Leadership Positions

Most interviewees and FGD discussants backed the findings from the questionnaire responses (76%) confirming that most females can be competent school leaders. They came with examples about the substantial qualities females can bring positive changes to their schools through their courage to decide, their punctuality, their general seriousness on school work, and their abilities to resolve conflicts. Thus, informants appreciated female decisiveness, assertiveness, and versatility. Moreover, females' nurturing qualities were added qualities to their organizational and consultation capabilities. A few interviewees however stressed that the professional capability of females greatly depends on their background, exposure, and education and thus female teachers who are well educated, supported, and encouraged can become very good school leaders.

Interest of Female Teachers to Become Elementary School Leaders

It was difficult for the FGD discussants to decide about the interest-disinterest level of female teachers in school leadership roles. This was also partly indicated by the responses in the questionnaire to indicate 40-60% to the respective interest-disinterest shares of their views. It is thus wise for the school system to notice, as a few among the FGD discussants remarked, that female disinterest in leadership roles is a "hidden interest" behind social, organizational, and personal discouragements that prevail in and around their school environments. Interviewees as well as FGD discussant suggested that the school system in its diverse policy levels should intentionally promote females accompanied by consistent educational and economic incentives.

Practices that Support Females to Hold Elementary School Leadership Positions

The findings of the study have revealed that there is a general insolvency to encourage females to hold school leadership positions at all levels of the school system. Rarely are female teachers identified, assessed, mentored, and encouraged to be leaders for elementary schools. The school system should bring to the fore exemplary female teachers and female school leaders to allow other females to develop the interest to becoming school leaders. One persistent finding disclosed in the study was the need for the school system to develop highly qualified teachers for elementary schools and use the pool for identifying adequate qualities and numbers of females to become school leaders. The lack of training at school level and at Mendefera school system level was also criticized by the questionnaire respondents to the extent of 93% and 84% respectively. Females more than males tend to become more models for female teachers to rise to school leadership titles and that the very few female leaders should be encouraged by higher levels of the school system to develop formal mentoring arrangements in schools. Moreover, very few schools have been able to allow female teachers to use day care facilities in their respective schools that can encourage them to hold leadership roles.

Strategy to Encourage Female Teachers to Become School Leaders

Participants in the study generally agree that they have no knowledge of either a clear policy or initiatives taken by all levels of the school system to encourage females to hold leadership positions in schools in general and elementary schools in particular. Participants also mentioned that the general view of the national government to view females as equals should be demonstrated in practical terms to encourage females to hold leadership roles in schools. Further than the obstacles mentioned in the above sub sections of this section, a strong barrier for female teachers to hold leadership ranks the study explored is that the two-fold social and professional responsibility of female teachers should be well understood by the school system to encourage them to arrange strategic and practical measures that allow females to become school leaders.

Recommendations

The need to enhance female teachers' educational and political maturity is mainly the responsibilities of the school system by empowering them to break socio-cultural, organizational, and individual obstacles. Females should be allowed to get pertinent education in pedagogy and school leadership that enables them to access to becoming competent teachers and leaders in elementary schools. Their capability can be the main source for developing their interest and commitment to prefer and decide to become school leaders. Such decisiveness that is empowered by competence can ensure the credibility of their respective communities to trust in them. Female teachers should also understand that waiting for their schools to donate leadership positions to them in the mid of all the socio-cultural and organizational barriers discussed in this study is fruitless and hence they should toil to look for collaborative self-learning alternatives to develop pedagogic and school leadership skills and attitudes.

In general, however, the study was able to come up with the following recommendations for enhancing and sustaining the role of female teachers to hold leadership positions in elementary schools in the Mendefera subzone:

- Encourage and support all elementary schools in the subzone to conduct a collaborative action research that is coordinated by the education office of the subzone
- Based on the outcomes of the research,
    - develop a clear profile of both female and male teachers and elementary school leaders on their situation, capability, and interest to hold leadership positions
    - come up with strategic and practical measures that the subzone should take to enhance the position of female teachers to hold leadership positions

Collaborate with pertinent levels of the school system and other national partners to begin and sustain mechanisms that improve the position of female teachers in general and youth in particular to hold leadership ranks in elementary schools and the subzone education office.

As mentioned in the recommendations, the subzone education office should invite the involvement of all the elementary schools within its jurisdiction to conduct a collaborative and action-oriented research in each school. The primary intervention step should thus be to conduct a series of workshops for potentially effective female and male teachers on how the collaborative action research should be conducted and supported in the schools. Constituting a team of teacher researchers will thus be important. The outcomes of this research project can become an important ground for concretizing and enabling the process of developing the conduct of the collaborative action research in each school. Expectedly, the process of encouraging and conducting collaborative action research in the elementary school will demand the establishment of a research team within the subzone with committed and competent members. The team is expected to initiate and sustain the likely challenges involved in the collaborative action research process both at subzone and school levels and must include females as its members in the subzone and school levels.

References

Aburdene & Naisbitt, (1992) *Mega Trends for Women* (New York: Villard Books Estler),

S.E. (1975). "Women as leaders in public education." *Journal of Women in Culture and Society*, 1(2), 363-86.

G. Baker (Ed.), (1994), *Handbook on the community college in America: Its history, mission, and management*. (Westport, CT: Greenwood Press).

Grady, M. L. (1992). *Women in Educational Administration: Certified by not Employed. Educational Considerations*, 20(1), 33-36.

Harris et al., 2003 Harris, A. & Ch, Chapman. (2002). "Leadership in Learning" *International Electronic Journal*, for, volume 6, No 9, Retrieved September,10/2011, from (http;//www.acs.ucalqary.cal-jeill).

Lane & Whitaker, (1990) "What is "a woman's place" in educational administration?" *Education Digest*, 56(3), 12-16.

Leliftu 2014, "Females Participation in Educational leadership in secondary schools of Ilu Aba Bora Zone."

Liyod, K.R. (2009). Women in Management. Harvard business Review 69(3). Ortiz & Marshalls, Women in educational administration. In N.J.

Porat, K.L. (1991). Women in administration: The difference is positive. Clearing House, 64(6), The Clearing House: A Journal of Educational Strategies, Issues and Ideas, Volume 64, Issue 6 (1991)

Shakeshaft, C. (1989). Women in Educational Administration. Newbury Park, CA: SAGE Publications.

Schmuck, P.A. (1980). Changing women's representation in school management: A systems perspective. In S.K. Biklen & M. Brannigan (Eds.), Women and educational leadership, 239-259. Lexington, MA: Lexington.

Tallerico, M., & Burstyn, J. (1996). Retaining women in the superintendency: The location matters. Educational Administration Quarterly, 32(Suppl.), 642-665.www.wpnet.com/cgibin/epwtop/

Developing Education through the Collaborative Interpretation of Research-based insights

Abraham Tadesse[1], Desalegn Zerai[1], Elina Lehtomäki[2], Hanna Kontio[3], Josephine Moate[1], Khalid Idris[4], Mussie Tewelde[1], & Yonas Mesfun Asfaha[5]

[1]University of Jyväskylä, Finland,
[2]University of Oulu, Finland,
[3]University of Helsinki, Finland,
[4]Eötvös Loránd University, Hungary
[5]Asmara College of Education, Eritrea

Abstract

The Eritrean education system faces formidable challenges in implementing inclusive and engaging educational experiences. Research-based insights of practitioners in the field seem to have limited role in informing practice and policy in developing education. In this study, the authors addressed key educational challenges through a collaborative interpretation of research findings from diverse sites of educational practice in Eritrea. Through the meta-interpretation of recently published qualitative research, perspectives and collaborative interpretations, this study contributed research-based insights into educational practices in Eritrea and outlines key areas for further development. The meta-interpretation of findings from 12 recent studies by the authors focused on areas of learner-centred interactive pedagogy, English medium education, differentiated instruction (inclusive education), and teacher education where the gaps between policy and practice remain wide. Framed within contemporary agendas for educational development and theories of teaching, the authors uncover practices and possibilities in implementing advocated learner-centred teaching approaches, linking conditions and experiences of English language education, inclusive teaching strategies in schools, and pedagogic interventions in teacher education. The synthesis study shares insights for educating teachers and the role of qualitative and practitioner researches in educational development in Eritrea.

Keywords: education development; collaborative meta-interpretation; educational research; Eritrea

Developing Contextual Knowledge in Education

One of the African Union's Continental Strategy key objectives is "revitalizing the teaching profession to ensure quality and relevance at all levels" (African Union, 2016, p. 22). Furthermore, the Strategy clearly deems the value of contextual knowledge in teacher preparation. In the review of the Strategy implementation across the continent Heleta and Bagus (2021, p.175) conclude, however, that:

> without quality, affordable and accessible higher education that develops competent and well-rounded graduates and engages in relevant research, struggling countries will remain dependent on unreliable outside aid and assistance. These countries will also continue to be passive consumers of knowledge produced elsewhere that does not consider local challenges, realities and needs and will not be able to take charge of their own destiny, improve people's lives and develop sustainably.

Moreover, as Idris, et al. (2017) noted, in order to bring about positive change and worthwhile learning, contextual challenges need to be recognised and explicitly addressed in teacher education programs and practices. In most of the Sub-Saharan Africa (SSA) contexts, constructivist education ideals have dominated education policy discourses (Vavrus et al., 2011) while their implementation were either frustrating (Schweisfurth, 2011) or unsustainable and 'expertise'-driven and not necessarily complimentary to bottom-up initiatives (e.g. Barnes et al., 2018). Practice and improvement orientation in education in SSA contexts also seem to have been associated with implementing what are known as small-scale research approaches, such as action research by teachers, which are widely phrased yet least embodied practice particularly by higher-education based teacher educators (Worku, 2017). Additionally, teachers rarely get opportunities in developing an understanding on how their situated practices could mediate idealized and real teaching practices (Barak, 2024). This is partly because professional development experiences tend to be either fragmented in nature (Idris et al., 2017) or top-down in approach (Schweisfurth, 2011). Accordingly, developing improvement-oriented contextualized knowledge in and during practice seem to be riddled with a transmissive telos of formal education (Harber, 2014) and academic traditions of higher education institutions which do not often bridge research and practice (Raelin, 2007).

Developing context relevant education and teaching requires pedagogic and research initiatives which centre voices and practices of the main agents, mainly students, teachers, and school leaders. For Alexander (2008), pedagogy is "the act of teaching together with the ideas, values and beliefs by which that act is informed, sustained and justified" (p. 4). In this definition, however, the act of teaching is dependent on the surrounding discourse for its meaning. If education is understood as the process of learning, then the act of teaching can be considered as facilitation or guidance, or indeed unnecessary if students are able to learn independent of teachers. If, however, education is understood as a practice that goes beyond the process of learning and necessitates teaching, then the integrity of teaching can be seen in a new light and the act, indeed art, of teaching can contribute to educational discourse and educational practice. In point, Biesta (2022) contests current proposals to base education on the interests of students alone as this promotes an ego-centric model for education. In place of an ego-centric model, Biesta (2022) proposes an eco-centric model for education which requires education beyond student interests. This is not to suggest that education should diminish student interest, but rather that education should not be limited to what students already know, imagine or care to know.

Further research in developing contextual knowledge in education afforded classroom dynamics and interactions as contexts for improvements and a basis of scholarship in moving perennially vexing topics forward. For example, establishing an inclusive, workable policy for language education (cf. Asfaha et al., 2024), building on school-related factors in adapting inclusive practices (Ainscow, 2016), and calls for pedagogies of enactment in revitalising the roles of teacher educators (Vanassche, 2022).

A view of transforming "teacher education and professional development from course-based, individual endeavours to lifelong, collaborative, and teacher led processes" further stirs orientations towards practice and practitioner-centred development initiatives in the context of reported teacher shortages world-wide (UNESCO, 2023, p. 31). It also appears that the language used in capturing and charting educational aspirations may be insufficient if they fall short of engaging with the intentional, reflective, and relational nature of educational practice (Biesta, 2015). Hence, embeddedness and embodiments of the seemingly lofty goals as lifelong learning and teacher-led research need to be understood as site-bound (Kemmis et al., 2014) and prefigured by the affordances and constraints of educational contexts (Barak, 2024). Accordingly, a biographical and narrative approach towards teacher education research practices (Smith, 2024) would not only facilitate in rendering practitioners' own change initiatives (Lehtomäki et al., 2014) but also frame and develop a language *of* and *for* practice. This paper builds on and draws inspiration from discourses of research to engage, improve, and empower practitioners (Kemmis, 2022; Lehtomäki et al., 2014) and possibilities for transformation through research insights which acknowledges the complexities and conditions for developing context relevant knowledge for educational practices in Eritrea.

In response to the need for developing contextually relevant knowledge, this study syntheses recent research findings on education in Eritrea. Through a meta-interpretation of recently published qualitative research this study contributes research-based insights into educational practices in Eritrea and outlines key areas for further development. Framed within the previously discussed contemporary educational agenda and theories of teaching, the authors uncover practices and possibilities for strengthening learner-centred interactive pedagogy, language education, inclusive teaching strategies, and teacher education. The overarching research task is to gain a better understanding of education as a practice within the particular context of Eritrea. The synthesized qualitative insights of educational practices and implications for policy point to the dire need for reimagining teacher education. Hence, in the discussions underpinned by the following research questions, insights for educating teachers are drawn.

1. What educational practices can be gleaned from the synthesis of 12 contemporary studies?
2. What do the qualitative research findings say about the policy and its implementation in Eritrea?

Synthesizing the studies

The initiative for synthesizing recent research findings on education in Eritrea was supported by the Global Innovation Network for Teaching and Learning (GINTL) at the University of Helsinki, Finland under a project titled 'generating better practices in education through international partnerships in doctoral education'. The project of a 9 months period (November 2023- June 2024) aimed at 'synthesising and publicising the key findings from four doctoral dissertations on education development in Eritrea; and connecting the individual research projects to the broader frames of internalization and decoloniality in higher education' (GINTL poster, Annex 1). The project comprised of four doctoral candidates and four supervisors from the universities of Jyväskylä, Oulu, and Helsinki in Finland and Eötvös Loránd University in Hungary. Since the four doctoral candidates where in a manuscript writing stage based on their published articles, the project was also seen as platform in collaborating to synthesize broader connections and implication to educational development in Eritrea. The four PhD projects are qualitative studies. As typical of qualitative educational studies they involved in-depth description of rich data collected through a variety of instruments and several analytical methods. Data and participants include in-depth interviews with individual teachers and students across the entire education system, elementary, middle, secondary, and higher education. Besides data was collected through video recordings of classroom instruction. Action research and self-study interventions and reflective practices were other research approaches.

The project supported a two-day workshop conducted at the University of Helsinki on January 24-25, 2024 among the doctoral candidates and supervisors in developing and supporting a meta-synthesis process. The focus was on what the individual studies share about education and teaching. The discussions for synthesis were based on the findings of published and submitted studies rather than the insights from the compiled dissertations. In line with this focus, a meta-interpretations approach (Weed 2008) was adopted in which the "underlying assumption of synthesis is that such procedures can add to the body of knowledge, and in this tradition the aim of meta-interpretation is to make conceptual links across previously unconnected areas of "undiscovered public knowledge" (p. 26). Accordingly, different tasks were designed during the workshop in (a) exploring interconnections and tensions within and among the 12 research studies, and (b) identifying what the findings within and across the doctoral research themes say about related education policies and their implementation in the Eritrean context. For example, questions as 'What incidental insights can be gleaned from the different themes'? 'How do these insights resonate with your findings' facilitated in generating a common 'table of findings' and a 'policy table' in capturing essential insights of respective studies and developing common themes (Tables 1& 2).

After this workshop, it was agreed by the participants (authors) to work on and update the tables, situate synthesized insights within education development discourses, and develop a conceptual framework for a journal article. A commonly shared Google Drive was created for this purpose where the tables were constantly and iteratively updated, an article draft was being developed, and relevant materials were shared. For the purpose of this international conference on Eritrean studies, it was agreed to focus on presenting synthesized findings on educational practice and policy implementation and implications to teacher education development. We conclude this presentation by highlighting the value of collaborative, qualitative, and practice-based research in education development in Eritrea.

Educational Practices in Context

Identification of key findings of the 12 studies facilitated in generating synthesized and overarching themes on educational practices ranging from elementary schools to higher education learning and teaching contexts. These synthesized themes include: teachers, students and teacher educators as mediators of education, influencing present practices shaping future possibilities, and creating conditions for changing practices (Table 1). In this section, these themes are discussed in turn.

Table 1

Key findings and synthesized themes

Tewelde et al., 2023a	Zerai et al., 2023a	Tadesse et al., 2021	Idris 2023	Synthesized theme of the findings
Systemic conditions can compromise pedagogical goals, activities & relationships	Eritrean teachers show positive attitudes towards differentiated instruction and some teachers can creatively provide DI	Positive teacher attitudes towards an innovation are not enough to introduce significant change	Mediating student-teachers' learning challenges with atypical reflective requirements	Teachers, students and teacher educators as mediators of education as a shared endeavour
Assigned teachers seem to have little capacity to develop pedagogical conditions that promote language learning	Eritrean teachers construct different meanings to DI, 78% positive, 22% negative	New approaches in education also rely on the willingness of students	Learning while practising reflection	
Many students struggle to be agentic in their language studies under critical conditions.	Expecting teachers to enact DI without support can significantly limit what teachers can do	Transforming a vision into practice requires investment, understanding & partnerships	Language related tensions while facilitating reflective practices	

Tewelde et al., 2023a	Zerai et al., 2023b	Tadesse et al., 2023b	Idris et al., 2024	Synthesized theme of the findings
Language learning is an emotional-filled experience. Emotions can be positive, negative or ambivalent.	Eritrean teachers use a variety of question modification strategies to differentiate instruction	LCIP policies are made into practices through teachers	Pedagogizing research while embedding reflective practices during a teacher education course	Actions and responses in the present help future possibilities to take shape
Emotions mediate experiences and perception of self in the present and future	Teachers' question modification strategies represented reactive not proactive responses	Teachers have to simultaneously negotiate with what has been, what should be and what can be in this moment	Exemplifying professional learning in and through a practice of modelling	
The emotions of individuals are in response to, not determined by, surrounding conditions.	Teachers can act sensitively & spontaneously even in mixed ability, linguistically diverse, and large classrooms	Using a different lens (DT) can help bring potential areas for development into view	Exemplifying professional learning in and through a practice of modelling Idris et al. (under review)	

Tewelde et al. (under review)	Zerai et al. (under review)	Tadesse et al. (under review)	Idris et al. (under review)	Synthesized theme of the findings
Students can become highly agentic. Students' use of language learning strategy informs the formation of their agency within the English Language Education environment.	Eritrean teachers display several pedagogical tensions in their narratives and actual classrooms instructions	Dialogic teaching breaks traditional practices by broadening the pedagogical repertoire of teachers	Creating conditions for developing reflective practices through self-study research approach	Creating conditions for changing practices
Students developed different agentic orientations to use the affordances in the environment and embrace the systemic and didactic conditions within the ecosystem.	Eritrean teachers adapt their instructions by utilising individual and small groups tasks to accommodate diversity	Creating active engagement and deeper learning through a possible link between dialogic and learner-centred teaching	Practices of teaching and learning as contexts for teacher educators' professional learning in teacher education	
	Equipping teachers with continuous training on DI help them handle the pedagogical tensions they encounter	Dialogic teaching fosters learner-centred practices	Practices of teaching and learning as contexts for teacher educators' professional learning in teacher education	

The studies explored critical education practices of implementing learner-centred interactive pedagogy (LCIP) in secondary schools, challenging conditions of English language education (ELE), inclusive teaching practices in elementary and middle schools, and teacher education pedagogy. Tadesse et al.'s (2021) study showed teachers' cautious (not necessarily negative) perceptions on the implementation of LCIP in large, diverse, and under-resourced classrooms. Zeria et al. (2023a) study presented positive meanings teachers constructed on differentiated instruction (DI), which was synonymized with implementing LCIP among the study participants. On the other hand, Tewelde et al.'s (2023a) findings shared restrictive and insensitive teacher roles within a challenging ELE ecosystem in a higher education setting. Idris's (2023) self-study on embedding reflective practice among experienced student teachers in a teacher education program, however, uncovered tensions which scrutinized and sensitized his facilitations. Teachers and teacher educators as mediators of education - between the system and the students, between the curriculum and studying activities, between policy and practice, between innovations and actualities – is a central theme across these studies conducted in diverse educational settings and communities.

The studies directly addressed how bridging schooling and teacher education practices could empower teachers' mediating roles to meaningfully engage their students. For example, Tewelde et al.'s (2023a) observation on the limited capacity of English teachers implicated the lack of preparation of such teachers within a challenging ELE ecosystem. The study also shared uncritical and colonial stances of participants about 'mastering' English which indicated the need for preparing language teachers who appreciate the ideological nature of ELE in postcolonial and multilingual contexts (Asfaha et al., 2024). Similarly, existing positive attitude of teachers on DI is an untapped resource because the flexible pedagogical approaches and self- reflective understandings of the teachers may help the wider community to see the practice of teaching from a different perspective and incorporate such elements into teaching processes. In poorly resourced education systems, this is particularly remarkable and points to the key role of teachers in improving the quality of education (Zerai et al., 2023a, p. 15).

Idris's (2023) study showed how teacher educators could learn in and from their teaching to meaningfully influence the mediating roles of student teachers. The study showed how embodying idealized practices, i.e., reflective and learner-centred practices, in teacher education classrooms could challenge and revitalize teachers' assumed roles and actions.

Qualitative analysis of video-recorded classroom interactions, interviews, and interventions of the studies underscored what Kemmis (2022) calls as 'happeningness' of educational practices in those activities "occur only in the present, although they are oriented towards the future and in response to the past" (p.29). For example, Tadesse et al.'s (2023) identified dialogic moments amidst a conventional teacher-led interactions which showed instances of implementing LCIP. This study showcased that quality interactions do *happen* in challenging classroom contexts. In a teacher education classroom context, Idris et al.'s (2024) study recounted a process of a teacher educator's modelling as conditions for improved facilitations of reflective practice among a group student teachers. "Modelling the process of professional learning revealed how excelling in the very practices that were being promoted amongst the [student teachers] was pedagogically supportive, as it exemplified visible professional learning" (p.12). Similarly, Zerai et al.'s (2023b) study stressed teachers' resourcefulness in enacting a more inclusive question modification strategies in large and under resourced mathematics and science classrooms. Tewelde et al.'s (2023b) findings echoed (presence or lack of) teacher resourcefulness in their study of how students' emotional experiences in ELE settings shaped their future selves as "the negative emotions provoked by the distal language policy [...] can be softened or reinforced through the actions of the proximal teacher or family" (p.11).

These findings that uncovered the happeningness of educational practices in Eritrea highlighted the essential nature of teaching, teachers, and teacher educators (Biesta, 2022), on the one hand, and how their roles could be supported and revitalized (Demoz, 2024) on the other. The promising educational practices that the studies have empirically verified were in most a factor of teachers' and teacher educators' *presence* in attending to learning needs and intentionality and reflectivity on their actions. For instance, a teacher's use of humour invoked not only a more open dialogue but employed the students' first language in a meaning making interactions (Tadesse et al. 2023(b)). Such happeningness of practices hold potent potential for educational development within the existing school and teacher education affordances and constraints. The potential of dialogic pedagogy as a practice of LCIP (Tadesse, 2025), differentiating and questioning strategies as embodiments of inclusive education (Zerai, 2024), role of emotions and ELE ecosystems in developing language education (Tewelde, Forthcoming), and researching teacher educators' practices as a form of professional and scholarship development (Idris, 2025) are key development areas that were identified through the studies. These potentials are also rich grounds for reimagining teacher education within and beyond colleges of education and school settings.

The studies further suggested how conditions could be created for changing teaching practices in schools and teacher education institutions. Tadesse et al.'s (2023b) study suggested the significance of broadening pedagogic repertoires of teachers and turning dialogic opportunities into quality interactions in whole-class and customarily teacher-led classroom environments. During such interactions, the studies highlighted teachers' and a teacher educator's need for managing pedagogic tensions while adapting inclusive teaching practices (Zeria, 2024), "English as the medium of instruction which appeared to hinder how learners could express their thoughts, ideas and opinions" (Tadesse et al., 2023b, p.10), and mediating verbal (mix of local and English) and written (English) reflective practice of student teachers (Idris, 2023). In a study of professional development of teachers in Israel, Barak (2024) emphasized the potential of such pedagogic tensions in supporting teachers' concept development while mediating ideal practices (as LCIP) with classroom realities. Idris's (Forthcoming) study employed Kemmis's (2022) theory of practice architectures to show how a changing course arrangement were changing the learning and facilitation practices during a semester's (16 weeks) course. The study stressed how a researching teacher educator in practice could prefigure the value of learning from experience in a teacher education course.

It seems research by teachers and teacher educators is a strategic undertaking in not only surfacing pedagogic tensions but also in learning to manage them. The role of professional development has been suggested by the studies in, for example, managing tensions related to DI (Zeria, 2024) and enhancing professional competence of teachers for dialogic pedagogy (Tadesse et al., 2023b). However, such calls may result in professional development practices which are either intermittent or fragmented when they are not meaningfully integrated into the schools' and teacher education programs and are 'expertise' dependent (Barnes et al., 2018). A realistic approach for professional development in and for practice based on the studies is the partnerships that teacher education institutes could create with schools and the central role of teacher educators in this process. Knight (2024) suggested "that teacher educators are especially well placed to help teachers bridge between practical experience and wider forms of knowledge" (p.1). The lived experiences of teachers and existing assumptions in terms of learning and teaching in schools should be the basis for designing and enacting teacher education (Demoz, 2024). Within teacher education practices itself, teaching-led research have a real potential in nurturing a teacher-researcher stances among student teachers and the development of a teaching-research nexus to form the basis for teacher education programs (Idris et al., 2024).

Policy Implementation

In this section, we discuss what the qualitative studies say about four educational policies, in line with the themes of the doctoral projects, and their implementation. The studies addressed the policy areas of English as the language of instruction beyond elementary schooling; special and inclusive education; learner-centred interactive pedagogy; and teacher education (Table 2). The studies point to gaps between policy and existing practices in all of the stated educational policy areas. Yet, our intention is not to dwell on those 'gaps' but to discuss shared insights on how the official policies are being enacted and negotiated and point to promising practices for bridging the existing gaps. ELE is among the centrally challenging issues raised in the studies while facilitating meaningful, engaging, and inclusive educational experiences. This is in the backdrop of mother tongue education as a 'contested resource' (Tewelde et al., 2023b) on the one hand and English proficiency as a desirable (Idris, 2023) and even idealistically conflated with performance compared with 'native' standards (Tewelde et al., 2023a) on the other. Accordingly, the studies point to ways in which the linguistic diversity and the multilingual policy affordances in Eritrea are mobilized for meaningful learning while improving English proficiencies starting from basic education levels. This should not be seen as a far-out undertaking considering the emerging scholarship of translanguaging pedagogies on the significance of employing multiple languages in learning and teaching (Asfaha et al., 2024).

The views and sensitivities teachers display on the diversity of their students is another relevant finding with policy implications. While Tewelde et al.'s (2023b) study emphasized the importance of teachers' sensitivity on the 'emotional tenor' in ELE classrooms, Zeria et al.'s (2023a,b) study showed the care and resourcefulness teachers displayed in attending to learning diversities. Tadesse et al.'s (2021; 2023b) study show practical dilemmas teachers face in large class sizes as they often resorted to frontal teaching methods at the expense of attending to diverse learning needs. These studies point to what extent and how teachers were prepared for diversity and whether teacher education programs take stock of the caring and resourceful teacher qualities as positive schooling realities in fostering inclusive education practices. Beyond teaching and learning subjects and 'mastering' English, there are relational and subjective educational intricacies (Biesta, 2022) that the teachers and teacher educators were not necessarily prepared for. Attending and learning to manage such intricacies are what makes schooling experiences *educational* (Kemmis et al., 2014). As Demoz (2024) puts it, there is a need for 'reframing contents for educating teachers' in including pedagogic, cultural, and character areas in line with the interest of students in schools and the social, national, and global milieu in which we live in.

Table 2
What do the qualitative research findings say about the policy and its implementation?

Educational policy	Tewelde et al.	Zerai et al.	Tadesse et al.	Idris et al.	Shared insights
English as the language of instruction beyond elementary schooling	Although English is the language of education, many students struggle to attain mastery of English even at tertiary level. This struggle affects their overall educational achievement.	Teachers experience problems in communication with their students in middle schools as the medium of instruction is English. Teachers' use of code switching to local languages links to mother tongue (language) education policy, multiling-ualism and teachers' sensitivity to and their effort to accommodate linguistic minority groups (Zerai et al, 2023b, Zerai et al., n.d.)	LCIP policy implies centredness of language and instruction.	Student-teachers were inspired to improve their written English during the reflective assignments. This shows the value teachers attach to their English proficiency.	Limitations on English proficiencies throughout the school and teacher education system.
	Despite the positive policy of mother-tongue based education in elementary grades, the transition to the use of English has significant implications for learning other subjects. The focus on grammar in English lessons, and emphasis on accuracy and fear of mistakes does not prepare students for using English in studying other subjects (Tewelde et al., 2023a).		English as a medium instruction is considered as a barrier to classroom interaction in multilingual classroom contexts. (Tadesse et al., 2021).	Teacher educators struggle in mediating supporting deeper reflective practice and improving proficiency in written English. Professional learning of teacher educators is vital in this regard.	
	Excluding different mother tongues from English lessons or using mother tongues to compensate for lack of understanding means that students don't learn how to use their multilingual resources in their studies (Tewelde et al., 2023a).		Collaborative learning and classroom discussions are central to LCIP; however, in contexts where teachers and students are not proficient in the medium of instruction, classroom interactions tend to shift to local languages in pursuit of better teaching. (Tadesse et al., 2023)		

Educational policy	Tewelde et al.	Zerai et al.	Tadesse et al.	Idris et al.	Shared insights
Special and inclusive education	Negative emotions of students and the struggles teachers face has a negative impact on the overall educational trajectory(Tewelde et al., 2023b).	Eritrean teachers hold a strong desire, readiness, and positive attitude toward supporting all individual students in the classroom through DI, and they attach personally and emotionally to their students.	Teachers' focus on the whole class lecturing rather than individual students' needs undermines students' learning	Inclusive and creative representation of reflective experiences needed: e.g., audio-visual and arts-based reflections. Exploring innovative and engaging learning experiences.	Building on pedagogy of care for inclusive educational practices
	Support and opportunities for personal freedom can transform negative emotions into positive resources (Tewelde et al., 2023b).	Teachers experience confusion and they lack conceptual clarity regarding the meanings and definitions of Inclusive Education and Differentiated Instruction. The challenges are related to lack of guiding principles of Differentiated Instruction and Inclusive Education in the school curricula as well as the Teacher Education Preparation programs.		Under-developed pedagogy for special need student teachers in teacher education need due attention	
		Teachers' voices in student learning are not considered. They are only required to transfer knowledge without inadequate in-service training and professional development. Teachers were found to show interest and sensitivity towards individual needs of their students (Tadesse et al., 2021) as a result, teaching methods were modified to satisfy students' needs (Tadesse et al., 2023)			

Educational policy	Tewelde et al.	Zerai et al.	Tadesse et al.	Idris et al.	Shared insights
Learner-centred interactive pedagogy	Teachers' low language proficiency and large class sizes can make in class communication unidimensional and limit opportunities for practice, pacifying learners. (Tewelde et al., 2023a)	The teachers tried their level best to make their classroom more interactive through open question, changing methodologies like field work, practical hands on exercises, group work and pair work, peer teaching, and student individual work (Zerai et al; 2023a, 2023b, Zerai et al; n.d.)		Pedagogy of modelling as an important notion in showing students the challenges, uncertainties, and possibilities of a learner-centred and interactive classroom. Focus on teacher educators' learning.	Building on seeds of engaging and interactive pedagogy in schools and teacher education institute(s)
	Multilingual linguistic repertoires can enable students & teachers to devise practice opportunities through translanguaging and giving them a sense of autonomy to transform the environment. (Tewelde et al.)	Teachers believe that DI is synonymous with LCIP, they show positive attitudes towards individual students (Zerai et al, 2023a), and they create positive classroom environments that are student friendly (Zerai et al.,n.d.)	Teachers' positive attitudes and learner-centred practices are hindered by a number of barriers, such as class sizes, failure to address individual needs, authoritarian culture and a lack of professional development (Tadesse et al., 2021).	Pedagogizing research as an important practice for showing improvements and development of relevant scholarship in and for teacher education. Practice based research and research based practice in teacher education.	
			Professional development, educational innovation investment and partnerships are required to bring vision to actual practice (Tadesse et al. 2021).		

Educational policy	Tewelde et al.	Zerai et al.	Tadesse et al.	Idris et al.	Shared insights
Teacher education	Assigned teachers seem to have little capacity to develop pedagogical conditions that promote language learning.	Teachers indicated the disconnect between teacher education, MOE policy and classroom practices due to lack of proper training of DI and IE (Zerai et al., 2023a)	Connecting curriculum and pedagogy is paramount in teacher education.	Introducing course management practices in teacher education, e.g., moving from practice of course outline (content oriented) to course design (process oriented)	
	Students can be agents in their own development when they can recognise and use different opportunities in the environment.	Incorporation of the guiding principles of DI and IE in the school curricula as well as Teacher Education programs in HEI is required (Zerai et al , 2023a, 2023b, Zerai et al., n.d.)	Teacher training should be dynamic and practical modelling LCIP and dialogic teaching incorporating opportunities for self reflection and adapt classroom realities (Tadesse et al., 2023)	Improvement and practice oriented research to be an integral part of teacher education programs.	Schooling contexts and practices as grounds for (re)imagining the education of teachers
	The policy-practice gap challenges educators when introducing change, and under-qualified teachers struggle to reform the conditions of education and are inhibited to attempt to change (Tewelde et al., 2023a).	Strengthening teachers' potentialities and expertise through ongoing in-service teacher training programmes that are research-based teaching practices (Zerai et al, 2023b).		Developing capacity of teacher educators in researching their practices and reading proficiencies, particularly literature on practice and critical theories in education.	Focus on teacher educators' practices
	For teacher educators, the findings highlight the importance of understanding the emotional tenor of the language classroom, to help students utilize their agency and to recognize the fundamental role teachers have in (re)forming the conditions of ELE in Eritrea (Tewelde et al., 2023b)	Teachers' innovative activities, creativity in implementing different strategies, and their resourcefulness give direction for future PD programmes, and utilising teacher potentialities in designing practice based activities in Teacher Education curricula (Zerai et al., 2023a; Zerai et al., n.d.)			

The studies show that the implementation of LCIP in school and teacher education practices have faced structural (e.g. teacher preparation and support, ELE system), material (e.g., class-sizes), and cultural (e.g., frontal teaching) challenges despite a generally positive or even idealized conceptions of school communities on the approach. Yet, the studies also shared seeds of interactivity for meaningful and inclusive learning that indicate the promising appropriation and recontextualization of the LCIP policy in diverse settings. Tadesse's (2025) studies has, for example, culminated in stressing the potential of dialogic pedagogy in strengthening LCIP in secondary schools. Idris (Forthcoming) was also convinced about the potential for practicing and modelling LCIP by building a collegial and committed community of practice and through intentionally integrating research to revitalize processes of teaching in teacher education programs. As the principles of LCIP also have close resonances with the role of education in Eritrea's contemporary history for liberation (Idris, Forthcoming), we argue it's high time to capitalize on and upscale these modest experiences in the school system. Hence, the question should not be whether implementing LCIP is realistic but about creating conditions in and for schools in matters which engage, challenge, motivate, and educate students.

These discussions bring us to the central role teacher education institutes could play in mediating policy and practice. Teachers and teacher educators were discussed across the studies as central agents for changing (or perpetuating untoward) practices. Hence, their characteristics and quality of preparation should be a policy priority. The findings indicate two integrated priorities which require due focus: strengthening teacher education institution(s) and teacher educators' practices. There is an urgent need to reconstructing the customary theory into practice model for teacher preparation in affording school practices as essential grounds for imagining teacher education programs and practices. Organizing teacher education around the traditional subjects being taught in schools seem to overlook the essence of teacher education which is mainly about 'learning to understand what teaching implies' (Demoz, 2024). Teacher educators', who are often not identified as such in Eritrea, sensitization (not to mention preparation) to this issue of what teaching implies becomes central in this regard. Accordingly, the work and professional identity of teacher educators is of utmost significance in the process of influencing prospective teachers who would work in a complex ELE ecosystem, with students of diverse learning needs, and are expected to implement innovative pedagogies of engagements. We believe the policy and research focus that is being afforded for the work and professional development of teacher educators in the Global North contexts (e.g. Vanassche et al., 2021) could be a timely resource to draw from in the Eritrean context.

Conclusion

This study was framed by discourses on the need of generating contextual knowledge for developing education. The practices and possibilities analysed in the 12 studies reaffirmed the essential roles of teachers and teacher educators not only as mediators but also as enablers of education. The central issue, then, becomes how teachers and teacher educators are prepared, supported, and managed. These tasks are not only left to teacher education and higher education establishments though such institutes *orchestrate* the preparation and support of the essential roles of teachers and bring forth practice-based, dynamic, and developmental insights. This initiative reiterates the value of collaboration in developing contextual insights for improving educational practices. It also highlights the importance of connecting seemingly disparate educational issues in gaining a better and broader understanding for improvements in line with the existing educational needs. In this, qualitative studies in education have a significant potential in uncovering complex educational practices, surfacing teachers' and students' voices, and building circles of learning professional communities.

References

African Union. (2016). Continental education strategy for Africa: 2016–2025. Addis Ababa: African Union.

Ainscow, M. Diversity and Equity: A global education challenge. *NZ J Educ Stud*, 51, 143-155. DOI 10.1007/s40841-016-0056-x

Alexander, R. (2008). Culture, dialogue and learning: Notes on an emerging pedagogy. In N. Mercer, & S. Hodgkinson (Eds.), *Exploring talk in schools: Inspired by the work of Douglas Barnes* (pp. 91–114). Sage.

Asfaha, Y.M., Spotti, M., & Idris, K. (2024). Translanguaging pedagogies in the global south: Review of classroom practices and interventions. In C. McKinney, P. Makoe, & V. Zavala (Eds), *The Routledge Handbook of Multilingualism 2nd Edition*, pp. 252-268. Routledge.

Barak, M. (2024). Educational ideals and classroom realities: Developing teachers' concepts of dialogic pedagogy in real-world contexts. *Teaching and Teacher Education*, https://doi.org/10.1016/j.tate.2023.104401

Barnes, A. E., Zuilkowski, S. S., Mekonnen, D., & Ramos-Mattouss, F. (2018). Improving teacher training in Ethiopia: Shifting the content and approach of pre-service teacher education. *Teaching and Teacher Education*, *70*, 1-11. https://doi.org/10.1016/j.tate.2017.11.004

Biesta, G. (2015). What is education for? On good education, teacher judgement, and educational professionalism. *European Journal of Education, 50*(1), 69–87. https://doi.org/10.1111/ejed.12109

Biesta, G. (2022). *World-centred education: A view for the present*. Routledge.

Demoz, B. (2024). *A Professional Pursuits for Educating Teachers: Managing Complexity with Complexity***.** [Unpublished manuscript]

Harber, C. (2014). *Education and international development: Theory, practice and issues*. Symposium books.

Heleta, S., & Bagus, T. (2021). Sustainable development goals and higher education: leaving many behind. *Higher Education*, 81, 163–177. https://doi.org/10.1007/s10734-020-00573-8

Idris, K.M., Asfaha, Y. M. & Ibrahim, M. A. (2017). Teachers' voices, challenging teaching contexts and implications for teacher education and development in Eritrea. *Journal of Eritrean Studies*, III (1), 31-58.

Kemmis, S. (2022). *Transforming practices: Changing the world with the theory of practice architectures*. Springer.

Knight, R. (2024). Teacher educators as knowledge brokers: reframing knowledge co-construction with school partners. *Professional Development in Education*, https://doi.org/10.1080/19415257.2024.2360461

Kemmis, S., Wilkinson, J., Edwards-Groves, C., Hardy, I., Grootenboer, P., & Bristol, L. (2014). *Changing practices, changing education*. Springer.

Lehtomäki, E., Janhonen-Abruquah, H., Margaret T. Tuomi, M. T., Okkolin, M., Posti-Ahokas, H., & Palojoki, P. (2014). Research to engage voices on the ground in educational development. International Journal of Educational Development, 35, 37-43. http://dx.doi.org/10.1016/j.ijedudev.2013.01.003

Raelin, J. A. (2007). Towards an Epistemology of Practice. *Academy of Management Learning & Education*, *6*(4), 495–519. https://doi.org/10.5465/amle.2007.27694950

Schweisfurth, M. (2011). Learner-centred education in developing country contexts: From solution to problem? *International Journal of Educational Development*, *31*, 425–432.

http:// doi:10.1016/j.ijedudev.2011.03.005
Smith, K. (2024). Evidence-based, evidence-informed or evidence-ignored teacher education? The role of research in teacher education. In V. Symeonidis (Ed.), *Enhancing the value of teacher education research: Implications for policy and practice* (pp. 23-41). BRILL.
UNESCO. (2023). *Global report on teachers: Addressing teacher shortages (Highlights).* UNESCO.
Vanassche, E. (2022). Four propositions on how to conceptualize, research, and develop teacher educator professionalism. *Frontiers in Education, 7.* https://doi.org/10.3389/feduc.2022.1036949
Vanassche, E., Kelchtermans, G., Vanderlinde, R. & Smith, K. (2021). A conceptual model of teacher educator development: An agenda for future research and practice. In R. Vanderlinde, K. Smith, J. Murray & M. Lunenberg (Eds.), *Teacher Educators and their Professional Development: Learning from the Past, Looking to the Future* (pp. 15-27). Routledge.
Vavrus, F., Thomas, M., & Bartlett, L. (2011). *Ensuring quality by attending to inquiry: Learner-centred pedagogy in sub-Saharan Africa.* UNESCO CBA International Institute for Capacity Building in Africa.
Weed, M. (2008). A potential method for the interpretive synthesis of qualitative research: issues in the development of 'meta-interpretation'. International journal of social research methodology, 11(1), 13-28.
Worku, M. Y. (2017). Improving primary school practice and school–college linkage in Ethiopia through collaborative action research. *Educational Action Research, 25*(5), 737-754. http://dx.doi.org/10.1080/09650792.2016.1267656

Synthesized studies

Zerai, D., Eskelä-Haapanen, S., Posti-Ahokas, H., & Vehkakoski, T. (2023a). The meanings of differentiated instruction in the narratives of Eritrean teachers. *Pedagogy, Culture & Society, 31*(3), 419–437. https://doi.org/10.1080/14681366.2021.1914712
Zerai, D., Eskelä-Haapanen, S., Posti-Ahokas, H., & Vehkakoski, T. (2023b). The use of question modification strategies to differentiate instruction in Eritrean mathematics and science classrooms. *Education Sciences. 13,* 284. https://doi.org/10.3390/educsci13030284
Zerai, D., Kontio, H., Eskelä-Haapanen, S., & Vehkakoski, T. (2024). Differentiation-related tensions in the thinking and instruction of Eritrean elementary and middle school mathematics teachers. [Manuscript under review]. Department of Education, University of Jyväskylä.
Idris, K. M. (2023). Tensions of Embedding Reflective Teaching Practices in Teacher Education in Eritrea: A Self-Study on Facilitation Experiences. *Studying Teacher Education,* 19(3), 251-270. https://doi.org/10.1080/17425964.2023.2202012
Idris, K. M., Posti-Ahokas, H., & Lehtomäki, E. (2024). Developing a teaching-research nexus in teacher education: Recounts of an educator's self-study. *Teaching in Higher Education, 30*(4), 898-914. https://doi.org/10.1080/13562517.2023.22988
Idris, K. M., Posti-Ahokas, H., & Lehtomäki, E. (under review). Creating conditions for reflective practice in teacher education: Insights from a self-study research in a challenging context.
Tadesse, A., Eskelä-Haapanen, S., Posti-Ahokas, H., & Lehesvuori, S. (2020). Eritrean teachers' perceptions of learner-centred interactive pedagogy. *Learning, Culture and Social Interaction, 28,* 100451. https://doi.org/10.1016/j.lcsi.2020.100451
Tadesse, A., Lehesvuori, S., Posti-Ahokas, H., & Moate, J. (2023). The learner-centred inter-

active pedagogy classroom: Its implications for dialogic interaction in Eritrean secondary schools. *Thinking Skills and Creativity, 50*, Article 101379.

Tadesse, A., Lehesvuori, S., Posti-Ahokas, H., & Moate, J. (under review). Advancing learner-centred interactive pedagogy through dialogic teaching: An intervention with Eritrean secondary school teachers.

Tewelde, M., Moate, J., & Posti-Ahokas, H. (2023a). Exploring the conditions of English language education through the experience of Eritrean Sophomore English language learners and teachers: An ecological perspective. *Apples – Journal of Applied Language Studies.* 17 (1), 104 – 123. https://doi.org/10.47862/apples.120265

Tewelde, M., Moate, J., & Posti-Ahokas H. (2023b). Investigating the Emotional Experiences of Sophomore Students in English Language Education in Eritrea from an Ecological Perspective. *Educ. Sci.*13, 1181. https://doi.org/10.3390/educsci13121181

Tewelde, M., Moate, J., & Posti-Ahokas H. (under review). Exploring the language learning strategies employed by successful sophomore students in English Language Education in Eritrea from an Ecological perspective.

Doctoral manuscripts

Zerai, D. (2024). *Towards inclusive education: Eritrean elementary and middle school mathematics and science teachers' views on and implementation of differentiated instruction,* (JYU Dissertations No. 798) [Doctoral Dissertation, University of Jyväskylä]. University of Jyväskylä. http://urn.fi/URN:ISBN:978-952-86-0192-0

Tadesse, A. (2025). *Strengthening learner-centred interactive pedagogy through dialogic teaching in Eritrean secondary schools*.[Doctoral Dissertation, University of Jyväskylä]. University of Jyväskylä. https://urn.fi/URN:ISBN:978-952-86-0506-5

Idris, K.M. (2025). *Improvements from within : a self-study on teaching practices in a teacher education context in Eritrea.* [Doctoral Dissertation, University of Oulu]. University of Oulu. https://urn.fi/URN:NBN:fi:oulu-202502051464

Tewelde, M. (anticipated autumn 2025) Investigating the language learning pathways of sophomore English students in Eritrean Higher Education: an eco logical perspective, Doctoral dissertation.

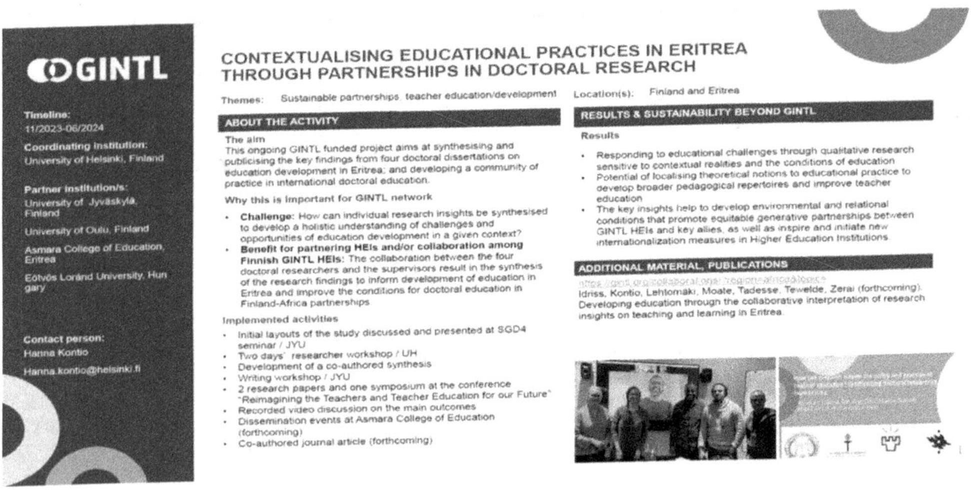

Annex 1: GINTL poster

Collaborative Action Research in Education: A Catalyst for Professional Development and Policy Advancement in a Demanding Context

Amanuel Yosief

Department of General Education, Ministry of Education, Eritrea.

Abstract

This study contends that the fragmentation between intentions and programmes in teacher education (TE) and schools have been critically affecting the quality of practices operating within the educational landscape in Eritrea. It delves into the impact of a TE course 'collaborative action research' (CAR) offered to in-service teachers in their school practices. It was informally modelled by a group of teacher educators (TEs) to enable practitioners to navigate along demanding educational contexts, make efforts to transform practices, inform and advance education policy through data-driven voices and actions. It employed a follow-up study using in-depth interviews, questionnaire for reflections, and critical observations. It was conducted between June 2019 and January 2023 to qualitatively ensure rigour and perceptiveness of situations, illuminate why circumstances are as they are, and envisage how challenges could be transformed. The inductive examination of data rendered in this study gave rise to six emergent themes which beget profounder implications. This study indicates CAR as a catalyst for strengthening TEs' and school teachers' professional relationships, developing firm engagement, and promoting legitimate and transformative practices. By offering insights into practitioners' actions, challenges, and accomplishments, it contributes to school practices and inspires teachers, educators, policymakers, and researchers to embrace CAR for positive transformation in education.

Keywords: collaborative action research, practitioners, professional transformation and development, education policy, teacher education, Eritrea.

Introduction

This study primarily reveals the outcomes of collaborative action research (CAR) used as a model and designed to link a teacher education program (TEP) and schools in Eritrea. It argues that the fragmentation between TEP and school practices in Eritrea have been critically affecting quality. There has been disconnection and separation between colleges and schools as two different learning arenas that overrun TEPs for years (Sjølie & Østern, 2020). This study was thus initiated to improve practices through creating relationships among teacher educators (TEs) and school teachers thereby inform policy using practitioners' voices and actions. Such intention was thought to be realised through framing a model that are characterised by bottom-up approach. Models of partnership are often established more informally which appear to be more flexible in structure and implementation (Livingston, 2019). Although the concept of university and school partnership and what a partnership model of teacher education means in practice remains unclear (ibid.), this study proposes CAR as a catalyst for practitioners' connection, learning, professional transformation, and development.

This study divulges that CAR is participatory by its nature, involving practitioners as active agents in the research process. CAR is a form of participatory research, which takes place conjointly with teachers and researchers (Heron, 1996). Diverse contexts can benefit from CAR which can also provide framework of creating a community of practice (Bruce & Easley, 2000). CAR deepens teacher researchers' commitment to improving their teaching through overcoming obstacles such as a lack of skill in research methods, valuing teachers as researchers, and researchers as learners, and developing and testing theoretical frameworks (Catherine et al., 2011). This collaborative and reflective approach allows practitioners to draw upon their expertise and experiences, bridging the gap between theory and practice, and fostering a culture of continuous improvement. Within a demanding context, which may encompass factors such as complex educational challenges, CAR becomes a promising and valuable means for schools and colleges in their profession engagements. CAR as a model is a foremost approach to continuous professional development (Mitchell et al., 2009; Altrichter, Posch & Somekh, 1993) that involves transformation through participation in collaborative endeavours (Rogoff, 2003).

Teachers as practitioners must be encouraged to reflect critically on their practice to construct new knowledge for effective professional development (Darling-Hammond & McLaughlin, 1995). Professional development is an ongoing process that engage teachers to transform their conceptions and practices and find new ways of fulfilling the needs and interests of their own contexts through CAR (Castro & Martínez, 2016), which ultimately empower them. Empowerment is identified as a key term in action research to enable practitioners achieve professional autonomy through professional development (Cohen et al., 2018). Therefore, since engaging in CAR as professional development is beneficial to pre-service or in-service teachers, adequate time must be given to the formulation and analysis of developing research questions, ensure training in how to undertake action research, and a system of rewards and accountability (Levin & Rock, 2003) for teachers.

As future teachers, in-service teachers (ISTs) are expected to develop intentional, theoretical, and collaborative expertise with others to be effective teachers (Day, 2019). TEs *know* the need to engage meaningfully with schools but their research practices tend to be within the university setting (Kitchen et. al., 2019). The overall existing situation of teachers' and TEs' cultural dependency on "expert knowledge" and external theory however might have been a factor for limiting capabilities in promoting knowledge base about native classroom requirements through action research (Barnes et al., 2018, p. 10). The depth of learning that come about through inquiry and its quality are related to the support and coaching teachers acquire. This study was a project plotted to assess the impact of a course initiative facilitated in a TEP on school practices through grasping a model that was primarily collaborative, context-sensitive, inquiry-based, action-oriented, and improvement-aimed. It insists that there is a potential for contravening the distance between TEPs and schools, to change practice, and practitioners' professional development recognizing that CAR is educationally and socially an impactful process.

Background and Research Question

The author had managed a three credit-hour semester course titled 'CAR' offered to ISTs from February to June 2018 in the currently defunct Asmara Community College of Education (ACCE). Four critical colleagues (CCs) were supportive in the course design, facilitation, and mentoring the ISTs (N=93). Together, we strongly believed that efforts to systematise practices are key to quality learning experiences. During the facilitation of the course, we and ISTs were attentively engaged in making practices more systematic, evidence-based, and transformed. The author and CCs had supported ISTs in carefully and critically examining why they do what they do (Vanassche et. al., 2019) through empowering them to develop the notion of "independent professionalism" (Leung, 2009 as cited in Yuan & Burns, 2017, p. 746). ISTs had worked hard during the course enactment and were fervent to bring about individual and collective improvements in their respective schools.

The CAR projects the ISTs developed in ACCE also ensured the continuous reconstruction of knowledge developed through engagement in inquiry and problem-solving approaches. Assessment of the ISTs' CARs conducted between June 2018 and June 2019 witnessed pledges, volition, readiness, and motivation to initiate further actions changing their future school practices. The process and outcome of the 16 CAR projects the ISTs brought into being with the support and mentorship of the author and CCs proved that ISTs were able to change their attitude regarding professional practices. The likelihood of change depends on the quality of collaboration to bring cultural change that is also a prerequisite for effective structural change (Hargreaves, 2019). Such endeavour was undertaken believing that change is possible, and making a difference involves collective acts (Clegg et. al., 2016).

We had prompted the ISTs to be emotionally perceptive to confront the predictable and unpredictable school situations. Unlike the pre-service graduates, the ISTs were aware of certain challenges and majority of them had the volition and readiness to tackle them. A few ISTs had worries regarding the realisation of CAR in their respective schools. They were concerned with the possibility of whether the necessary support to initiatives would be acknowledged by their school communities. We deemed that friction may arise between the needs and interests of individual teachers, the schools, or the society at large. The ISTs conceived that the culture of learning and teaching in their previous schools was far from the CAR practices they learned in ACCE which could affect identity formation and development. Yuan and Burns (2017) unveiled that teachers who are working in a discouraging environment with different contextual impediments might experience low identity development.

The most worrying part of the intended tasks in schools was the indispensable support and mentorship the ISTs gained while they were studying in ACCE. In fact, the space and time barrier had become a challenge for the author, CCs and all former ISTs (currently teachers). These missing opportunities restricted some of the former ISTs from the close follow-up and collaborations bestowed, especially for those who worked in remote areas. Nevertheless, we hoped to work hard in schools individually and collectively and contribute to the improvement using school-based CAR. It was meant to ensure firm engagement and ownership, redress prevalent challenges, and promote improved practices through creating "partnership including in research and other forms of scholarly activity" (Clegg et. al., 2016, p. 235). It was geared toward finding discrepancies that existed between what was endeavoured and developed in a TEP and what really happened in the schools. With such a background, this study aims to address an overarching question, "What can CAR influence in developing teacher practitioners professionally and in informing education policy?"

Methodology

This study focuses on actions to inductively understand context realities thereby draw theoretical underpinnings pertinent to practitioners and policies. The literature used in this study are not meant to guide and direct it but support patterns or categories identified (Creswell, 2014). There are however arguments by those sceptics who hold the view that the quality of theory generated through action research is questionable (Yuan & Burns, 2017), which is "in effect a scenario that disempowers and de-professionalises teachers as agents of educational change" (Elliot, 2015, p. 19). Practical experiences however indicate that action researchers can contribute to the development of quality theory (Whitehead & McNiff, 2006).

This study was thought up to be longitudinal to make sense of the prevailing education backdrops cautiously. It was done through a close follow-up study to explore and understand what happened after the ISTs (participants) left the TEP and re-entered the schools. Close-up research was conducted to ensure depth and understanding of situations, explain why things are as they are, and see how they might be changed (Clegg et al., 2016). Follow-up study helps discover the relationship between the world of programmes and the world of work. It is an investigation of consequent development of participants after taking part in a programme. In follow-up study, the same teachers would receive a follow-up to find out how they are doing (Lodico et al., 2006). It allows researchers to document the rate at which teachers are leaving the field (ibid.). This kind of study has relevance to the education decision makers and others interested in human resources management (Lamoure, 1996).

We were able to frequently meet and conduct reflective "talks" (Bullough & Smith, 2016, p. 338) with the former ISTs and the extended time of this study allowed to deeply sense the nature of school practices. We were intently meeting with participants, visiting, and observing schools to acquire evidence for the contextual realities. The chosen qualitative research approach that also used questionnaire, interviews, and document analysis allowed for a comprehensive exploration of practitioners' endeavours in CAR within demanding educational contexts.

Data Collection and Participants

Data was drawn from the study conducted between June 2019 and January 2023 in selected regions (N=3) and schools (N=10). The former ISTs (N=93) whose average age was 27, had studied a two-year diploma program at the former ACCE. Identifying the characteristics of ISTs and recognising them as a resource was an essential element of this study. About 50 percent of them were from the country's southern region and more than 77 percent were Tigrinya (one of the nine ethno-linguistic groups in Eritrea). They all were engaged in teaching, ranging from one to 26 years of service, in various elementary and middle schools in the country. As qualitative researchers, we found ourselves in close contact with participants to understand the meanings of their experiences (Maharaj, 2016).

The study arranged meetings with the former ISTs beginning from June 2019. Informal learning conversations with colleagues, visits to schools or teacher education institutions were values that would improve learning and teaching. Conducting meetings with former ISTs was however challenging for many of them were working in remote areas of the country. Participants who were coveted to reconnoitre the process and outcome of the study were selected. The question of how participants were selected was important for transparency and quality in qualitative approaches (Normann, 2017). For that reason, purposive sampling seemed more appropriate to develop the follow-up study that considered variables consisting of sex, age, year of experience, region and sub region. In purposive sampling, sample elements judged to be typical or representative were chosen from the population (Ary et al., 2010). Spending ample time in schools to acquire valid data and understand situations in a much more reliable ways demanded time and other ineluctable expenses. The unprecedented Covid-19 pandemic and the subsequent lock-down also made visits and meetings difficult. Later in the academic year 2021/22, meetings and observations for data collection purposes were however much doable.

We adopted a questionnaire that included open-ended items, which dealt with the role of CAR, reflections on TEP's practices, prevalent school challenges and opportunities while doing CAR, and viable suggestions for what ought to be done. Data using the questionnaire was collected on January 2022. Asking the participants to write gave them the opportunity for more reflection time but as writing is cognitively demanding, rich data is difficult to be gathered compared to talking (Normann, 2017). So, to generate more data, interview items almost analogous to the questionnaire items were prepared to better understand the attitudes and emotions of the participants through inquisitory questions. The interviews were conducted with ten teachers (former ISTs) and five school leaders on June 2022. Critical observation (beginning from 2019) in the selected elementary and middle schools was organised to explore diverse issues. The study observed the participants, involved desk work to analyse documents, conversed with teachers in various schools, and different events (Ferguson, 2016). This study acknowledges potential limitations, such as the subjective nature of qualitative research and the generalisability of findings to broader contexts. It however asserts that individual or subjective aspects of research are at the same time social and objective because the individual is a socio-cultural being with experiences that are relatable, social, generalisable, and theoretical (Rådesjö, 2017).

Data Analysis

Stories from the data collected were constructed through in-depth reading of reports made by the former ISTs in the ACCE and school-based CAR. Author's and CCs' feedback and insights, and review of notes were also significant. As a research group, we offered each other good support and insight while reading and reviewing the texts (Kramp, 2004). The analytic process had cautiously followed the steps discussed by Normann (2017). MS Excel 2019 spreadsheet helped group and code the meaning and messages of texts and see peculiarities, patterns, and themes. Repeated and idiosyncratic ideas, views, suggestions, and needs were classified and paraphrased, which were combined with author's and CCs' records, insights, literature and conclusions.

Emergent issues

Many of the information collected were meaningful and significant in developing themes quintessence for transforming and education policies in the country. The thorough examination of data gave rise to six emergent issues structured and presented as themes and reports respectively, which principally depicted CAR as a catalyst for professional development through thinking and doing and informing what need to be considered while framing education policy.

CAR as agent of professional advancement

Approached former ISTs working in various regions of the country reasserted that CAR is an essential tool that ensures practice advancement. They repeatedly mentioned the benefits of CAR among those teachers who teach the same subjects as an example. Kihwele (2020) found that lesson study has been a research tool for teachers who teach the same subjects to collaboratively work in identifying problems, planning a lesson, teaching that lesson, and producing solutions to the problem identified, which tenders them pedagogical skills, change their attitudes, beliefs and grow professionally. The abiding issue here, according to Day (2019), is "whether teachers have the time and energy to do so, especially in the increasingly intensive, emotionally challenging, and some might claim, toxic policy environments which exhaust and demoralize many" (p. 502). Participants' responses however showed variations in the extent to which they practised CAR in schools.

This study had seen the processes of the CARs underway by participants who came to learn that CAR practically becomes challenging while initiating tasks in a demanding environment. It establishes that the complexities of CAR are vividly sensed in a situation where practitioners happen to experience complex socio-cultural, economic, and educational constitutions whereby the idea of ensuring advancement in learning becomes difficult. The respondents stated various whys and wherefores regarding the challenges they faced while doing CAR. Nevertheless, they, time after time, uttered and univocally affirmed that CAR is an indispensable tool of professional advancement.

Continuousness of TEPs practices as school add-on

Participants argued that extrication characterise TEPs' and schools' practices in Eritrea. The absence of policies, principles, and strategies that ensure the togetherness of TEPs and schools is creating havoc in practices. It is observed that a bona fide nexus between TEP and school practices are non-existent and there is no strategy that connects both. Characteristically, there has been a breakage where TEPs release and send-off teachers without creating an environment of connectedness of practice for learning improvement and development. The effect of this existent severance has been limiting teacher quality. The Ministry of Education (MoE) has been making efforts to upgrade teacher professional qualifications. However, the process is found to be limited, sporadic, and decelerated, which led to the reigning of weak sense of teacher professionalism.

Participants informed that TEs need to continuously learn and develop qualities to support teachers not only in TEP grounds but also in schools. Such ideas might be ambitious and difficult to realise. This study perceived that the collaborative commitment to create professional links and relationships among TEs and teachers ensures mutual professional development. Participants' voices principally demonstrate the need for a paradigm shift in the nature and direction of TEP policies and practices. Mechanisms such as changing the pedagogics of TEPs using CAR results can ensure practice continuum for better learning and teaching in schools. Participants had practically sensed that teachers need hope, encouragement, support, and professional opportunities which indicated the need to ensure TEP practices continuity through CAR. Such suggestions can be accomplishable by creating networks of teacher educators (TEs) and teacher professionals who endeavour to develop qualities of incessant inquiries through collaboration and communication. This study discerns that the MoE had been providing training for teachers regarding action research for more than a decade. Nevertheless, the efficiency of such undertakings does not seem to be substantial, which calls the need for rethinking about strategies of professional development schemes.

School leadership as decisive factor in professional development

Many of the participants demanded the creation of a conducive working environment that encourages teachers to use schools as a space for professional learning and development. This study deduced that the school leadership can play a role in safeguarding teachers' emotional state. Those participants who initiated tasks in schools at their own pace and control were encouraged and supported by their school leaders. Others were found to be affected by their leaders' propensities in initiating tasks relevant for learning and teaching. They elucidated the prevalent challenges in their schools that created difficulty in managing the processes of learning and teaching. They sought encouragement from the school, sub region, and region because they found it difficult to do extra tasks without support, recognition, and incentives. This study observed discrepancies among teachers who work in private schools owned by the religious institutions (currently defunct) and public government owned schools. Those teachers who worked in the private schools reflected positive support from the school leaders that made them do CAR which was beyond the ordinary undertaking.

Teaching and research are complementary and not contradictory to each other (Smith & Flores, 2019) but teacher practitioners face difficulties in using research-based knowledge (Wahlgren & Aarkrog, 2020). This study validates that school leaders who initiate research activities beyond the conventional duties promote teachers' and learners' learning. The school leaders are however preponderantly observed doing school activities including assigning teachers to teach subjects, controlling the discipline of learners, taking teachers' and learners' attendance, managing the assessment processes, reporting school performances all uninformed by research. They consider such activities as their primary duty and responsibility though this does not imply that what the school leaders do in the schools are trivial.

Collaboration as indispensable attribute of professional development

Collaboration as an important essence of learning was repeatedly cited by the participants. This study grasped that school leaders can create a community of professionals who manifest collaborative demeanour to ensure collective betterment. In spite of this, some school leaders aspire to "colonize collaborative efforts with purposes and activities other than one's teachers might initiate themselves" (Hargreaves, 2019, p. 7). Under excellent leadership, teachers could make sense of the complexities and ambiguities of their professional works through working together (ibid.). This study confirms that teachers were observant of their school milieu where they understood the collaborative role of parents, learners, teacher, school leadership, media etc. Although participants acknowledged the value of collaboration as indisputable in promoting better learning and teaching, some teachers tend to work individually. They also shared a scenario regarding parental role in the learning and teaching processes, where learners are observed having no chance of getting additional help at home.

Participants proposed ideas that solve such prevailing challenges through the collaboration of all members of the society. Schools understand that professional collaboration leads to increased student achievement (ibid.). Participants were aware of the multidimensional issues that are affecting their learners' performance. This study has recorded voice of teachers who vividly illustrated that it is through the concerted efforts of the school community that improvement and change in the learners' achievements occur. They also described the social aspect of learning while discoursing about the challenges and opportunities of their contexts' cultural diversities in the school performance. Participants witnessed how they manage the diversities that exist in their school and the benefits they are gaining from. Such teacher respondents' taste of the social demeanour and its role in learning and teaching is a crucial element of understanding the pedagogics of classroom behaviour and in the contravention of the isolation between schools and the community in redressing challenges.

Teacher professionalism as critical issue of quality education

This study contends that MoE has been recruiting uncertified teachers. Participants have unequivocally emphasised on teacher professionalism as an essential aspect of quality in education. Professionalism is connected to the profession's internal quality, authority, values and autonomous practices (Hargreaves, 2000). They firmly argued that without the development of teacher professional qualities, desired improvement and change is difficult to achieve. They also attested that those teachers who are not professionally trained show leniency, miss classes, become absent, and do only "ordinary" activities. This study found that many teachers who are pedagogically uneducated have been recruited by MoE to teach subject matters that might or might not be related to their speciality. Furthermore, some teachers who have been teaching for several years were observed exhibiting professional limitations.

Participants insisted that teacher recruitment criteria exercised by the MoE have not been valuing volition, interest, quality, and professional ethics. Another challenge noticed by the respondents was that teachers have not been professionally growing coupled with being socially and economically destitute. This study found that teachers have been deprived of necessities, rewards, and incentives that are currently the source of frustrations to many, which are practically limiting the morale, motivation, and quality to contribute aptly and adeptly. Teaching is primarily a practice that encourages practitioners' intellectual growth (Noddings, 2003). A step towards the education and professional development of teachers through various in-service programmes like summer training, distance education, regional and schools-based professional training, though currently adjourned, have become issues of the past. The participants took a firm stand that teachers should not linger in one level and must boost their capacity and career levels through learning. The MoE had previously made commendable efforts to provide various scholarships opportunities to teachers. By doing so, the MoE was indeed able to certify many uncertified teachers.

Participants asserted that engagement in various forms of continuous professional development schemes is one way of strengthening teacher professionalism. Many teachers in the country are unnoticed for such opportunities that has off-puttingly affected their morale and commitment. They attested that the processes and opportunities of professional development initiatives in their respective contexts were erratic. Nevertheless, critical challenges identified in this study in ensuring teacher professionalism was the development of obligatory behaviour, self-initiatives and self-thoughtfulness of teachers to learn, develop professionally, and improve performances. Such missing qualities mentioned by respondents are found to be bottlenecks to teacher professionalism.

Noddings (2003) asserts that most young teachers enter teaching professions because they want to make a difference in the lives of the learners, but altruism or service is usually perceived as a sign of weakness with respect to professional status. Participants described that low motivation and commitment profoundly affects the quality of education. They asserted that the provision of quality education cannot be realised without adequate supply of highly motivated and well-educated professional teachers. This study acknowledges that the existing realities limit teachers' capabilities to prepare themselves for the challenges of work in the country within a continuous social, economic, and technological changes. Respondents demanded professional development opportunities to qualify, motivate, and retain teachers. Ping et. al. (2018), however, reason that professional development undertakings are primarily intrinsic rather than extrinsically motivated. Conversely, results of a study by Coombs et. al. (2020) indicated "extrinsically motivated teacher candidates (e.g., motivated by perceived job security) experienced greater gains in their learning than their intrinsically motivated counterparts" (p. 3).

Socio-Economic status as teacher position determinant

This study found the working environment of schools as highly demanding. Participants saw a gap between what is theoretically perceived in TEPs and what practically exists in schools. Teachers voiced that an obstacle that blocked them from performing their job proficiently is the existing socio-economic life. They admitted that all professionals in the country have similar income, but they pleaded that teaching requires more care and investment. Exacerbated by the social attitudes toward teaching, teachers depicted themselves as economically underprivileged. They mentioned that MoE should work hard to create opportunities for teachers' professional development that shall equally augment their economic and social position and tackle perceptions that de-level teachers. This study insists that teachers in remote areas of the country need opportunities and privileges. Teacher respondents reflected that teacher attrition in their schools is high. Those teachers working in urban areas are also observed engaging themselves in other jobs including moonlighting. Such situations are deteriorating the moral, commitment, motivation, and pride of teachers. The hope and determination of some teachers, however, make our education promising.

Implications and conclusion

This study was a tracer study aimed at understanding how former ISTs were influenced and what kind of learning and teaching practices were prevailing after they re-joined their schools. This study revealed that TEs are not using their own position and acts to influence student teachers. It was highly relevant to see how action research cycles could be sustained in schools in shaping teaching practices. Various meeting occasions were set with some of the former ISTs that had multifaceted implications to professional behaviours and the general education systems in the country. The readiness, influences, and relationships created were encouraging. However, TEs needed to have the confidence and self-critical stance to capably respond to and engage with diverse perspectives, influences, and relationships.

This study realises that what former ISTs voiced out were primarily about the professional behaviour of TEs and teachers. All the issues they reckoned, expounded, and illustrated during the CAR processes and afterwards essentially inform how most ISTs (teachers) and TEs behaved in the TEP and the schools. Moreover, they have been courageous to raise critical issues of TEP and school learning and teaching dynamics that also included self-criticisms of their own experiences. This study helped develop understanding the dynamic nature of collaborative learning. It perceives that such experiences have wider annotations and implications and might potentially exemplify realities of other TEs and teachers in the country and beyond.

Every aspect of the processes and outcomes of the whole study also have meaning to the pedagogy of TEPs and education policies. The issues identified are associated with the essences of teacher professionalism, TEs tendencies, thoughts, and practices, TEP purposes and functions, as well as policy orientations and implementations. They urge all to bear responsibility and act relentlessly to ensure learning improvements and changes. It was found that what TEs, ISTs, and teachers voiced and acted can lay the ground for professional practices in TEPs and the schools. The entire process informed that TEs need to have "stronger voices and engage in the debates on education and teacher education so that they can capably contribute in educational reforms.

The government of the state of Eritrea (GoSE) well regards that a key input in human resources development is education (MoE, 2011). Furthermore, the commitment of the GoSE toward quality education is vividly manifested in the various MoE policy documents. The analysis made on the education system in Eritrea, however, indicates challenges regarding the quality of teachers. This contextual extrapolation stipulates the need to enhance teacher professionalism in teacher education programmes and schools through highly focusing on developing indispensable teacher professional skills and characters.

This study reaffirms that quality of learning and teaching and the success of learners is primarily dependent on teacher quality. The insinuation of this is that educational provision will have to be linked to the overall human resources development strategy. This provision has also to contend with the deeply adverse legacy of war, including the challenges of quality access to education. In such a context, the role of professionally qualified teachers, school leaders, supervisors, support systems in materialising the objectives of education in the country is enormous. Although the key to improved student learning is quality teaching (Smith & Flores, 2019), how the quality of teaching is achieved is, however, "at the heart of the teacher gap" (Ovenden-Hope, 2020, p. 135).

Quality is at the heart of any education and teacher education system in which it influences what student teachers learn, how well they learn, and what benefits they draw from their education and training (Zewdu, 2017). Regrettably, one of the major flaws of the current situation in Eritrean education system is the acute shortage of qualified teachers, which is an impediment to the goal of ensuring the provision of equitable and quality education to all citizens. The MoE is working to repurpose teacher policies and committedly ensure their realisations thereby creating a conducive working environment for the education professionals. This study assumes that research practices such as CAR promote the quality of teachers and improves school practices and learners' success. The MoE's previous interventions to develop teachers using different modes should be reviewed and reformed. This study advocates for a strong link between TEPs and school practices to promote TEs, teachers' and school learners' learning and development. It attests that connections can be made possible through strategies of making TEP courses continuous and relevant to schools. It insists that TEs' and school teachers' professional development should be coordinated, sustained, diversified, and accredited. They should give teachers the opportunity to professionally develop, improve practices, and guarantee them to hold apposite roles in the society. One of the most satisfactory and rewarding element of professional *doings* of the former ACCE ISTs was the criticality in reflecting the learning and teaching situations in their schools. Such ideas, beliefs, and experiences shared by the former ISTs were excellent indicators of the education realities, which were also indicators and directives to the solution of the prevailing education challenges.

References

Altrichter, H., Posch, P., & Somekh, B. (1993). Teachers investigate their work: an introduction to the methods of action research. London: Routledge.

Ary, D., Jacobs, L. C., Sorensen, C. & Razavieh, A. (2010). Introduction to Research in Education (8th ed.). Cengage Learning represented. Canada: Nelson Education.

Barnes, A. E., Simmons, Z., Mekonnen, D. & Ramos-mattaoussi, F. (2018). Improving teacher training in Ethiopia: shifting the content and approach of pre-service teacher education. *Teaching and Teacher Education,* 70, 1-11.

Bruce, B. & Easley, J. (2000). 'Emerging communities of practice: collaboration and communication in action research'. *Educational Action Research*, 8 (2): 243-259.

Bullough Jr., R. V. & Smith, L. K. (2016). Being a Student of Teaching: Practitioner Research and Study Groups. In Loughran, John., Hamilton, Mary Lynn (Eds.), *International Handbook of Teacher Education* (Vol. 2, pp. 305-352). Springer.

Castro, A. Y. & Martínez, G. L. (2016). The role of collaborative action research in teachers' professional development. *PROFILE Issues in Teachers' Professional Development*, 18(1), 39-54.

Catherine D. B., Flynn, T. & Stagg-Peterson, S. (2011). Examining what we mean by collaboration in collaborative action research: a cross-case analysis. *Educational Action Research*, 19:4, 433-452

Clegg, S., Stevenson, J. & Burke, P. (2016). Translating close-up research into action: a critical reflection. *Reflective Practice,* 17:3, 233-244. DOI: 10.1080/14623943.2016.1145580

Cohen, L., Lawrence M., and Keith M. (2018). Research Methods in Education, 10017. 8th ed. New York, NY: Routledge.

Coombs, A., DeLuca, C. & MacGregor, S. (2020). A person-centered. analysis of teacher candidates' approaches to assessment. *Teaching and Teacher Education,* 87. https://doi.org/10.1016/j.tate.2019.102952

Creswell, J. W. (2014). Research design qualitative, quantitative, and mixed method approaches. (4th ed.) CA: Sage.

Darling-Hamnmond, L. & McLaughlin. (1995). Policies that support professional development in an era of reform, Phi Delta Kappan, 76(8), 587-604.

Day, C. (2019). Policy, teacher education and the quality of teachers and teaching. *Teachers and Teaching*, 25:5, 501-506, DOI: 10.1080/13540602.2019.1651100

Elliot, J. (2015). Educational action research as the quest for virtue in teaching. *Educational Action Research*, 23:1, 4-21, DOI: 10.1080/09650792.2014.994017

Ferguson, K. (2016). Lessons learned from using shadowing as a qualitative research technique in Education. *Reflective Practice*, 17:1, 15-26, DOI: 10.1080/14623943.2015.1123683

Hargreaves, A. (2000). Four ages of professionalism and professional learning. Teachers and Teaching: History and Practice, 6(2),151-182.

Hargreaves, A. (2019). Teacher collaboration: 30 years of research on its nature, forms, limitations and effects. *Teachers and Teaching*, DOI: 10.1080/13540602.2019.1639499

Heron, J. (1996). Co-operative Inquiry: Research into the Human Condition. Sage Publications.

Kihwele, J. (2020). Affordances and Constraints of Implementing Lesson Study for Teachers' Professional Development: A Review. *African Journal of Teacher Education.* Vol. 9. https://doi.org/10.21083/ajote.v9i0.5731

Kitchen, J., Berry, M. & Russell, T. (2019). The Power of Collaboration, *Studying Teacher Education,* 15:2, 93-97. DOI: 10.1080/17425964.2019.1628560

Kramp, M. K. (2004). Exploring Life and Experience Through Narrative Inquiry. In Kathleen de Marrais & Stephen D. Lapan (Eds). *Foundations for Research Methods of Inquiry in Education and the Social Sciences*. Lawrence Erlbaum associates, publishers. Mahwah, New Jersey, London.

Lamoure, J. (1996). Follow-up and Tracer studies. Paris: International Institute for Educational Planning, (Learning Modules 1-7), UNESCO.

Levin, B. B., & Rock, T. T. (2003). The effects of collaborative action research on preservice and in-service teacher partners in professional development school. *Journal of Teacher Education,* 54, 135-149.

Livingston, K. (2019). Reflections on teacher education: developments and challenges, *European Journal of Teacher Education*, DOI: 10.1080/02619768.2020.1705653

Lodico, M. G., Dean, S. T., Katherine, V. H. (2006). Methods in educational research: From Theory to Practice. Jossey-Bass. San Francisco, CA.

Maharaj, N. (2016). Using field notes to facilitate critical reflection. *Reflective Practice*, DOI: 10.1080/14623943.2015.1134472

Mitchell, S., Reilly, R. & Logue M. E. (2009). 'Benefits of collaborative action research for the beginning teacher' Teaching and Teacher Education, 25(2): 344-349.

MoE. (2011). National Curriculum Framework. Asmara: Eritrea. Ministry of Education.

Noddings. N. (2003). Is Teaching a Practice? *Journal of Philosophy of Education*, 37 (2), 241-251.

Normann, A. (2017). Becoming a phenomenologically skilled researcher – an account of a journey started. *Reflective Practice*, DOI: 10.1080/14623943.2017.1307720

Ovenden-Hope, T. (2020). The teacher gap. *Journal of Education for Teaching*, 46:1, 134-136. DOI: 10.1080/02607476.2019.1692560

Ping, C., Schellings, G. & Beijaard, D. (2018). Teacher educators' professional learning: A literature review. *Teaching and Teacher Education*, 75, 93–104.

Rådesjö, M. (2017). Learning and growing from 'communities of practice': autoethnographic narrative vignettes of an aspiring educational researcher's experience. *Reflective Practice*, DOI: 10.1080/14623943.2017.1361917

Rogoff, B. (2003). The cultural nature of human development. New York: Oxford University Press.

Sjølie, E. & Østern, A. L. (2020). Student teachers' criticism of teacher education - through the lens of practice architectures. *Pedagogy, Culture & Society*, DOI:10.1080/14681366.2020.1729842

Smith, K. & Flores, M. A. (2019). Editorial: Teacher educators as teachers and as researchers. *European Journal of Teacher Education*, DOI:10.1080/02619768.2019.1648972.

Vanassche, E., Kidd, W. & Murray, J. (2019). Articulating, reclaiming and celebrating the professionalism of teacher educators in England. *European Journal of Teacher Education*, DOI: 10.1080/02619768.2019.1628211

Wahlgren, B. & Aarkrog, V. (2020). Bridging the gap between research and practice: how teachers use research-based knowledge, *Educational Action Research*, DOI: 10.1080/09650792.2020.1724169.

Whitehead, J. & McNiff, J. (2006). Action Research Living Theory. London: Sage.

Yuan, R. & Burns, A. (2017). "Teacher Identity Development through Action Research: A Chinese Experience." *Teachers and Teaching,* 23(6), 729–749. DOI:10.1080/13540602.2016.1219713.

Zewdu, A. A. (2017). Challenges in Ethiopian Teacher Education Pedagogy: Resistance Factors to Innovative Teaching-Learning Practices. *African Journal of Teacher Education,* Vol. 6. DOI https://doi.org/10.21083/ajote.v6i0.3624

A Systematic Literature Review on Teachers' Perceptions and Challenges in Integrating STEM Curriculum in Education

Samson Kifletsion Elias,[1] Francies Stonier[1] & Irene Musengamana [1]

Southwest University, Chongqing, P.R. China
Emial: Samsonkifletsion2@gmail.com

Abstract

It is critical to comprehend teachers' perceptions, attitudes, preparation, and beliefs regarding the methods and implementations of STEM (Science, Technology, Engineering, and Maths)-integrated curricula if schools are to offer high-quality STEM education. Teachers have a crucial role in helping students to improve their critical and creative thinking skills since they have personal perspectives and experiences that might affect how they integrate STEM. This study uses a variety of existing literature to clarify what is known about teachers' opinions, beliefs, and attitudes regarding STEM curriculum integration in education. Empirical publications that matched the study's objectives and were published between 2013 and 2023 in English in scholarly journals (Springer and Web of Science databases) met the inclusion criteria. Twenty-four peer-reviewed scholarly articles were included in this systematic review after quality assessment. Finding themes within the data required the application of thematic analysis. The results show that while teachers hold favourable perceptions toward STEM education, there are still some obstacles to teachers' implementation of STEM including pedagogical difficulties, curriculum difficulties, structural difficulties, student anxiety, assessment anxiety, and a lack of teacher support. Teachers believed that peer collaboration, high-quality curricula, administrative support, prior experiences, and efficient professional development would all benefit their efforts to adopt STEM instruction. Conclusions and practice recommendations include quality in-service instruction over STEM pedagogy best practices and district support of collaboration time with peer teachers. Recommendations for future research are given.

Keywords: STEM curriculum, systematic literature review, teacher challenges, teacher perception

Introduction

Globally, there is an emphasis on providing exceptional educational opportunities, particularly in emerging fields like school-level STEM education. STEM, which integrates science, technology, engineering, and mathematics, promotes cross-curricular learning by applying real-life situations to solve problems (Herro & Quigley, 2017). STEM education is essential for economic prosperity worldwide (Croak, 2018). Understanding teacher perceptions, attitudes, beliefs, readiness, and practices is crucial for enhancing teacher development and fostering a creative society. This approach can create authentic learning experiences, inspiring students to pursue STEM careers essential for future employment. This study examines in-service teachers' beliefs, understanding, and valuation of STEM pedagogical approaches, their impact on students' interest in STEM, and the challenges teachers face in implementing STEM practices.

The STEM integrated method has been marketed primarily as a tool for schoolteachers to use to build a foundation of math and science-related knowledge while motivating students to express their ideas in a range of creative ways using the arts (Spyropoulou et al., 2020; Conradty & Bogner, 2020). Not only that but to lay the foundation for future academic success, students taught using STEM-integrated techniques must exhibit critical thinking and innovative approaches to solving problems. Therefore, to better prepare students for future problems children in the twenty-first century need to develop their critical-thinking abilities (Dell'Erba, 2019; Miller & Knezek, 2013). More recently, STEM education—which combines STEM courses with the arts—has emerged as a fresh strategy for enhancing students' capacity for innovative thinking, engagement, problem-solving, and other cognitive benefits (Bahrum et al., 2017; Watson & Watson, 2013).

STEM role in the Sustainable Development of Countries

The significance of quality STEM education and its application has become increasingly recognized across various countries, leading to a heightened focus on the relationship between STEM (science, technology, engineering, and mathematics) education and sustainable development in educational institutions (Hsiao & Su, 2021; Wang et al., 2018). This emphasis stems from the understanding that STEM is crucial for fostering sustainable development by enhancing future generations' capabilities (Khadri, 2022). Through STEM education, students are expected to develop essential skills such as innovation, creativity, critical thinking, scientific problem-solving, and sustainability mindsets. Moreover, it promotes interdisciplinary learning approaches that tackle complex economic, social, and environmental issues, enabling students to actively engage in their communities and improve their quality of life.

Education for sustainable development (ESD) aims "to empower and equip present and future generations to meet their needs using a balanced and integrated approach to the economic, social, and environmental dimensions of sustainable development" (Leicht et al., 2018). Today's students must acquire the knowledge and skills necessary to address sustainability challenges for the future (Sinakou et al., 2019). ESD encourages the development of the knowledge, abilities, beliefs, and worldviews needed for individuals to support sustainable living practices, enabling them to reflect on and engage with their surroundings (Hoffmann & Siege, 2018). In light of the growing global attention on STEM, there is a pressing need for effective teaching methodologies and implementation strategies in early education settings. The STEM educational framework has spurred the development of new teaching competencies, addressing scientific reasoning, integrative skills, and workforce readiness among educators (Roberts et al., 2018). In a survey conducted in Uzbekistan, educators emphasized that the primary goal of STEM education is to cultivate students' talents for future society (Y. Lee, 2021), with the secondary objective focusing on improving scientific classroom instruction, highlighting a tendency to view STEM as a means to enhance science teaching rather than embracing its holistic potential across various subjects (Y. Lee, 2021).

Teachers' Perception of STEM Practices

Teachers' pedagogical content knowledge and perceptions—comprising beliefs, conceptions, and attitudes—significantly impact their classroom goals, efficacy, and professional development (Crawford & Capps, 2018; García-Carrillo et al., 2021). These perceptions, which influence science education, often evolve into entrenched, traditional attitudes that are difficult to change (Hutner & Markman, 2016; El-Deghaidy & Mansour, 2015; Chan et al., 2019; Fives & Buehl, 2014; Schommer-Aikins, 2004). Several factors influence how teachers perceive and implement STEM integration, including their training, beliefs about the nature of science, background, prior experience, pedagogical knowledge, curriculum, assessment-based systems, time, school resources, and parental involvement (DeJarnette, 2018; Herro & Quigley, 2017; Quigley & Herro, 2016). Additionally, factors such as student behaviour, ability (Jamil, Linder, & Stegelin, 2018), reflections on classroom outcomes (Hawari & Noor, 2020; Jamil et al., 2018), feelings of control (Zaher & Hussain, 2020), and age (Alghamdi, 2023) further affect perceptions. Alghamdi (2023) and other researchers emphasize that by reflecting on their classroom practices, teachers can shift their perceptions.

According to Richardson (1996), the type and content of teaching practices can reshape teachers' self-perceptions, as they interpret and analyze student actions through their conceptual frameworks (Van Es & Sherin, 2002). Research indicates that teachers with more subject knowledge are generally more adaptable and receptive to student ideas (Shulman, 1986). However, traditional views of deductive instruction can impede objective classroom observation and assessment (Meschede et al., 2017). STEM education aims to apply scientific knowledge in specific contexts, linking inquiry with problem-solving in the engineering design process (Y. Lee, 2021; Kennedy & Odell, 2014). STEM also fosters cognitive and emotional learning (Conradty & Bogner, 2019). Despite substantial research on STEM education, awareness remains low in developing nations. Given their critical role in the success of STEM, understanding teachers' opinions and beliefs about it is essential. Studies in South Korea have explored teachers' attitudes toward STEM at its introduction (Hong, 2017; Kim et al., 2018; Park et al., 2016; Lee, 2020).

Research indicates that teachers' perceptions of STEM education vary by school level, gender, and teaching experience. For example, Park et al. (2016) found that elementary school teachers generally hold more favorable views of STEM than middle and high school teachers, while male teachers often exhibit more positive attitudes than their female counterparts. In the UAE, the shift toward STEM education aims to develop a competitive, knowledge-based economy (Said, 2016). Despite various initiatives to promote STEM integration (Suwarma & Kumano, 2019), effective implementation remains a topic of ongoing discussion, particularly among science teachers (Kelley & Knowles, 2016). Furthermore, teachers' knowledge, professional development experiences, and self-efficacy significantly impact their beliefs and practices regarding STEM education (Thibaut et al., 2018; Srikoom & Faikhamta, 2018; Sturtevant & Wheeler, 2019). In Uzbekistan, while in-service teachers generally exhibit positive attitudes toward STEM, but rarely integrate it into their teaching.

Barriers to STEM curriculum integration in education

Numerous studies indicate that deeply rooted views and perceptions hinder many teachers from adapting to innovative curricular approaches (El-Deghaidy & Mansour, 2015). Research shows that teachers' perspectives and skills play a significant role in the integration of STEM into their teaching (Herro, Quigley, & Cian, 2019; Marzano, 2007). Although teachers' beliefs about STEM are critical for pedagogical reasoning, these beliefs can often obstruct the actual implementation of STEM in classrooms (Herro et al., 2019; D. Kim & Bolger, 2017). Despite the potential benefits of STEM education, many teachers encounter difficulties in effectively integrating it into their practices (Herro & Quigley, 2017; Wu, 2022). If teachers perceive STEM pedagogies as unhelpful for teaching or learning, they may become reluctant or unable to employ them meaningfully (Affouneh et al., 2020).

STEM education faces several challenges, including cultural, governmental, financial, and inequality issues, along with insufficient adult supervision during STEM activities, which can lead to negative behaviors among children (Ejiwale, 2013). Many children also lack access to digital technologies due to poverty and cultural constraints (Edwards, 2013). The complexity and ambition of implementing STEM education contribute to teaching challenges related to its conceptualization and teacher training (Mejias et al., 2021). Factors such as curricular organization and teacher preparation further hinder the integration of STEM education into classrooms (Affouneh et al., 2020; Margot & Kettler, 2019). To address these obstacles, professional development (PD) programs are essential for enhancing teachers' attitudes, skills, and knowledge, focusing on the selection and integration of necessary curricular components, instructional strategies, and materials for effective STEM education (DeJarnette, 2018). However, despite various initiatives, there remains a limited understanding of STEM education's implementation in classrooms and how teachers, as key policy agents, value and conduct it. Previous studies (Alghamdi, 2023; Hong, 2017; Inomjonovna, 2023; Y. Lee, 2021) have emphasized promoting STEM careers but have overlooked the involvement of teachers, which remains under-researched (Marginson et al., 2013; Herro & Quigley, 2017). Additionally, notable differences in perceived obstacles exist based on school level and gender, with high school teachers and female teachers reporting greater concerns regarding increased workloads and the use of advanced technology (Jho et al., 2016).

Research Questions

The purpose of the study is to review the literature on teachers' opinions of STEM curriculum implementation and practice. This study also makes an effort to clarify what is known about teacher readiness, attitudes, and views toward the advancement of STEM education. The following inquiries were used to examine the literature:

1. How does existing literature articulate teachers' perceptions and value of STEM education, including their readiness for utilizing STEM education for sustainable development?
2. What are the teachers' perceived challenges and difficulties in implementing STEM-integrated curricula as their teaching methodology?
3. What are practical suggestions on what would enhance teachers' efforts to implement STEM-related curricula in their classrooms?

Methods

To explicitly and systematically analyze published research articles on teachers' perceptions and their readiness for STEM education, a systematic literature review was conducted. A systematic review of literature is a transparent and replicable method of literature review that answers research questions based on accountable methods with explicit criteria that include or exclude previous studies on a certain topic (Gough et al., 2012). Data of systematic review are extracted from the results and findings of existing research studies, and it aims at systematically synthesizing the existing knowledge on a particular research topic and discovering gaps in the literature (Møller & Myles, 2016).

To ensure credibility, consistency, and transparency, this systematic review was conducted under the guidelines of the Preferred Reporting Items for Systematic Reviews and Meta-Analyses (PRISMA). The review methods were organized according to 4 phases 1) Literature search, 2) Literature source, 3) Selection criteria, and 4) Data extraction and analysis. In particular, the PRISMA consists of a detailed checklist addressing all the major sections of a systematic review, providing an evidence-based foundation for transparency in identifying, selecting, appraising, and synthesizing the studies being reviewed (Moher et al., 2009).

Literature

For this systematic search, the authors developed a search strategy to identify the existing empirical studies that attempted to explore the current status of STEM education, particularly the perceptions of teachers of STEM education, and teachers' readiness for the implementation of STEM curriculum in education for sustainability. The search keywords included predefined search terms related to "STEM or STEM education", "teacher perception", "teachers' beliefs and attitude", "teachers' readiness", "challenges and difficulties", "sustainable development" and variations of these terms. The search was conducted in the months of July and August 2023, and the search results were strictly limited to peer-reviewed scholarly articles written in English and published in the last 10 years (from 2013 to 2023).

For this systematic review, Southwest University- China's library website was used to access the necessary information for this search. A comprehensive literature search was tailored to 2 major established electronic databases: Web of Science and Springer considering how they are recognized for their rigor and importance in research, particularly in the educational sphere. The initial search of this study resulted in 1103 studies, with more than 73% sourced from the Web of Science, and the remaining studies were extracted from the Springer database.

Selection criteria

This systematic review utilized the 2020 PRISMA (Preferred Reporting Items for Systematic Review and Meta-Analysis Protocols) statement for including and excluding research articles (Page et al., 2021). Articles retrieved from the literature search were screened based on the relevance of their titles and abstracts to the research topic; if an abstract was absent, the full text was examined. The selected articles underwent a full-text review for appropriateness, with this process conducted by two researchers. The focus was on empirical studies exploring teachers' perceptions of STEM education and their readiness to implement it for sustainable development, considering studies that presented empirical data and/or employed qualitative or quantitative methodologies, with a search span limited to the last 10 years, excluding studies published before 2013. Eligibility criteria such as language, publication period, document type, and country of focus were applied using database filters, followed by data export to an Excel file. Duplicate studies were eliminated by comparing titles and author names, leading to the removal of studies that did not meet the eligibility criteria after careful consideration. Ultimately, 24 studies were retained for this systematic review, as indicated by an asterisk in the references list, with Figure 1 illustrating the screening process in detail.

Figure 1 PRISMA flowchart of the studies screening process

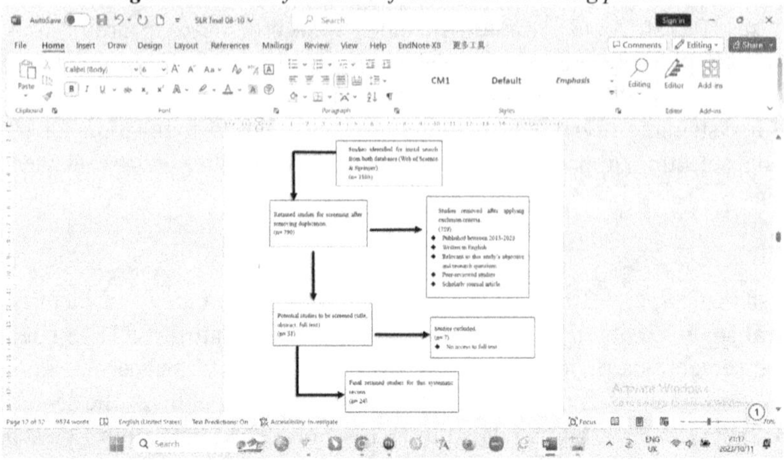

Data Extraction and Analysis

Relevant data, such as the author(s) of the study, publication year, country of focus, research methods and design, participants and sample size the study employed, thematic focus, the purpose of the study, and major findings were extracted from the selected articles, and studies were coded based on this information. This systematic review employed a thematic analysis to identify common themes and patterns across the studies. The identified themes were categorized into positive and negative perceptions, challenges faced, support and professional development needs, and possible recommendations for successful STEM education implementation for sustainable development.

Results

As seen in Figure 2 below, a fascinating phenomenon was discovered when the publication years of each of the 24 retained articles were examined. The first article on the subject was published in 2013, after which there were more pieces on the subject published each year until there were five articles released in 2021.

Figure 2 *Publication distribution time of studies on teachers' perceptions of STEM education.*

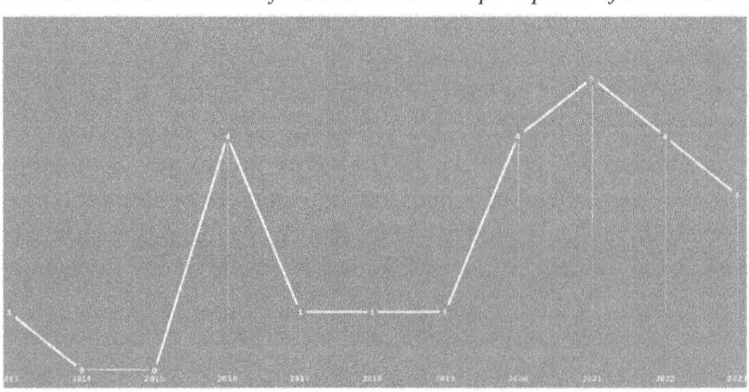

This issue is significant outside of the United States because 23 of the articles that were kept were undertaken outside there. Countries involved in these studies are; South Korea, Saudi Arabia, Vietnam, Slovakia, South Africa, Egypt, Uzbekistan, Spain, Australia, China, UAE, and Greece. Seven of the retained studies employed a qualitative technique, five mixed methods, and ten quantitative methods to examine the data gathered. In addition, the table below (see Table 1) provides a summary of the articles that were retained for review and also included in the reference as (*).

***Table 1** Summary of studies included in the systematic review.*

No	Author (s)	Year of publication	Location	Study design	Main findings	Sample size
1	Ogegbo et al.	2020	South Africa	Qualitative (Semi-structure interview & class observation)	Teachers value STEM integration and have a positive attitude, but its implementation is hindered by insufficient parental involvement, school support, technological resources, and teacher training.	(n=8)
2	Jho et al.	2016	South Korea	Qualitative-case study (Interview, observation, videotaping &documentation)	Teachers emphasized the importance of open-mindedness, community involvement, and including non-science educators for successful STEM education.	(n=10)
3	Yumi L.	2021	Uzbekistan	Qualitative (Surveys, interviews, documentation)	Limited awareness of STEM education and its importance, coupled with teachers' instructional challenges, hampers its effective implementation.	(n=57)
4	Teresa et al.	2023	Spain	Qualitative-longitudinal case study	Teachers design and facilitate STEM projects, benefiting from flexible planning, but face challenges with time management and students' inadequate STEM skills.	(n=2)
5	Faiza et al.	2017	USA	Mixed methods	The effectiveness of STEM education depends on teachers' beliefs, with challenges including time constraints, limited resources, and varying focus on assessment over practical application.	(n=41)
6	Ahlam A., A.	2022	Saudi Arabia	Quantitative (Survey)	Teachers view STEM education positively for young children but have moderate beliefs about its implementation and limited knowledge of its principles.	(n=245)

7	Jessica et al.	2023	USA	Quantitative (Online surveys)	Teachers recognize the importance of STEM for all young children, including those with disabilities, but face challenges like limited resources, funding, time, and evidence-based practices. Positive attitudes and enhanced science knowledge are crucial for effective STEM inclusion in early childhood education.	(n=280)
8	Zhan et al.	2020	China	Mixed methods (Surveys, observation, and interviews)	Teachers generally understand engineering and its education, but struggle to effectively integrate engineering practices into their science teaching due to limited knowledge. The low level of integration and the absence of high-quality STEM training programs are significant challenges for educators.	(n=1185)
9	Nguyen et al.	2020	Vietnam	Quantitative	The study categorizes STEM education into understanding, competencies, and difficulties, showing that teachers' perceptions significantly influence their practices and vary by experience and education, with all groups facing similar implementation challenges.	(n=186)
10	Hamad et al.	2022	UAE	Mixed methods	The study categorizes STEM education into understanding, competencies, and difficulties, revealing that teachers' perceptions, influenced by their experience and education, affect their practices, with all groups facing implementation challenges despite novice teachers' positive views.	(n=3 Qual.) (n=61 Quan.)
11	Park et al.	2016	South Korea	Quantitative (Online survey)	Most male and experienced teachers have positive perceptions of STEM education; however, its implementation in Korea faces challenges such as limited class time, increased teacher workloads, and inadequate administrative and financial support. To enhance STEM education, significant changes like government support and national curriculum redesign are essential.	(n=705)

12	Nguyen et al.	2016	Vietnam	Quantitative (Online &offline surveys)	Vietnamese teachers, who initially lacked experience in STEM and engineering-based learning, reported transformative perceptions after participating in professional development training, correlating positively with the time spent on STEM knowledge and lesson planning.	(n=150)
13	Young and Seung	2021	South Korea	Quantitative (Online survey)	teachers generally have a positive perception of STEM education and instructional materials, influenced by school changes, collaborative resources, and their own instructional expertise and self-efficacy.	(n=650)
14	Hanaa O. Khadri	2022	Egypt	Mixed methods	science teachers and those with higher qualifications demonstrate greater understanding and competencies, novice teachers also hold positive views of STEM, with consistent implementation -	(n=52)
15	Wan et al.	2013	China	Qualitative	Science teachers reported to hold positive attitudes toward STEM education. Their perceptions were largely influenced by their conception of science.	(n=24)
16	Simoncini et al.	2018	Australia	Quantitative (Survey)	Teachers' understanding of STEM in early childhood education is linked to subjects, play-based and hands-on learning experiences, and mind development, with a consensus on its importance for children's social-emotional growth and the utility of the Early Childhood STEM Habit of Mind Framework for planning integrated learning experiences.	(n=117)
17	Yau et al.	2022	Hong Kong	Qualitative (Phenomenological)	The study categorizes teachers' conceptions of science and technology into areas like technology bridging and competence cultivation, highlighting the challenge of a shortage of qualified science and ICT teachers, and recommending the implementation of STEM education in both technical and non-technical schools.	(n=28)

18	Katarina K.	2019	Slovakia	Q-Sort	The study identifies three types of teachers' perceptions regarding the effectiveness of science and STEM education: students as curious independent thinkers, the necessity of foundational knowledge for individual growth, and the importance of teachers creating stimulating learning environments that encourage interaction.	(n=5134)
19	Kim & Bolger	2016	South Korea	Mixed methods	While most teachers have a positive perception of STEM education and recognize its curricular importance, they struggle to implement it; however, their attitudes improve significantly with effective STEM lesson plan design and targeted pre- and post-survey interventions.	(n=119)
20	Canan et al.	2023	-	Qualitative	The findings indicate that teachers have positive attitudes towards STEM in early childhood education, recognizing its benefits for enhancing their teaching skills, improving students' skill development, and applying curriculum content in practice.	(n=134)
21	Shaw et al.	2021	Vietnam	Qualitative	Most Vietnamese teachers lack confidence and understanding in engaging early students in STEM education, highlighting the need for long-term training using Innovative Learning Environments (ILE) to enhance teaching approaches.	(n=105)
22	Catherine & Franz	2020	-	Quantitative	Teachers' attitudes influenced teachers' STEM creativity. There was a significant positive effect of STEM creativity on teachers' motivation. The integration of teachers' PD into school life is a necessity to foster creativity in classrooms.	(n=550)
23	Burke et al.	2021	Australia	Quantitative	Teachers require more guidance in using diverse science and technology teaching strategies for hands-on learning, with in-service teachers needing fewer professional development opportunities focused on collaboration and pedagogy, yet both groups benefit from training in knowledge construction and teaching strategies.	(n=303)

24	Papadakis et al.	2021	Greece	Quantitative	While some pre-service and in-service teachers hold positive attitudes towards Educational Robotics as part of STEM education, others express negative attitudes and doubts; overall, teachers' knowledge and understanding of Educational Robotics are positively correlated with their attitudes toward STEM education.	(n=201)

As in the figure below, there is a change through time in the topic or area of teachers, perception, conception, and attitude toward STEM education. For the first four years, the study was on the effect of teachers' STEM conceptions and attitudes towards it, and the role of teacher professional development in transforming teachers' perception of STEM was conducted. In the second phase which took three years the study was conducted on the role of teachers' beliefs in fostering STEM implementation, what skills do novice teachers need most about STEM. And the third phase the studies conducted was on the role of STEM education integration and association between teachers' perception and teaching practices towards STEM (see Figure 3).

Figure 3 *Timeline of change in study topics/area of teachers' perception, conceptions, and attitudes toward STEM education.*

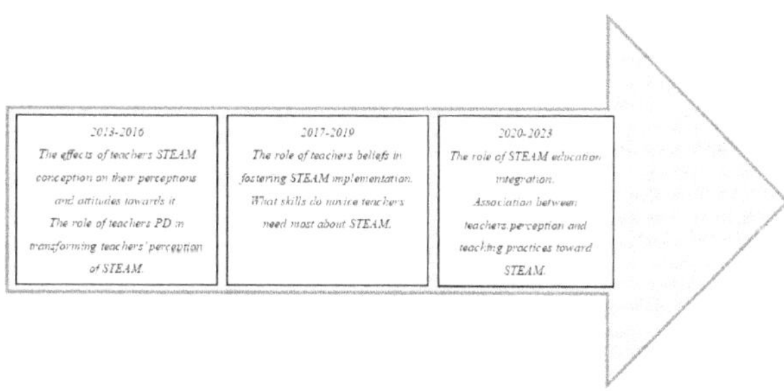

How are teachers' perceptions of using STEM pedagogy described by the available research?

Through the research that has been already conducted some understanding is gained of the teachers' attitudes about STEM education. To condense the body of knowledge concerning the following codes, several themes emerged: instructors' demographics, application activities, professional development and training for instructors, parental participation, accessibility to technology resources, and STEM awareness.

Table 2 Teacher demographics and their perception of STEM education

Teacher demographics and their perception of STEM education	
Gender	❖ Positive views about STEM held by male teachers than females (Amsarby, 2023;)
Age	A positive correlation between teachers' age and their perceptions of STEM education (Park et al., 2016; Kotuľáková, 2021). ❖ A positive attitude toward STEM education in the classroom grew with the age of the teachers (Thi To Khuyen et al., 2020; Wan et al., 2013).
Experience	❖ Novice teachers (1-5 years of teaching): less value of STEM education (Alghamdi, 2023; Kotuľáková, 2021). ❖ Moderately experienced teachers (6-10 years of teaching) hold the least familiarity with STEM education and are likely to be doubtful about students' ability to learn STEM (Alghamdi, 2023; Lupión-Cobos et al., 2023). ❖ Experienced teachers with above 10 years of teaching experience hold more favorable perceptions of STEM and they value STEM education (Alghamdi, 2023; Kotuľáková, 2021).
Expertise area	Teachers who hold degrees in science and technology hold positive attitudes towards and value STEM education than their colleagues whose expertise is in social sciences and other subjects (Amsbary et al., 2023; Alghamdi, 2023).
STEM Perceptions and Readiness	❖ The higher teachers value STEM education, the higher readiness for implementation, and vice versa is true (Lupión-Cobos et al., 2023; Park & Cho, 2022).

What are the challenges and barriers that teachers identify in implementing STEM pedagogy?

Six categories can be used to group the issues teachers describe as problems and barriers to STEM education: priorities for pedagogy or instruction barriers, curricular problems, structural barriers, student concerns, assessment concerns, and teacher vision and support. The summary of the teachers' challenges and barriers in implementing STEM education in their classrooms with the dimensions and subdimensions but also the support that they expect to be provided are in Table 3 below.

Table 3 Summary of key findings in the systematic review

Dimensions	Sub-dimensions	Articles
A. Challenges and barriers to STEM integration		
Pedagogical or instructional Challenge	= a fundamental shift away from teacher-led instruction to student led-instruction = a view of instruction that aligns with the philosophy of the STEM curriculum authors. = mismatch between the teachers' pedagogy and the curriculum pedagogy = incorrectly or inadvertently misinterpret the STEM developer's expectations = meeting the diverse needs of all learners, particularly those with disabilities and various cognitive abilities	Ogegbo et al. (2020) Yau et al. (2022) Yumi L. (2021) Young & Seung (2021) Jessica et al. (2023) Kim &Bolger (2016) Nguyen et al. (2016) Canan et al. (2023)
Curriculum Challenge	* Teachers had apprehension about following someone else's curriculum plan * Teachers felt miscommunications between various STEM teachers' perceptions of each other's domain led to feelings of anger and subsequently caused the interdisciplinary curriculum to fail	Zhan et al. (2020) Canan et al. (2023) Yumi L. (2021) Hamad et al. (2022)
Structural Challenge	= Teachers felt the confines of class scheduling prohibited the interdisciplinary nature of STEM lessons. = Various teachers teaching their specific subjects were not conducive to interdisciplinary work = The structure of student schedules, and lack of flexibility in them = Lack of control over the pacing of the curriculum and the sequence of instruction = At the district level, teachers felt administrative and financial support could be a challenge to STEM implementation = Lack of technology resources available to students. = The way education is organized and evaluated at the state level.	Young & Seung (2021) Yau et al. (2022) Ogegbo et al. (2020) Faiza et al. (2017) Jessica et al. (2023) Kim & Na, (2022) Lee et al. (2019)
Student concern	*= Teachers underestimate students' abilities to solve STEM problems *= Many of these teachers did not believe their students were competent enough in content areas to apply these skills to self-directed STEM problems *= Teacher beliefs about their students may be related to their implementation fidelity of STEM curriculum	Katarina K. (2019) Canan et al. (2023) Teresa et al. (2023) Simoncini et al. (2018) (Steele&Ashworth, 2018) (Papadakis et al., 2021)
Assessment Time Knowledge	= Teachers perceive that lack of quality assessment tools, planning time, and knowledge of STEM disciplines. = Not enough formative assessments. = Challenging concerns about group grading. = increased workload associated with STEM programming and planning. = lack of subject matter knowledge concerning STEM content = Lack of instructional resources leads to loss of confidence =lack of technology resources available to students	Faiza et al. (2017) Yumi L. (2021) Teresa et al. (2023) Zhan et al. (2020) Ahlam A., A. (2022) Park et al. (2016) Hamad et al. (2022) Wan et al. (2013) Shaw et al. (2021) Ogegbo et al. (2020) Nguyrn et al. (2016) Kim &Bolger (2016)

B. Teachers' expected support		
Collaboration	* Teachers believe that a culture of collaboration would increase the viability of STEM programs * Not only create an atmosphere that enhances preparation for STEM lessons but also model a team approach to students. * Partnerships with university faculty and accessing the expertise in their community * Teachers felt intentional time was necessary throughout the school year for various disciplines to meet together to plan to bridge the disciplines * A technology-enabled network could be an effective means of collaborating between content area teachers * Providing time and opportunities for collaborative planning and open communication between teachers may be critical to successful implementation.	Hamad et al. (2022) Young & Seung (2021) Park et al. (2016) (Lee, 2021)
Curriculum	= Teachers believe that the availability of a quality curriculum would enhance the likelihood of success of STEM initiatives = The importance of a flexible curriculum that is engineering-based = The curriculum must be flexible enough to be used with various ability levels and educational environments while still being focused on the engineering design process = Teachers believed this type of curriculum increased their belief in themselves, or self-efficacy, to teach STEM = Teachers expressed a desire for specific, ready-made STEM problems they could use in their classrooms immediately. = There must be fidelity in the implementation of the curriculum such that the teachers utilize the expectations and goals intended by the curriculum designer.	Simoncini et al. (2018) Hamad et al. (2022) Park et al. (2016)
Administrative support	* Teachers perceive school district support, guidance, and flexibility were necessary for STEM initiatives * A supportive administrator or administrative team is important when teachers are implementing STEM pedagogy. * Teachers believed guidance and constant dialog with administrators is needed to successfully utilize STEM programs * Teachers also felt that districts need to help parents and students understand course offerings and what STEM courses will teach them.	Jho et al.(2016) Jessica et al. (2023) Park et al. (2016)
Prior Experiences	= Teachers perceive that previous experience using student-centered, inquiry models of instruction facilitate success in a STEM initiative = Confidence in STEM pedagogy increased because of these prior experiences.	(Jamil et al., 2019) Nguyen et al. (2016) Papadakis et al. (2021)

Professional development	= Teachers believe that well-organized and frequently available professional learning opportunities would facilitate successful STEM initiatives = Effectiveness of STEM education was learning opportunities for teachers to increase their ability to effectively integrate STEM content into their curriculum. = Significant increases in their confidence, knowledge, and efficacy to teach STEM after attending professional development programs	Ogegbo et al. (2020) Zhan et al. (2020) Simoncini et al. (2018) Catherine &Franz (2020) Burke et al. (2021) Nguyen et al. (2016) Katarina K. (2019) Shaw et al. (2021)

Teachers' Preceptions of STEM Education on sustainable development

The necessity for education that fosters a sustainable world is increasingly recognized, as it not only enhances individual lives but also contributes to sustainable development (Jho et al., 2016; Sinakou et al., 2019; Hsiao & Su, 2021). Sustainable development, defined as meeting present needs without compromising the ability of future generations (Nam et al., 2020), seeks to balance economic growth with environmental and social integrity (Z. Wang, 2008). STEM education plays a pivotal role in global educational reforms by integrating science, technology, engineering, and mathematics, thereby emphasizing the interconnectedness of these disciplines (Sinakou et al., 2019; H. Park et al., 2016; Wu, 2022). This approach not only addresses legislative goals but also fosters competencies in mathematics and science (Marginson et al., 2013). The learner-centered methodology of STEM enhances students' self-direction, problem-solving, and collaboration skills (Hamad et al., 2022), while promoting innovation through the resolution of real-world problems via integrated STEM applications (S. Kim et al., 2018; García-Carrillo et al., 2021).

In Vietnam, teachers have pinpointed several key development issues to tackle through STEM instruction, with over 80% selecting environmental pollution as a focal point to enhance students' environmental awareness and understanding of monitoring tools (Nam et al., 2020). Following this, sustainable consumption and production were prioritized, emphasizing the importance of promoting sustainable usage through design-based learning. Recycled materials ranked third, alongside sustainable agriculture, food production, and renewable resources (Hsiao & Su, 2021). Teachers perceive STEM education as a contemporary approach that bolsters the role of secondary education in addressing national development challenges (Hamad et al., 2022).

Discussions, Recommendations for Practice

The required resources must be made available so that instructors and STEM programs may assist students' development while working to cultivate STEM talent. Teachers require top-notch curricula that adhere to district and state standards and include formative assessment strategies that teachers can utilize to gauge their pupils' conceptual understandings (H. Park et al., 2016). It is also essential that the group of teachers who will be using the curriculum participate in professional development that allows them to apply STEM principles and methods in real-world settings.

The need for more time to collaborate with the other STEM instructors must be made clear to the district administration. The incorporation of STEM would be facilitated by allowing instructors during the school year to plan classes and prepare lessons with colleagues. Teachers will want time to collaborate as they transition from teaching single subjects to teaching cross-curricular courses. To successfully incorporate this multidisciplinary approach to thinking and learning, teachers need to be inspired to collaborate and come up with creative solutions. Regarding teacher support and needs for STEM education, the school and district must have an open and transparent channel of communication.

Recommendations for future research

More research demonstrates the effectiveness of STEM programs with low-ability and diverse student populations would be beneficial in motivating teachers, as they believe that all pupils can gain from STEM education and will be encouraged to incorporate STEM activities as they witness student achievement in their classes. Investigating the differences in STEM perceptions between elementary and high school educators could provide insights into STEM integration (H. Park et al., 2016). Future research should utilize representative data to understand the effective conditions for STEM implementation and evaluate its impact on students' interest in science and mathematics. Studies should also take into account factors such as classroom variations, instructional methods, and school collaborations to assess transformative perceptions and implementation (Nam et al., 2020), with the reform process including ongoing testing and evaluation (Hamad et al., 2022). Furthermore, future studies should explore the effects of the STEM approach on programming and educational robotics in teachers' instruction while considering the interactions between teachers and students (García-Carrillo et al., 2021).

As engineering design emerges as a key focus, examining teacher educators' and teachers' Technological Pedagogical Engineering Knowledge (TPAEK) is essential, indicating a need for further research in this area (Chai, 2019). Lastly, research that incorporates teacher interviews and observations of the design process will enhance understanding of how history and STEM can collaborate through integrated STEM curricula, providing insights into the real-world challenges teachers face when merging history and STEM in the curriculum, thus benefiting from theories of curriculum integration to investigate various models of STEM integration, including history (W. Park & Cho, 2022).

Conclusions

The study's 24 findings should serve as a guide for STEM programs in schools and districts. These 24 findings, which were found in several studies, could help to direct future professional development and teacher efforts to boost teachers' efficacy in STEM integration and instruction for sustainable development. Teachers appear to support STEM education, believing that it enhances student learning outcomes and prepares students for the future. If educators had more confidence, STEM activities would likely be more successfully incorporated in the classroom. There is no doubt in the research that higher levels of confidence enhance classroom performance, which will enhance student learning.

In conclusion, equitable and high-quality education for sustainable development, sustainable lifestyles, and social equity may help to accomplish sustainable development goals. However, it ought to have gotten more consideration in national and international sustainable development agendas. The implementation of inclusive STEM education necessitates innovation and change in pedagogical approach, curriculum creation, student assessment methodologies, school management structure, and initiatives for teacher support. For schools in rural locations, serving ethnic minorities, underprivileged pupils, and places of social unrest, more funding is also needed for facilities, infrastructure, and technology.

References

References with (*) are these 24 articles of literature we systematically reviewed.

*Abdulwali, H., Naem, M., & Aljallal, M. (2019). Saudi Arabian science and mathematics teachers' attitudes toward integrating STEM in teaching before and after participating in a professional development program. *Cogent Education, 6* (1), 1580852. In.

*Amsbary, J., Yang, H.-W., Sam, A., Lim, C.-I., & Vinh, M. (2023). Practitioner and Director Perceptions, Beliefs, and Practices Related to STEM and Inclusion in Early Childhood. *Early Childhood Education Journal,* 1-19.

Affouneh, S., Salha, S., Burgos, D., Khlaif, Z. N., Saifi, A. G., Mater, N., & Odeh, A. (2020). Factors that foster and deter STEM professional development among teachers.*Science Education, 104*(5), 857-872. https://doi.org/10.1002/sce.21591

*Alghamdi, A. A. (2023). Exploring early childhood teachers' beliefs about STEM education in Saudi Arabia. *Early Childhood Education Journal, 51*(2), 247-256.

Bahrum, S., Wahid, N., & Ibrahim, N. (2017). Integration of STEM education in Malaysia and why STEM. *International Journal of Academic Research in Business and Social Sciences, 7*(6), 645-654. http://dx.doi.org/10.6007/IJARBSS/v7-i6/3027

Capraro, R. M., & Han, S. (2014). STEM: The education frontier to meet 21st-century challenges. *Middle Grades Research Journal, 9*(3), XV.

Chai, C. S. (2019). Teacher professional development for science, technology, engineering, and mathematics (STEM) education: A review from the perspectives of technological pedagogical content (TPACK). *The Asia-Pacific Education Researcher, 28*(1), 5-13. https://doi.org/10.1007/s40299-018-0400-7

Chan, K. K. H., Rollnick, M., & Gess-Newsome, J. (2019). A grand rubric for measuring science teachers' pedagogical content knowledge. *Repositioning pedagogical content knowledge in teachers' knowledge for teaching science*, 253-271. https://doi.org/10.1007/978-981-13-5898-2_11

Chen, Y.-L., Huang, L.-F., & Wu, P.-C. (2021). Preservice preschool teachers' self-efficacy in and need for STEM education professional development: STEM pedagogical belief as a mediator. *Early Childhood Education Journal, 49*, 137-147. https://doi.org/10.1007/s10643-020-01055-3

*Conradty, C., & Bogner, F. X. (2019). From STEM to STEM: Cracking the code? How creativity & and motivation interact with inquiry-based learning. *Creativity Research Journal,31*(3),284-295. https://doi.org/10.1080/10400419.2019.1641678

Conradty, C., & Bogner, F. X. (2020). STEM teaching professional development works Effects on students' creativity and motivation. *Smart Learning Environments, 7*, 1-20. https://doi.org/10.1186/s40561-020-00132-9

Crawford, B. A., & Capps, D. K. (2018). Teacher cognition of engaging children in scientific practices. *Cognition, metacognition, and culture in STEM education: Learning, teaching and assessment*, 9-32. https://doi.org/10.1007/978-3-319-66659-4_2

Croak, M. (2018). The effects of STEM education on economic growth.

DeJarnette, N. K. (2018). Implementing STEM in the early childhood classroom. *European Journal of STEM Education, 3*(3), 18.

ell'Erba, M. (2019). Preparing Students for Learning, Work, and Life through STEM Education. Policy Brief. *Education Commission of the States.*

Edwards, S. (2013). Digital play in the early years: A contextual response to the problem of integrating technologies and play-based pedagogies in the early childhood curriculum. *European Early Childhood Education Research Journal, 21*(2), 199-212. https://doi.org/10.1080/1350293X.2013.789190

Ejiwale, J. A. (2013). Barriers to successful implementation of STEM education. *Journal of Education and Learning (EduLearn),7*(2),63-74. https://doi.org/10.11591/edulearn.v7i2.220.

El-Deghaidy, H., & Mansour, N. (2015). Science teachers' perceptions of STEM education: Possibilities and challenges. *International Journal of Learning and Teaching, 1*(1), 51-54.

Fives, H., & Buehl, M. M. (2014). Exploring differences in practising teachers' valuing of pedagogical knowledge based on teaching ability beliefs. *Journal of Teacher Education, 65*(5), 435-448. https://doi.org/10.1177/0022487114541813

*García-Carrillo, C., Greca, I. M., & Fernández-Hawrylak, M. (2021). Teacher perspectives on teaching the STEM approach to educational coding and robotics in primary education. *Education Sciences, 11*(2), 64. https://doi.org/10.3390/educsci11020064

Gough, D., Thomas, J., & Oliver, S. (2012). Clarifying differences between review designs and methods. *Systematic reviews, 1*(1), 1-9. https://doi.org/10.1186/2046-4053-1-28

*Hamad, S., Tairab, H., Wardat, Y., Rabbani, L., AlArabi, K., Yousif, M., . . . Stoica, G. (2022). Understanding science teachers' implementations of integrated STEM: Teacher perceptions and practice. *Sustainability, 14*(6), 3594. https://doi.org/10.3390/su14063594

Hawari, A. D. M., & Noor, A. I. M. (2020). Project-based learning pedagogical design in STEM art education. *Asian Journal of University Education, 16*(3), 102-111. https://doi.org/10.24191/ajue.v16i3.11072

Herro, D., & Quigley, C. (2017). Exploring teachers' perceptions of STEM teaching through professional development: implications for teacher educators. *Professional Development in Education, 43*(3), 416-438. https://doi.org/10.1080/19415257.2016.1205507

Herro, D., Quigley, C., & Cian, H. (2019). The challenges of STEM instruction: Lessons from the field. *Action in Teacher Education, 41*(2), 172-190. https://doi.org/10.1080/01626620.2018.1551159

Hoffmann, T., & Siege, H. (2018). What is Education for Sustainable Development (ESD)? *Human Development, 1*(8), 1-6.

Hong, O. (2017). STEM education in Korea: Current policies and future directions. *Science and Technology Trends Policy Trajectories and Initiatives in STEM Education, 8*(2), 92-102.

Hsiao, P.-W., & Su, C.-H. (2021). A study on the impact of STEM education for sustainable development courses and its effects on student motivation and learning. *Sustainability, 13*(7), 3772. https://doi.org/10.3390/su13073772

Hutner, T. L., & Markman, A. B. (2016). Proposing an operational definition of science teacher beliefs. *Journal of Science Teacher Education, 27*(6), 675-691. https://doi.org/10.1007/s10972-016-9480-5

Inomjonovna, R. I. (2023). STEM EDUCATION IS ONE OF THE MAIN TRENDS IN THE WORLD. *Journal of New Century Innovations, 21*(2), 27-32.

Jamil, F. M., Linder, S. M., & Stegelin, D. A. (2018). Early childhood teacher beliefs about STEM education after a professional development conference. *Early Childhood Education Journal, 46*, 409-417. https://doi.org/10.1007/s10643-017-0875-5

*Jamil, F. M., Linder, S. M., & Stegelin, D. A. (2018). Early childhood teacher beliefs about STEM education after a professional development conference. Early Childhood Education Journal, 46, 409-417.

*Jho, H., Hong, O., & Song, J. (2016). An analysis of STEM/STEM teacher education in Korea with a case study of two schools from a community of practice perspective. *Eurasia Journal of Mathematics, Science and Technology Education, 12*(7), 1843-1862. https://doi.org/10.12973/eurasia.2016.1538a

*Juškevičienė, A. (2020). STEM teacher for a day: a case study of teachers' perspectives on computational thinking. Informatics in Education-An International Journal, 19(1), 33-50.

*Kotuľáková, K. (2021). Identifying teachers' beliefs before CPD training focusing on an inquiry-based approach in science education. *Research in Science Education, 51*(1), 183-211.

Kelley, T. R., & Knowles, J. G. (2016). A conceptual framework for integrated STEM education. *International journal of STEM education, 3*, 1-11. https://doi.org/10.1186/s40594-016-0046-z

Kennedy, T. J., & Odell, M. R. (2014). Engaging students in STEM education. *Science Education International, 25*(3), 246-258.

*Khadri, H. O. (2022). Becoming future-proof STEM teachers for enhancing sustainable development: A proposed general framework for capacity-building programs in future studies. *Prospects, 52*(3-4), 421-435. https://doi.org/10.1007/s11125-021-09588-0

Kim, D., & Bolger, M. (2017). Analysis of Korean elementary pre-service teachers' changing attitudes about integrated STEM pedagogy through developing lesson plans. *International Journal of Science and Mathematics Education, 15*, 587-605. https://doi.org/10.1007/s10763-015-9709-3

Kim, S., Lee, Y., & Heungdeok-gu, C. (2018). A study on teachers' practices of STEM education in Korea. *International Journal of Pure and Applied Mathematics, 118*(19), 2339-2365.

*Kim, Y.-H., & Na, S.-I. (2022). Using structural equation modelling for understanding relationships influencing the middle school technology teacher's attitudes toward STEM education in Korea. *International Journal of Technology and Design Education,* 1-32.

*Lee, M.-H., Chai, C. S., & Hong, H.-Y. (2019). STEM education in the Asia Pacific: Challenges and development. In (Vol. 28, pp. 1-4): Springer.

Lee, S.-S. (2020). Research on the developmental aspects of the STEM education program development in Korea. *Journal of Engineering Education Transformations, 34*(2), 33-44.

*Lee, Y. (2021). Examining the Impact of STEM Education Reform on Teachers' Perceptions of STEM in Uzbekistan. *Asia-Pacific Science Education, 7*(1), 34-63.

Leicht, A., Heiss, J., & Byun, W. J. (2018). *Issues and trends in education for sustainable development* (Vol. 5): UNESCO publishing.

*Lupión-Cobos, T., Crespo-Gómez, J. I., & García-Ruiz, C. (2023). Challenges and Opportunities to Teaching Inquiry Approaches by STE (A) M Projects in the Primary Education Classroom. *Journal of Baltic*

Science Education, 22(3), 454-469.
Marginson, S., Tytler, R., Freeman, B., & Roberts, K. (2013). STEM: country comparisons: international comparisons of science, technology, engineering and mathematics (STEM) education. Final report. In: Deakin University,
Margot, K. C., & Kettler, T. (2019). Teachers' perception of STEM integration and education: a systematic literature review. *International journal of STEM education, 6*(1), 1-16. https://doi.org/10.1186/s40594-018-0151-2
Marzano, R. J. (2007). *The art and science of teaching: A comprehensive framework for effective instruction*: Ascd.
Mejias, S., Thompson, N., Sedas, R. M., Rosin, M., Soep, E., Peppler, K., . . . Bell, P. (2021). The trouble with STEM and why we use it anyway. *Science Education, 105*(2), 209-231. https://doi.org/10.1002/sce.21605
Meschede, N., Fiebranz, A., Möller, K., & Steffensky, M. (2017). Teachers' professional vision, pedagogical content knowledge, and beliefs: On its relation and differences between pre-service and in-service teachers. *Teaching and teacher education, 66*, 158-170. https://doi.org/10.1016/j.tate.2017.04.010
Miller, J., & Knezek, G. (2013). *STEM for student engagement.* Paper presented at the Society for Information Technology & Teacher Education International Conference.
Moher, D., Liberati, A., Tetzlaff, J., Altman, D. G., & Group*, P. (2009). Preferred reporting items for systematic reviews and meta-analyses: the PRISMA statement. *Annals of Internal Medicine, 151*(4), 264-269. https://doi.org/10.7326/0003-4819-151-4-200908180-00135
Møller, A., & Myles, P. (2016). What makes a good systematic review and meta-analysis? *BJA: British Journal of Anaesthesia, 117*(4), 428-430. https://doi.org/10.1093/bja/aew264
*Nam, N. H., Van Hien, N., Van Bien, N., Trang, N. T. T., & Minh, T. H. (2020). Transformative perceptions of in-service teachers towards STEM education: The Vietnamese case study. Вопросы образования(2 (eng)), 204-229.
*Ogegbo, A. A., & Aina, A. (2020). Early childhood development teachers' perceptions on the use of technology in teaching young children. South African Journal of Childhood Education, 10(1), 1-10.
Page, M. J., McKenzie, J. E., Bossuyt, P. M., Boutron, I., Hoffmann, T. C., Mulrow, C. D., . . . Brennan, S. E. (2021). The PRISMA 2020 statement: an updated guideline for reporting systematic reviews. *International journal of surgery, 88*, 105906.
*Papadakis, S., Vaiopoulou, J., Sifaki, E., Stamovlasis, D., & Kalogiannakis, M. (2021). Attitudes towards the use of educational robotics: Exploring pre-service and in-service early childhood teacher profiles. Education Sciences, 11(5), 204.
*Park, H., Byun, S.-y., Sim, J., Han, H.-S., & Baek, Y. S. (2016). Teachers' perceptions and practices of STEM education in South Korea. *Eurasia Journal of Mathematics, Science and Technology Education, 12*(7), 1739-1753. https://doi.org/10.12973/eurasia.2016.1531a
*Park, W., & Cho, H. (2022). The interaction of history and STEM learning goals in teacher-developed curriculum materials: opportunities and challenges for STEM education. *Asia Pacific Education Review, 23*(3), 457-474. https://doi.org/10.1007/s12564-022-09741-0
Quigley, C. F., & Herro, D. (2016). "Finding the joy in the unknown": Implementation of STEM teaching practices in middle school science and math classrooms. *Journal of science education and technology, 25*, 410-426. https://doi.org/10.1007/s10956-016-9602-z
Richardson, V. (1996). The role of attitudes and beliefs in learning to teach. *Handbook of research on teacher education, 2*(102-119), 273-290.
Roberts, T., Jackson, C., Mohr-Schroeder, M. J., Bush, S. B., Maiorca, C., Cavalcanti, M., . . . Cremeans, C. (2018). Students' perceptions of STEM learning after participating in a summer informal learning experience. *International journal of STEM education, 5*(1), 1-14. https://doi.org/10.1186/s40594-018-0133-4
Said, Z. (2016). Science education reform in Qatar: Progress and challenges. *Eurasia Journal of Mathematics, Science and Technology Education, 12*(8), 2253-2265. https://doi.org/10.12973/eurasia.2016.1301a
Schommer-Aikins, M. (2004). Explaining the epistemological belief system: Introducing the embedded systemic model and coordinated research approach. *Educational psychologist, 39*(1), 19-29. https://doi.org/10.1207/s15326985ep3901_3
*Shaw, P. A., Traunter, J. E., Nguyen, N., Huong, T. T., & Thao-Do, T. P. (2021). Immersive-learning experiences in real-life contexts: deconstructing and reconstructing
Shulman, L. S. (1986). Those who understand: Knowledge growth in teaching. *Educational researcher, 15*(2), 4-14.
Sinakou, E., Donche, V., Boeve-de Pauw, J., & Van Petegem, P. (2019). Designing powerful learning en-

vironments in education for sustainable development: A conceptual framework. *Sustainability, 11*(21), 5994. https://doi.org/10.3390/su11215994

Spyropoulou, C., Wallace, M., Vassilakis, C., & Poulopoulos, V. (2020). Examining the use of STEM Education in Preschool Education. *European Journal of Engineering and Technology Research.* https://doi.org/10.24018/ejeng.2020.0.CIE.2309.

Srikoom, W., & Faikhamta, C. (2018). Assessing in-service teachers' self-efficacy and beliefs about STEM education. *Journal of Education, 12*(4), 169-186.

*Steele, A., & Ashworth, E. L. (2018). Emotionality and STEM Integrations in Teacher Education. *Journal of Teaching and Learning, 11*(2), 11-25.

Stohlmann, M., Moore, T. J., & Roehrig, G. H. (2012). Considerations for teaching integrated STEM education. *Journal of Pre-College Engineering Education Research (J-PEER), 2*(1), 4. https://doi.org/10.5703/1288284314653

Sturtevant, H., & Wheeler, L. (2019). The STEM faculty instructional barriers and identity survey (FIBIS): Development and exploratory results. *International journal of STEM education, 6*, 1-22. https://doi.org/10.1186/s40594-019-0185-0

Suwarma, I., & Kumano, Y. (2019). *Implementation of STEM education in Indonesia: teachers' perception of STEM integration into the curriculum.* Paper presented at the *Journal of Physics:* Conference Series. https://doi.org/10.1088/1742-6596/1280/5/052052

*Thi To Khuyen, N., Van Bien, N., Lin, P.-L., Lin, J., & Chang, C.-Y. (2020). Measuring teachers' perceptions to sustain STEM education development. *Sustainability, 12*(4), 1531. https://doi.org/10.3390/su12041531

Thibaut, L., Knipprath, H., Dehaene, W., & Depaepe, F. (2018). The influence of teachers' attitudes and school context on instructional practices in integrated STEM education. *Teaching and teacher education, 71*, 190-205. https://doi.org/10.1016/j.tate.2017.12.014

Van Es, E. A., & Sherin, M. G. (2002). Learning to notice: Scaffolding new teachers' interpretations of classroom interactions. *Journal of technology and teacher education, 10*(4), 571-596.

*Wan, Z. H., Wong, S. L., Wei, B., & Zhan, Y. (2013). Focusing on the classical or contemporary? Chinese science teacher educators' conceptions of the nature of science content to be taught to pre-service science teachers. Research in Science Education, 43, 2541-2566. DOI: 10.1007/s11165-013-9372-z

Wang, H.-H., Moore, T. J., Roehrig, G. H., & Park, M. S. (2011). STEM integration: Teacher perceptions and practice. *Journal of Pre-College Engineering Education Research (J-PEER), 1*(2), 2. https://doi.org/10.5703/1288284314636

Wang, X., Xu, W., & Guo, L. (2018). The status quo and ways of STEM education promoting China's future social sustainable development. *Sustainability, 10*(12), 4417. https://doi.org/10.3390/su10124417

Wang, Z. (2008). National humiliation, history education, and the politics of historical memory: Patriotic education campaign in China. *International Studies Quarterly, 52*(4), 783-806. https://doi.org/10.1111/j.1468-2478.2008.00526.x

Watson, A. D., & Watson, G. H. (2013). Transitioning STEM to STEM: Reformation of engineering education. *Journal for Quality and Participation, 36*(3), 1-5.

Wu, Z. (2022). Understanding teachers' cross-disciplinary collaboration for STEM education: Building a digital community of practice. *Thinking Skills and Creativity, 46*, 101178. https://doi.org/10.1016/j.tsc.2022.101178

Zaher, A. A., & Hussain, G. A. (2020). *STEM-based active learning approach to selected topics in electrical/computer engineering.* Paper presented at the 2020 IEEE Global Engineering Education Conference (EDUCON). https://doi.org/10.1109/EDUCON45650.2020.9125367

Reframing Contents for Educating Teachers: Pursuits for Professional Engagements

Berhane Demoz

Ministry of Education

Abstract

Teacher education in Eritrea dates back to the early 1940s. Yet, its history and its impact to the quality of school education will require professionally thoughtful, collaborative, and sustainable engagements in research. This should penetratingly mean to strategically and practically reframe teacher education in general and *initial teacher education* (ITE) in particular to lay the foundations for nurturing responsible citizens. In line, Musset (2010) argued that any education reform that does not take serious consideration about the indispensability of teacher education is predestined to gross failures, and countries like Eritrea require a school system guided by unfaltering teachers. The paper is an extract of one chapter in a recent manuscript, by the present author, entitled *Professional Pursuits for Educating Teachers: Managing Complexity with Complexity*. The study echoes the pressing needs that educational establishments and their partners are required to rethink in developing educationally grounded, historically vital, practically relevant, necessarily collaborative, resourcefully applicable, and sustainably engaging ITE *contents*. The main data sources were the writers' persistent taking of notes over four decades of his teaching experience in general and the last two decades in particular accompanied by close reflective discussions with colleagues for more than ten years, and consistent collaborative action research engagements with learners and colleagues. These sources were closely enriched by relevant literature in content management. The outcomes of the process brought to bear general frames for educating reliable teachers grounded on *school practices* as implied by the educational needs of school learners, *pedagogy* that should closely accompany school practices, vital professional identities, and core areas of knowledge. It emphasizes that coloniality has been a systematic project for removing our contexts, and a pedagogy towards decoloniality must be our project (Reyes, 2022, p. 1244) in reframing our teacher education programs.

Keywords: school practice, quality ITE, pedagogy, professional identity, content management, Eritrea

Background

Teacher education in Eritrea dates back to the early 1940s. Yet, its history and its impact to the quality of school education will require professionally thoughtful, collaborative, and sustainable engagements in research. This should penetratingly mean to strategically and practically improve teacher education in general and *ITE* in particular with approaches that involve all essential stakeholders as organized and guided by the Ministry of Education and the National Higher Education and Research Institute in the country. The intention of such collaborative and sustained research endeavor should target the *context and practice* of the levels and types of schools we have in our country with the focus on the role of quality teachers in respective types of schools.

The enormity of the endeavor is vast but we can prioritize the process considering the gravity of the swift erosion of the quality of school education to educate responsible and resourceful young citizens on the one hand, and the availability of our human, material, and financial resources on the other. Currently, the demand for weaving teacher education into the '*tapestry*' (Cochran-Smith, 2000, p. 2, emphasis added) of the country's diverse political, social, economic, and educational context has never been as urgent. Countries like Eritrea require high quality school system guided by unfaltering teachers considering the country's targeted political victimization by the west and Ethiopian regimes. In line, Musset (2010) argued that any education reform that does not take serious measures about the indispensability of teacher education is predestined to gross failures.

Objectives and Research Questions

The paper is an extract of one chapter among the 13 chapters of a recent manuscript, by the present author, entitled *Professional Pursuits for Educating Teachers: Managing Complexity with Complexity*. The study echoes the pressing needs that educational establishments and their partners are required to rethink, reframe, and reposition the critical place of ITE content in developing educationally grounded, historically vital, practically relevant, necessarily collaborative, resourcefully applicable, and sustainably engaging ITE *contents* or curricula in the country.

The specific objective of the paper is to elaborate the connotations of content in the process of preparing teachers in general and the importance of seriously considering the context and practices in our schools for developing quality and relevant contents for ITE. The intention is to provide a general framework that will be largely relevant for *entering into a thorough process of research* for designing relevant organization and composition of contents based on school practices that must be recognized by pertinent higher education establishments with the following main research questions:

- *Content*: Why is the connotation of relevant content crucial in preparing teachers?
- *Experience*: What are the limitations of our experiences in ITE?
- *Content management*: How should we visualize the value of content management?
- *Research*: What must be the place of research for profound engagement?

Methodology

The methodology we followed for helping us visualize the reframing process for conducting penetrating research efforts in ITE contents included looking back, looking here and now, and looking forward by using the qualitative data sources we had at our disposal that we collected and organized from our notes and reflective experiences, collective experiences of diverse participants, and collaborative engagements in action research. The main data sources for responding to the objectives and research questions were the author's persistent taking of notes over four decades of his teaching experience in general and the last two decades in particular accompanied by close reflective discussions with colleagues for more than ten years. These data sources were accompanied by consistent collaborative action research engagements with learners and colleagues at the Faculty of Education in the University of Asmara, at the College of Education in the Eritrea Institute of Technology, and at the Asmara Community College of Education. Besides, his active involvement with teams to support schools and educational offices in sub-regions and regions have contributed to recognize the critical influences of the contexts and practices of these educational institutions in teacher education programs.

The immense data gathered during the years have enabled the author, his colleagues, and learners to enter into thoughtful collaborative research endeavors not only to come up with diverse outcomes for improving ITE programs but also to develop essential materials for teacher education and research about action research (meta-research) in teacher education. Moreover, these endeavors and experiences have been closely supported by relevant and timely literature search and reviews.

Findings and Discussion

The outcomes of our efforts for years allowed us to develop a general design for reframing contents in ITE programs that should seriously value the educational needs of school learners for educating reliable teachers grounded on *school practices and programs* as implied by:

- the *educational needs* of school learners,
- the *pedagogy* that should closely accompany school practices,
- the influence of *professional identities* that should prevail in schools, and
- the need for *core areas of knowledge* essential for preparing teachers.

We have come to learn that these grounds are fundamental for *decolonizing* our ITE. Our experiences confirmed that "coloniality has been a systematic project of removal, [and hence] a pedagogy of and towards *decoloniality must also be our project* of "RE"s: resisting, refusing, rehumanizing, restoring, reimagining, etc." (Reyes, 2022, p. 1244, emphasis added) our teacher education programs.

Connotation of content in preparing teachers

The meaning of content for our purpose is the study areas that learner teachers should deliberately learn in ITE programs to develop professional sayings, beings, and doings through discrete or integrated forms of learning areas that are relevant for the purposes of learning to teach (e.g. while they are in the ITE programs) and/or teaching to learn (e.g. while they are teaching in schools). It is the input of a variety of experiences shared or portrayed, openly or tacitly, by the programs intended to prepare teachers for beginning their professional teaching in schools.

Understanding content in ITE should both include the process of learning and the process of teaching. It should mainly match the educational and social needs of learner teachers with higher meaning and worth to their expected roles in schools that clearly target school learners and their society because contents for *learning to teach* and *teaching to learn* are both about a professional pursuit for quality learning in order to teach. Our experiences proved to us to never take teaching for granted. Instead to honesty begin to respect what it means to teach by comprehensively considering the part contents play for professional empowerment of teachers to learn to teach and to teach to learn on the one hand, and to develop professional becoming and doing on the other. The role of educators hence becomes critical in sharing and exemplifying (as opposed to simply lecturing) essential educational experiences to ensure that learner teachers retain, remember, retrieve, relate, reveal, and remold their contributions to new situations.

Selecting contents in preparing teachers for professional process requires adequate deliberations about what it really means to the practice of preparing capable teachers. Its vitality lies in its multiplicity unlike its customary or common meaning that often mystifies even colleges to focus on its detached meaning of representing only the ingredients of *subject matters* taught in colleges. Even then many colleges of education provide much less percent of their credit hours to educational courses or contents for preparing teachers to teach. In fact, our colleges of education had the following characteristics that fell short to grounding legitimate behaviors and functionings of teachers to *serve schools*:

- programs that clearly lack visionary and comprehensive aims and process of selecting candidates to join the teaching profession
- programs with minimum links with schools to openly seek to esteem school contexts, active involvement of school leaders and teachers, and the school curricula
- programs with least values to processes that allow learner teachers to closely connect and practice teaching in schools and classrooms during their stay at the colleges
- programs that did not ensure solid educational base both in content and time for teachers to work in KGs and primary schools despite the relative importance of such schools for allowing learners to assume a solid base for further learning
- programs that provide educational courses at the end of the program thus alienating subject matter courses from educational courses implying that the latter was only superfluous 'subject matter'
- programs allotted less credits for educational studies as compared to subject matters
- programs that lacked earnest and rigorous assessment process, often using tests and exams, for education related courses
- programs that procduce *lecturers and instructors* who never wanted to be called teachers
- program courses that were irrelevant to teaching practices in schools and were highly theoretical often founded on western cultures and literature.

In reality, however, *not only strict-subject-wise-prepared graduates* but also *incomprehensively prepared graduates of teachers education* fail to become effective teachers *in real situation of schools*. Pedagogically unprepared teachers happen to become not only unsuccessful teachers but also inflict unlimited damages on the future of school learners and tend to mostly leave the profession very early thinking that they were subject matter 'experts'. Either way, former waste in the colleges *massively replicates* the latter waste in schools.

The above arguments imply that content in preparing teachers should be strategically and comprehensively perceived to develop a set of learning areas in subprograms, each requiring fair share with close focus to intra- and inter-subprogram linkages. This influences teacher education initiatives to assume cohesive, enlightened, balanced, open, and consistently progressive programs to recognize school realities. Besides a lot of socio-cultural and civic wisdoms can be integral parts of the comprehensive contents because teachers must be the proprietors and proponents of such deep wisdoms of educational insights. Professional preparation that value school practices should be highly esteemed primarily by college of education leadership to ensure that every teacher begins with firm initial capacities to contribute to early quality learning, enjoy teaching as a vocation, and promote the development of responsible citizenry.

Figure 1 emphasizes that learning to teach of, for, and in school engagements requires intensive preparation for developing relevant capacities that is grounded on school context in general and school practices in particular through consistent efforts of learning about learning and teaching. It highlights that our focus should be on how learning and teaching should be well understood by learner teachers accompanied by why they are there in the profession.

***Figure 1** Learning to Teach in, of, and for Professional Process*

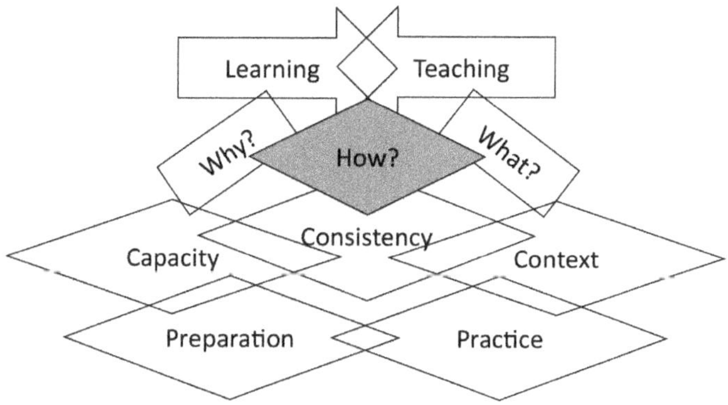

General Frameworks of Contents in Preparing Teachers

Content is said to derive from the Latin word '*contentum*' (plural '*contenta*'), meaning substance or substances enclosed or contained in something. Content has thus a relative meaning in that we can talk about successive content within a content, like for example life science being a content within natural sciences (Figure 2) or micro-biology within biology. In substance, therefore, content for preparing teachers for professional practice is wide by essence and level of the program(s) we intend to develop. Selecting essential sets of subprograms that contain strategic and practical contents within the main program(s) for *teacher preparation* is what makes the program(s), the subprograms, and contents relevant, resourceful, and judicious in many professional aspects of learning to teach. Subprograms should then be packages that institutes or colleges must develop to ensure qualities that learner teachers will pursue to be ready to face professional challenges in real school practices that *contain* effective, dynamic, and profound *course ingredients*.

General Framework of Contents

Content for teachers' professional practice is wide in essence, by school level, by school type, etc. and its selection should be:

- aim-sensitive, national, contextual, and practical
- research-based and participatory
- activate the roles of CoE in national development
- recognize the key position of ITE in higher education

Such decisions in selecting content, as we argued earlier, should be research-based, participatory, and witty to ensure fair selection of process and content from the infinite reservoir of national and international experiences. The challenge for choosing qualitatively relevant blends among and within the subprograms can be turned into opportunity if policy, institutions, and educators become fair enough to *searchingly* respect that teaching is meant to learn, to relate, to care, to guide, to inquire, and to lead one-self and learners while serving schools and communities. This will benefit educational institutions to become thoughtful about why this or that content is *indispensable for teachers* to prepare for professional process and to continue to refine and adapt their resourceful contribution to learners, communities, and themselves in practice, of practice, for practice in schools. In point, it is wise learning from the critiques Kosová and Kovaľcíková (2023) make on Slovakian teacher education system about the underestimation of pedagogical and psychological components that negatively affect any anticipated paradigm shift in teacher professionalism (p. 77).

Parallel to searching to decide what and which subprogram areas of content are essential to prepare teachers, is the need for the fair inclusion of the diverse relevant components of experiences (what Bloom also calls the taxonomy of knowledge). These should include the implicit necessity, unity, and balance of *knowledge, skills, attitudes, and creative application* of subprograms and contents within the subprograms (Figure 2). For example, knowledge can include information, theories, facts, figures, evidences, views, realizations, etc. Skills can include articulations, relationships, manipulations, talents, competences, etc. Attitudes can include approaches, outlooks, manners, stances, mindset, behaviors, etc. Creativity can include initiatives, imaginations, bright ideas, originalities, wits, resourcefulness, etc. These components should be balanced in each content depending on why the contents contribute to teachers' proactive, consistent, capable, and contextual involvement in schools supported by relevant pedagogic practices.

Another parallel challenge in the selection process for preparing teachers is the organization of the contents into whether they should be *discrete or integrated* (Figure 2). The challenge of selecting discrete subject, for example biology, may not be that difficult but the challenge of integrating biology with pedagogy (Biology Education) becomes more difficult in the selection and decision process of contents. Besides and based on the aims of integration, the balance of *which* contents to include between or among the main areas to be integrated becomes educationally vital. Here then preparing teachers for a school level, like preparing teachers for the foundational grades and schools respectively becomes more challenging than preparing for middle or secondary school grades, implying that the preparation of teachers for the foundational (often erroneously identified as 'lower') grades or levels demand thorough development of programs, subprograms, and integrated courses.

Figure 2 *General Scene of Content for Preparing Teachers*

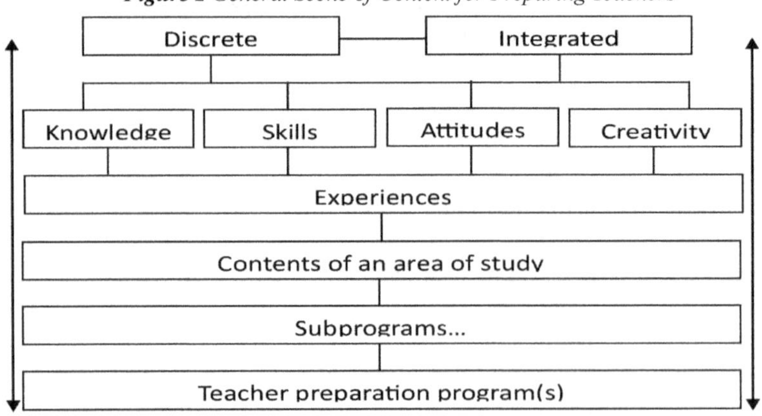

Figure 2 shows a general landscape of content for preparing teachers in defining our programs to include well-designed subprograms. Each subprogram should contain essential areas of studies that are relevant to school practices with adequate balance of the four areas of experience: knowledge, skills, attitudes, and creative adaptations. The study areas can be then organized in either discrete or integrated experiences depending on the educational needs of learners in schools.

Essential Subprograms in Teacher Preparation

It is not practical to draw one and inflexible teacher preparation program for all school systems that is universally applicable and acceptable. Neither is it also possible to do so for all types (e.g. vocational schools) and levels (kindergarten) of schools. The aggregate teacher education program should understandably contain subprograms that are classified into systematic, related, logical, and also relevant to the aims and processes of preparing teachers. In the same way, courses and their contents should make up the subprograms. Subprograms in a teacher preparation program may differ depending on national intentions, context, and practices in school types and school levels. In general, the professional requirements for preparing teachers greatly vary in time, extent, and mainly in quality; quality implying that a program and its subprograms are relevant and rigorous enough to ensure good education in pertinent schools. Depending on these influences, subprograms can also differ in number not only among school types and levels but also among institutions along with influences of availabilities of expertise and other resources.

Our experiences in cataloging the main subprograms in an ITE program for primary and general secondary schools has for long been shaped by colonial influences. Menter (2023) stresses, much of the scholarship on teacher education hitherto has had a strongly western or northern focus and it is time to address such imbalance if teacher education can function as a strategic means for school systems to actively contribute to avoid political upheavals, inequalities, migration, and climate change (p. v).

Often, we design some good educational aims and strategies but we vastly fail in implementing our aims by habitually getting trapped to accommodate colonial forms and contents of teacher preparation programs. Such pasted programs had either very short-term implementation arrangements that inherently restrict the preparation of effective teachers demanded by our schools (e.g., the elementary school teacher preparation of one year), or programs greatly dominated by subject matters and western oriented educational studies (e.g., secondary school teacher education programs). Mignolo and Walsh (2018) note that western coloniality, colonialism, and colonization "have impacted indigenous peoples through a global project of domination over ways of thinking, knowing, valuing, feeling, doing, being, and becoming" (in Reyes, 2022, p. 1238). Such evil intentions "as violent projects of indigenous dispossession, exploitation, miseducation, separation, eradication, and erasure" (Reyes, 2022, p. 1238) should be well examined "in order to cultivate radical sites of praxis toward *decoloniality*" (Ibid., emphasis added) in education in general and teacher education in particular.

During the last two decades we have tried to reflect on what subprograms can be more relevant to begin to transform the contents we have so far been *entirely trapped in spontaneous, short-lived, and unresponsive measures*. We were able to learn that the issue of content in teacher preparation programs should strictly suit to the national aims and strategies on the one hand and the types, levels, extent, timeliness, and number of subprograms and contents on the other. Making professional preparation of teachers rigorous and well-knit may take time but we need to be firmly committed to begin NOW to develop expertise and other resources to ensure well-primed teachers. The challenge is therefore forked: rethinking the role of colleges in national development with close ties to our country's hostilities by the west, and rethinking the place of teacher education in pertinent colleges.

The diverse inquiries into the programs we thus far have been making have allowed us to learn from practice that we need to reconsider about what subprograms will be necessary for preparing teachers in KGs, primary schools, secondary school, etc. Figure 3 provides the basic categories of subprograms that we learned from practice during the last two decades or so. The Figure illustrates about five categories of subprogram frames that are essential for well-developed programs for preparing teachers. We may require to adapt to other important influential factors like context, school type, and school level of the programs but these subprograms are fundamental for preparing teachers for all school levels and types. As mentioned in section 4.2, subprograms mean the main frames or divisions of a teacher preparation program that contain further divisions within them. Our findings reveal, as also shown in Figure 3, that preparing teachers must primarily be grounded on school programs and school pedagogy that are openly and tacitly marinated with culture and character in all study areas of teacher education.

***Figure 3** Basic Categories of Learning-teaching Subprograms for Preparing Teachers*

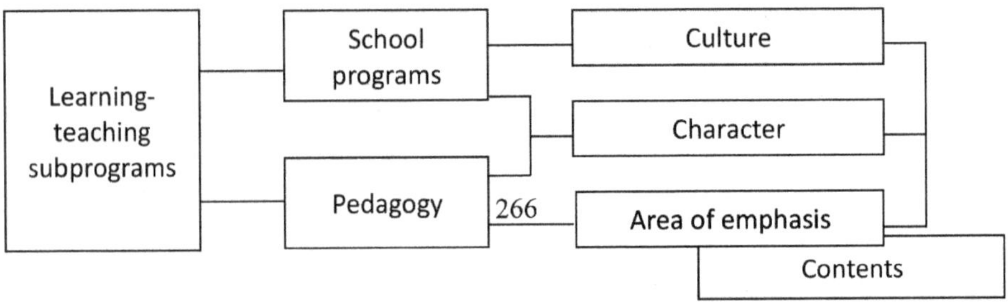

A key issue to note is that subprograms should not be confused with the organization structure of teacher education institutes or colleges. Institutes or colleges may choose to develop diverse organizational structures for their smooth functionings as long as relevant subprograms are well positioned, coherent, and integrated in their composition.

Inclusion of School Programs

Actual practices indicate that institutes meant to prepare teachers are often 'distant' from school programs and situations though much is said about the importance of such relationships in theory. The consequences we have observed is that schools look at the institutes with displeasure that leads to reluctance to cooperate even during the intermittent and negligible initiatives the institutes take to visit schools or demand for some sort of collaboration. Year after year, we have heard both schools and novice teachers assigned to teach in schools that the years spent in our institutes are not that worthy as compared to the experiences in schools, especially in the degree programs. The programs in preparing teachers in our colleges should thus earnestly consider moving towards the praxis of decoloniality implying that we recognize the essential merits of:

- cooperating closely with school communities for sustained sharing of mutual benefits by avoiding distance from schools
- embracing and integrating school practices into college programs.
- We thus need to clearly identify at least two main domains of professional capabilities:
- quality learning: to engage school students in quality learning
- quality teaching: to ensure teachers demonstrate high standards of behavior and commitment.

One essential virtue that can convince schools to acknowledge and collaborate seriously with colleges of education is when the colleges make school programs a fundamental part of their teacher education programs, subprograms, and courses. This can be a win-win state of professional fulfilment: institutes will work with schools, schools will work with institutes, and learner teachers will work for both. Essential virtues that derive from the necessity of making school programs very important sources of teacher preparation include:

- schools tend to seek to closely work with the institutes in critical issues of pre-preparation phases like selecting candidates for the institutes
- schools will develop institutionalized relationships with the institutes for improving programs in the institutes and schools as well
- schools tend to continue to work with the institutes in in-service programs both in terms of creating the situation and in directly participating as facilitators
- schools will develop trust in the institutes and develop sense of ownership of college programs
- schools will take part in assessing, planning, implementing, monitoring, and researching initiatives in diverse educational and related programs
- schools will provide their facilities free or at a very negligible cost to the institutes.

Moreover the inclusion of contents directly related to school programs allow learner teachers:

- to own contents related to develop adequate conceptual and practical meanings as related to the knowledge of working in schools and classrooms
- to appreciate contents related to developing adequate practical and conceptual values of skills to teach learners in schools and enjoy the vocation early on
- to merit and assess attitudes and viewpoints of teachers on the ground and learn from them
- to learn resourceful applications of what it means to teach in contexts that are often flexible
- to assess the relevance and quality of the lessons and experiences in their institutes by relating it to school contexts and communities.

Though selectivity of necessary experiences and practices is essential, comprehensive inclusion of school programs and their crucial activities should be the basis for picking objectives, contents, and processes of teacher preparation programs. Subprograms that relate and apply intimately to schools like educational needs of school leaners, school development and management issues, involvements of communities, school curricular, co-curricular, and extra-curricular issues, learners' assessment into the programs can make teacher preparation highly relevant. For example, Brunei's (a 'small' country in South East Asia) teachers' standards framework are described by two main domains of learning and teaching competencies i.e. (i) for keeping learners absorbed in learning, and (ii) for keeping teachers in practical professional engagements in teaching (Harun et al., 2023, p. 35). The Brunei Ministry of Education (2019) suggested the inclusion of school related programs for developing teacher in the following standards and competencies within two domains:

Domain 1 Develop professional knowledge and skills

- Standard 1 Ability to engage students in intellectually challenging learning
- Standard 2 Ability to base learning on well-designed content and methods
- Standard 3 Ability to suit work to student's individual needs
- Standard 4 Ability to use assessment and reporting effectively

Domain 2 Professional engagement

- Standard 5 Ability to adopt and refine professional practice
- Standard 6 Ability to demonstrate high standards of personal and professional conduct (in Harun, 2023, pp. 35-36).

The initiative to make subprograms closer to school realities, like Brunei's case, is challenging in the sense that it requires efforts of the institutes to sustain professional relationships with the school system and schools. Experience has demonstrated that it is relatively easy for institutes to stay remote but its consequences to quality teacher preparation are damaging ethically, educationally, socially, economically, and politically. Institutes should learn to *practically* accept this fact to make their institutional aims fit to that of the schools. Such relationships can actually deepen, to the extent of jointly developing relevant theories of education for enriching college programs and subprograms.

Inclusion of Pedagogic Contents

Pedagogy is still a dynamically progressing aspect of educational studies and practice. Honestly, it may be the often-uttered term in educational institutions and schools but unfortunately the least practiced not only by most individuals in these institutions but also among the institutions. Such unawareness often leads both institutions and schools to enter into practices that emphasize subject matter contents rather than *how* to understand and use subject matters in life and vocations like teaching. One of the prime aims of teacher preparations programs must be to emphasize on the indispensability of pedagogy for effective teaching in multiple ways but mainly in understanding and using:

- pedagogy as a science and an art: pedagogy as having wide and deep content in itself and requiring the inclusion of adequate and relevant contents to expose learner teachers to its indispensability
- pedagogy of disciplines: pedagogy as relative and specific to the process of teaching and learning of discrete or inter-disciplinary areas of knowledge with emphasis on school subject matters
- pedagogy of self-teaching: often with the same meaning as self-learning that is essential for empowering both teachers and learners to understand the concepts and practices of self-reliance and self-regulation through reflective practices and inquiries in learning and teaching.

The principle of learner center interactive pedagogy (LCIP) is vital to the principles of successful education in schools with due concern to making learners conscious about their environment. About two decades ago schools were required to implement the principle of LCIP in the country. LCIP "emphasizes skills, such as enquiring, analyzing, evaluating, problem-solving and critical thinking which require learners to interact and participate in classroom discourses by drawing on their own experiences" (Abraham et al., 2023, p. 1). In practice however, it has proved to be 'difficult' to generally implement mainly because teachers were not well-prepared, and teachers in duty have not been well acquainted to its principles and practices. In LCIP, majority of learners should be the target for acquiring learning beyond minimum level of achievements. The relationships that are constructed in LCIP should be planned, implemented, and assessed by teachers for further improvement by encouraging learners to become proactive owners of the learning process.

Inclusion of Cultural Contents

The broad and positive meaning of education lies in its role to generate and propagate material, social, and ideological cultures to and from its pertinent citizens. As education shares culture, it in turn develops culture that serves to consistently improve the intentions, contents, and processes of conveyance thus crafting its own traditions. Over its lifespan, schooling has been able to construct organizational varieties that define its type depending on the services it provides to its respective communities. Most have been called general and many provide specific intentions. Speaking about the general categories of schools, their service should be more of propagating broad sets of culture that fits their main intent of being 'general'. This implies that they should provide not only scientific categories of disciplines but also cultural disciplines, packages, or organizational settings that widen and enrich learner teachers' vision, thinking, taste, behavior, ability, and resourcefulness to engage in professional process in real school practice.

The prevalence and inclusion of contents related to culture is not debatable at least in principle. In practice, however, we find it 'demanding'. Incorporating culture in teacher education programs and contents cultivates soft talents like developing ideology of self-reliance and identity of confidence in one's overall disposition. Cultural contents should be integrated to include values and ways of life to enrich vision, taste, behavior, and abilities that are Eritrean in multiple ways, formal-nonformal or compulsory-optional as contained in disciplines, activities, and college ambiances. A good opportunity for teachers to be exposed to the broad meaning of culture and its implication to identity is during their preparation. Some relevant and representative materials (e.g., pottery), social (e.g., tradition), and ideological (e.g., self-reliance) components can be included to expose and allow learner teachers to relevant cultural contents. Such values, beliefs, customs, civility, civilizations, and ways of life can be integrated in pertinent disciplines, subject areas, or institutional subprograms.

Depending on the school system requirements, teacher preparation programs should define the value and amount learner teachers should take including optional cultural domains like material art (e.g., sculpture) and non-material art (e.g., drama), visual art, performance art, and music. The focus of providing such cultural domains in teacher preparation programs may have branched agendas for learner teachers in priming them to develop knowledge, skills, and attitudes for themselves and at the same time for developing such experiences to enrich teaching, and to encourage and organize cultural engagements in their respective schools later. Teacher preparation institutes that encourage learner teachers to engage in cultural experiences tend to also cultivate several hidden talents of their learners thus becoming fortunes not only for the institutes but mainly for school learners.

Inclusion of Contents in Character

Successful teaching as a vocation mainly depends on decent and professional relationships of teachers primarily with their learners and also with colleagues, parents, and other partners. The development of decent professional characteristics that define teachers' integrity and identity should begin as early as the first semester in the institutes mainly portrayed by three architypes of decent character:

- demonstrated by colleges' thorough organizational set up
- exemplified by the leadership, educators, and staff in the colleges
- implied and educated in all study areas that emphasize character.

The organizational philosophy and principles of institutions that prepare teachers are portrayed mainly in both tacit and open images and contents that learner teacher observe and acquire to pursue the teaching profession. Developing and continuously enriching these codes of character and culture as essential subprograms are mandatory educational decisions to introduce and sustain learner teachers' identity and character both during their stay at the institutes and during their duty phases. The tacit educational values are demonstrated in the environmental integrities of the overall programs of the institutes to treat and respect the learner teachers and the communities around them. These are often strong elements that can influence learner teachers to understand what it takes to become professional teachers. The language used in its multiple formats, the traditions and routines practiced, and the experiences revealed to bring efforts together in utilizing scarce resources such as time inherently transmit the sophistication of the institutes in modelling essential qualities among the learner teachers.

Disclosed or latent characteristics of educators at individual, group (e.g., departments), or collective levels are important sources of character development both for valuing the staff as professionals and individuals, and for making efforts to seep through, cultivate, and multiply exemplary characters for themselves and for their learners, colleagues, and school partners. The frequency of meeting and working with leaders and educators is often high and the level of appreciating to work with them highly depends on the atmosphere created by the leaders and educators to allow learner teachers to thrive in learning to become effective teachers. Educators must thus build solidarity with learner teachers as well as with school teachers through prioritizing significant mental and emotional energy (Razfar & Troiano, 2022, p. 155).

The role of contents as educated and implied in all disciplines in general and in pertinent disciplines that learner teachers wish to emphasize for their professional career have immense influence in developing coherent character among learner teachers. Physical education educators who do not model and emphasize the need for fit body for becoming fit teachers is practically an inconsistent educational process. Concurrently, the pedagogy used to empower the active involvement and absorption of disciplines is often as important as the contents in developing the dispositions necessary for learner teachers to use similar processes for articulating and sharing related and other contents later in their pertinent schools. Fairly, therefore, educators should understand that content should be for influencing positive behavior and that its worth should be implied in the *contents* accompanied by consistent process that charges learner teachers with responsible characters because "teachers teach not just subjects, but people" (Kim, p. 1).

Inclusion of Areas of Study

Inclusion of areas of study or emphasis highly depends on the possible future roles of learner teachers in schools. Study areas should be viewed as essential educational division of work in schools and demand serious endeavors to make them qualitatively relevant by revolutionizing:

- respect for the groups of population served
- value for the types and levels of schools
- transformation of the role of educators and leadership
- esteem for the role of time in learning to teach.

A country may choose to allow prospective teachers to be prepared for about six years after an accredited high school or equivalent levels. It may accordingly also choose to entitle recruits in the programs as 'learner teachers' right from the beginning of their entry into the professional process. Six years may not be a mandatory duration because duration depends on the professional integrity of institutions and educators to take the aims of quality education to excellence through mobilizing committed efforts of learner teachers and partners. In such an environment the selection and upkeep of learner teachers to choose areas of emphasis or 'subjects' as a career can be very important for them in schools. It is then very crucial that institutes that prepare teachers consider relevant subject areas for teachers. Obviously, this will depend on the focus of the institutes to prepare teachers for different types and levels of schools, and it also depends on the extent of the general and groups of population to be served. Institutes may also use streaming of areas of study depending on the educational demands of school types and levels.

Conclusion: Implications and Relationship of Subprograms

Regardless of the diversity of structure that institutes choose, presence and accommodation of the five categories of subprogram packages we have discussed are fundamental in *how to manage contents* for preparing teachers for effective professional pursuits. Colleges of education should visualize the indispensability of content management within and among the five categories of the framework we discussed. Viewing content management as a challenging section of educating teachers can greatly help in selecting, positioning, and relating contents. Two issues should be noted among the relationships of subprograms whatever position they hold in the overall program. The first is the need for a careful study in placing a certain course content of a subprogram in an organization structure of an institute. This should lead us to the second issue of the *intra and interrelationship* of subprogram contents that there may not be one and only one position for a content of a subprogram in the structure because it depends on the aim and scope of the contents as related to the organizational structure. One of the concerns should be in how to avoid unnecessary redundancy and incompatible positioning of contents in the aggregate program of an institution.

The issues discussed are also the basic essentials of teacher preparation. Institutes may choose to widen and deepen their contents depending on the services they provide to their respective education communities. Quality, relevance, balance, focus, and flexibility are also important organizational elements in positioning contents. Apparently, availability of resources in general and human resources in particular influence the positioning of contents. Our main principle along the process of such an urgent responsibility in educating teachers should be to prioritize the interest of learners in schools and ensure the coherence of content to learn (formal or non-formal curriculum) with content to teach (see Figure 1, school programs and practice) with adequate regard to the influence of hidden meanings of organizational set ups and the positioning of contents in content management.

The challenges in prioritizing the interest of learners in schools in teacher education programs imply that we give precedence to conduct penetrating research in the following issues of content management:

- How truly do we appreciate the value and meaning of teaching?
- What are the consequences of detachment of colleges of education from schools?
- How deep have the damages inflicted by coloniality been?
- What are the critical professional abilities for teachers to engage school students in active quality learning?
- What are the vital abilities for teachers to develop professional engagement and integrity?
- What should be the role of educators in selecting relevant teacher education contents and programs?

References

Abraham Tadesse, Lehesvuori, Sami, Posti-Ahokas, Hanna, Moate, Josephine, (2023), The learner-centered interactive pedagogy classroom: Its implications for dialogic interaction in Eritrean secondary schools, *Thinking Skills and Creativity, Elsevier,* 50 (2023) 101379, pp. 1-12.
https://doi.org/10.1016/j.tsc.2023.101379

Cochran-Smith, Marilyn. (April 2000). "*The outcomes question in teacher education, AERA Vice Presidential Address for Division K: Teaching and Teacher Education.*" AERA Annual Meeting. http://www2.bc.edu/~cochrans/mcsvpaddress.html

Harun, Zurina, Chong, Maureen Siew Fang, Hana, Siti, Muiz, Tajuddin, Abdul Shamsiah Zuraini Kanchanawati, Kifle, Hazri and Azmi, Siti Suhailla. (2023). Raising the profile of teacher professionalism: approaches and strategies in Brunei Darussalam. In Wang, Yan, Halász, Gábor, Guberman, Ainat, Baghdady, Ahmed, and Mcdossi, Oded, (Eds), *Research, Policymaking, and Innovation: Teacher and Education Development in Belt and Road Countries* (pp. 23-40). Singapore: Springer.
https://doi.org/10.1007/978-981-19-4349-2

Kim, Jiwon. (2025), Introduction to Interdisciplinary Education. In J. Jiwon and S. Zak, (Eds.), *Teaching and Teacher Education: Developing Teacher Competencies for Interdisciplinary Instruction (*pp. 1-9). Lanham: Rowman & Littlefield.
https://lccn.loc.gov/2024038055

Kosová, Beáta and Kovalˇcíková, Iveta. (2023). Striving for policy coherence and implementational consistency teacher education in Slovakia. In Wang, Yan, Halász, Gábor, Guberman, Ainat, Baghdady, Ahmed, and Mcdossi, Oded, (Eds), *Research, Policymaking, and Innovation: Teacher and Education Development in Belt and Road Countries* (pp. 63-78). Singapore: Springer.
https://doi.org/10.1007/978-981-19-4349-2

Lampert, Jo, Kaukko, Mervi, Wilkinson, Jane and Garcia-Carrion, Rocio (2024). *Disadvantage*. London: Bloomsbury Academic, Bloomsbury Publishing Plc.

Menter, Ian. (2023). Preface. In Menter, Ian, (Ed), *The Palgrave Handbook of Teacher Education Research,* (pp. v-vii). Gewerbestrasse: Palgrave MaCmillan.
https://doi.org/10.1007/978-3-031-16193-3

Musset, P. (2010), "Initial Teacher Education and Continuing Training Policies in a Comparative Perspective: Current Practices in OECD Countries and a Literature Review on Potential Effects." OECD Education Working Papers, No. 48, OECD Publishing. doi:10.1787/5kmbphh7s47h-en

Razfar, Aria and Troiano, Beverly. (2022). *Action Research in STEM and English Language Learning: An Integrated Approach for Developing Teacher Researchers* New York: Routledge, Taylor & Francis Group. doi:10.4324/9781351001168

Reyes, G. T. (2022). Pedagogy of and towards decoloniality. In Michael A. Peters (Ed.),*Encyclopedia of Teacher Education* (pp. 1238-1244). Singapore: Springer.
https://doi.org/10.1007/978-981-16-8679-5

Russell, Tom. (2024). Constructing a personal pedagogy of teacher education. In B. M. Butler, and M. S. Bullock, (Eds), *Understanding a pedagogy of teacher education: Contexts for teaching and learning about your educational practice* (pp. 11-29). New York: Routledge Taylor & Francis Group. DOI: 10.43

Students' Learning Motivation and Engagement in Secondary Schools: A Case in Maekel Region

Bersabeh Daniel[1] and Rodas Tekle[1]

[1] Ministry of Education

Abstract

Interest and active engagement in learning among students is critical for quality learning. This paper examined the dispositions of students' motivation and engagement in classroom-based activities in 10 secondary schools at Maekel Region in the academic year 2019/2020. A descriptive survey with a mixed approach that combined both qualitative and quantitative techniques was employed. Data was collected through focus group discussions and multi-scaled questionnaire. The study involved 244 secondary school teachers and 69 student-teachers of Asmara College of Education. The results showed that students' motivation to learn and engage as indicated by their class participation and diligence to accomplish academic activities fell below average. The factors for identifying and characterizing students' drive and active participation were based on teachers' observation and teachers' perspectives. Lack of role models, students' misperception about the worth of education, and the influence of the school curriculum, and practices to implement the curriculum were among the main factors behind the inconsistencies. The study is expected to serve as an empirical essay to guide research-based intervention programs aimed at responding to contemporary students' motivation and engagement in active learning. It can also be a showcase to imply the need for enriching teachers, research practitioners, and policy makers' responsibilities in the Eritrean school system.

Keywords: academic engagement, learning motivation, secondary school, Eritrea

Introduction

Formal education aims at developing the potential and capacity of learners in multi-dimensional ways. Schools being the means, play vital role in preparing the young generation for daily life and workplace by providing and cultivating the necessary knowledge, skills, and attitudes. They are expected "to cultivate confident and compassionate students who become successful learners, contribute to their communities, and serve society as ethical citizens" (Bialik et al., 2015, p. 1). Learners need to develop humane attitudes or character, academic competence and culture that their respective society values for meaningful interactions, purposeful living, and acceptable involvement.

Students' keen interest and engagement is critical in the learning-teaching process. Students' learning motivation promotes earnest and meaningful task engagement in classroom learning. According to Schlechty (as cited in Dary et al., 2016), highly engaged students show outstanding task involvement, perseverance in the face of challenges and triumph in achieving their goal. If such impetus is missing in school-based learning, quality education can hardly be ensured. Hence, the kind of learning motivation and quality of engagement of students in schools determine the success or failure of a school system.

Students commit tremendous amount of time to schooling. The significance of such investment is questionable if it lacks keen motivation and profound engagement in school task. We observe that quite a number of secondary school students seem to appear in class just for the sake of attendance. Doing homework, individual assignments and participating in classroom discussions are becoming the tasks of the very few. For the sake of marks, and the hope of avoiding failure in a grade, not few would copy from those who completed their homework and assignments. Unconcerned about their academic tasks, many students misuse technological gadgets in and out of classrooms. Besides, schools are concerned by inappropriate behavior such as missing classes on purpose, and disrespecting teachers as well as school authorities. Many parents incessantly complain that their children are not obedient to them, do not study nor check their exercise books and have less interest in learning. These observed situations are results of disengagement and therefore demand curious minds and adequate attention for deeply probing into the matter.

All in all, students' disregard for education has often become our daily experience as teachers, and we had been wondering what reasons lie behind the occurrences. It is not professionally appreciated to simply observe as it unfolds and silently ignore the concern as it ruins the future of the young learners and upcoming generations of our society. The invaluable sacrifices that our country paid for earning its independence would be hard to sustain if the current demotivation and disengagement of students remain rampant in schools.

Henceforth, the findings of this study are presumed to bridge the literature and current contextual knowledge gap in the area and serve as an empirical document to guide research-based intervention programs in response to the contemporary students' learning motivation and engagement in the Eritrean school system. Furthermore, they can give relevant insights to parents, teachers, school leaders, counselors, research practitioners, policy makers, and other stakeholders to understand how important their influence is for students' learning motivation and engagement. The main purpose of the study is thus to find out the current state of students' learning motivation and engagement in 10 Eritrean secondary schools located in the Maekel region and to empirically understand the factors influencing it.

Research Questions

What is the current state of students' learning motivation and engagement in the secondary schools? What factors influence the secondary school students' learning motivation and engagement?

Literature Review

Student engagement is an essential part of good student character and is described as "...the degree of attention, curiosity, interest, optimism, and passion that students show when they are learning or being taught, which extends to the level of motivation they have to learn and progress in their education" (Hidden Curriculum, as cited in Dary et al., 2016). The process of learning barely gets successful without the appropriate engagement of learners. A classroom where its learners are engaged is lively, teachers and students get motivated to learn tirelessly, and show promising results in acquiring the knowledge and skills provided. Bialik et al. (2015) found an interesting link between curiosity and students' engagement with class material. Dary et al. (2016) also articulated student learning engagement and the characteristics of engaged students as follows:

> *...student engagement occurs when young people have invested themselves, their energy, and their commitment to the learning environment, both within and outside the classroom. They willingly put forth the required effort to find a level of personal success academically, socially, and emotionally...care about others' successes as well, including both their peers and the adults around them...contribute meaningfully to the school and classroom climate [and] internally understand that their presence matters (p. 5)*

Scholars have tried to pinpoint the concepts of motivation in relation to students' learning. Bayraktar (2015) defined the concept of motivation as, "the power that gives energy and direction to the behavior of individuals thereby enabling them to move towards a specific goal" (p. 1080). Similarly, Yilmaz et al. (2017) noted that motivation is indispensable for drawing students in academic activities.

Yet, it is important to first be able to identify the characteristics of a motivated and engaged student. Alina Ioana and Alexandru Filip (2022) identified, "a student motivated to learn is *curious*, *involved in school activities*, and *persistent*" (p. 66). Engagement can be identified by recognizing the different types mentioned by Fredricks et al. (2004), the first being behavioral engagement, refers to, "a student's participation in academic, social, and extracurricular pursuits" (p. 60). The second and third are emotional and cognitive engagements which simultaneously indicate how the students respond and think towards and about the school as well as education in general (Fredricks et al., 2004).

Any teacher can testify that students' learning motivation and engagement are very important aspects of a learning-teaching environment without which quality learning and thereby quality education could not happen. Li and Xue (2023) recognize that, "student engagement promotes learning quality and performance" (p. 3). Another finding by Carini et al. (2006) also proved that expedient results of learning, including critical thinking, have been found to be positively associated with student engagement. Students' motivation and engagement apart from being the necessities of academic characters are corner stones for learning and assimilating knowledge supporting the ultimate goal of the teaching and learning process. Such concepts have driven many curious minds to the arena raising endless questions as why some students are engaged while others are not, what factors affect students' motivation, and if the factors apply only inside the school or are there other factors. Hardre´ et al. (2006) highlighted the complexity that "student outcomes are not the result of simple cause–effect relationships, but of systemic interactions of factors that include the characteristics that teachers and students bring to the instructional context, as well as their institutional and cultural contexts" (p. 189).

Potential factors for Motivation and Engagement in Learning

Several factors that happen within the school have the potential of motivating students to basically appear in school and to stay engaged in the academic activities. Therefore, considering that the school is a place where educational service is provided, it is important to enquire the quality of its service. Yilmaz et al. (2017) explained how researches have underscored that the quality of service in education is central to students' learning motivation. Annamdevula and Bellamkonda (2016) noted an existing relationship among, "perceived service quality, student satisfaction, loyalty and motivation" (p. 19). Teachers and school leaders tend to point their fingers towards students, their parents and other external conditions, attributing the lack of motivation and engagement without even looking at what is happening inside the school and the quality of educational service provided. Several researches have therefore verified that teachers, peers, the physical environment of the school, the quality of learning materials, resources and activities that exist within the physical environment are some of the most important in-school factors affecting students' learning motivation and engagement as reviewed in the literature.

Teachers have significant power in controlling and directing the learning-teaching atmosphere especially the students' learning motivation and engagement when their power is used appropriately. Williams and Williams considered the teacher as one of the key ingredients impacting students' motivation (2011). Simply stating that the teacher is responsible might be too general and thus, it is necessary to clarify which teachers' aspects act as factors related to increasing or decreasing students' learning motivation and engagement. Tanveer et al. (2012) mentioned that teacher's behaviour is a factor that influences student engagement and motivation. Li and Xue (2023) categorized the factors associated with teachers that influence student engagement into promoting factors like, "positive teacher behaviour, the teacher-student relationship and partnership, as well as teaching factors" and hindering factors like, "negative teacher behaviour" (p. 4). As an example of the negative teacher behavior, Assor et al. (2005) showed that, "directly controlling teacher behaviors (DCTB), such as giving frequent directives, interfering with [learner]'s preferred pace of learning, and not allowing critical and independent opinions appear particularly harmful because they lead to a-motivation that is intertwined with anger and anxiety" (p. 397). Tanveer et al. (2012) noted that teachers' teaching style influences learning motivation of students.

Some of the teacher factors considered germane to students' engagement are teacher support (Ali & Hassan, 2018). Yilmaz et al. (2017) similarly proved that student motivation is affected by teacher's positive feedback which is one way to support students. Teacher interaction and their relationship with the students influence students' engagement. Hardre´ et al. (2006) revealed that most students consider their individual interaction with teachers as highly valuable and those, "students who regard teachers' one-to-one interpersonal style as being supportive of their choices and learning, were more engaged and worked harder" (p. 202). Quin (2016) found an association between better quality teacher-student relationships and several indicators of student engagement such as, "higher levels of psychological engagement, academic achievement, school attendance, reduced levels of disruptive behaviors, suspension, and dropout" (p. 29).

Hardre´ (2001, as cited in Hardre´, 2006) supported the notion about teachers' instructional decisions that they, "can promote or reduce students' motivation for learning and achievement, in terms of … teachers' design of the classroom learning environment" (p. 191). Another aspect of the instructional decisions to be made by the teacher is the teaching method or process. Teaching method or process, as defined by Williams and Williams (2011), "is the way in which content is presented, that is, the approach used for instruction" (p. 11). Bayraktar (2015) and Yilmaz et al. (2017) supported that the method a teacher uses and selects affects how well students will be motivated. Another considerable factor influencing students' learning motivation and engagement which needs a critical decision by the teacher is learning material or resource. Hofferber et al. (2016) found that the intrinsic motivation that happened as a result of autonomy-supportive teaching behavior still depended on whether the teaching materials were interesting. Furthermore, Laing (2011) showed that providing students with balanced activities in terms of its level of difficulty which are easy and hard improves students' motivation.

Although teachers hold on the larger portion of the factors that influence students' learning motivation and engagement, several other physical and social aspects of the school environment have significant roles. A critical social factor within the school environment although not restricted to the inside is the influence peers have. Ali and Hassan (2018) identified that peer factor is related to good engagement of students and noted, "acceptance and rejection of peers influences the engagement of students as peers will encourage the development of student social relationships, build a sense of comfort, and inspire them to be active in school" (p. 2167). Hardre´ et al. (2006) showed how the perception of peer support affected student engagement, "students who regard peers as being supportive … were more engaged and worked harder" (p. 202).

Considering the physical aspect, school factors are mainly related to the school environment and school contexts. Ali and Hassan (2018) noted:

> Context of the school environment is a critical factor that contributes to student engagement. Context refers to the occurrence of an event and the context of the environment of each school is different. The school environment has had a strong influence on student participation and creating a "sense of belonging" in the students towards the school (p. 2165)

In spite of the fact that the service the school provides is an indispensable factor that leads to either heightened or reduced learning engagement and motivation of students, that is not necessarily always a secluded factor. Therefore, besides all the in-school factors mentioned above, there are also diverse out of school factors that are not that easy to pinpoint or even control. Researchers have studied the role of environment, parents and community, as well as the students themselves in determining the students' learning motivation and engagement. According to Ali and Hassan (2018), family factor is one main factor that affects students' engagement in the process of learning. Furthermore, Li and Xue (2023) put it that some of the key factors which affect students' motivation and engagement are "students' learning and thinking abilities, behavior, and individual personality characteristics" (p. 4). Similarly, according to Laing (2011), *self-efficacy* fosters "willingness to take on challenging tasks, increased effort, persistence despite obstacles, less anxiety, flexibility in learning strategies, and self-monitoring" (p. 10). Students' self-efficacy also affects students' engagement on academic tasks. As Petsche (2009) noted, it influences students' judgement on what activities they put effort and on what activities they become less engaged. The personality characteristic the student possesses is also a great factor for the motivation and engagement. There are students who believe their success, failure and opportunities lie outside themselves; if students don't believe they have control over their situation, they will be less likely to be motivated nor will they take control of their learning process. This clearly applies to *attribution theory*, which refers to the characteristic students possess. The credit they give toward success and failure affects the effort they exert on any activity (Petsche, 2009).

Research Methodology

This study is an excerpt from a senior project paper with the title *Fostering students character development and learning motivation in Eritrean secondary schools*. It was conducted in the year 2020 in the central (*Maekel*) region of the country,'involving a total of 10 secondary schools located in 4 subzones. The study employed descriptive survey research design combining mixed approach of qualitative and quantitative method.

The participants of the study were secondary school teachers of *Maekel* region and two batches of the student-teachers of the post graduate diploma program at Asmara college of education. The student-teachers were selected for the reason they had years of teaching and administrative experiences in secondary schools throughout the country. 283 teachers participated as respondents to the questionnaire, whereas 30 student-teachers of the post graduate program participated in three focus group discussions (FGD).

Convenience sampling method was used for selecting the secondary schools. The central region being the most convenient, it consists of four sub-zones. The secondary schools from the three sub-zones, namely; Berik, Gala-Nefhi and Serejeqa were selected randomly. As in the case of Asmara subzone, seven of the 12 secondary schools were randomly selected for filling the questionnaires. The secondary schools were selected through block stratification and random sampling from each cluster. All the members of the first batch PGD program also participated in filling the questionnaire. The FGD participants from the second batch PGD program were selected using snowball sampling in order to meet the inclusion criteria of teachers with more than five years of experience in teaching.,

Two types of instruments were developed for data collection. These are the guiding questions that are used in the FGD as well as the numbered scale questionnaire. The questionnaire was pilot tested in one secondary school that was not included in the sample. The Cronbach's alpha results of the pilot test and the actual data collection are 0.948 and 0.971 respectively. The rating of the statements used in the analysis were based on Likert's scale.

The data was collected following formal communication between the Asmara College of Education and the MoE of *Zoba Maekel* branch office regarding the legitimacy of the study. The PGD program participants were contacted in the Asmara College of Education and the school management team along with the school teachers were consulted in their workplaces. The FGD were audiorecorded and meticulously transcribed. The questionnaires were distributed at the college and in the secondary schools. The data was analysed qualitatively using thematic analysis and supported by a quantitative data analysed descriptively using SPSS and Microsoft Excel. For the sake of participants' anonymity, the participants are represented by alphabets A-Z in the case of direct quotes of the qualitative data in the discussion section.

Results

The quantitative and qualitative results in connection with the two research questions, the current state of affairs of students' learning motivation and engagement in the secondary schools and the factors affecting students' learning motivation and engagement in the secondary schools are presented below.

Figure 1:*Status of Students' Learning Motivation and Engagement*

The findings show the status of the students' motivation and engagement in their academic tasks. The academic indicators assessed from the perspectives of students' learning engagement and motivation, diligence to accomplish tasks and their classroom participation are below average.

Figure 2: *Factors Influencing Students' Learning Motivation and Engagement*

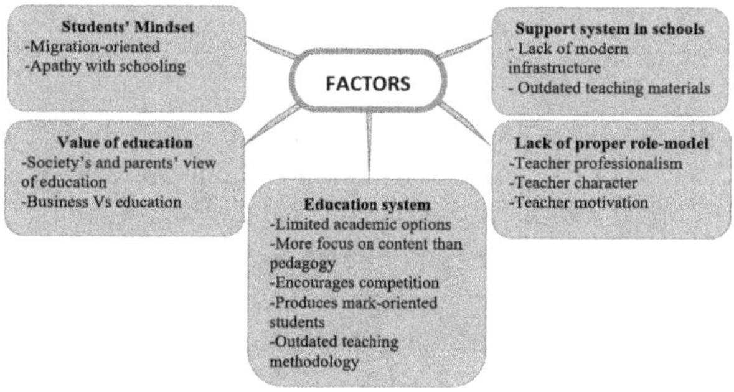

The above thematically organized qualitative findings represent the factors that were impacting the learning motivation and academic engagement of the secondary school students. From the participants' perspective, students' mindset, lack of proper role model in teachers' side, perceived value of education, education system and lack of modern support system in schools in specific are possible factors affecting secondary school students' learning motivation and engagement as shown in the figure.

Discussion

Current Status of Students' Learning Motivation and Engagement in the Secondary Schools

Academically motivated students are eagerly involved in academic tasks. They are emotionally and cognitively part of the learning and teaching process. This internal inspiration cultivates accepted students' behaviors and fosters student characters essential for academic engagement. Participant S stated, "the level of student engagement varies across different variables including gender, type of subjects, location of the schools and students' residential areas." The participants reported that secondary school students' engagement in their learning is not satisfactory. The quantitative data supports the findings in the qualitative data and shows similar pattern of the students' learning motivation and engagement. The learning engagement, motivation, class participation and diligence to accomplish assignments fell below average indicating they are not in a good shape (Figure 2). In the secondary schools, late coming, missing classes, absenteeism without any permission, drop outs etc. are becoming common scenes. As the teachers reported, these minor and major misbehaviors are the reflections of the students' increasing disengagement and diminishing motivation in the learning process. Similarly, the above point is supported in research by Bayraktar (2015):

> students with low motivation [exhibit behaviors such as] ... listening to courses not carefully, directing the attention to their friends or outside events, when facing difficulty in the homework no willingness to search answers, appearing in classes indifferent to the topic, drawing their attention to the course is difficult by the teacher etc. (p. 1081).

Respondent M stated that "not more than 15% to 20% of students are engaged and actively participate in the learning process while others come to class without doing their homework, assignments and fail to actively contribute their input in the school matter."

Factors Influencing Students' Learning Motivation and Engagement in Secondary Schools

Students' Mindset

The teachers are deeply concerned about the current mindset of the students. As the participants have reported, there is observed deteriorating students' interest in schooling. Students perceive a gloomy future through education. Instead, they are having a mindset of going abroad, opening their own business and getting rich through shortcuts. The idea of migration had been refreshed with the 2018 political status of Eritrea with Ethiopia when the borders were opened. The participants reported that it has affected students' motivation and interest in learning.

In effect, students are coming to school for the sake of attendance or due to their parents' pressure. Once they are in class, they become passive and disengaged in any school matter. The students do not seem to have a vision to succeed through academic endeavors. The current study also found out that the teachers are highly concerned about the type of character the students are developing in the schools. In their view, the important virtues like hard work and dedication to learn among the student community are deteriorating from time to time. Students are becoming negligent and just having a prey eye for shortcuts. Respondent C has expressed the above points by saying:

> If you are to participate and engage in [the learning and teaching process] first you have to have a vision, then interest and finally you have to work hard; but I am afraid our students are either missing one or all from the list (FGD).

Participants expressed that there is prevailing idea in the secondary school students that financial stability could not be achieved through education. As a result, the students do not value education and look for ways to earn more money and make a living. The human capital theory of educational development by Harber (2014) noted that the rate of return depends on the years individuals spend in education, that is, "the more educated earn more than the less educated" (p. 54). The controversial side of the methodological problem in calculating the rate of returns and the increasing unemployment rates of the educated may explain such students' mindset, but the idea of measuring education only in terms of economy and devaluing its other importance is a point to be critically taken.

Students who come from families of low socio-economic status, give less value to education and prioritize any type of work which would contribute to their immediate survival. Students are less likely to be engaged in doing homework and assignments when they have part-time jobs outside class and are even less likely to be actively engaged in class when they are physically and mentally tired from their jobs.

From the many factors, participants mentioned student factors as influencing factors for students' learning motivation and engagement. The participants mentioned the goals students have, their self-efficacy and self-perception influence students' level of engagement and motivation. Similarly, the goal achievement theory articulated students' goals are equally important as their self-determination and added, "both sets of characteristics help to explain the reasons why students engage (or fail to engage) in school-related tasks" (Hardre´ et al., 2006, p. 19). Participants also mentioned that students perceive that they lack certain skills and cognitive ability, as a result it affects their self-esteem and motivation.

Lack of Role Model

When students actively interact in school, home or community; there is a tendency to influence their thoughts, feelings, actions and behaviors. The school community specially teachers hold the upper hand to influence the students. Therefore, teachers are required to show teacher professionalism as well as humane characters as role models for students. Participants stressed that teachers should be exemplary in the way they talk, teach, interact and dress. They also stated that for students to be influenced positively, teachers should teach devotedly and act responsibly. But participants reflected their concern about the professional commitment of the contemporary teachers as compared to the past.

Moreover, the result indicates that lately the ineffective teacher training and the deployment of graduates from non-teaching fields to teach in schools is adding up to the challenges. Graduates from teaching as well as non-teaching fields of specializations are assigned for the responsibility after graduation to fill gaps created by shortage of teachers in the schools. This is nurturing lack of interest to teach on the teachers' side. On-job training and refresher courses like psychology and pedagogy that considerably help teachers further understand human behavior and learning are missing. The trainings are vital for teachers who graduate from non-teaching fields. Such programs have great importance in the teaching profession to positively influence the skills and attitudes of the graduates who are assigned to serve as teachers without any clue of the teaching profession. The experience would help them develop indispensable skills to guide their students and positively change their own views towards teaching profession. The situation calls the college of teacher education to work diligently in order to positively impact the professional identity of teachers.

In the absence of proper training for the would-be teachers, the teaching profession is becoming "a *must* part time job" as respondent K has noted. The graduates assigned to teach are not giving teaching their whole time and energy. The situation affects how they prepare their lessons and teach in class. All the participants expressed their concern that currently teachers all around the country are getting less motivated to teach. If the teachers lack teacher professionalism as in the above-mentioned aspects, then the students are lacking role models. In the absence of teachers who can motivate and inspire students to learn and be fully engaged, it is easy to assume the nature of the learning motivation and their engagement in schooling. Participant X said, "if we the teachers are not motivated, how can we expect the students to be so?"

The quality of education is dependent on the quality of the teacher who closely works with students and therefore considerable investment in the provision of qualification to teachers should be made. The quality of education can then significantly improve. Since the teacher is a primary figure that directly works with students, the psychological wellbeing of the teacher greatly affects the whole learning and teaching process. When teachers have a positive state of wellbeing, they are more likely to perform their job well and to share positivity to students. By reinforcing teachers in a way that increases their psychological wellbeing, schools can aid students into having hope and vision thereby enhancing their motivation to learn. Respondent H commented, "the government and schools should motivate teachers so that we can have motivated students". This can have a positive effect on the academic engagement.

Misperception of the Value of Education

The worth students, teachers and the society in general give to education affects how much time, resources and interest they pour into it. Participants reiterated that due to the low value the society is giving to education, the students' engagement and their thirst to learn is highly being affected. The participants explained that parents specifically and the society in general are lately measuring the value of education in terms of money and rate of return.

The society's observation of the educated individuals being unable to gain financial stability by earning salary, they conclude that being educated is worthless and unable to get them out of poverty. One participant said "teachers hear some parents saying Mr. X's son has a degree but has done nothing so the family is still poor but Mr. Y's son after he has left the country, he has changed his family's living within two years". Literally, if parents frequently say this to their children, it conveys a message that education can get you nowhere regarding financial freedom encouraging the students to migrate to other countries prioritizing work than education. This type of attitude of the society on education is becoming a contributing factor for students' less engagement and motivation in school.

Although the major focus of education is dynamic and has been changing according to the era it serves, which at some point in time was related to rate of return, the main long-lived aim of education shouldn't change. Our limited parameters on the outcome of education being rank, money, prestige has affected how we perceive the value of education as a society. It has made people measure it as valuable only if it serves the above-mentioned purpose. As Anugerah (2018) noted, "education is a humanitarian issue that should be approached from a human development itself" (p. 43). The parameter on the value of education should extend to how well students are becoming ethical, moral, creative, adaptive and collaborative. Education should attend and enable the learners to face challenges and complexities of our world by helping them develop an attitude of growth, development and learnability.

Lack of Modern Support System and Resources in the School

As humans, anything that is fancy and beautiful to the eye captures our attention and increases our interest in that particular object. This analogy also applies to the connection between students' learning motivation and school environment and type of resources in a school. Capturing school environment, modern classroom equipped with advanced facilities stimulates students' motivation to attend school and makes learning enjoyable. The teachers earnestly commented that the lack of modern infrastructure, less advanced way of teaching, unattractive classroom structure, inconvenient desk arrangement for learnercentered pedagogy play its stake for the decreasing learning motivation and engagement. Similar to this finding, Li and Xue (2023) put the support of learning resources as part of the factors promoting student engagement. What happens inside the school, how the students are treated, how relevant the academics appears to be, how the learning-teaching process is carried on, how enjoyable the in-school experience is and so many other instances have an incredible potential of motivating students to basically appear in school in the first place as well as to stay engaged in the academic activities.

Education System

Respondent B stated that pointing fingers at the students' and blaming their decreasing motivation and engagement is not fair. The problem is also related to not getting what interests them and not being served with what they need. The existing curriculum is inflexible in that it favors certain types of intelligence like arithmetic and verbal intelligence. All students may not have the same interest and intelligence but the existing system in its pedagogy and assessment style may be favorable to students with the above listed intelligence. It seems to marginalize students with musical, artistic, kinesthetic and other inclinations. The education system should provide a ground to explore and utilize variety of skills which students possess. A welcoming and non-judgmental school atmosphere which accepts every student for who they are regardless of their ability, socio economic status, gender and race creates a comfortable aura and an atmosphere of acceptance and belongingness. This type of atmosphere gives students the advantage to be creative, exploratory and imaginative without the fear of making mistakes and failure; facilitating their learning motivation and engagement.

The type of the teaching methodology used in the schools has taken its role in the demotivation of learners and their disengagement from purposeful learning. The 21st century puts a lot of demand on teachers, parents and the schools to cultivate students who can be productive and successful in the ever-changing technological era. The current findings revealed that the demand of modern and practical ways of teaching is not being met. Educational practices and resources in the schools are not adjusted to the needs of the 21st century learners and lack the quality and advancement to cultivate the learners' intellect and character simultaneously as needed. The existing education system overemphasizes on academics and marks more than behavioral change. Moreover, the current teaching methodology encourages rote learning and memorization. Authentic learning which enables students to practically associate their learning in class with their daily life is missing. The phenomenon compels students to end up being silently excluded, less engaged, and increasingly disinterested in learning. Students' relative engagement in individual activities and tasks that contain marks while giving less value to group activities explains the type of engagement in the schools as remarked by the participants. The participants further noted that the type of assessment and grading system is encouraging individualistic and competitive mindset in the students. The practice discourages their internal motivation and interest for lifelong learning.

Conclusion

According to the study, students' learning motivation and engagement as indicated by the diligence to complete school tasks and participation fall below average. This shows that students are showing a decline in learning motivation and engagement which are typical elements of their academic characters. Students' mindset, lack of exemplary role models, misperception of the value of education, lack of students' support system and resources in the school and the current education system are among the major factors that are influencing the students' learning behaviors. The multifaceted factors are adversely impacting the students' learning motivation and engagement. The situation requires an urgent intervention in terms of teacher education, curriculum revision, public awareness campaigns, positive school environment promotion, home-school engagement, raising students' motivation and change of mindset related efforts in the secondary schools.

References

Ali, M. M., & Hassan, N. (2018). Defining Concepts of Student Engagement and Factors Contributing to Their Engagement in Schools. *Creative Education, , 9*, 2161-2170. https://doi.org/10.4236/ce.2018.914157

Alina Ioana, C., & Alexandru Filip, P. (2022). Curriculum Integration and Learning Motivation of the Secondary School Students. *JOURNAL OF EDUCATION, SOCIETY & MULTICULTURALISM*. https://doi.org/https://doi.org/10.2478/jesm-2022-0018

Annamdevula, S., & Bellamkonda, R. S. (2016). Effect of student perceived service quality on student satisfaction, loyalty and motivation in Indian universities: Development of HiEduQual. *Journal of Modelling in Management, 11*(2). https://doi.org/http://dx.doi.org/10.1108/JM2-01-2014-0010

Anugerah, E. (2018). The analysis of character education in Indonesia. *International Journal of Humanities, Art and Social Studies, 3*(4), 39-46.

Assor, A., Kaplan, H., Kanat-Maymon, Y., & Roth, G. (2005). Directly controlling teacher behaviors as predictors of poor motivation and engagement in girls and boys: The role of anger and anxiety. *Learning and Instruction, 15*, 397-413. https://doi.org/10.1016/j.learninstruc.2005.07.008

Bayraktar, V. (2015). STUDENT MOTIVATION IN CLASSROOM MANAGEMENT AND FACTORS THAT AFFECT MOTIVATION. *Turkish Studies, 10*(3), 1079-1100. https://doi.org/http://dx.doi.org/10.7827/

Bialik, M., Bogan, M., Fadel, C., & Horvathova, M. (2015). Character Education for the 21st Century: What Should Students Learn?

Carini, R. M., Kuh, G. D., & Klein, S. P. (2006). STUDENT ENGAGEMENT AND STUDENT LEARNING: Testing the Linkages. *Research in Higher Education, 47*(1). https://doi.org/10.1007/s11162-005-8150-9

Dary, T., Pickeral, T., Shumer, R., & Williams, A. (2016). *Weaving student engagement into the core practices of schools: A National Dropout Prevention Center/Network position paper*. Clemson, SC: National Dropout Prevention Center/Network. Retrieved from www.dropoutprevention.org/resources/major-research-reports/student-engagement/student-engagement-2016-09.pdf

Fredricks, J. A., Blumenfeld, P. C., & Paris, A. H. (2004). School Engagement: Potential of the Concept, State of the Evidence. *Review of Educational Research, 74*(1).

Harber, C. (2014). *Education and International Development: Theory, Practice and Issue*. Symposium Books Ltd.

Hardre´, P. L., Chen, C.-H., Huang, S.-H., Chiang, C.-T., Jen, F.-L., & Warden, L. (2006). Factors Affecting High School Students' Academic Motivation in Taiwan. *Asia Pacific Journal of Education, 26*(2), 189–207. https://doi.org/10.1080/02188790600937326

Hofferber, N., Basten, M., Großmann, N., & Wilde, M. (2016). The effects of autonomy-supportive and controlling teaching behaviour in biology lessons with primary and secondary experiences on students' intrinsic motivation and flow-experience. *International Journal of Science Education*, 1-20. https://doi.org/http://dx.doi.org/10.1080/09500693.2016.1229074

Laing, K. A. (2011). Factors that Influence Student Motivation in the Middle and High School French Language Classroom.

Li, J., & Xue, E. (2023). Dynamic Interaction between Student Learning Behaviour and Learning Environment: Meta-Analysis of Student Engagement and Its Influencing Factors. *Behavioral Sciences, 13*(59), 15. https://doi.org/https://doi.org/10.3390/bs13010059

Petsche, A. (2009). *Factors that influence student motivation* (Publication Number 1326) University of Northern Iowa].

Quin, D. (2016). Longitudinal and Contextual Associations Between Teacher–Student Relationships and Student Engagement: A Systematic Review. *Review of Educational Research, 20*(10), 1-43.

Tanveer, M. A., Shabbir, M. F., Ammar, M., Dolla, S. I., & Aslam, H. D. (2012). Influence of Teacher on Student' Learning Motivation in Management Sciences Studies. *American Journal of Scientific Research*(67), 76-87.

Williams, K. C., & Williams, C. C. (2011). Five key ingredients for improving student motivation. *Research in Higher Education, 12*(1), 1-23.

Yilmaz, E., Şahin, M., & Turgut, M. (2017). Variables Affecting Student Motivation Based on Academic Publications. *Journal of Education and Practice, 8*(12), 112-120.

Law

Environmental Implications of and Liability for Acts of Violence in Maritime Spaces: The case of the *Rubymar*

Munir Omer,[1] Deborah Dawit,[1] Luwam Mogos[1] & Siphora Yohannes[2]

[1]Ministry of Justice, State of Eritrea
[2]University of Law, The United Kingdom
Email: erifalkon@gmail.com

Abstract

On February 18, 2024, the Yemeni belligerent group Houthi launched an attack on the oil tanker *Rubymar*, causing extensive hull damage and a significant oil spillage. This incident has severe environmental and socio-economic implications, threatening marine life, coastal ecosystems, and the livelihoods of local communities across the Red Sea region. Yemen's prime minister, Ahmed Awad Bin Mubarak, described it as "an unprecedented environmental disaster," warning that neighbouring Red Sea states would also face the adverse consequences of the spillage. The incident highlights the urgent need for regional and international cooperation to address environmental harm resulting from conflicts and to safeguard sustainable development.

This paper explores responsibility and liability mechanisms under international law and highlights the importance of collective efforts in addressing marine pollution during armed conflict. It examines state responsibility, strict liability for vessel owners, and obligations under international humanitarian law (IHL). The analysis considers Yemen's potential liability under state responsibility despite limited territorial control and evaluates exceptions in strict liability frameworks for vessel pollution caused by armed attacks. Furthermore, it discusses the accountability of non-state actors, specifically the Houthis. The study emphasizes the need for active cooperation on regional level to address transboundary environmental harm in the Red Sea.

Adopting a doctrinal methodology, complimented by case-based analyses and policy-oriented research, this study highlights how international and regional cooperation can strengthen environmental accountability and support sustainable development. The findings advocate for collaborative legal and policy frameworks, analysing legal mechanisms to mitigate marine pollution and promote long-term ecological sustainability. This study tries to give its fair share of contribution by addressing the intersection of international environmental law, conflict and regional cooperation and finally proposing collective strategies to ensure marine ecosystem resilience and sustainable development.

Keywords: sustainable development, international cooperation, marine pollution, state responsibility, Red Sea

Introduction

In the aftermath of attack on the *Rubymar* tanker, various legal articles in journals and book chapters written about the incident focused on the protection of freedom of navigation and safety of internet cables within the sea, protecting the smooth flow of trade as well as other legal aspects of economic nature.[1] Some articles tried to illustrate the legality of the Houthi's actions according to international law, and some gave analyses as to the legality of United States of America (U.S.), the United Kingdom (U.K.) and other countries' military strikes against the Houthis in Yemen in the aftermath of the incident.[2] Unfortunately, the environmental aspects of the incident from a legal point of view were not given enough emphasis. This is not surprising as environmental concerns have long been sidelined during conflicts. One can refer to the ongoing Israeli aggression in Gaza and the environmental catastrophe it has caused; in just 120 days of 2024, up to 652,000 tons of CO_2 were emitted, exceeding annual emissions of 26 countries, alongside alleged white phosphorus use and rising solid waste.[3] This environmental disaster has hardly made the mainstream news. The authors believe environmental protection concerns resulting from conflicts, whether in maritime or other spaces, must be given due recognition and be legally analysed and questioned. Understandably, if human life and development come first, but sustainability demands attention to the environment too.[4]

In this chapter, the authors analyze the environmental implications of the *Rubymar* incident and try to explore ways to hold perpetrators accountable under international law. The lack of reputable publication regarding the environmental implications of such violence in maritime spaces, especially when a non-state armed group assessing the relevant international conventions in relations to the facts of the *Rubymar* is involved, indicates to the originality of this research paper. While analyzing laws as well as decisions of international tribunals, the authors also discuss the immediate actions needed for mitigating the environmental catastrophe that might result from the *Rubymar* sinking.

The main argument which will be echoed throughout this paper is that existing international legal frameworks for environmental protection are fragmented and insufficient for ensuring liability for environmental harms caused in relation to armed violence, particularly in critical regions like the Red Sea. Addressing these gaps requires enhanced international and regional cooperation to unify efforts, mitigate risks, and establish robust mechanisms for sustainable maritime environmental governance. In a nutshell, this paper tries to address three research questions. What are the potential environmental impacts of the attack on the *Rubymar* ship? Who bears legal responsibility for the actual and potential environmental

1 Knights, Michael. *Assessing the Houthi War Effort Since October 2023*. Washington Institute for Near East Policy, 2024.; Rogers, Peter, and Hansen, Julia. *Economic and Security Ramifications of Maritime Incidents in Conflict Zones*. Journal of Maritime Law 42, no. 2 (2024): 112–127.; International Maritime Organization (IMO). *Freedom of Navigation in the Red Sea Amidst Rising Tensions*. IMO Repository, 2024.

2 Haight, B. W. *US Strikes Against the Houthis: A 'Splendid Little War' of Self-Defence*. American University National Security Law Brief, 2024.; Henderson, Christian. *US and UK Military Strikes in Yemen and the Jus Ad Bellum*. International and Comparative Law Quarterly 73, no. 2 (2024): 215–230.

3 Otu-Larbi, Frederick, Benjamin Neimark, Patrick Bigger, Linsey Cottrell, and Reuben Larbi. *A Multitemporal Snapshot of Greenhouse Gas Emissions from the Israel-Gaza Conflict*. *ResearchGate*, 2024.

4 Hopwood, Bill, Mary Mellor, and Geoff O'Brien. *Sustainable Development: Mapping Different Approaches*. Sustainable Development 13, no. 1 (2005): 38–52.

harm caused by the attack on *Rubymar* ship? How does the incident impact Eritrea's path in achieving sustainable development?

The authors have narrowed the scope of this research to exploring responsibility for environmental harm caused by the *Rubymar* incident. At least from legal point of view, it will not entertain discussions related to economic and other implications of the incident. But such implications will be briefly deliberated upon when addressing sustainable development implications of the incident in part III of this paper.

This paper will be organized in five chapters. Chapter one gives introduction as to the objectives and content of the paper. Chapter two discusses the environmental and economic significance of maritime spaces like the Red Sea as well as environmental implications of the *Rubymar* incident. Chapter three will then assess the facts of the case in the lens of relevant international conventions as well as decisions. Chapter four will then delve into the various sustainable development goals Eritrea is pursuing and how the *Rubymar* incident might affect their attainment. Chapter five will give conclusion and recommendations.

Background of the *Rubymar* incident

On March 2nd 2024, *Rubymar*, reportedly a UK-owned bulk carrier (but later confirmed that it had no current links to the UK), sank in the Bab el-Mandeb Strait, a waterway linking the Red Sea and the Gulf of Aden, after being struck by the Houthi anti-ship ballistic missile on February 18th 2024.[5] The Houthi leader, Mohammed Ali al-Houthi posted on social media site X, that there is still a chance to prevent further damage if UK Prime Minister, at the time, Rishi Sunak and his government act quickly and allow aid into Gaza and stated that they are responsible for the *Rubymar* attack because of their support for the war on Gaza.[6] The *Rubymar* vessel was linked to an address in a residential block of flats in Southampton. This link, in the eyes of the Houthis made the *Rubymar* a "British" vessel, and despite being widely reported as British-owned, the only tie that the *Rubymar* had to the UK was its insurance.[7]

The vessel was reported to have been carrying approximately 22,000 (note that the number varies from one new source to another) metric tons of ammonium phosphate-sulphate fertilizer as well as 200 tons of heavy fuel oil and 80 tons of marine diesel.[8] The ship sank after 14 days of taking water and almost immediately started leaking oil. The former head of the Yemeni Public Authority for Environmental Protection, Abdul Qadir Al-Khraz, emphasized that the 20,000 tonnes of the artificial fertilizer have severe environmental repercussions, should they dissolve and contaminate the water. Once the fertilizer dissolves, it disrupts the balance of the marine ecosystems triggering cascading effects through the food web, and this will affect various species that depend on these ecosystems and, in turn, po-

5 Richard Meade, *Sunken Rubymar's Only UK Link Was Its Disputed Insurance Cover*, Lloyd's List, 2024, Accessed: July 7, 2024 https://www.lloydslist.com/LL1148459/Sunken-Rubymars-only-UK-link-was-its-disputed-insurance-cover.

6 Sarah Shamim, *Could Houthi-Sunk Ship Rubymar Spell Eco-Disaster for the Red Sea?*, Al Jazeera, March 7, 2024, Accessed July 4, 2024 https://www.aljazeera.com/news/2024/3/7/could-houthi-sunk-ship-rubymar-spell-eco-disaster-for-the-red-sea.

7 See *Sunken Rubymar's Only UK Link Was Its Disputed Insurance Cover.*

8 International Maritime Organization (IMO), *MV Rubymar - Call for Contributions of Oil Pollution*, June 25, 2024, Accessed 16 august, 2024 https://wwwcdn.imo.org/localresources/en/MediaCentre/Documents/Circular%20Letter%20No.4890%20-%20Mv%20Rubymar%20-%20Call%20For%20Contributions%20Of%20Oil%20Pollution.pdf.

tentially impacting the very livelihoods of coastal communities.[9] Abdusalam al-Jaabi, of the Yemeni government's Environmental Protection Agency, also warned that this would impact 78,000 fishermen and their families – up to half a million people.[10]

Understandably, this incident was labelled a major maritime disaster with far-reaching implications for water pollution and the marine ecosystem. As per reports on the 5th of March 2024, the *Rubymar* lies approximately 30km West of Yemen's Red Sea Port of Mocha,[11] which means that the *Rubymar* is close in proximity to Eritrea's Assab port. This is a concern for Eritrea as marine pollution spreads easily. A 29km fuel slick was reported, and the cargo poses an environmental risk in the Red Sea.[12] Due to the movement of the winds, it could be assumed that Yemen will not be the only affected area as the fertilizer will spread throughout the Red Sea and cause damage to marine organisms.

In response to the attack on *Rubymar*, the Yemeni government established an emergency response unit to address the vessel when it was stranded and leaking. However, despite the numerous warnings to take timely action, the UN experts, ambassadors from the European Union, the US, Britain and China arrived to assess the situation after the *Rubymar*'s sinking and failed to take actions that yield any tangible results.[13]

The prevalence of attacks on vessels in the Red Sea, especially in the past year, has had an effect on the route that vessels have started to take. Global insurers have doubled their premiums for vessels that take the route through the Red Sea, through the Bas al-Mandab strait and some even refusing to underwrite Red Sea coverage.[14] The increase in attacks over the past year have resulted in vessels taking a lengthy detour around the Cape of Good Hope, which adds around 3,500 nautical miles (4,000 miles/6,500km) and 10-12 days sailing time each trip, which results in an additional $1million worth of extra fuel.[15] Therefore, the use of more fuel per trip means that these ships are going to emit more carbon into the atmosphere, creating more environmental repercussions. For example, a large container ship's journey from Shanghai to Hamburg, emits 38% more CO_2, or 4.32 million kilograms[16] if it goes

9 Greenpeace MENA , *Greenpeace MENA on the Sinking of the Rubymar Ship: Immediate Action Must Be Taken to Avert an Environmental Catastrophe in the Red Sea*, 2024. Accessed 7 September, 2024 https://www.greenpeace.org/mena/en/sinking-of-the-rubymar-ship/.

10 AFP correspondents with Amanda Mouawad. *Yemen Faces 'Environmental Disaster' as Sunken Ship Threatens Red Sea*. Al-Monitor, 2024. Accessed 17 October, 2024 https://www.al-monitor.com/originals/2024/03/yemen-faces-environmental-disaster-sunken-ship-threatens-red-sea.

11 See *Sunken Rubymar's Only UK Link Was Its Disputed Insurance Cover.*

12 Ibid.

13 Rasanah International Institute for Iranian Studies. *The environmental disaster wrought by the sinking of MV Rubymar in the Red Sea*. March 14, 2024. Accessed August 30, 2024.https://rasanah-iiis.org/english/wp-content/uploads/sites/2/2024/03/The-Environmental-Disaster-Wrought-by-the-Sinking-of-MV-Rubymar-in-the-Red-Sea.pdf.

14 Jonathan Saul and Carolyn Cohn. *Red Sea Insurance Costs Soar as Houthi Shipping Threats Loom, Sources Say*. Reuters, September 19, 2024, Accessed 11 October 2024https://www.reuters.com/world/middle-east/red-sea-insurance-costs-soar-houthi-shipping-threats-loom-sources-say-2024-09-19/.

15 Chris Baraniuk. *Red Sea Crisis: What It Takes to Reroute the World's Biggest Cargo Ships*, www.bbc.com, January 21, 2024, Accessed 11 October 2024 https://www.bbc.com/future/article/20240119-red-sea-crisis-how-global-shipping-is-being-rerouted-out-of-danger.

16 Richa Naidu, *How Red Sea Disruptions Are Driving up Carbon Emissions*. The Japan Times, June 2024. Accessed 2 August, 2024. https://www.japantimes.co.jp/environment/2024/06/01/climate-change/red-sea-shipping-co2-emissions/.

around Africa instead of through the Suez Canal.[17] As of mid-September, the average daily transits through the Suez Canal stood at 29 compared with about 80 in October 2023.[18]

Environmental and economic significance of Maritime Spaces

Maritime spaces are indispensable for both environmental health and economic prosperity. Environmentally, these areas are biodiversity hotspots, home to a vast array of marine life, including fish, coral reefs, and marine mammals, which are crucial for maintaining ecological balance.[19] They play a vital role in regulating the Earth's climate by absorbing significant amounts of carbon dioxide and heat, thus mitigating the impacts of climate change.[20] Additionally, maritime spaces are integral to nutrient cycling, including the nitrogen and phosphorus cycles, which are essential for the growth of marine organisms and the overall health of marine ecosystems.[21] They also help in breaking down and dispersing pollutants, although this capacity is increasingly overwhelmed by human activities.[22] The toxic components of oil can persist in the environment for decades, causing chronic harm to marine species and ecosystems.

Economically, maritime spaces are vital for several reasons. They support fisheries and aquaculture, providing a significant source of food and livelihoods for millions of people worldwide.[23] Oceans are also the highways of global trade, with over 80% of international trade by volume carried by sea, making maritime spaces critical for the global economy.[24] Ports and shipping lanes are essential for the transportation of goods and raw materials.[25] Coastal and marine tourism generate significant revenue and provide employment opportunities through activities such as diving, snorkelling, and beach tourism.[26] Furthermore, maritime spaces are rich in energy resources, including oil, natural gas, and renewable energy sources like offshore wind farms, which are crucial for meeting global energy demands.[27] The seabed contains valuable minerals, including rare earth elements essential for modern technologies, making deep-sea mining an emerging industry with significant economic potential, though it poses environmental risks.[28]

---

17 Ibid.

18 John Power, *Houthi Red Sea Attacks Still Torment Global Trade, a Year after October 7*. Al Jazeera, October 5, 2024, Accessed 12 October, 2024. https://www.aljazeera.com/economy/2024/10/5/a-year-after-october-7-houthi-red-sea-attacks-still-torment-global-trade.

19 Jernelöv, Arne. *The Threats from Oil Spills: Now, Then, and in the Future. Ambio* 39, no. 5 (2010): 353-366. https://doi.org/10.1007/s13280-010-0085-5.

20 Ibid.

21 Soyer, Baris, and Andrew Tettenborn, eds. *Pollution at Sea: Law and Liability*. London: Informa Law from Routledge, 2012.

22 See *The Threats from Oil Spills: Now, Then, and in the Future.*

23 Ibid.

24 Klein, Natalie. *Maritime Security and the Law of the Sea*. Oxford: Oxford University Press, 2011.

25 United Nations. *Applying the Law of the Sea to Protect International Shipping. UN Chronicle*, February 28, 2024. Accessed 2 June, 2024 https://www.un.org/en/un-chronicle/applying-law-sea-protect-international-shipping.

26 Platjouw, Froukje Maria, and Alla Pozdnakova, eds. *The Environmental Rule of Law for Oceans: Designing Legal Solutions*. Cambridge: Cambridge University Press, 2023.

27 Rothwell, Donald R., and Tim Stephens. *The International Law of the Sea*. 3rd ed. Oxford: Hart Publishing, 2023.

28 Ibid.

The economic implications of marine pollution include substantial costs for clean-up and compensation for affected communities and industries.[29] Effective legal mechanisms are necessary to address these economic burdens.[30] Ensuring the safety and security of maritime spaces is vital for maintaining global trade and economic stability.[31] Threats like piracy and environmental disasters can have far-reaching economic consequences.[32]

A certain study conducted by Krueger *et al.,* recently investigated the resilience of the reef-building coral Stylophora pistillata from the northern Red Sea to environmental stressors like high temperatures and reduced pH, simulating future ocean conditions. Coral reefs globally are threatened by rising ocean temperatures and acidification, which often lead to coral bleaching.[33] Corals in the northern Red Sea resist bleaching and harm despite high temperatures and future pH levels. This is due to evolutionary adaptation from migrating through warmer southern regions, favouring heat-tolerant species. They currently live below their optimal temperature, suggesting warming may improve their performance until reaching a tolerance limit. This shows the Red Sea may serve as a refuge for corals amid climate change. However, the study emphasizes the need to reduce local stressors like pollution and habitat destruction to protect these resilient coral communities, as their survival under global stressors will also depend on the mitigation of local environmental impacts.[34] The stakes can never be higher, and unless the international community averts Red Sea marine pollution from vessels like *Rubymar*, it risks losing the very coral reefs which can resist the adverse effects of climate change.

Assessing Facts of the Case through the Lens of Applicable International Laws

Spillage and release of oil and other dangerous chemicals resulting from armed conflict presents unique challenges that distinguish it from other forms of marine pollution. This paper examines the *Rubymar* ship incident in the Red Sea, and the applicability of international legal frameworks such as MARPOL and SUA. Under the doctrine of *lex specialis,* SUA's provisions may take precedence in this context. Additionally, the paper explores avenues for state responsibility, particularly concerning Yemen, and the potential liability of non-state actors like the Houthis. But first a brief remark will deal with the CLC Convention and its relevance to the *Rubymar* incident.

The Shortcomings of the CLC

The CLC establishes that shipowners are strictly liable for oil pollution damage caused by their ships, meaning they are responsible regardless of fault. But at the same time, it gives

29 Chen, Huiping. *The Liability and Compensation Mechanism under International Regulations of Marine Pollution from Ships*. Dordrecht: Springer, 2013.

30 See *Pollution at Sea.*

31 See *Maritime Security and the Law of the Sea.*

32 Kraska, James. *Contemporary Maritime Piracy: International Law, Strategy, and Diplomacy at Sea.* Santa Barbara, CA: Praeger, 2011.

33 Ibid.

34 Krueger T, Horwitz N, Bodin J, Giovani M-E, Escrig S, Meibom A, *Fine M.2017 Common reef-building coral in the Northern Red Sea resistant to elevated temperature and acidification.* R. Soc. open sci.4: 170038. Accessed July 10, 2024 http://dx.doi.org/10.1098/rsos.170038.

an exception where the shipowner is not liable, such as in situations damage is caused by acts of war, hostilities, or insurrections.[35] The *Rubymar* case falls under the exceptions of the convention because the Houthis carried out an armed violence on the *Rubymar* ship that can be construed as a hostility.

The Shortcomings of MARPOL

The *Rubymar* incident, involving oil spillage and potential release of 21,000 tonnes of ammonium phosphate-sulphate as a direct result of violence, highlights the complexities of addressing such pollution under existing legal frameworks. Unlike accidental or operational discharges typically covered by MARPOL, the intentional or incidental pollution from armed conflict necessitates a different legal approach. MARPOL's primary focus is on peacetime operations, as evidenced by its Annex I and Annex III, where the former sets regulations for the prevention of pollution by oil through technical measures and the latter preventing pollution by harmful substances carried by sea in packaged form.[36] However, these provisions do not explicitly consider scenarios stemming from armed conflict. In this context, the limitations of MARPOL become apparent. While it remains a cornerstone of international marine environmental protection, its effectiveness is diminished in the face of deliberate acts of violence. This gap underscores the need for supplementary frameworks, such as SUA, which specifically address unlawful acts against maritime navigation.

SUA CONVENTION

The SUA Convention, adopted to suppress unlawful acts against maritime navigation, offers a more appropriate legal framework for incidents like the *Rubymar* attack. SUA provisions are particularly designed to address violence at sea so they have greater relevance and can precede in their application, without prejudice to the more general maritime law in cases involving violence at sea. Article 3 defines offenses such as seizing a ship by force, using violence against persons on board, and placing devices that may destroy or damage a vessel—acts that apply directly to the *Rubymar* case.

SUA outlines jurisdiction for prosecution. Article 6 allows a state to exercise jurisdiction if the offense occurs on a ship flying its flag or involves its nationals. Since *Rubymar* sailed under Belize's flag, Belize may assert jurisdiction. Yemen may also do so if its nationals carried out the attack. The convention promotes international cooperation on prosecution and extradition. However, it does not cover liability for environmental harm caused by such acts, leaving a gap in its relevance to pollution from armed violence.

State Responsibility and the Case of Yemen

The *Trail Smelter* Arbitration case established a key principle of state responsibility for transboundary environmental harm.[37] The arbitration tribunal held that Canada was responsible for air pollution that crossed into the United States, requiring it to prevent future harm and

35 International Maritime Organization (IMO). *International Convention on Civil Liability for Oil Pollution Damage (CLC)*. Accessed September, 2024. https://www.imo.org/en/About/Conventions/Pages/International-Convention-on-Civil-Liability-for-Oil-Pollution-Damage-%28CLC%29.aspx.

36 International Maritime Organization (IMO). *International Convention for the Prevention of Pollution from Ships (MARPOL), Annex I: Regulations for the Prevention of Pollution by Oil and Annex III: Prevention of Pollution by Harmful Substances Carried by Sea in Packaged Form.* London: International Maritime Organization, 1973/78.

37 Trail Smelter Arbitration (*United States v. Canada*), 3 R.I.A.A. 1905, 1938.

compensate for damages.[38] This case, along with a range of other case judgements given by international tribunals, made the prevention of transboundary harm of the environment a customary international environmental law that States should adhere to. This shows the duty of states to prevent harm to other states' environments, a principle applicable to Yemen if it can be shown that the state failed to prevent the *Rubymar* attack. The International Court of Justice's (ICJ) *Nicaragua Case* is also another case which addressed state responsibility and the use of force.[39] The ICJ found that the United States' actions in supporting the Contra rebels and mining Nicaraguan harbours violated international law and amounted to an unlawful use of force.[40] This case accentuated the complexities of attributing state responsibility during conflicts but reinforces the principle that states must refrain from actions that cause harm to other states.

Similarly, Principle 2 of the Rio Declaration on Environment and Development also establishes that states are responsible for ensuring that activities within their jurisdiction do not cause environmental damage to other states or areas beyond national jurisdiction.[41] This principle supports the argument that there is an obligation by states to regulate both state and non-state actors, so that they don't cause transboundary environmental harm. In such case, Yemen would be held accountable if it is proved that significant environmental transboundary harm has accrued as a result of the sinking of the *Rubymar* tanker. Efforts to hold non-state actors accountable have often started to involve a combination of diplomatic, economic, and legal strategies to mitigate harm and encourage adherence to international environmental standards.[42] Sanctions and international pressure, imposed on Armed Groups in the Democratic Republic of the Congo (DRC) to hold them accountable for illegal logging and mining for e.g. is argued by many academics and practitioners as an alternative recourse to deal with non-state armed groups like the Houthis.[43] Efforts by the international community to hold states and individuals accountable for environmental harm resulting from armed conflict have generally yielded limited success. The UNEP states that a significant exception is the case of Iraq, which was held liable for environmental damages from the 1990-1991 Gulf War, leading to billions of dollars in compensation.[44]

The United Nations Convention on the Law of the Sea (UNCLOS) provides for responsibility in cases where there is pollution of the marine environment. However, it is clear that the convention is a state-centric treaty, intended primarily for peacetime operations.[45] Consequently, the articles stop short of explicitly regulating actions of non-state armed groups that could affect the marine environment and put it at risk of pollution. Even though, from UNCLOS's perspective, states are arguably expected to regulate the acts of entities within their territory, including armed groups, the current situation in Yemen would make it problematic.

---

38 Ibid, para. 15.

39 International Court of Justice. *Case Concerning Military and Paramilitary Activities in and Against Nicaragua (Nicaragua v. United States of America), Merits. ICJ Reports* 1986, 14, para. 172–205 and 187–190.

40 Ibid, 292-293.

41 United Nations. *Rio Declaration on Environment and Development.* United Nations Conference on Environment and Development, 1992.

42 Brauch, Hans Günter, Úrsula Oswald Spring, Czeslaw Mesjasz, John Grin, Patricia Kameri-Mbote, Béchir Chourou, Pal Dunay, and Jörn Birkmann, eds. *Globalization and Environmental Challenges: Reconceptualizing Security in the 21st Century*. Berlin: Springer, 2008.

43 La Rosa, Anne-Marie, and Carolin Wuerzner. *Armed Groups, Sanctions, and the Implementation of International Humanitarian Law*. International Review of the Red Cross 90, no. 872 (2008): 393–423.

44 United Nations Environment Programme (UNEP). *Protecting the Environment During Armed Conflict: An Inventory and Analysis of International Law*. Nairobi: UNEP, 2009, 4.

45 United Nations. *United Nations Convention on the Law of the Sea (UNCLOS).* 1833 UNTS 3, December 10, 1982. Articles 192, 194, 197, 207–212, 301.

Yemen's responsibility under international law is contingent on its ability to exercise control over the areas from which the Houthi attack originated. The doctrinal framework, grounded in Article 8 of the International Law Commission's Articles on State Responsibility[46], requires proof that Yemen directed or had effective control over the Houthis or the territory in question. However, given the conflict-ridden nature of Yemen, where significant regions are under Houthi control, the state might be lacking the territorial authority or organizational influence necessary to meet the effective control threshold. As Crawford emphasizes, a state's inability to control actors within its territory complicates attributing responsibility for internationally wrongful acts.[47] While Yemen may bear residual obligations to mitigate environmental harm under treaties such as UNCLOS or MARPOL, the lack of effective control over the perpetrators and the coastal areas absolves it of direct accountability for the *Rubymar* incident. This highlights the limits of state responsibility in fragmented territorial contexts under international law.

Non-State Actor Liability: The Case of the Houthis

The role of non-state actors, such as the Houthis in the *Rubymar* incident, necessitates an examination of their liability under international law. While international law traditionally focuses on state actors, there is growing recognition of non-state actor liability, especially in armed conflicts. Holding the Houthis accountable involves proving their direct involvement in the *Rubymar* incident and the environmental damage caused. The International Criminal Court's *Al Mahdi* case shows that non-state actors can face prosecution for international crimes. Al Mahdi was convicted for destroying cultural heritage, setting a precedent for individual accountability in cases causing serious harm, including environmental damage.[48]

In his book *Prosecuting Environmental Harm Before the International Criminal Court*, Gillett argues that severe environmental harm caused intentionally during conflict can fall within the ambit of international criminal law, particularly under Article 8(2)(b)(iv) of the Rome Statute of the International Criminal Court, which criminalizes acts during armed conflict that cause "widespread, long-term, and severe damage to the natural environment."[49] Applying this standard, the deliberate targeting of the tanker, which led to a substantial oil spill in the Red Sea, could potentially meet this threshold.[50] However, as Gillett notes, proving intent and establishing the severity of the harm remain critical challenges in prosecuting such cases.

Furthermore, the Rome Statute lacks explicit provisions for prosecuting non-state actors like the Houthis, although states that fail to prevent such harm within their jurisdiction may face secondary liability.[51] Yemen's limited control over Houthi-held regions complicates attribution, but the doctrine of command responsibility under international criminal law could hold individual Houthi leaders accountable if it is demonstrated that they ordered

46 Crawford, James. *The International Law Commission's Articles on State Responsibility: Introduction, Text, and Commentaries.* Cambridge: Cambridge University Press, 2002, 121.

47 Ibid, 110–115.

48 Prosecutor v. Ahmad Al Faqi Al Mahdi, *Judgment and Sentence*. ICC-01/12-01/15, International Criminal Court (ICC), September 27, 2016.

49 Gillett, M. *Prosecuting Environmental Harm Before the International Criminal Court.* Cambridge: Cambridge University Press, 2022.; *Rome Statute of the International Criminal Court*, Article 8(2)(b)(iv), 1998.

50 Sands, Philippe, and Jacqueline Peel. *Principles of International Environmental Law*. 3rd ed. Cambridge: Cambridge University Press, 2018, 850–860.

51 See the *International Law Commission's Articles on State Responsibility*, 110–115.

or had knowledge of the attack and failed to prevent it.[52] This paper concludes that the *Rubymar* case has highlighted the gaps existing in international criminal frameworks, as Gillett emphasizes the need for clearer legal standards to address environmental crimes in conflict zones effectively.[53]

Finally, as the *Rubymar* incident involved an armed violence, recourse was made to discuss the Geneva Conventions and their Additional Protocols, along with customary IHL and their applicability with respect to the incident. The writers of this paper believe that while the environment is considered a civilian infrastructure during armed conflicts, further clarifications made by the International Committee of the Red Cross's (ICRC) Guidelines on the Protection of the Natural Environment Under International Humanitarian Law[54] as well as International Law Commission's (ILC) Draft Principles on the Protection of the Environment in Relation to Armed Conflicts[55] are directly applicable to the *Rubymar* case, as long as the facts of the incident warrants the incident to be considered as an armed conflict *per se*, hence subject to IHL.

The applicability of IHL to the situation involving Houthi attacks on vessels in the Red Sea depends on whether the conflict meets the thresholds of an armed conflict under Common Article 3 of the Geneva Conventions and Additional Protocols.[56] For IHL to apply, the situation must constitute either a non-international armed conflict (NIAC) or an international armed conflict.[57] The Houthis are a non-state armed group (NSAG) engaged in ongoing hostilities in Yemen against the Saudi-led coalition and other factions, which has been widely recognized as a NIAC.[58] Their coordinated missile and UAV attacks against shipping vessels like *Rubymar* is likely to satisfy the organization and intensity criteria outlined in customary IHL and international judgements.[59] The organization criterion,[60] examines whether the Houthis possess the command structure necessary to sustain military operations, which they demonstrably do, given their sophisticated maritime operations.[61] The intensity criterion[62] considers whether the hostilities in the Red Sea go beyond sporadic acts of violence. The frequency and scale of Houthi attacks, combined with their broader strategy to disrupt interna-

52 See *Rome Statute of the International Criminal Court*, Article 28.

53 See *Prosecuting Environmental Harm Before the International Criminal Court*, 180.

54 International Committee of the Red Cross (ICRC). *Guidelines on the Protection of the Natural Environment in Armed Conflict under International Humanitarian Law*. Geneva, 2020.

55 International Law Commission. *Draft Principles on the Protection of the Environment in Relation to Armed Conflicts*. UN Doc. A/74/10, 2022.

56 ICRC. *Commentary on the Geneva Conventions of 1949: Common Article 3*. Cambridge: Cambridge University Press, 2016.

57 Ibid.

58 M. Khan and Z. Ullah. *The Yemen Armed Conflict in the Perspective of International Humanitarian Law.* Global Legal Studies Review, 2020, pages 15–20.

59 Prosecutor v. Tadić, *Decision on the Defence Motion for Interlocutory Appeal on Jurisdiction*, ICTY Appeals Chamber, 2 October 1995, para. 70.

60 Ibid.

61 Knights, Michael, Abdulaziz Al-Gabarni, and Charles Coombs. *The Houthi Jihad Council: Command and Control in 'the Other Hezbollah'*. CTC Sentinel 15, no. 2 (2022): 13.

62 See *Decision on the Defence Motion for Interlocutory Appeal on Jurisdiction*

tional trade and maritime security, could meet the requisite threshold, extending the existing NIAC in Yemen to the maritime domain under Articles 1 and 3 of Additional Protocol II.[63]

Key provisions from IHL provide further guidance on the legality of such actions. Article 13 of Additional Protocol II protects civilians and civilian objects, including commercial vessels, during NIACs, prohibiting attacks unless they contribute to military objectives. Under Article 35(3) and Article 55 of Additional Protocol I, which reflects customary international law, the natural environment and civilian infrastructure must also be protected from "widespread, long-term, and severe damage," a provision applicable to both international and non-international conflicts. The Houthis' justification for their actions—targeting vessels in retaliation for Israel's actions—raises the question of whether these attacks constitute a legitimate military objective under IHL or violate the prohibition against indiscriminate attacks.[64]

Assuming that these thresholds are met, the situation would satisfy IHL applicability criteria. The Houthis, as a highly organized NSAG, possess the command structure and operational capacity to sustain hostilities.[65] The scale and frequency of attacks in the Red Sea, combined with their deliberate targeting strategy, support the intensity criterion. These actions are not isolated but are part of a broader NIAC already recognized in Yemen. Consequently, the Red Sea operations can fall under the legal framework of IHL, highlighting the Houthis' obligations as an organized armed group. However, the legality of specific attacks must be further scrutinized to determine whether the targeted vessels were military objectives or civilian objects protected under IHL. The need for detailed factual analysis of individual incidents to assess compliance with IHL principles is acknowledged.

The ICRC states in the guideline that the days when the environment was an overlooked victim of conflict are over and that today the global community recognizes that human well-being and planetary health are deeply interconnected.[66] Victories achieved at the planet's expense are hollow, and militaries that ignore this reality risk both their own future and that of humanity.[67] Based on the discussions above, there is a responsibility to avoid causing disproportionate harm to the environment during armed conflict. In case of failure to do this, recourse can be made to holding the perpetrator responsible, though the enforcement mechanism is still limited.

Impact of the *Rubymar* incident on Eritrea's path to Sustainable Development

The concept of "sustainable development" is relatively recent in the context of global environmental governance. The term gained prominence between 1987 and 1992, evolving

63 Protocol Additional to the Geneva Conventions of 12 August 1949, and relating to the Protection of Victims of Non-International Armed Conflicts (Additional Protocol II). Adopted on 8 June 1977, Articles 1 and 3.; Patrick Maluki and Faith Njoki. *Maritime Security Concerns in the Emerging Global Order: An African View on the Red Sea.* Journal of International and Area Studies, 2024, pages 29–32.

64 See *Additional Protocol I (1977) to the Geneva Conventions*, Article 52(2).

65 See *The Houthi Jihad Council: Command and Control in 'the Other Hezbollah', 14.*

66 Obregón Gieseken, Helen, and Vanessa Murphy. *The Protection of the Natural Environment under International Humanitarian Law: The ICRC's 2020 Guidelines.* International Review of the Red Cross 105, no. 924 (2023): 1180–1207.

67 Ibid.

from the Brundtland Commission's report into the main framework of global environmental governance, finalized at the 1992 Rio Conference. The Rio Declaration framed sustainable development as a balance between economic growth and environmental protection.[68]

Sustainable development tries to make sense of the interactions of three complex systems: the world economy, the global society, and the Earth's physical environment. Environmental protection is important for ensuring that future generations can enjoy a healthy and sustainable environment. A country with strong economic growth, but high levels of inequality and poverty or one with good social and human well-being indicators, but a polluted environment, or vice-versa, does not constitute the sustainable development that is envisioned. The vision of the concept of sustainable development is to have a holistic view of development, where economic, social and environmental factors are all taken into consideration. This paper does not debate the vagueness of sustainable development but notes its broad consensus and calls for future research. It examines the environmental aspects of the *Rubymar* case in relation to Eritrea's path to sustainable development, understood as a normative concept reflected in international instruments. This paper takes this into consideration and tries to examine the environmental aspects of the *Rubymar* case in relation to Eritrea's path to sustainable development. By sustainable development, it must be understood that the term is taken not as a mere concept but as a normative one as is reflected in the existing international instruments.

The experience Eritrea has with environmental protection and conservation dates back to the days when customary laws started to be codified, way before the creation of the state. Although basic and focused on minor matters like tree cutting seasons and liability, these customary laws reflect the people's longstanding care for the environment.[69] Afterwards, Eritrea went through different periods, such as colonization by different countries with different agendas followed by the struggle for independence period and all the way to post-independence challenges. All of these had significant impact on the environment and efforts made to protect and conserve it. However, the core passion for protecting the environment continues to be reflected to this day with Eritrea signing various international agreements related to the protection of the environment, such as the Convention on Biodiversity (CBD) and United Nations Convention to Combat Desertification (UNCCD). The authors of this paper would like to underline here that protecting the environment for the sole reason that it is an obligation under various international environmental laws has never been the primary reason for Eritrea. Such reason compliments Eritrea's own efforts to protect the environment for the present and future generation that has been passed from generation to generation. Emphasizing this fact is relevant, because it must be put into context when discussing how the *Rubymar* case affects Eritrea's path to sustainable development.

Progress Towards the SDGs

The challenges Eritrea faces can be examined from the U.N.'s Sustainable Development Goals perspective as well as international laws which have implications for sustainable development in general. As has been discussed extensively above, we have seen how the *Rubymar* case affects the overall development process in various ways

68 Rajamani, Lavanya, and Jacqueline Peel, eds. *The Oxford Handbook of International Environmental Law*. 2nd ed. Oxford: Oxford University Press, 2021, 286.

69 Shabait. *Environment: A Non-Trivial Matter*. August 27, 2022. Accessed July 4, 2024. https://shabait.com/2022/08/27/environment-a-non-trivial-matter/.

In terms of the SDGs, the most affected goals would be Decent work and Economic Growth (SDG 8), Life below Water (SDG 14), Peace, Justice and Strong institutions (SDG 16), Partnerships for the Goals (SDG 17), Life on Land (SDG 15), Climate action (SDG 13) while other goals such as No poverty (SDG 1), Zero hunger (SDG 2), Decent health and Well-being (SDG 3), Clean Water and Sanitation (SDG 6), Industry, Innovation and Infrastructure (SDG 9) as well as Responsible Consumption and Production (SDG 12) are also to some extent affected. The U.N. once stated that the ocean facilitates 90% of global trade, underpins the internet, and generates jobs for over 3 billion people worldwide.[70] The significance of protecting the ocean from threats that would hamper its usage has drastic effect that is plain to the eyes. We shall briefly discuss the effects on the main SDGs while taking into consideration Eritrea's context and reality.

Life below Water (SDG 14)

In its 2022 inaugural voluntary national review report, it is stated that Eritrea's rich array of minerals, natural resources, and diverse ecosystems, alongside its unspoiled Red Sea coastline, presents significant opportunities for economic progression.[71] This biodiversity offers prospects for economic expansion and holds significant cultural and heritage value for Eritrea. To ensure the sustainable management of its land and marine environments, Eritrea has implemented extensive conservation efforts in collaboration with government entities, development partners, and community-based organizations. Acknowledging the essential role of biodiversity and ecosystems for social, economic, and environmental gains in sustainable development, the Eritrean government has shown strong commitment to the CBD. The updated National Biodiversity Strategy and Action Plan (NBSAP) guides the systematic allocation of resources toward biodiversity conservation and sustainable livelihoods, with a goal of maintaining a stable natural environment that can support societal well-being through 2040.[72] The *Rubymar* incident threatens sea life and sustainable development due to over 20,000 tonnes of ammonium nitrate risking toxic leaks.[73]

Decent Work and Economic Growth (SDG 8)

The Eritrean marine and coastal ecosystem include diverse habitats like coral reefs, mangroves, sea grass beds, and flats that support fisheries and offer tourism potential. These intertidal zones are ecologically and economically significant, although the Eritrean Red Sea, with limited nutrient supply and low primary productivity, differs from many global oceans due to minimal upwelling and lack of permanent river inflows.[74] Eritrea's coastal regions, particularly the Dahlak archipelago, create ideal conditions for fisheries, with over 600 recorded fish species and 154 stony coral species in waters conducive to coral growth.[75] Mangrove forests cover approximately 70 km² along 380 km of coastline, particularly dense around Assab Bay and several islands, while sea grass beds, hosting species like sea turtles

70 United Nations. *Video: Ocean 20: Building a Sustainable Blue Economy.* United Nations Sustainable Development Goals. Accessed August 2, 2024. https://sdgs.un.org/news/video-ocean-20-building-sustainable-blue-economy-49912

71 Eritrea. *Voluntary National Review Report on the Implementation of the 2030 Agenda for Sustainable Development.* 2022, 54.

72 Eritrea. *National Biodiversity Strategy and Action Plan,* Version 2. 2015, 52. Accessed July 12, 2024. https://www.cbd.int/doc/world/er/er-nbsap-v2-en.pdf

73 See *Greenpeace MENA on the Sinking of the Rubymar Ship: Immediate Action Must Be Taken to Avert an Environmental Catastrophe in the Red Sea*

74 Eritrea. *State of the Coast Report, 2006-2007,* 3-4. Accessed August 7, 2024. http://www.eritreaembassy-japan.org/data/State_of_the_Coast_2006-2007_FULL.pdf

75 Ibid.

and dugongs, are abundant near islands such as Baradu and Norah.[76] Those mentioned facts, among other even more important ones, are sufficient enough to understand Eritrean marine environment's importance for attaining the decent work and economic growth goal. The *Rubymar* incident also poses a threat to employment in coastal areas, impacting fishing, tourism, and trade. Pollution from ammonium nitrate could harm marine life, leading to diminished fish stocks and reduced income for fishermen. A decline in water quality could deter tourists once Eritrea's tourism grows, harming jobs in hospitality, guiding, and boat operations. This would also impact businesses tied to fishing and tourism, causing job losses in Red Sea communities.

Peace, Justice and Strong Institutions (SDG 16)

Worldwide, achieving comprehensive peace is essential, yet the path to global peace and security has grown increasingly challenging.[77] In fact, even though subject to argument, the *Rubymar* incident is a spill over of another hurdle to peace and security currently facing the international community, i.e. the Israeli aggression towards Palestine. The Houthis have repeatedly stated they will persist in targeting ships associated with Israel as a show of support for Palestinians until Israel halts its military actions in Gaza.[78] Houthi spokesperson Mohammed Abdul Salam argues that international and European navigation remains safe, provided no hostile operations take place, dismissing any need to militarize the Red Sea.[79] Instead, he emphasized that the international community's priority should be an immediate, comprehensive ceasefire in Gaza to address urgent humanitarian concerns.[80] Regardless of the authenticity behind the motive of the attacks on the shipping vessels as claimed by the Houthis, the important thing to note is that other geopolitical conflicts in surrounding areas has created more violence, with the achievement of peace as a SDG farfetched. This shows that the *Rubymar* incident also has direct implications on the peace and security of the globe.

Moreover, Target 16.1 of SDG 16 states the need to develop effective, accountable and transparent institutions at all levels. This includes, of course, the international level. Liability that might arise from the *Rubymar* incident would effectively be claimed if there is rule of law and strong institutions such as international tribunals ready to enforce it. The United Nations security council adopted resolution 2722, stating maintenance of international peace and security, and demanded that the Houthis immediately stop attacks on merchant and commercial vessels in the Red Sea. Such resolution stopped short of mentioning environmental concerns, arguably because more often than not it prioritizes immediate threats to peace and security, which in this case were deemed to be safety of navigation and the economic impact of the disrupted shipping routes.[81]It also didn't include language related to the conflict in the Gaza Strip, although China, Russian Federation, Algeria and Sierraleone voted in favour of it.[82] This gives a glimpse of how geopolitical power play might also affect the responses given to the *Rubymar* incident as well as achievement of global peace in general.

---

76 Ibid.

77 United Nations. *Goal 16: Peace, Justice, and Strong Institutions. United Nations Sustainable Development Goals.* Accessed September 4, 2024. https://sdgs.un.org/goals/goal16#progress_and_info

78 See *Yemen's Houthis 'Will Not Stop' Red Sea Attacks until Israel Ends Gaza War.*

79 Ibid.

80 Ibid.

81 United Nations. *Security Council Demands Houthis Cease Attacks on Vessels, Adopt Resolution 2722 (2024). United Nations Press*, January 10, 2024. Accessed August 2, 2024. https://press.un.org/en/2024/sc15561.doc.htm.

82 United Nations. *Security Council Addresses Regional Cooperation for Peace, Stability in Red Sea, Gulf of Aden.* Last modified February 12, 2024. Accessed July 7th, 2024 https://press.un.org/en/2024/sc15552.doc.htm.

Partnerships for the Goals (SDG 17)

Perhaps, it would be better to analyse this goal as a mechanism to solve the challenges that arise out of the *Rubymar* incident rather than explaining how it might be affected by the incident. SDG 17 calls for strengthening of the means of implementation and revitalizing the global partnership for sustainable development. It has been argued above that the *Rubymar* incident affects various sustainable development goals. Surrounding countries like Eritrea might be negatively affected in a way that makes it hard for them to achieve their SDGs. Hence, regional and global partnership becomes necessary to assist developing countries in preventing environmental harm as well as mitigating the effects of such violence. A good instance where such co-operation bear fruitful outcomes can be given. The deteriorating FSO Safer, carrying over 1.1 million barrels of oil near Yemen's coast has long raised concerns among Red Sea nations and beyond. Eritrea consistently alerted the United Nations to the imminent risk of the ship breaking apart and urged swift action to remove the oil.[83] Last year, the oil was successfully transferred to a different vessel, averting the threat of a spill in the Red Sea.[84] Such successful cooperative actions taken to avert environmental catastrophes whether in marine or land environment shows that cooperation can actually help achieve sustainable development. An oil spill from the Safer could have resulted in a severe humanitarian and environmental crisis.[85] A large spill would have likely devastated Yemen's north western coastline, including its Red Sea islands, especially Kamaran Island and it could have spread posing risks to neighbouring countries such as Djibouti, Eritrea, and Saudi Arabia.[86] But this was all averted as a result of prudent collaboration between different international organizations along with relevant actors.

Furthermore, in January 2020, Saudi Arabia launched the Red Sea Council, a multilateral framework of littoral countries.[87] The signing of the founding charter by the Foreign Ministers of the Council of Arab and African Coastal States of the Red Sea and the Gulf of Aden reflected the leaders' commitment to enhancing coordination and consultation regarding this crucial waterway, which holds significant economic, commercial, and investment value for the global economy.[88]

Eritrea is one of the countries which signed the charter, and showcased its desire for regional co-operation. In fact, the government of Eritrea released two policy documents in 2019, which clearly communicated Eritrea's "......policy perspectives and precepts on the matter [Red Sea]."[89] Basically, the documents can be paraphrased as follows, "Eritrea advocates for a robust, collaborative approach to ensuring peace and stability in the Red Sea region. This requires establishing a comprehensive security framework that is grounded in clear strategies and coordinated action plans, while emphasizing the importance of sover-

83 Eritrea Ministry of Information. *Eritrea's Statement at the Sixth Ministerial Level Session of the United Nations Environment Assembly.* March 1, 2024. Accessed August 2, 2024. https://shabait.com/2024/03/01/eritreas-statement-at-the-sixth-ministerial-level-session-of-the-united-nations-environment-assembly/.

84 International Maritime Organization. *Preventing an Oil Spill from the FSO Safer*. International Maritime Organization, Accessed August 2, 2024. https://www.imo.org/en/MediaCentre/HotTopics/Pages/FSO-SAFER-oil-spill-risk.aspx.

85 Ibid.

86 Ibid.

87 Saudi Press Agency. *The Signing of the Founding Charter by the Foreign Ministers of the Council of Arab and African Coastal States of the Red Sea and the Gulf of Aden.* Accessed June 6, 2024. https://www.spa.gov.sa/2019804.

88 Ibid.

89 Shabait. *The Geopolitics of the Red Sea.* Last modified January 19, 2024. Accessed July 8th, 2024 https://shabait.com/2024/01/19/64701/.

eignty and adherence to international law. Eritrea underscores the necessity of detailed threat assessments and effective categorization to address security challenges. It promotes the development of autonomous defence capabilities among littoral states, advocating for each state to possess its own naval forces rather than relying on external delegations. Additionally, Eritrea emphasizes collective efforts to manage tasks beyond individual capacities through coordinated support and cooperation with external partners based on mutual consent. Efforts also include environmental protection, prevention of pollution, and the formation of a "command and control" coordinating committee to enhance information sharing, joint exercises, and resource complementarity. Eritrea acknowledges that achieving these goals will require significant time, resources, and sustained commitment from all parties involved.[90]" The authors of this paper wholesomely agree to the policy and its contents as a foundation to act on the *Rubymar* incident for the sake of environmental protection as well as sustainable development in general. UNCLOS (to which Eritrea is not a party) also underscores the importance of international cooperation and shared responsibility in protecting the marine environment through several key provisions. Article 197 promotes global and regional collaboration, while Article 198 requires states to notify others of any imminent or actual pollution threats. Article 200 encourages states to engage in research, data sharing, and collaborative scientific efforts to address marine pollution. Additionally, Articles 204–206 mandate that states monitor environmental risks and assess the potential impacts of their activities, fostering proactive environmental stewardship. Apart from this, the authors would like to note Eritrea's absence in the Regional Convention for the Conservation of the Red Sea and Gulf of Aden Environment (known as Jeddah Convention.) The convention expresses in clear terms the commitment and the political will of the governments of the region to tackle challenges to the marine and coastal environments of the Red Sea and Gulf Aden through joint and coordinated efforts. Upon making preliminary research, assessing the potential signing of the convention is encouraged.

Conclusion and Recommendations

Establishing liability and responsibility for marine pollution from armed conflict involves navigating multiple legal frameworks and actors. While MARPOL's applicability is limited in this context, the SUA Convention provides a more targeted approach for addressing acts of violence at sea. However, SUA does not explicitly address environmental harm, highlighting the need for complementary legal frameworks. State responsibility, as seen in the case of Yemen, faces challenges due to lack of effective control, but international legal avenues remain viable through cases like the *Trail Smelter Arbitration* and the ICJ's *Nicaragua Case*. Non-state actors, such as the Houthis, can also be held accountable under evolving international legal standards, with precedents like the ICC's *Al Mahdi Case* providing a pathway for prosecution. By leveraging these legal avenues, the international community can enhance accountability for environmental harm resulting from armed conflict, promoting both justice and environmental protection. But then again, enforceability remains weak.

While the legality of the Houthis' attacks on shipping vessels is beyond the scope of this paper, it is important to note that such discussions would not significantly impact the issue of accountability for the environmental harm they have caused. The broader geopolit-

90 Shabait. *Press Release*. Last modified September 23, 2019. Accessed May 20, https://shabait.com/2019/09/23/press-release-84/; Shabait. *The Geopolitics of the Red Sea*. Last modified January 19, 2024. Accessed May 20, https://shabait.com/2024/01/19/64701/.

ical tensions in the Middle East, particularly the ongoing humanitarian crises and escalating conflicts, have undoubtedly contributed to spill over effects like the *Rubymar* attack. These incidents are emblematic of how regional instability and unresolved grievances can critically impact international waterways, affecting global trade and environmental security.

The deafening silence of the international community regarding such crises, particularly the catastrophic humanitarian situation in Gaza and the clear disregard for international law, exacerbates these tensions. When international law is selectively enforced or ignored, it creates a vacuum of accountability that non-state armed groups may attempt to fill, albeit in dangerous and unlawful ways. While these actions remain unequivocally illegal under international law and must be condemned to prevent anarchical recourse, they also serve as a stark reminder of the urgent need for international law to be respected and applied consistently among and between states.

This leads to broader reflections on the current rules-based international order, where double standards in upholding international law are increasingly visible to the global community. While this paper summarily concludes that the Houthis' actions constitute an offense against the marine environment, it also highlights the pressing need for systemic reassessment of international law to ensure its equal enforcement and to address the dissatisfaction many states and actors express toward its uneven application.

As to the *status quo*, no one seems to be in a rush to address the looming pollution from the sunken *Rubymar* vessel, located approximately 30 kilometres west of Yemen's Red Sea port of Mokha.[91] The authors would want to emphasize on the urgency of the matter. As discussed above, whereas the exact volume of oil spilled from the *Rubymar* until the present time has not been specified. For instance, the MV Wakashio oil spill off the coast of Mauritius in 2020, amounted to around 7300 barrels of oil, which is similar to the estimated oil leaked from *Rubymar*.[92] The spill occurred near environmentally sensitive areas, including the Blue Bay Marine Park reserve, home to diverse marine life and coral reefs.[93] The oil contamination threatened the survival of thousands of species in the region.[94] As discussed in relation with SDG 17 above, swift regional and international co-operation is necessary so as to help the littoral countries to the Red Sea to summon the support they need to avert the looming disaster. The littoral states themselves are also encouraged to discuss various matters such as clean-up, containment and restoration measures, with due consideration of the gravity of the implications it has on the protection of the marine environment from pollution.

Reference

Abu Bakr, Faisal. *UN Raises Funds to Clean Spill from Ship Houthis Sank in Red Sea. Al-Fassel*, July 25, 2024. Accessed September 10, 2024. https://alfasselnews.com/en_GB/articles/gc1/features/2024/07/25/feature-02.

AFP correspondents with Amanda Mouawad. Yemen Faces 'Environmental Disaster' as Sunken Ship Threatens Red Sea. Al-Monitor, 2024. Accessed 17 October, 2024 https://www.al-monitor.com/originals/2024/03/yemen-faces-environmental-disaster-sunken-ship-threatens-red-sea.

Al Jazeera. *UK Cargo Ship Hit by Yemen's Houthis Sinks in the Red Sea.* Al Jazeera, March 2, 2024. Accessed November 11, 2024. https://www.aljazeera.com/news/2024/3/2/*Rubymar*-cargo-ship-earlier-hit-by-houthis-has-sunk-yemeni-government-says.

Al Jazeera. Yemen's Houthis 'Will Not Stop' Red Sea Attacks until Israel Ends Gaza War. December 19, 2023, Accessed 13 August, 2024. https://www.aljazeera.com/news/2023/12/19/yemens-houthis-will-not-stop-red-sea-attacks-until-israel-stops-gaza-war.

Baraniuk, Chris. *Red Sea Crisis: What It Takes to Reroute the World's Biggest Cargo Ships.* BBC Future, January 21, 2024. Accessed October 11, 2024. https://www.bbc.com/future/article/20240119-red-sea-crisis-how-global-shipping-is-being-rerouted-out-of-danger. CBC News. *Cargo Ship Hit by Houthis Sinks, Spilling Oil and Fertilizer into Red Sea,* March 2, 2024. Accessed June 6, 2024. https://www.cbc.ca/news/world/houthi-attack-oil-spill-1.7132518.

BBC News. X-Press Pearl: The 'Toxic Ship' That Caused an Environmental Disaster, June 10, 2021, sec. Asia, Accessed on 12 July 2024 https://www.bbc.co.uk/news/world-asia-57395693.

Brauch, Hans Günter, Úrsula Oswald Spring, Czeslaw Mesjasz, John Grin, Patricia Kameri-Mbote, Béchir Chourou, Pal Dunay, and Jörn Birkmann, eds. *Globalization and Environmental Challenges: Reconceptualizing Security in the 21st Century.* Berlin: Springer, 2008.

Chen, Huiping. *The Liability and Compensation Mechanism under International Regulations of Marine Pollution from Ships.* Dordrecht: Springer, 2013.

CNN. *Mauritius Oil Spill Threatens Coral Reefs, Marine Ecosystems*, August 9, 2020. Accessed June 6, 2024. https://edition.cnn.com/2020/08/09/world/gallery/mauritius-oil-spill-2020/index.html.

Crawford, James. *The International Law Commission's Articles on State Responsibility: Introduction, Text, and Commentaries.* Cambridge: Cambridge University Press, 2002.

Eritrea Ministry of Information. *Eritrea's Statement at the Sixth Ministerial Level Session of the United Nations Environment Assembly.* March 1, 2024. Accessed August 2, 2024. https://shabait.com/2024/03/01/eritreas-statement-at-the-sixth-ministerial-level-session-of-the-united-nations-environment-assembly/.

Eritrea. *National Biodiversity Strategy and Action Plan*, Version 2. 2015, 52. Accessed July 12, 2024. https://www.cbd.int/doc/world/er/er-nbsap-v2-en.pdf

Eritrea. *State of the Coast Report, 2006-2007*, 3-4. Accessed August 7, 2024. http://www.eritreaembassy-japan.org/data/State_of_the_Coast_2006-2007_FULL.pdf

Eritrea. *Voluntary National Review Report on the Implementation of the 2030 Agenda for Sustainable Development.* 2022, 54.

Gillett, M. Prosecuting Environmental Harm Before the International Criminal Court. Cambridge: Cambridge University Press, 2022.

Greenpeace MENA. *Greenpeace MENA on the Sinking of the Rubymar Ship: Immediate Action Must Be Taken to Avert an Environmental Catastrophe in the Red Sea.* 2024. Accessed September 7, 2024. https://www.greenpeace.org/mena/en/sinking-of-the-*Rubymar*-ship/.

Haight, B. W. *US Strikes Against the Houthis: A 'Splendid Little War' of Self-Defence. American University National Security Law Brief*, 2024.

Henderson, Christian. *US and UK Military Strikes in Yemen and the Jus Ad Bellum. International and Comparative Law Quarterly* 73, no. 2 (2024).

Hopwood, Bill, Mary Mellor, and Geoff O'Brien. *Sustainable Development: Mapping Different Approaches. Sustainable Development* 13, no. 1 (2005).

ICRC. *Commentary on the Geneva Conventions of 1949: Common Article 3.* Cambridge: Cambridge University Press, 2016.

International Committee of the Red Cross (ICRC). *Guidelines on the Protection of the Natural Environment in Armed Conflict under International Humanitarian Law*. Geneva, 2020.

International Court of Justice. *Case Concerning Military and Paramilitary Activities in and Against Nicaragua (Nicaragua v. United States of America).* ICJ Reports 1986.

International Law Commission. *Draft Principles on the Protection of the Environment in Relation to Armed Conflicts*. UN Doc. A/74/10, 2022.

International Maritime Organization (IMO), MV *Rubymar* - Call for Contributions of Oil Pollution, June 25, 2024, Accessed 16 august, 2024.

International Maritime Organization (IMO). *Convention for the Suppression of Unlawful Acts Against the Safety of Maritime Navigation.* Adopted March 10, 1988.

International Maritime Organization (IMO**).** *International Convention for the Prevention of Pollution from Ships (MARPOL)*, Annex I: Regulations for the Prevention of Pollution by Oil. London: International Maritime Organization, 1973/78.

International Maritime Organization (IMO). *International Convention on Civil Liability for Oil Pollution Damage (CLC).* Accessed September 2024.

International Maritime Organization. *Preventing an Oil Spill from the FSO Safer*. International Maritime Organization, Accessed August 2, 2024. https://www.imo.org/en/MediaCentre/HotTopics/Pages/FSO-SAFER-oil-spill-risk.aspx.

Jernelöv, Arne. *The Threats from Oil Spills: Now, Then, and in the Future*. *Ambio* 39, no. 5 (2010).

John Power, Houthi Red Sea Attacks Still Torment Global Trade, a Year after October 7. Al Jazeera, October 5, 2024, Accessed 12 October, 2024. https://www.aljazeera.com/economy/2024/10/5/a-year-after-october-7-houthi-red-sea-attacks-still-torment-global-trade.

Klein, Natalie. *Maritime Security and the Law of the Sea.* Oxford: Oxford University Press, 2011.

Knights, Michael, Abdulaziz Al-Gabarni, and Charles Coombs. *The Houthi Jihad Council: Command and Control in 'the Other Hezbollah'. CTC Sentinel* 15, no. 2 (2022).

Knights, Michael. *Assessing the Houthi War Effort Since October 2023. Washington Institute for Near East Policy*, 2024.

Kraska, James. Contemporary Maritime Piracy: International Law, Strategy, and Diplomacy at Sea. Santa Barbara, CA: Praeger, 2011.

Krueger, T., N. Horwitz, J. Bodin, M.-E. Giovani, S. Escrig, S. Meibom, and M. Fine. *Common Reef-Building Coral in the Northern Red Sea Resistant to Elevated Temperature and Acidification. Royal Society Open Science* 4 (2017): 170038.

La Rosa, Anne-Marie, and Carolin Wuerzner. Armed Groups, Sanctions, and the Implementation of International Humanitarian Law. International Review of the Red Cross 90, no. 872 (2008).

M. Khan and Z. Ullah. *The Yemen Armed Conflict in the Perspective of International Humanitarian Law.* Global Legal Studies Review, 2020.

Meade, Richard. *Sunken Rubymar's Only UK Link Was Its Disputed Insurance Cover. Lloyd's List*, 2024. Accessed July 7, 2024. https://www.lloydslist.com/LL1148459/Sunken-*Rubymars*-only-UK-link-was-its-disputed-insurance-cover.

Naidu, Richa. *How Red Sea Disruptions Are Driving up Carbon Emissions. The Japan Times*, June 2024. Accessed August 2, 2024. https://www.japantimes.co.jp/environment/2024/06/01/climate-change/red-sea-shipping-co2-emissions/.

Obregón Gieseken, Helen, and Vanessa Murphy. *The Protection of the Natural Environment under International Humanitarian Law: The ICRC's 2020 Guidelines.* International Review of the Red Cross 105, no. 924 (2023).

Otu-Larbi, Frederick, Benjamin Neimark, Patrick Bigger, Linsey Cottrell, and Reuben Larbi. "A Multi-temporal Snapshot of Greenhouse Gas Emissions from the Israel-Gaza Conflict." *ResearchGate*, 2024.

Patrick Maluki and Faith Njoki. *Maritime Security Concerns in the Emerging Global Order: An African View on the Red Sea.* Journal of International and Area Studies, 2024.

Platjouw, Froukje Maria, and Alla Pozdnakova, eds. *The Environmental Rule of Law for Oceans: Designing Legal Solutions.* Cambridge: Cambridge University Press, 2023.

Prosecutor v. Ahmad Al Faqi Al Mahdi, Judgment and Sentence. ICC-01/12-01/15, International Criminal Court (ICC), September 27, 2016.

Prosecutor v. Tadić, Decision on the Defence Motion for Interlocutory Appeal on Jurisdiction, ICTY Appeals Chamber, 2 October 1995.

Protocol Additional to the Geneva Conventions of 12 August 1949, and relating to the Protection of Victims of Non-International Armed Conflicts (Additional Protocol II). Adopted on 8 June 1977.

Rajamani, Lavanya, and Jacqueline Peel, eds. *The Oxford Handbook of International Environmental Law.* 2nd ed. Oxford: Oxford University Press, 2021.

Rogers, Peter, and Julia Hansen. "Economic and Security Ramifications of Maritime Incidents in Conflict Zones." *Journal of Maritime Law* 42, no. 2 (2024).

Rome Statute of the International Criminal Court,1998.

Rothwell, Donald R., and Tim Stephens. *The International Law of the Sea.* 3rd ed. Oxford: Hart Publishing, 2023.

Sands, Philippe, and Jacqueline Peel. Principles of International Environmental Law. 3rd ed. Cambridge: Cambridge University Press, 2018.

Sarah Shamim. *Could Houthi-Sunk Ship Rubymar Spell Eco-Disaster for the Red Sea?* March 7, 2024. Accessed July 4, 2024. https://www.aljazeera.com/news/2024/3/7/could-houthi-sunk-ship-*Rubymar*-spell-eco-disaster-for-the-red-sea.

Saudi Press Agency. *The Signing of the Founding Charter by the Foreign Ministers of the Council of Arab and African Coastal States of the Red Sea and the Gulf of Aden.* Accessed June 6, 2024. https://www.spa.gov.sa/2019804.

Saul, Jonathan, and Carolyn Cohn. *Red Sea Insurance Costs Soar as Houthi Shipping Threats Loom, Sources Say*. Reuters, September 19, 2024. Accessed October 11, 2024. https://www.reuters.com/world/middle-east/red-sea-insurance-costs-soar-houthi-shipping-threats-loom-sources-say-2024-09-19/.

Shabait. *Environment: A Non-Trivial Matter*. August 27, 2022. Accessed July 4, 2024. https://shabait.com/2022/08/27/environment-a-non-trivial-matter/.

Shabait. *Press Release*. Last modified September 23, 2019. Accessed May 20, https://shabait.com/2019/09/23/press-release-84/

Shabait. *The Geopolitics of the Red Sea.* Last modified January 19, 2024. Accessed July 8th, 2024 https://shabait.com/2024/01/19/64701/.

Soyer, Baris, and Andrew Tettenborn, eds. *Pollution at Sea: Law and Liability.* London: Informa Law from Routledge, 2012.

The Maritime Executive, IMO Honors Crew of Product Tanker That Was Hit by a Houthi Missile. The Maritime Executive, July 10, 2024. Accessed 12 October 2024, https://maritime-executive.com/index.php/article/imo-honors-crew-of-a-product-tanker-hit-by-a-houthi-missile.

Trail Smelter Arbitration (United States v. Canada), 3 R.I.A.A. 1905, 1938.

U.S. Central Command. UPDATE: Fire Extinguished on M/v Marlin Luanda Following Houthi Anti-Ship Ballistic Missil, January 27, 2024. Accessed 13 October 2024.https://www.centcom.mil/MEDIA/PRESS-RELEASES/Press-Release-View/Article/3658448/update-fire-extinguished-on-mv-marlin-luanda-following-houthi-anti-ship-ballist/.

United Nations Environment Programme (UNEP). Protecting the Environment During Armed Conflict: An Inventory and Analysis of International Law. Nairobi: UNEP, 2009.

United Nations Environment Programme. *Oil Spill in Mauritius Calls for More Efforts to Safeguard Coral Reef Ecosystems*. UNEP, August 20, 2020. Accessed June 6, 2024. https://www.unep.org/news-and-stories/story/oil-spill-mauritius-calls-more-efforts-safeguard-coral-reef-ecosystems.

United Nations. Applying the Law of the Sea to Protect International Shipping. UN Chronicle, February 28, 2024. Accessed 2 June, 2024 https://www.un.org/en/un-chronicle/applying-law-sea-protect-international-shipping.

United Nations. *Goal 16: Peace, Justice, and Strong Institutions. United Nations Sustainable Development Goals*. Accessed September 4, 2024. https://sdgs.un.org/goals/goal16#progress_and_info

United Nations. *Rio Declaration on Environment and Development.* United Nations Conference on Environment and Development, 1992.

United Nations. *Security Council Addresses Regional Cooperation for Peace, Stability in Red Sea, Gulf of Aden*. Last modified February 12, 2024. Accessed July 7th, 2024 https://press.un.org/en/2024/sc15552.doc.htm.

United Nations. *Security Council Demands Houthis Cease Attacks on Vessels, Adopt Resolution 2722 (2024). United Nations Press*, January 10, 2024. Accessed August 2, 2024. https://press.un.org/en/2024/sc15561.doc.htm.

United Nations. *United Nations Convention on the Law of the Sea (UNCLOS).* 1833 UNTS 3, December 10, 1982.

United Nations. *Video: Ocean 20: Building a Sustainable Blue Economy.* United Nations Sustainable Development Goals. Accessed August 2, 2024. https://sdgs.un.org/news/video-ocean-20-building-sustainable-blue-economy-49912

Xiaokai Zhang et al. *The X-Press Pearl Disaster Underscores Gross Neglect in the Environmental Management of Shipping: Review of Future Data Needs. Marine Pollution Bulletin* 189 (April 1, 2023): 114728.

The Role of Traditional Law in the Development of the Eritrean Legal System

Salvatore Mancuso
University of Palermo.
Honorary professor of African law, Center for African Laws and Society, Xiangtan University.
Visiting professor, Somali National University
Email: salvatore.mancuso06@unipa.it

Introduction

Legal pluralism is probably the most important characteristic of African legal systems where the presence of different normative orders at the same time creates situations of conflict among them.[1] Such conflict is a result of the traditional approach to legal pluralism where the different normative orders are competing with each other under a state that purports to have the monopoly to produce laws. Unfortunately, very often such laws are a pure transposition of a rule borrowed from another (generally Western) legal system into the receiving one. The consequence is that quite often citizens have scarce confidence on state law and prefer to resort to other set of rules to which they are more familiar, even if such rules remain at informal level.[2]

Nevertheless, Africans continue to rely on local traditions which are still coexisting with the statutory law. Despite this, large parts of informal normative orders today are not recognized by the official state law, and the possible harmonization of such informal normative orders with state law has thus become an objective for present legal reform projects, as most African countries today are facing the same issue.[3]

In the African continent the official state law still follows the western pattern that

1 Literature on legal pluralism is endless. Among others, useful references can be: Anne Griffiths, "Legal Pluralism", in Reza Banakar, Max Travers, *An Introduction to Law and Social Theory*, (Portland: Hart Publ., 2002) 289; Ghislain Otis, "La gestion du pluralisme juridique: definitions et approche", in Ghislain Otis, Jean Leclair, Sophie Thériault (dir.), *La vie du pluralisme juridique* (Paris: LGDJ, 2022) 7; Jacques Vanderlinden, "Les droits africains entre positivisme et pluralisme", (2000) 46 *Bulletin des séances de l'Académie royale des sciences d'outre-mer*, 279; John Griffiths, "What is legal pluralism?", 18 (1986) *Journal of Legal Pluralism and Unofficial Law* (24) 1; Jacques Vanderlinden, "Villes africaines et pluralisme juridique", 42 (1998) *Journal of Legal Pluralism and Unofficial Law*, 245; Jan C. Bekker, Christa Rautenbach and Nazeem M. I. Goolam, *Introduction to Legal Pluralism in South Africa*, (6th ed., Durban: Lexis Nexis, 2018); Jacques Vanderlinden, "Return to Legal Pluralism: Twenty Years Later", 21 (1989) *The Journal of Legal Pluralism and Unofficial Law* (28) 149.

2 Marco Guadagni, *Il modello pluralista* (Turin: Giappichelli, 1996).

3 See more in Salvatore Mancuso, "Pluralismo e ordini normativi in Africa. Dal conflitto alla collaborazione", in Michele Graziadei, Mario Serio cur., *Regolare la complessità. Giornate di studio in onore di Antonio Gambaro* (Turin: Giappichelli, 2017) 157; Id., "Short Notes on the Legal Pluralism(s) in Somaliland", in Séan P. Donlan, Lukas Heckendorn Urscheler (eds.), *Concepts of Law* (London: Ashgate, 2014).

has been imported during the colonial occupation and that has never been adapted to reflect the local conditions of the people living in Africa. After having achieved independence, the African states kept the imported legal system, as it was the only common legal framework in territories where different people having different traditions and normative orders were arbitrarily put together. Such western-based state legal systems have since become part of the reality of African countries where Africans grow up and are educated. However, only those who can afford education fees and university studies abroad can get all the necessary information and "tools" to understand and use the official legal system. On the other side, the vast majority of Africans still does not identify with it, as local rules amongst the population differ considerably from the western-based legislation in key issues like land tenure, successions, family and personal law. As a result, in many African countries, only an infinitely small number of disputes is dealt with by the state courts.

This phenomenon is caused by different factors. First, there is still an issue of limited infrastructures of formal justice around the African countries out of the big cities and a general problem of access to justice for people living in rural areas. Secondly, the unpredictability of the official justice due to the fact that still large cases of corruption and clientelism contribute to the problem.[4] Furthermore, in most countries the language of the judiciary is the European one, inherited from the colonial period, which frequently is not spoken by the locals. Last but not least, the application of state law involves requiring compliance with a system of values and a coherent legal framework based on fundamentally different moral and religious ideas.[5]

The situation could be different if the traditional approach to legal pluralism is changed from the usual vision – based on the competition of the different normative orders – to a situation where the different normative orders cooperate to produce a system of rules that is closer to the local legal culture.

It has long been recognized that it is impossible to eliminate the African legal tradition by force. It is too strong and too rooted in people, even in those who live in cities and apply Western lifestyle models. Is it therefore possible to hypothesize cooperation between these normative orders? One could object that collaboration is not in fact possible, because the application of one legal order necessarily excludes the application of others. But this is not necessarily true.

Article 4 of the Mozambican Constitution – entitled, precisely, legal pluralism – has now become famous among experts as one of the most significant examples on the path to the formal recognition of legal pluralism. It contains a strong declaration of principle, but

4 With reference to this issue see the vast bibliography related to the Organization for the Harmonization of Business Law in Africa (OHADA), with particular reference to the reasons on which the idea of harmonizing business law in Africa has been based. The website www.ohada.com contains a considerable part of such bibliography. Reference can be made to Jean Claude Mutombo, *L'harmonisation en Afrique du droit des affaires à travers l'OHADA: Aspects institutionnels et normatifs* (Saarbrücken: Éditions Universitaires Européennes, 2019); Boris Martor, *Droit OHADA africain des affaires* (Paris: LGDJ, 2017); Justine Diffo Tchunkam (dir.), *L'OHADA au service de l'economie et de l'entreprise* (Cape Town: JUTA, 2014); Hygin Didace Amboulou, *Le droit des affaires dans l'espace OHADA* (Paris, L'Harmattan, 2014); Salvatore Mancuso, *Direito comercial africano (OHADA)* (Coimbra: Almedina, 2012); Paul-Gérard Pougoué, *Encyclopédie du droit OHADA* (Paris: Lamy, 2011); Paul-Gérard Pougoué, Yvette Rachel Kalieu Elongo, *Introduction critique à l'OHADA* (Yaoundé: Presses Universitaires d'Afrique, 2008).

5 D. Kohlhagen, "State Law and Local Law in Sub-Saharan Africa", in Alula Pankhurst and Getachew Assefa dir., *Grass-Roots Justice in Ethiopia. The Contribution of Customary Dispute Resolution* (Addis Ababa, Centre français des études éthiopiennes, 2008) 77; Salvatore Mancuso, "L'espressione diversa di un diritto diverso: lingua e diritto in Africa", in Barbara Pozzo (ed.), *Lingua e diritto: oltre l'Europa*, (Milan: Giuffré, 2014).

two observations deserve to be made. First, the State continues to assume the task of determining the extent to which the recognition of different normative orders can be carried out (*"na medida em que não contrariem os valores e os princípios fundamentais da Constituição"*); and, furthermore, it is not clear how this principle can be put into practice.

Collaboration can be based on a hierarchical relationship (of supremacy) or on equality. The first is the ethnocentric model (which in some ways could also be deduced from the wording of the aforementioned art. 4 of the Mozambican Constitution). Experience tells us that this approach does not pay off, and that the different legal orders should be recognized as having equal dignity. On the other hand, it is argued that at least a *grundnorm* is necessary to dictate the fundamental rules of this coexistence. But again, it can be observed how the very idea of the *grundnorm* is alien to the African legal culture that is more inclined to the circumstantial and non-predetermined approach.

Practice also suggests some examples that can indicate the path to take.

In Mozambique, the Court of First Instance of Gondola, in the central province of Manica, sentenced two young men arrested by the police while having "sexual" relations with a goat (…) to six months in prison for aggravated theft. The judge ruled that the two young men had committed the theft of the goat, as the act of having sexual relations with a goat is not foreseen and regulated by Mozambican law, even though it must be considered as contrary to the traditional morality and culture of the local population. The owner of the goat, who appeared in court, requested compensation for the damage suffered through the payment of the bride price (*lobolo*) on the basis of the "sexual" relations that had taken place and the fact that no one would ever buy that goat at the market again: the judge declared himself incompetent to decide on the request for compensation, referring the shepherd to the traditional authorities as the only body with the necessary legal competence to decide on the request.[6]

Beyond the "strangeness" of the episode, the way in which the state judge intended to resolve the compensation problem is certainly interesting.

This is not an isolated example. Again, in Mozambique, a district court judge, interviewed on the resolution of family disputes (which fall within the jurisdiction of these first instance courts), stated that these types of conflicts "should not be dealt with by a judge, but should be resolved within the family or in the neighbourhood".[7]

In Somaliland, elders often ask the ordinary judge to resolve disputes by themselves directly, and the judges favor this out-of-court solution based on the consent of the parties. In the civil sector, this can happen at any level of the trial, and the trick used is the use, depending on the level of the trial, of Articles 117 and 239 of the Somali Code of Civil Procedure. According to the aforementioned Articles, borrowed from the Italian code, the judge can invite the parties to find an out-of-court solution to the dispute; the interpretation given to these articles has always been in the sense of allowing the parties to find a solution to the dispute by applying the rules and methods of traditional law. Here, it is not the judge who applies traditional law: he simply takes note of the resolution of the dispute in application of traditional law by the competent authorities in the matter. The parties register the agreement reached with the official judge for the sole purpose of closing the file, with the execution of the agreement itself being left to the mechanisms provided for by traditional law. The same phenomenon also occurs in the criminal sector. Here, judges consider this method of resolving disputes an application of the principle of Islamic law according to which the offended

6 This case has been reported in the Correio da Manhã – March 3, 2010. See also https://macua.blogs.com/moambique_para_todos/2010/03/dois-jovens-condenados-pelo-tribunal-por-furto-qualificado.html.

7 L'episodio è riportato in Boaventura de Sousa Santos, *The Heterogeneous State and Legal Pluralism in Mozambique*, 40 (2006) *Law and Society Review* (1) 39-75.

person (or his heirs) have the right to choose whether the perpetrator of the unlawful act should be punished personally, or whether the matter should be resolved through the intervention of the qadi (who is nothing other than the traditional authority) and the payment of *diya*, and therefore through compensation. In this case, the judge accepts the extrajudicial decision taken by the elders (who play the role of the qadi), possibly imposing a symbolic penalty that represents the punishment of that part of the crime consisting in the offense against the State.[8]

And again, in South Africa the constitutional judge has introduced into the system the traditional concept of ubuntu (a Zulu and Xhosa word that expresses the bond between individual and community, for the translation of which it is necessary to refer to the expression *umuntu ngumuntu ngabanta* in Zulu or *umntu ngumntu ngabanye* in Xhosa, which can be translated as: "a person is a person only through the community") as a principle that must guide the formation and application of South African law as a fundamental value of society, although there is no trace of it in the constitutional text.[9]

One of the pillars of the peace and reconciliation process in Rwanda after the 1994 genocide has been resorting to the *gacaca* traditional dispute settlement system to solve most of the criminal cases connected to the genocide.[10] This Rwandan traditional institution has been transformed and adapted from its original framework – which was to cover small conflicts between family members and neighbors (including theft, property damages and physical violence) to cover crimes committed during the genocide with the aim of using a system where people could actively participate with a view to securing appropriate punishment and facilitating at the same time the re-establishment of unity among Rwandans and the rebuilding of the country.[11] The resort to the *gacaca* jurisdiction has been proposed by the Rwandan government as a quicker and more effective alternative to trials by state courts, being a process based more on forgiveness and reconciliation than on punishment. The *gacaca* courts have been substantially introduced in the court system and six laws have been enacted by the Rwandan parliament to deal with the different aspects of the system.[12]

Similarly, the 2000 Arusha peace and reconciliation agreement for Burundi expressly resorted to traditional cultural values (*ibanga, ubupfasoni* and *ubuntu*) and a traditional dispute resolution system (*ubushingantahe*) to restore and maintain peace in the country.[13]

---

8 Salvatore Mancuso, *Short Notes* cit. at footnote 3.

9 *State v. Makwanyane*, 1995 (3) SA 391 (CC).

10 The literature on the *gacaca* is today quite extensive. An interesting overview has been recently made by Roelof H. Haveman, "Gacaca in Rwanda. Customary Law in Case of Genocide", in Jeanmarie Fenrich, Paolo Galizzi, and Tracy E. Higgins (eds.), *The Future of Customary Law*, (Cambridge: Cambridge University Press, 2011) 387. For a more comprehensive approach see Paul Christoph Bornkamm, *Rwanda's Gacaca Courts* (Oxford, Oxford University Press, 2012). A response to the critics to the *gacaca* system can be found in Felix Mukwiza Ndahinda, Alphonse Muleefu, "Revisiting the Legal and Socio-political Foundations and (Western) Criticism of gacaca courts", in Thomas W. Bennett et al. (eds.), *African Perspectives on Tradition and Justice* (Cambridge: Intersentia, 2012).

11 Such position is well illustrated in Fanie Du Toit, *Reconciliation and Transitional Justice: The Case of Rwanda's Gacaca Courts*, (2011) Institute for Justice and reconciliation – Africa Programme, Occasional Paper 2.

12 See more in detail Roelof H. Haveman, *op. cit.*, at 398 and ff.

13 On the *ubushingantahe* system of dispute resolution see Léa Havard, Youssouf Ali, "Initiation au pluralisme juridique à travers l'étude des Bashingantahe", in Gilda Nicolau, Fabrice Hourquebie (dir.), *Cultures juridiques en quête de dialogue*, (Paris: Karthala, 2014) 239; Dominik Kohlhagen, "Le bushin-

Unlike the informal laws, state law (the formal one) is a law almost always imported or in any case inspired by a Western model, not deriving from the legal culture of African populations. Often even those who have studied, live and work in cities and are exposed to both models prefer the local one, considered more effective, quick and easy to manage, unlike the European one (in which they barely recognize themselves) slow, cumbersome, less practical and more expensive.

The following example allows us to understand the reasons.

In Ziguinchor (Senegal) a lady buys a plot of land, and of this transaction she has a deed signed by the seller, by herself as the buyer and by two witnesses (the last two having added their fingerprints). Later, the neighbor learns that the seller of the land has made a second sale of the same plot of land. Armed with her deed of sale, the lady goes to the *chef de quartier* to request his intervention. The latter summons the parties involved and on the appointed day everyone is present: the *chef de quartier*, the lady, her neighbor, the former owner, the two witnesses and the second buyer. A long discussion ensued, the result of which is that the seller promises to return the money from the second sale, with which the *chef de quartier* procures another plot of land to be allocated to the second buyer.[14] In this situation, the lady could certainly have appealed to the state judge and obtained the return of the plot of land, but going down this path would have determined a win-lose situation, where – in exchange for the assignment of the land to the lady – the second buyer would have been obliged to return the property, and the seller would have been ordered to pay legal costs in addition to the return of the sum received from the second buyer , not to mention the time required to reach this result. With the appeal to the local authority, a win-win situation is determined: the lady takes back the plot of land to which she is entitled, the seller is not burdened with additional charges, the second buyer still obtains a plot of land that he intended to purchase, and the seller of the second plot of land has found a buyer without any effort.

The example illustrates that cooperation between the different normative orders goes through dialogue and reciprocal integration, so that the different normative orders can help each other.

The Eritrean Situation

Turning now to the specific case of Eritrea, the territory of the Eritrean state can be summarily divided into three large geographical areas that separate the population from both an ethnic and a religious point of view: the plateau, inhabited by populations belonging to ethnic groups of Tigrinya matrix, of Christian-Coptic religion, carrying out a mainly sedentary agricultural activity, and that shared with the Ethiopian ones, among other things, a series of social institutions and rules, with minimal differences between the various realities; the lowland, where different ethnic groups are found, essentially devoted to pastoral and agro-pastoral activities with a significant presence of Muslim communities; and finally the coastal zone and the Danakil depression, a desert area crossed by the Afar populations of Sunni Muslim religion, who are nomadic shepherds by tradition. The country experienced the Italian colonization, followed by a short period of British administration after the Italian defeat in the Second World War. Eritrea was ruled by Ethiopia as a province before it became independent in 1991.

The two large groups that constitute the ethnic base of the Eritrean populations gave rise to numerous and different peoples who, in turn, have developed into equally different

---

gantahe au Burundi: transformations et reminiscences d'un concept judiciaire ancien", in Gilda Nicolau (dir.), *Dire le droit, render la justice* (Paris: Karthala, 2009) 113.

14 Gerti Hesseling, "Le droit foncier dans une situation semi-urbaine. Le cas de Ziguinchor", in B. Crousse, E. Le Bris & E. Le Roy (eds.), *Espaces disputés en Afrique noire*, (Paris : Karthala, 1986) 113-132.

lineages and under-lineages. This occurred following intersections and transformations in which the Hamitic and Semitic elements intertwined or overlapped and produced populations with sometimes different languages and traditions.[15]

Traditional law in Eritrea represents an essential element of the country's unique socio-cultural landscape, characterized by a deep historical meaning and a profound link with the usages and practices of its different ethnic groups. With its unique features as oral traditions, community decision-making and local practices, traditional law has traditionally served as a main means of conflict resolution and social regulation between various communities, especially in rural areas where formal legal institutions can be less accessible.[16]

Rooted in the rich tapestry of the cultural heritage of Eritrea, traditional legal practices are based on lifestyles, shared values and beliefs of its many ethnic groups. In particular, nine main ethnic groups have been generally recognized, namely Tigrigna, Tigre, Saho, Afar, Blin, Kunama, Nara, Bidawyet and Rashaida.[17] The traditional laws of the Eritrean people are known and transmitted through oral tradition. These traditional laws have been collected with greater depth for the populations residing on the plateau, in numerous writings mainly due to collections curated by Italian scholars during the colonial period.[18]

Eritrea has a variegated patchwork of traditional normative orders, also influenced by different religious beliefs. They are the fundamental component of the Eritrean legal pluralism, which includes – out of those traditional normative orders – rules of religious origin (Christian, mainly in the highlands, and Muslim, mainly in the lowlands); laws of Ethiopian origin, as amended after the independence of Eritrea; rules deriving from membership in regional organizations (like COMESA) or from international treaties to which the country is a party.

The Eritrean state is today under a process of modernizing its legal system by producing new sets of rules that supersede those inherited from the Ethiopian legal system after independence.

In 2015 the Eritrean government issued drafts of new Civil, Civil Procedure, Penal and Criminal Procedure codes. Here the choice seems to be to incorporate traditional principles and values in codes drafted using the Western pattern. In the Civil Code is said that "[…] continued efforts have been made to make it consistent with the values and principles of the Eritrean society […]"[19], while the Penal Code says, more explicitly, that "[f]irst and foremost, Eritrea is renowned for being a treasure-nation of age-old, self-advanced customary laws. These customary laws embodied, inter alia, principles of penal law, based on peaceful

15 Alberto Pollera, *Le popolazioni indigene dell'Eritrea*, (Licinio Cappelli Editore: Bologna, 1935). See also G. Cucci, "Le fonti del diritto indigeno d'Eritrea", in 7 (1933) *Rivista delle Colonie Italiane* (11) 853; Carlo Conti Rossini, *Principi di diritto consuetudinario dell'Eritrea*, (Rome: Tipografia dell'Unione Editrice, 1916).

16 Uoldeul Chelati Dirar, "Colonialism and the construction of national identities: The case of Eritrea", 1 (2007) *Journal of Eastern African Studies* (2) 256-276.

17 Senai W. Andemariam, *Ensuring Access to Justice Through Community Courts in Eritrea. Traditional Justice, Practitioner's Perspectives*, (2011) International Development Law Organization, 15.

18 The reference here is mainly to the texts by Carlo Conti Rossini, *Principi di diritto consuetudinario dell'Eritrea*, cit.; Felice Ostini, *Trattato di diritto consuetudinario dell'Eritrea*, (Officina Grafica-Corriere Eritreo: Asmara, 1956); and Ilario Capomazza, *Diritto consuetudinario dell'Acchelè-Guzai*, (Tip. E. De Angeli: Asmara, 1909). On the general characters of Eritrean traditional law see Marco Guadagni, "Eritrean Law Between Past and Future: An Introduction", in Lyda Favali, Elisabetta Grande, Marco Guadagni eds., *New law for new states: politica del diritto in Eritrea* (Turin: L'Harmattan Italia, 1998) 11-19; and also Gebre H. Tesfagiorgis, "Customary Laws in Eritrea", in Tesfa G. Gebremedhin, Gebre H. Tesfagiorgis eds., *Traditions of Eritrea: Linking the Past to the Future*, (Trenton: Red Sea Press, 2008) 1-36.

19 See the Prelude to the Drat Civil Code of 2015. A similar statement is also made in the short introduction to the code by the Eritrean Minister of Justice, Mrs. Fawzia Hashim.

settlement of certain criminal acts that we today have found to be fitting for incorporation into the penal law for independent Eritrea. It was, thus, imperative, over and above the 1991 amendments, that we review the Penal Code, which was currently in force, in light of the traditions of our societies as revealed in the customary laws. This required us to embody the workable penal principles of our customary laws into the new Code".[20]

On the procedural side, the Civil Procedure Code states that "[c]ountless meetings, correspondences, and studies of both customary law and practice in Eritrea and the experience of other nations culminated in the production of a preliminary draft Civil Procedure Code".[21] The Criminal Procedure Code is, again, more explicit by saying that "[a]n effort was also exerted in incorporating some acceptable and basic principles and perspectives in the age-long indigenous laws, traditions and practices similarly extant in the various communities of the Eritrean people as are related to criminal procedure".[22]

Even if all these codes have not yet entered into force and seem to serve as the draft basis for the enactment of new codes which take into more consideration of the Eritrean legal values, they already witness the attention that the Eritrean government dedicates to the legal culture of its people, even if the way in which all this should be done remains uncertain.

The issue that remains is the way in which common principles and values shall be extracted from the different traditional laws of Eritrea and how to embody them into a legislation prepared using a Western approach.

A Possible Way Forward

As it can be easily understood, this endeavor requires a proper methodological approach, which should move from preliminary research using the methodological tools of comparative law to identify common principles and to highlight differences to be "harmonized" in order to better understand how from those traditional normative orders common principles can be produced. The first assumption must be, however, that no individual legal system, not even a mixed system, should be taken as the standard. This avoids any risk of hegemonism, which could prejudice the project's feasibility.

In this regard the methodology adopted by the Common Core Project, itself based on the idea launched by Rudolf Schlesinger in his project at the Cornell Law School, is useful for our purposes.[23] It can even be more profitable in an African context if correctly used and contextualized.

The Common Core approach, utilizing the analysis of the legal formants acting in a given legal system[24], can provide an opportunity to take into consideration African elements and specificities that influence the "law in action" as well as the "living law", the phenomena of spontaneous or "underground" law existing beside (and sometimes competing with) official State law.[25]

One important difference with the Common Core Project should, however, be outlined. The declared aim of the original project is the search for commonalities and differenc-

---

20 See the Prelude to the Draft Penal Code.

21 See the introduction to the draft code by the Eritrean Minister of Justice, Mrs. Fawzia Hashim.

22 See the Prelude to the Draft Criminal Procedure Code.

23 See Rudolf B. Schlesinger (ed.), *Formation of Contracts, A Study of the Common Core of Legal Systems*, (2 volumes) (New York, London: Oceana Publications, 1968).

24 On the theory of legal formants see Rodolfo Sacco, "Legal Formants: A Dynamic Approach to Comparative Law", 39 (1991) *American Journal of Comparative Law*, 1 (Part 1) and 343 (Part 2).

25 See Mauro Bussani & Ugo Mattei, "The Common Core Approach to European Private Law", 3 (1997) *The Columbia Journal of European Law*, (3) 339 and ff., for the use of this methodology for the study of European private law.

es behind the different legal forms of the countries taken into consideration in order to obtain a reliable geographical map of the related legal framework. The research proposed here will be a useful instrument for legal integration among the different normative orders in the sense that reliable data should be obtained for use in preparing new common solutions that will be workable in practice.[26] Therefore, the research proposed for the Eritrean context should seek to discover common principles among the different Eritrean normative orders, this in order to offer to the Eritrean legislator the basis from which the insertion of the Eritrean values in the official law shall move.

Having identified the sectors where the integration of the Eritrean values shall be pursued, first of all a process of mapping the available resources will be necessary to have a reliable start point for the research. Then, the key tool for the research will be the questionnaire, with the aim of verifying and – if necessary – update the available information. The general guideline is to draft the questionnaires with a degree of specificity that requires the reporters to provide answers to the questions so that all of the circumstances which affect the law in the system that is being considered are addressed. This should guarantee that rules formulated in an identical way, but which may produce different applications, or even different interpretations from the doctrine, will not be regarded as identical. This should also allow one to see the elements that may play an official and declared role in one system, and that in another system may act in a more cryptic, unsystematic, and unofficial way. Identifying the role of such cryptic elements (the mind turns immediately to African traditional law) is crucial when determining the principles present in the operating rules.

The research could be complicated due to the interaction of the different normative orders. Therefore, this problem must be resolved by drafting the questionnaires and guiding the way to answering them. In addition, participants should not be limited to nationals who deal with their domestic law but should also include participants who are comparatists or scholars having at least some comparative law background, to support local researchers in the activity of tracing the common values. At the end, the researcher will handle a very large amount of information about the systems operating in the country.[27] Such information can thereafter be offered to the legislator who will decide which elements deserve to be incorporated into the state law.

In conclusion, Eritrean legal values are essential in the process of modernizing the country's legal system to produce laws in which people can recognize themselves. It shall never be forgotten that law is made for the people to use it: therefore, citizens must feel confident with the law, and this can happen only if the law reflects the values of the society. This requires both that the laws are inspired by the values of the people and that the different legal components present in the Eritrean society have the same dignity and cooperate with each other in a correct management of the country's legal pluralism.

Transforming Corporate Social Responsibility from Contractual Transplantation to Legislative Innovation in the Eritrean Mining Industry

Tomas Tesfamariam

Email: thomastsf756@gmail.com

Abstract

An Eritrean journal published to facilitate the then ongoing mining conference in 2014, noted that mining activities in Eritrea will offer good investment opportunities and pave the way to rapid sustainable economic growth and sustainable development. The Macro-policy provides a testing clause, an economic growth be parallel to the aim of social justice and that its national development objective be the creation of a modern, technologically advanced and internationally competitive economy. So, mining companies operating in Eritrea are, as part of their contractual obligation, required to employ internationally accepted standards/norms of operation. The mining companies have opted the CSR (Corporate Social Responsibility) tool as part of their corporate policies to answer to the pressure posed. Kofi Anan, former UN Secretary General described CSR as universal value binding together global corporations. CSR has great potential in mitigating environmental and socio-economic challenges, such as poverty eradication and environmental degradation through integrating businesses in nations developmental activities. Because many companies are found to be well-off in technology, know-how and financial capacity. However, application of CSR is hooked on the country's understanding of the constituents of a responsible business conduct. Therefore, arbitrariness of government regulation and action must be avoided.

The research will initially delve how responsible business conduct are currently depicted in Eritrean laws, policies, societal values and know-how, because CSR necessitates legal compliance as well as customary ethics. Moreover, this study will provide the framework by which the stakeholders of the operation are able to comprehend, explain, attribute, extrapolate, and predict what could happen in the future of the project, especially, the overshadowed stakeholders, nearby communities and diaspora to activate companies' responsible behaviour. The methodology of the study will comprise both primary and secondary data. Primary sources will include structured interviews with persons associated with each mining project; and distributing questionnaire as part of establishing a comprehensive understanding of responsible conduct. Secondary sources will include published research and reports made as part of their CSR obligation. Case-studies (comparative analysis) will also be part of discussions to propose an effective compliance program to facilitate implementation.

Keywords: Economic growth, Mining Sector, Social Justice, Standards of Operation, Sustainable Development.

Introduction

In 2002 a summit known as WSSD was conducted in South Africa to reaffirm international commitments to sustainable development.[1] This summit for the first time recognized the positive contribution of mining to sustainable development in its draft Johannesburg Plan of Implementation (hereinafter JPOI).[2]

Mining companies, as influential multinational enterprises (MNEs), make significant impact on the environment and economic growth of countries. Locally, the mining industry brings important economic and social benefits for Eritrea covering one of the country's largest GDP shares. Moreover, with the two new projects (the Colluli and Asmara project) set to begin, their prospective contributions is expected to boost its share significantly.[3] The industry can also offer export revenues, employment and infrastructural development for the country. However, this industry is at the same time making undesirable impacts on the environment related to land use and environmental pollution.

The idea of Corporate Social Responsibility (CSR), which is a modernized version of social contract, is one approach adopted to face these challenges. This version of responsibility arose in response to a growing public sentiment to hold global mining companies accountable for their social and environmental externalities. Besides, multinational mining companies are also using this notion to respond to the pressures and step-up their efforts.[4] Thus, the CSR is commonly understood as a broad concept which requires factorization of social and environmental concerns by businesses to seek to either avoid or rectify harmful effects of their undertakings beyond what the law simply requires.[5]

Mining Operations and CSR – Regulatory Approaches

The notion of CSR has different meanings in the environment of global mining industry.[6] It is argued by many that CSR is associated with the broad agenda of sustainable development, while some view it as responsible mining which considers CSR as a 'method whereby diversity of community demands and the need to protect the environment is ever-present with the aim to make profit.'[7]

These two definitions have two major implications. First, they imply that mining companies are required to carefully stabilize the competing interests among all stakeholders. Second, it suggests that mining CSR is simply a synergetic interaction among businesses, ecosystem and host communities.[8]

1 Madeleine Busacca, 'Corporate Social Responsibility in South Africa ' s Mining Industry: Redressing the Legacy of Apartheid', Claremont McKenna College Senior Theses (2013), at 23.

2 Ibid, at 24, JPOI, para. 44.

3 UNDP, 'Analysis of the Potential Contributions of Colluli Potash Project to Sustainable Development Goals in Eritrea', at 5.

4 Andrew Ngawenja Mzembe and Yvonne Downs, 'Managerial And Stakeholder Perceptions of an Africa-Based Multinational Mining Company's Corporate Social Responsibility (CSR)', the Extractive Industries and Society 225 (2014), at 230.

5 Angessa Eticha Asela, 'The Need to Adopt Corporate Social Responsibility Strategy in Ethiopia, a research thesis submitted to the Arsi University School of Law, (2020), at 1. See also Kofi Annan, Address to World Economic Forum, Davos, (1999), at 3. Available at: https://www.un.org/sg/en/content/sg/speeches/1999-02-01/kofi-annans-address-world-economic-forum-davos. (visited on January 10, 2024). Kofi Anan in his words also noted that CSR is the cement binding together our global corporations, since they are values people all over the world will recognize as their own.

6 See Andrew above n 4, at 227.

7 Ibid.

8 Ibid.

For an effective elucidation of CSR in the business culture of a country, a three-tier test has to be utilized. The first test involves examination of the voluntary and non-voluntary actions undertaken by the mining companies to address their social and environmental impacts and improve the social welfare. [9] Second is the means and degree of engagement with various stakeholders, usually termed as stakeholder relations. The last test emphasizes the steps mining companies undertake deal with the changing expectation of the society.[10]

Approaches to regulate CSR may be divided into self-regulation and mandatory regulation. Mandatory regulation may include mandatory due diligence, and mandatory philanthropy. CSR regulation may also be viewed from governance structure perspective.

Self-Regulation

The first strategy used to adopt CSR is self-regulation, such as introducing codes of conduct and establishing internal principles by businesses for responsible environmental and social conduct.[11] Other alternatives also shade some light with an aim to improve community relations and steadily lessen effects on the environment and society. These possibilities emphasize incorporating social objectives into a company,[12] causing a shift from voluntary to mandatory regulatory regimes. This move was further accelerated in several countries because of accidents in Ranza Plaza, Bangladesh and Shell's environmental disaster in Nigeria, which were a result of unregulated corporate behavior and defective corporate reporting.[13]

The International Organization for Standardization (ISO) and the International Chamber of Commerce (ICC) take the understanding that CSR is a voluntary undertaking taken by businesses implying that businesses responsibility extends beyond legal responsibility.[14] Nonetheless, many other international instruments were found to impose an obligation to respect rules of human rights and environment, thus making self-regulatory tools adopted by mining companies obligatory to consider minimum thresholds incorporated in legislations.[15] But the argument continues that a self-regulating CSR tool largely remains a voluntarily adopted normative rule developed by companies in the form of explicit formal rules like code of conduct to reap long-term benefits through greater customer satisfaction.[16]

Many countries fail to endorse voluntary CSR standards through their supply chains, because they are concerned with its potential to oust a domestic legislation. This is mainly due to the close association these standards have with market outcomes, offering less strict rules. Nevertheless, it should be duly noted that voluntary regulations serve as an essential step before full CSR integration.[17]

---

9 Tomas Frederiksen, 'Corporate Social Responsibility and Political Settlements in the Mining Sector in Ghana, Zambia and Peru', 74 ESID Working Paper (2017), at 9. Available at : https://ssrn.com/abstract=289313. (visited on April 20, 2024)

10 Ibid.

11 Carla Vintró and Josep Comajuncosa, 'Corporate Social Responsibility In the Mining Industry: Criteria and Indicators', University of Colombia, Volume 77, 31 (2010).

12 Ibid, at 34.

13 Michael Hopkins, *Corporate Social Responsibility and International Development Is Business the Solution* ? 1st ed. (Bath: Bath Press,2007), at 128. Oxfam in the report titled "Behind the mask: the real Face of CSR" conducted in 2004.

14 See Angessa, above n 5, at 25.

15 Ibid.

16 Tasew Abitew, 'Approaches to Regulating Corporate Social Responsibility in Ethiopia: The Case of Manufacturing Companies', research thesis submitted to College of Law and Governance Studies School of Law Masters of Business Law, Addis Ababa University (2021), at 27.

17 See Comajuncosa, above n 11.

Mandatory Regulatory Approaches

Mandatory regulation refers to regulations imposed by government laws. Laws and regulations play an essential role to organize company role in protecting environment and ensure its sustainability. Laws protect interests of disadvantaged and marginalized groups. Rules incorporated in state legislations may vary from restructuring taxes and fees imposed on foreign and national companies to policies and procedures to protect the environment and its components.[18] Moreover, CSR legislations may also appear to take the form of environmental, labour, and/or human rights law.[19]

The emergent varieties of mandatory approaches go beyond requiring CSR, demanding either companies' responsibility to observe due diligence duties; engage in philanthropic activities, to deliver conducive governance restructuring by including it as part of corporate duties under the available corporate laws.[20] These approaches signal the need for companies and their operators to effect a change in their organizational behavior and pave tne way for an introduction of new ethical standards and business relations.

CSR due diligence was initially developed as a form of soft corporate law to support a regulatory response to sustainability challenges. But now it is emerging as a key tenet of the transition of the economic system towards sustainable development.[21] This regulatory regime is the most common form of legislation where companies are required to declare their non-financial performances to express their due diligence as a general corporate risk management process i.e. requiring the companies to detect their environmental and social risk associated with their business operation and to execute plans to prevent harm.[22]

The plans must clearly identify the risks, make regular assessment and establish competent procedures to mitigate the risks. In case of non-compliance with such rules of due diligence, a civil action may be pursued. This right is widely recognized by international human rights treaties,[23] which call for remedies for victims. Domestic laws are also urged to deliver proper remedies. EU countries are the most active nations with this practice, with an energetic participation of Non-Governmental Organizations (NGO's), because these organization continuously advocate for an inclusion of civil liability for corporate misconduct.

The best example of this regulatory framework is France's 'Vigilance Law' which imposes legal liability in case of damage. The disastrous event in Ranza Plaza textile factory in Bangladesh was a boost to giving life to the 'Vigilance Bill'.[24] The bill removed the general CSR duty and reduced legal obligation of the companies and provided companies to be liable for any harm resulting from their operation. The main criticism for this legislation arises from the difficulty it poses to prove causation, which is demoralizing for victims.[25]

---

18 Raed Awashreh, 'Legalization of Corporate Social Responsibility in Dealing with Environmental Issues "Examples from Moroccan Arab Countries"', A'Sharqiyah University (2017), at 6. Available at: https://www.researchgate.net/publication/333166393 (visited on March 3, 2024).

19 Li Wen Lin, Mandatory Corporate Social Responsibility Legislation Around The World, Mandatory Corporate Social Responsibility Legislation Around The World, Emergent Varieties And National Experiences, 429, at 430.

20 Ibid, at 431.

21 Avni Shrivastava, 'Corporate Social Responsibility and Legal Accountability: Bridging the Gap, International Journal of Technical Science and Exploration (2022), at 43.

22 See Li Wen above n 19, at 431.

23 Ibid.

24 Ibid, at 436.

25 Ibid, at 439.

This type of CSR can be somehow depicted as corporate charity. The essence of this CSR approach is to require companies allocate certain percentage of their profit to designated CSR programs such as for building schools and hospitals. Corporate philanthropy can enhance reputations of managers or directors in their social circles and provide them with numerous benefits. The proponents of this idea also believe this program is not a 'tax' or 'charity', but a compensation on the cost they couldn't fully avoid. However, unless the cost of giving is at least balanced by reduced compensation, such spending on charity will signify a company cost.[26]

Companies' philanthropic activities are challenged based on the argument that such vision of a CSR is more restrictive and associated with public relation than CSR;[27] and diminishes the focus on accountable management of risks and delay.[28] In contrast, various corporations believe it is just another way of increasing corporate tax.[29] This approach is dominant in Africa.[30]

Structural Approaches

Countries adopt different governance structures to regulate CSR.

Mandatory structural governance of CSR adopts a structural way of executing mandatory CSR, making legal changes in the corporate governance to include stakeholders other than shareholders.[31] South Africa is the most known nation for this kind of mandatory regime. The abolition of apartheid was the reason for such introduction to bring about affirmative changes, obliging companies to establish an inclusive committee within the corporate governance.[32] The problem with this approach lies in the probability to make structural changes in the governance structure, which largely depend on the will of the government.

Another structural approach is general mandatory CSR duty. CSR duty under corporate law shows a state-centered approach, referring to the general legal duty of companies to act in a socially responsible manner. However, to equate CSR to a legal duty, the law has to really offer actionable standards since attempting to just widen the scope of a legislation and to include CSR will be insufficient.[33] The main criticisms of this approach lie in its nature of being inherently inconsistent with the basic essence of CSR, since CSR is fundamentally an inspirational idea, and it is very difficult to legislate aspirations.[34] China and Indonesia are most known for this approach.

All things considered, hard law regulation on CSR fail to provide the sovereignty and incentives for an efficient response given the plurality of the issues the companies face. Furthermore, the force of the approaches is prone to be weakened by politics and the flexible nature of CSR.[35]

---

26 William O. Brown, Eric Helland, Janet Kiholm Smith, Corporate Philanthropic Practices, 12 Journal of Corporate Finance 855 (2006), at 856.

27 Ibid.

28 B The Change, 'CSR in India is now a law'. Available at: https://bthecahnge.com/csr-in-india-is-now-a-law-2502 (visited February 4, 2024).

29 Anel Karnani, 'Mandatory CSR In India: A Bad Proposal'. Available at https://ssir.org/articles/entry/mandatory-csr-in-india-a-bad-proposal (visited February 4, 2024).

30 Clotilde Marielle, 'Why adopt a regulatory and legal framework to facilitate the growth of CSR?', https://Sofercom.com/en/news-insights/why-adopt-a-regulatory-and-legal-framework-to-facilitate-the-growth-of-csr.html (visited on September 23, 2023).

31 See Li Wen above n 19, 440.

32 Ibid.

33 Ibid, at 443.

34 See Karnani, above n 29.

35 See Li Wen above n 19.

Regulatory Approaches to CSR in Eritrea

Generally, there is not much awareness of CSR in Eritrea. Nor does the government factor Responsible Business Conduct (RBC) in its procurement decisions. The case of responsible conduct in Eritrea is barely entertained in court cases.[36] However, Eritrea's various legislations and regulatory standards include some fundamental elements of the notion of business responsibility. Such instruments, in addition to imposing legal obligations, also stipulate ethical responsibilities that business institutions should consider.

Furthermore, the attraction of CSR is that it is a system approach, meaning that the problem is defined and the system's boundary is delineated so that all the important factors in resolving the problem are involved in the deliberations.[37] Thus, countries aiming to use CSR to ensure sustainability from companies employ different approaches depending on their political and social background but not necessarily on the democracy level of a country.[38] Therefore, under this section the paper will examine how the available approaches can be utilized in Eritrea without looking for much change in the current legal system.

Self-regulatory Approaches to CSR

A standalone CSR policy is not adhered to by mining companies in Eritrea. It is contended that their compliance with the law and social and environmental management plan (SEMP) are essentially adequate to address environmental and social issues related to their actions. SEMP is part of the commitments made by companies to enter the Eritrean market. This may create liability for the companies based on Article 308(2) and Article 309 (1) (c) of the Transitional Commercial Code of Eritrea[39] (TCoCE), since the companies simply assume the commitments made by their founders. Such claim will be based on the accuracy of promises when companies fail to live up to the commitments made. However, the only recourse for the public for injury will be based on the rules of extracontractual liability because most of the elements and statements in the commitments made remain confidential.

Moreover, Eritrean mining companies claim practicing CSR when they do anything from minor charitable donation to devoting considerable resources to alleviate societal problems. However, this is not enough because these companies should also worry about losing market demand resulting from negative public perceptions, as well as being responsive to stakeholder demands for CSR.[40] Companies try to escape responsibility for any of their environmental and social impacts, because such responsibilities are externalities and only bring cost and reduce profit. Problems also occur in the implementation process, because the regulations may be paternalistic, and they often undermine local communities' skill to identify their own needs and priorities. Such situations will allow or encourage companies to pursue their own vision of good conduct.

Institutional Hurdles and Ideological Constraints

The Eritrean Standard Institution (ESI) provides standards for various actors in certain Er-

36 Interview with Mr. Sium Tekle, Director General at the Ministry of Justice conducted on 23 February 2024, from 4:00 – 4:30 pm.

37 See Hopkins, above n 13, at 8.

38 Hui Wen, 'Global Corporate Social Responsibility Reporting Regulation: Drivers and Impacts on Sustainable Development', University of Illinois at Urbana-Champaign November (2017), at 35.

39 Provisional Government of Eritrea, Gazette of Eritrean Laws, Volume 1, No.2, Proclamation 6/1991, Proclamation to establish Transitional Commercial Code of Eritrea.

40 Mulus Wijaya Kusuma, 'Corporate Social Responsibility: Concepts and Practices in Indonesia', 137 (2018), at 138.

itrean economic sectors and deals with nationwide standardization.[41] But it does not consider voluntary standards, because it is mainly assumed that the companies lack a set of standards and may run the risk of being disqualified by international standards.[42]

There are no public capital markets inside Eritrea and only a small number of publicly traded companies operate inside the country. Moreover, CSR as a management tool, is oriented to achieving excellence in the long-term and is characterized by the establishment of a new business culture based on criteria related to legitimacy, responsibility, and delegation of power.[43] Yet, the existing practices need to be compatible with the guiding principles of the market approach.

Mandatory Approaches to CSR Regulation

At national level there are two possibilities of how CSR can be regulated, one is through indirectly bringing substantial changes to government actions or directly addressing it to companies through rules on company law.[44]

A strong emphasis on philanthropic initiatives by companies may call within the mining industry for a broader approach to CSR issues indicating that these companies are committed to deflecting threats from disruptive stakeholders, as well as to deliberate approach to maintain positive relations with the host communities.[45] However, this approach may be found compatible with the National Environmental Management Plan of Eritrea (NEMP-E),[46] if a special tax was to be attributed to an environmental protection, which to some extent can safeguard an environmentally ethical ambition._

Furthermore, any kind of philanthropic activities in Eritrea outside government initiatives are seen with caution, although the regulatory authorities claim they are not against social contributions coming from private enterprises in principle.[47] This is because they do not see it as their primary responsibility.[48] Moreover, the authorities show great support for voluntary actions for internal controls on environmental pollution in mining companies, and they believe it is the key duty of the government to improve community welfare. The applicability of this approach in Eritrea, moreover, misses the mark due to its high probability of becoming a window-dressing strategy, where regulations are not effectively complied with.

The existing Eritrean corporate rules seem to equate the CSR duty with directors'

41 Provisional Government of Eritrea, Gazette of Eritrean Laws, Volume 5, No.9, Proclamation 75/1995, A Proclamation To Provide For The Establishment of The Eritrean Standards Institution, preamble.

42 Ibid, Art. 7(1) "...prepare compulsory Eritrean standards to products, practices... in the economic sector and follow up the implementation of same."

43 See Comajuncosa above n 11, at 35. Suggest that "the credibility, quality, transparency, and accountability of policy-making processes, public institutions, the legal regulatory climate, and sector governance are major determinants of how successfully countries can channel their resource wealth into sustainable development. *Emphasis added*

44 Christiana Ochoa, Corporate Social Responsibility and Firm Compliance: Lessons from the International Law International Relations Discourse, 9 Santa Clara J. Int'l L. 169 (2011). https://digitalcommons.law.scu.edu/scujil/vol9/iss1/6.volume-9|Issue1-Article-6-1-1-2011. If disclosure laws require firms to provide a wide range of accurate information about firm fundamentals, actual mandatory disclosure levels may never fall below voluntary ones (such as CSR disclosure).

45 See Andrew, above n 4, at 227.

46 Government of Eritrea, The National Environmental Management Plan For Eritrea(NEMP-E), (1995).

47 NEVSUN Resource LTD., 'Corporate Social Responsibility Report', 2014, at 4. Available at: www.nevsun.com at 22.

48 Interview With Mr. Abel Yohannes, Member of The Impact Assessment Unit of the DOE, Conducted On 9 October 2023, from 10:30 am -11:35.

fiduciary duties imposing due diligence duty.[49] Such due diligence act can be considered as an act complying with the fundamental rule of operation. For this approach to be utilized the rules must have broad perspective when involving undertakings with significant adverse impact. This is because CSR is less regarded as corporate obligation rather than merely part of directors duty.[50] However, with the scope of the existing due diligence rules in Eritrea being vague, its pertinence becomes rather problematic.

Mandatory CSR due diligence is another approach. This variety requires a mining company to identify social and environmental risks associated with the operation, establish and execute plans to prevent harm resulting from identified as well as unidentified risks. The due diligence obligation is similar with contents of SEMP of the companies and the principle of precautionary approach laid by the environmental management proclamation.[51] In these instances, the companies are required to employ, in addition to the National Environmental Assessment Procedures and Guidelines (NEAPG), international tools essential to address the risks to which they are exposed.

The application of such instruments, however, cause difficulty in the ability of victims (from the public) to prove causation in Eritrea. Meanwhile, not only are the terms of the obligations confidential but the standards are sophisticated for the existing legal system of Eritrea to adjudicate. Nevertheless, perceiving its nebulousness to some degree the rules of extracontractual liability, which make liable an entity for any harm caused, can be utilized.

Mandatory Governance Structure

Boards of Directors of mining companies in Eritrea comprise government authorities and company managers. Therefore the possibility of biased decisions can be simply expected, since both the d. constituents of the companies aim is for profit. For example, the government apart from its regulatory duties, also assumes the status of main shareholder through its mining entity called the Eritrean National Mining Company (ENAMCO) with the foreign companies as stated in Proclamation 157/2006 establishing the ENAMCO.[52] Besides, both the regulating ministries and the members of ENAMCO are appointed by the head of the state.

Therefore, it will be necessary to introduce a new law, or amend the existing ones, to reconfigure the composition of the board of directors of the mining companies, if this approach is to be pursued. However, as Philip Jennings: "Companies and governments overwhelmingly want the public both to believe in the ethical corporation and at the same time do not want to provide new legal backing for tighter ethical behavior."[53] Therefore, to avoid waste of time and the cost of enacting a new law, it is better if the available outlooks of good practices in the country are utilized proficiently. This will improve the business culture and corporate practices as a substitute for new rules aimed to reduce noncompliance resulting from decentralized regulations.

49 See Proclamation 6/1991, above n 39. Art. 362 and 364.

50 See Li Wen Lin, above n 19.

51 Provisional Government of Eritrea, Gazette of Eritrean Laws, Volume 25/2017, No.2, Proclamation 179/2017, The Eritrean Environmental Protection , Management and Rehabilitation Framework (hereinafter EPMRF), Art. 5 sub (8) and (9). Also, Interview with Dr. Seife Berhe, Executive Director of Andiamo Exploration, conducted on 15 September 2023, 25 September 2023 from 3:30-4:00 pm,

52 Provisional Government of Eritrea, Gazette of Eritrean Laws, Volume 15, No.5,. Proclamation 157/2006, Eritrean National Mining Corporation Establishment Proclamation.

53 Philip J. Jennings, UNI General Secretary, Union Network International, http:// www.union-network.org/uniindep.nsf/ .

Nationally Recognized Best Practices

In its newly introduced NEMP-E, the Department of Environment contemplates its plan to provide for the basics of what it considers best practices that should be promoted for environmental sustainability. For instance, endorsing multi-stakeholder approach to monitor and clear EIA reports on environmental issues through awareness enhancing programs and direct stakeholders' engagement in environmental conservation and compliance with global waste management best practices are some of the plans put in place.[54]

Multi-Stakeholder Engagement

In laying the groundwork for an inclusive engagement in which part of the society is included as a stakeholder, three distinguishing factors of decision-making process used in political arena from a typical market-controlled CSR can be utilized. First, instead of disbursements prone to unevenness, to consider preferences through votes. Second, conduct voting where the voter's preference may be considered. Third, involve an intermediary which can be a government agency or other independent organ. This movement can pave the way for a voluntary accord to be reached whereby the communities can play a decisive role in the smooth enforcement of environmental laws. It may create close bondage with the companies and the regulating body that will work to easily detect malpractices and forward solutions. This engagement as well can boost companies' activities.

Local Community

There is an inherent tension between the nature of industrial resource extraction and commitments to CSR, a tension that local communities in resource extraction areas can sense intuitively. This was seen in the implementation process of the Community Assistance Project (CAP). In the first attempt, CAP was not implemented; this deprived Bisha mining share company of a valuable mechanism to address potential risks to human rights, [55] as well as to contribute to national, regional and local development goals.[56] However, after interviews with community leaders and Sub-Zoba administrators were conducted about the basis for the implementation of the pilot projects prioritizing community needs and regional development priorities, the implementation was a success. Moreover, as similar experiences suggest in countries like Indonesia, local communities can have strong impact on local development and enhancement of human capital in their localities.

Traditional Civil Associations

A common denominator for traditional Eritrean civil society associations is that they take as their main objective the social welfare of their members, related with the aim of enhancing social solidarity and contributing to improve economic constraints.[57] A small local civil association is far more likely able to understand the real need of the people it represents, than an established and distant state organ.

54 DOE of MLWE, Draft Environmental Report for Reviewing and Updating NEMP-E 1995, Assessment report adopted on 7 September 2023,at 24.

55 Nevsun Resources Ltd., 'Human Rights Impact Assessment of The Bisha Mine In Eritrea(HRIA)', Lkl International Consulting INC. at 35. It also includes avoidance of discriminating activities and employment of grievance mechanism.

56 Ibid.

57 Redie Bereketeab, 'Conceptualizing Civil Society in Africa: The Case of Eritrea', Article in Journal of Civil Society (2009), at 8.

These organizations tend to employ less people and resources compared to a governmental agency to realize a similar objective. This can be for several reasons. But it is mainly associated with the deeply rooted idea of voluntarism in these organizations, which encourages people or private actors to devote the desired time and energy to provide solution to public problems proficiently. Additionally, resolutions made by such associations are forwarded on free and voluntary basis, where they can lessen the costs of the government. For instance, these societies have years of experience in distribution of limited resources which can aid in the distribution of benefits to be forwarded. In other words, their engagement in discussions associated to their sacred connection with the land can reduce societal dissatisfactions related to land use, relocation, and other socially disruptive behaviors the industry could cause.

Civic Organizations

Apart from the traditional civil organizations, there are several civil organizations supporting government social and economic projects. The existing civic organizations in Eritrea have reinforced the government in its national reconstruction and nation-building programs after independence. The Eritrean government is supportive of such organizations and urges them to play active role in the implementation of the country's policies and projects. This is essential where the government cannot fully satisfy demands of domestic actors. However, when Civil Society Organizations (CSOs) engage, they can make a calculated use of international alliances and intergovernmental organizations to alleviate the burden. For instance, there is the belief that the National Union of Eritrean Youth and Students (NUEYS), National Union of Eritrean Women (NUEW) and National Corporation of Eritrean Workers (NCEW) are CSOs and they function as such. Thus, given their available manpower and some specific experiences, they may enhance the bargaining power of local communities, particularly of the community groups they represent.

Diaspora

An effective CSR-based business model in the extractive sector is prone to remaning largely window dressing that serves a strategic purpose for mollifying public concerns inherently related with the destructive nature of extractive industries.[58] In this instance the Eritrean diaspora, with their exposure and the influence they enjoy, may strengthen the capacity to put considerable pressure on the companies and the government.

Disclosures of Reports

Companies are required to disclose social and environmental information to the public. This process does not mandate any substantive change in corporate behavior. Such disclosures serve as an indirect regulation where interested parties can interact with the companies. To realize this engagement, the reports should be prepared in a manner that is easy to understand for the host communities in order for them to comprehend, extrapolate and predict actions of the companies.

These reports can range from a simple public relations statement to a detailed and in-depth examination of the company's environmental performance, policies, practices and

58 Keith Slack, 'Mission impossible?: Adopting a CSR-based business model for extractive industries in developing countries', 37 Oxfam America Resources Policy 179 (2012), at 179–184. Available at: www.elsevier.com/locate/resourpol, (visited on September 4, 2023).

future direction.[59] Moreover, regardless of its specific objectives or targeted audience, it is essential for any environmental or social report, that the document remains relevant, reliable, comprehensible and comparable. For the beginning, it may be enough for the reports to focus primarily on the analysis on how the mining infrastructure is affecting the environment such as the company's performance in terms of pollutants, waste management and environmental investment.[60]

Conclusion and Recommendation

The adoption of CSR initiatives for mining companies in Eritrea largely remains voluntary, while the legal obligations with regard to the dimension of obligation clauses produce limited effect. Realizing these limitations the study assessed the viability of potential regulatory approaches for CSR and proposes solutions. The solution is to simply embrace the values of mandatory regulations of minimum standards with voluntary initiatives considering engagement of affected communities. This is essential in promoting responsible undertaking, because only when CSR activities adopt the existing values are the standards of socially responsible measures can be effective, whereas the existing level of the standards only ensure the fulfillment of the general responsibility and common minimum requirements.

Implementation of CSR in Eritrea is prone to becoming patronizing when companies undermine knowledge and skills of local communities to identify their own needs and priorities. This can lead to the lack of adaptation to local conditions, as companies appear to enforce their own vision of community 'good'. In terms of performance, CSR outcomes currently have a negligible contribution to society's welfare, because the perception of their social performance is also related to the high expectations formed around the mining companies. State intervention is a vital factor to rehabilitate and accelerate the economic growth of Eritrea. Companies cannot be relied upon to protect the society and environment without government necessary intervention.

It is essential to establish/improve private and public partnership linking the mining companies, Small and Medium Enterprises, government and local social organization. Such partnerships can aid in creating and enhancing the Small and Medium Enterprises contribution through setting an advisory office. Such partnerships may provide a platform for the transfer of their skill in managerial competence and entrepreneurial spirit that consider environmental and social responsibilities.

Lastly, in today's political economy, it is more likely that corporate policies and strategies will be adopted primarily to enhance the wealth of shareholders, not to promote social justice. Therefore, a fundamental restructuring of policies and thinking about the role of mining companies in community developmental programs ought to be made to conceptualize social responsibility as social capital.

References

Michael Hopkins, *Corporate Social Responsibility and International Development Is Business the Solution* ? 1st ed. (Bath: Bath Press,2007).

Andrew Ngawenja Mzembe and Yvonne Downs 'Managerial and stakeholder perceptions of an Africa-based multinational mining company's Corporate Social Responsibility (CSR)',(2014).

Mulus Wijaya Kusuma, 'Corporate Social Responsibility: Concepts and Practices in Indonesia', 137 (2018).

Raed Awashreh, 'Legalization of Corporate Social Responsibility in Dealing with Environmental Issues "Examples from Moroccan Arab Countries"', A'Sharqiyah University (2017). Available at: https://www.researchgate.net/publication/333166393 (visited on March 3, 2024).

Li Wen Lin, Mandatory Corporate Social Responsibility Legislation Around The World, Mandatory Corporate Social Responsibility Legislation Around The World, Emergent Varieties And National Experiences, 429 (2020).

William O. Brown, Eric Helland, Janet Kiholm Smith, Corporate Philanthropic Practices, 12 Journal of Corporate Finance 855 (2006).

Redie Bereketeab, 'Conceptualizing Civil Society in Africa: The Case of Eritrea', Article in Journal of Civil Society (2009).

UNDP, 'Analysis of the Potential Contributions of Colluli Potash Project to Sustainable Development Goals in Eritrea'.

Avni Shrivastava, 'Corporate Social Responsibility and Legal Accountability: Bridging the Gap, International Journal of Technical Science and Exploration (2022).

Giovanni Azzone, Michael Brophy, Giufiano Noci, Richard Welford and William Young, 'A Stakeholders' view of Environmental Reporting', 30 Long Range Planning, 699 (1997).

Philip J. Jennings, UNI General Secretary, Union Network International, http:// www.union-network.org/uniindep.nsf/

Hui Wen, 'Global Corporate Social Responsibility Reporting Regulation: Drivers and Impacts on Sustainable Development', University of Illinois at Urbana-Champaign November (2017)

Carla Vintró and Josep Comajuncosa, 'Corporate Social Responsibility In the Mining Industry: Criteria and Indicators', University of Colombia, Volume 77, 31 (2010).

Angessa Eticha Asela, 'The Need to Adopt Corporate Social Responsibility Strategy in Ethiopia, a research thesis submitted to Arsi University School of Law, (2020).

Madeleine Busacca, 'Corporate Social Responsibility in South Africa's Mining Industry: Redressing the Legacy of Apartheid', Claremont McKenna College Senior Thesis, (2013).

Tasew Abitew, 'Approaches to Regulating Corporate Social Responsibility in Ethiopia: The Case of Manufacturing Companies', Addis Ababa University, College of Law and Governance Studies School of Law Masters of Business Law (2021).

Interviews:

Interview with Dr. Seife Berhe, Executive Director of Andiamo Exploration, conducted on 15 September 2023, 25 September 2023 from 3:30-4:00 pm, 10 October 2023 from 4:00-4:20 pm, and 4 March 2024 from 10:00-10:25 am.

Interview with Mr. Abel Yohannes, member of the Impact assessment unit of the DOE conducted on 9 October 2023, from 10:30 am -11:35 am and 28 November 2023, from 10:00-10:30 am.

Interview with Mr. Sium Tekle, Director General at the Ministry of Justice conducted on 23 February 2024, from 4:00 – 4:30 pm.

Reports:

DOE of MLWE, Draft Environmental Report for Reviewing and Updating NEMP-E 1995 abridged, Assessment report adopted in 07 September 2023.

NEVSUN Resource LTD., 'Corporate Social Responsibility Report', 2014, at 4. Available at: www.nevsun.com.

Nevsun Resources Ltd., 'Human Rights Impact Assessment of The Bisha Mine In Eritrea(HRIA)', Lkl International Consulting INC.

Websites:

B The Change, 'CSR in India is now a law'. Available at: https://bthecahnge.com/csr-in-india-is-now-a-law-2502 (visited on February 4, 2024).

Anel Karnani, 'Mandatory CSR In India: A Bad Proposal'. Available at https://ssir.org/articles/entry/mandatory-csr-in-india-a-bad-proposal (visited February 4, 2024).

Clotilde Marielle, 'Why adopt a regulatory and legal framework to facilitate the growth of CSR?', https://Sofercom.com/en/news-insights/why-adopt-a-regulatory-and-legal-framework-to-facilitate-the-growth-of-csr.html (visited on September 23, 2023

Provisional Government of Eritrea, Gazette of Eritrean Laws, Volume 5, No.9, Proclamation 75/1995, A Proclamation To Provide For The Establishment of The Eritrean Standards Institution

Provisional Government of Eritrea, Gazette of Eritrean Laws, Volume 1, No.2, Proclamation 6/1991, Proclamation to establish Transitional Commercial Code of Eritrea

Provisional Government of Eritrea, Gazette of Eritrean Laws, Volume 15, No.5,. Proclamation 157/2006, Eritrean National Mining Corporation Establishment Proclamation.

Provisional Government of Eritrea, Gazette of Eritrean Laws, Volume 25, No.2, Proclamation 179/2017, The Eritrean Environmental Protection, Management and Rehabilitation Framework.

Legal Transplantation in Eritrea: History, Practice, and Prospects

Tomas Yohanes Abraha
Minister of Justice
Email: tomasjohannes54@gmail.com

Abstract

Legal transplantation has been a central feature of Eritrea's legal development, from customary systems to codified laws introduced through colonial rule, federation with Ethiopia, and post-independence reforms. This practice has produced both successes and failures depending on how well transplanted laws align with local contexts. This paper examines the history and practice of legal transplantation in Eritrea, analyzing motivations, outcomes, and critical factors influencing effectiveness. Through historical analysis and comparative evaluation, the study focuses on two examples: the Tobacco Control Proclamation and the Regulation on Ozone-Depleting Substances (ODS). These case studies illustrate the contrasting impacts of cultural compatibility, institutional readiness, and international support on legal implementation.

The findings reveal that successful legal transplantation in Eritrea depends on contextual adaptation, stakeholder participation, and enforcement capacity. Laws that lack these foundations tend to remain unenforced, despite formal adoption. The paper concludes by offering practical recommendations to guide future law reforms toward sustainable and functional legal development.

Keywords: Legal transplantation, Eritrea, law reform.

Introduction

Legal transplantation[1]—the movement of laws or legal institutions from one jurisdiction to another—has long been a central yet controversial concept in comparative law. The term gained prominence through the work of Alan Watson, who argued that most legal development occurs through the borrowing and adaptation of foreign legal ideas, regardless of cultural boundaries. According to Watson, the law is largely autonomous and can be transplanted from one legal system to another with minimal disruption. In his view, societies often adopt foreign laws not because of legal necessity, but because of perceived prestige or modernization value.[2]

However, this optimistic view has been challenged by scholars such as Pierre Legrand and Gunther Teubner. Legrand argues that legal norms are deeply embedded in the cultural, linguistic, and social frameworks of the societies that produce them, and therefore cannot be effectively transferred without losing their original meaning and function. He insists that

1 Legal transplantation is also referred to as *legal reception, legal borrowing, legal diffusion*, or *legal importation.*

2 Alan Watson, *Legal Transplants and Law Reform* (1976), 79-84.

what is transferred is often only the form of law—not its substance. Teubner extends this critique by introducing the concept of "legal irritants," suggesting that transplanted laws do not effortlessly integrate into the recipient legal system but instead create friction, resistance, and unintended consequences, which may eventually lead to hybrid legal forms.[3] These divergent perspectives illustrate that legal transplantation is not a uniform or neutral process. Rather, it is shaped by power dynamics, historical context, institutional capacity, and the socio-cultural fabric of the recipient country. In many cases, transplanted laws are adopted symbolically to align with international expectations or to facilitate access to global institutions, while remaining underenforced or unobserved at the local level.

Eritrea's legal history offers a compelling case study in the complex realities of legal transplantation. From pre-colonial customary laws to Italian colonial legislation, from imposed Ethiopian codes to post-independence legal reforms, Eritrea's legal system is a layered product of both voluntary and involuntary borrowings. Legal transplantation has occurred under varying motivations: colonial administration, legal harmonization, treaty compliance, ideological alignment, and practical necessity. However, the success of these transplants has been uneven. Some laws have been effectively domesticated and implemented, while others have failed due to lack of contextual relevance or institutional support.

Despite the central role of legal transplantation in shaping Eritrea's legal framework, scholarly attention to this phenomenon has been limited. This paper seeks to fill that gap by providing a comprehensive analysis of legal transplantation in Eritrea—past and present. It aims to identify the historical conditions, political motivations, and socio-cultural dynamics that have shaped both successful and failed transplants. Through this, it highlights the need for future legal reforms in Eritrea to be not only informed by global standards but also carefully adapted to local realities.

The Process of Legal Development in Eritrean Customary Laws

Eritrean customary laws largely evolved in direct attachment to the daily practices and livelihoods of the people. While not comprehensive for addressing all the complexities of 21st-century life, these laws encompassed a wide array of issues including domestic relations (betrothal, marriage, emancipation, divorce), inheritance, land rights, torts (damage done by and to animals, insults, bodily injury), adultery, and procedural matters such as civil procedure and rules of evidence, including witness and oath standards. Studying these developments is critical for understanding Eritrea's longstanding tradition of incorporating external legal concepts while preserving indigenous values.

Friederike Kemink concluded that "the majority of those that are known as Eritrean customary laws are translations, often of religious works not originating in the region. Few texts can truly be said to be indigenous sources reflecting the culture of the local population."[4] This writer challenges the Kemink claim, arguing that although religious influence exists, it does not deny the local innovation and development inherent in these laws. For example, Afar customary law was notably weakened after the community's conversion to Islam, especially concerning personal status, marriage, divorce, and family relations. In contrast, Bilen customary law remained largely secular.[5] This distinction offers compelling evidence

3 Kahn-Freund, *On Uses and Misuses of Comparative Law* (1974), 1-27

4 Friederike Kemink, 'The Tegreñña Customary Law Codes', Paideuma: Mitteilungen zur Kulturkunde, 37, 55, (1991), at 67.

5 Lyda Favali and Roy Pateman, *Blood, Land and Sex: Legal and Political Pluralism in Eritrea*, 1st ed. (Indiana Uni-

of legal systems that evolved independently of religious prescriptions, indicating a degree of indigenous legal autonomy.

There are also visible signs of expected domestic cross-influence among Eritrean customary laws. In some cases, customary law was involuntarily imposed across regions. For instance, Degiat Hailu of *Tseazega* compelled the people of *Loggo Chewa* and *Kabasa Chewa* in southern *Hamasen* to abandon their local customs and instead follow the code of *serate Habteslus Gebre-Kristos*.[6] On the other hand, voluntary transplantation of legal norms also occurred. The residents of *Zeban Seraw* and *Endadeqqo,* who had immigrated from Agame in Tigray, settled in an area where *Mai-Adgi* customary law was dominant. When they were denied participation in a conference for amending this law—on grounds that they were not descendants from the area—they responded by declaring their own legal code, which was essentially a transplant of the *Meem Mehaza* law with slight modifications.[7]

Harmonization also played a key role in the evolution of Eritrean customary laws. A notable example is *Hgi Shewate Anseba*, formed when seven previously distinct groups in the Anseba region came together to draft a single, consolidated code. Similarly, *Ser'at Adgena Tegeleba* emerged from a harmonization and standardization of various customary laws in the Akkele Guzai area. In some instances, foreign legal norms were adopted directly into customary law. For example, the *Adkame Mlgae* law incorporated provisions from Italian civil law concerning paternity. The following case is one example:

> Traditionally, a woman's oath alone could establish paternity—even in cases where the alleged father had no physical contact or was deceased. This practice was challenged by Italian authorities, who ruled that claims extending beyond the typical pregnancy period were invalid. This standard was subsequently codified into the 1948 *Adkame Mlgae* legal code.[8]

The European Colonial Rule and Federation Period

During more than fifty years of Italian colonial rule, Eritrea's legal system was heavily influenced by Italian models. Italian legislation served as the primary source of formal law. However, rather than pursuing a policy of full legal assimilation, the Italians implemented a strategy that recognized and selectively incorporated local customary laws into the colonial legal system. In the second decade of colonization, the drafting of new codes and the expansion of Italian state law were undertaken, but these laws were only applicable "as far as conditions of the locality permit," and subject to exceptions defined by particular colonial statutes.[9] This conditional application reflects an early awareness of the incompatibility between European legal constructs and local realities. The Italian approach left room for contextual adaptation, even within a colonial framework.

When the British Military Administration took over Eritrea during World War II, it adhered to its obligations under the 1907 Hague Convention, which prohibited occupying powers from making substantive changes to local institutions and laws. Accordingly, the British did not undertake substantial legal reforms in Eritrea. Changes made under their rule were

versity Press, 2003) at 7-8.

6 See Frederik Kemink, above n 4, at 61

7 Ibid.

8 Dennis J. Duncanson, 'Sirat Adkeme Mlgae – A Native Law Code of Eritrea', Journal of the International African Institutte, XIX, 2, (1949), at 141-149.

9 See Fevali and Pateman, above n 5, at 44.

minimal and piecemeal, generally limited to adjustments for administrative convenience or translation into English. The British retained the existing Italian legal system and made no significant legal transplantations during their administration.

Following the United Nations' Federal Act of 1950, Eritrea was federated with Ethiopia, leading to the 1952 Eritrean Constitution. This constitution was modeled on Western democratic systems and has been described as an archetype imposed by the United Nations Commissioner, with limited public participation in its drafting process. According to Rosen, the 1952 constitution "had no roots in the populace."[10] This characterization underscores the disconnection between the legal framework and Eritrean social context. Despite its foreign roots, the 1952 Constitution maintained provisions for the respect of customary law, especially in matters related to personal status, family law, and succession. Eritrean nationals retained the right to live under their respective customs in these areas, which highlights the enduring significance of customary law even during externally influenced legal reforms.[11]

The Legal Reform of Ethiopia (1955–1965)

Following a decade of gradual erosion of the 1952 Eritrean Constitution, Ethiopia formally annexed Eritrea in 1962. This move extended Ethiopian laws to Eritrea, making the 1955 Ethiopian Constitution and subsequent legal codes as the governing law in the empire, including Eritrea.[12] Importantly, by this time Ethiopia had already undertaken an ambitious legal codification process from 1955 to 1965. The legal history of this reform process is important for Eritrea since these laws application was extended to Eritrea after the annexation. Moreover, after independence, Eritrea adopted these laws voluntarily, with some amendments, as transitional codes which continue to apply at the time of writing of this paper.

Several reasons have been proposed for Emperor Haile Selassie's push for legal codification. Chief among them was the desire to replace diverse customary, ethnic, and religious legal systems with a uniform legal order based on state law. Moreover, some scholars argue that codification was part of the emperor's broader political aim to consolidate imperial authority. It did not make concessions to non-Abyssinian Ethiopians, particularly the non-Orthodox Christian *gebbars* (farmers) of imperial Ethiopia's peripheries.'[13]

The rationale behind these Codes can also be explained through the views of the drafters of the new codes. According to Graven, for instance, the drafting process of the Penal Code was most concerned with giving Ethiopia a set of laws that would draw the country into the modern world.[14] It aimed to modernize Ethiopia by introducing laws that assumed the existence of developed legal institutions and practices, most of which did not exist at the time. The Civil Code, prepared by the renowned French comparativist René David, was also a transplant based on foreign legal traditions. David himself stated that his goal was to draft "a civil code for Ethiopia, not the civil code of Ethiopia,"[15] implying the deliberate

10 Haile Semere, 'The Roots of the Ethiopia-Eritrea Conflict: The Erosion of the Federal Act,' Journal of Eritrean Studies, 1, 1, (1996), at 1–18.

11 Eritrean Constitution of 1952, Chapter 5, Article 36. Reproducing the text of the 1952 constitution would add color here

12 William H. Ewing and Beyene Abdi, *Consolidated Laws of Ethiopia*, vol. I, eds. (Addis Ababa, Ethiopia: Haile Sellassie I University, 1972), at 39.

13 Hailegabriel G. Feyissa, 'Complexities of Legal Codification Imperial Ethiopia', Third World Approaches to International Law Review, Issue 1, 152, (2020), at 153.

14 Graven, 'The Penal Code of the Empire of Ethiopia,' Journal of Ethiopian Law, 1, (1964) at 267 – 79.

15 Rene David, 'A Civil Code for Ethiopia: Considerations on the Codification of the Civil Law in African Countries,' Tulane Law Review, 37, (1963).

detachment from local legal customs.[16] Although David's original draft included adaptations of Ethiopian customary law principles, these elements were ultimately rejected by because of a lack of concern from the Ethiopian legal commission. Many commission members lacked adequate knowledge of customary law and were overly eager to Westernize the system. As a result, the codes had limited compatibility with local Ethiopian and Eritrean social realities. Schiller described the outcome as a "fantasy law," disconnected from the lived experiences of most Ethiopians and Eritreans.[17]

Nevertheless, some local elements, such as customary rules regarding family law, succession, and land tenure, were incorporated. However, the general approach remained foreign in both form and substance. According to Redden, customary laws particularly in the areas of family law, succession law, and land laws was taken in to account in shaping the code along the lines suited to Ethiopia. In exception to those few considerations, the civil code was officially promulgated as an original work of the drafter with radical changes in substance which were incompatible to the Ethiopian society.[18] Thus, this policy choice of legal transplantation in Ethiopia had brought a complete disruption of the institutions most closely valued by members of traditional society.[19]

To conclude, the legal penetration[20] of the new laws was questionable by many Ethiopian writers. The widespread application of these codes in Eritrea had minimal impact. Customary law persisted robustly, and the foreign legal codes were largely ineffective in displacing existing social and legal norms. The codification process was also criticized for being secretive and rushed which was typically alien to the Eritrean society, in which laws are made through deeply rooted and inviolable sacred public participation right of community.

EPLF Laws During the Armed Struggle

From the beginning of its founding in 1970 until 1977, the Eritrean People's Liberation Front (EPLF) operated without a comprehensive legal system, aside from a Military Law based heavily on the Chinese People's Liberation Army code of conduct. This choice likely stemmed from the military training background of EPLF leadership in China.[21]

At its first Congress in 1977, EPLF introduced its first legal reforms: amendments to the Military Law and the enactment of a Marriage Law. The latter was a near-literal Tigrinya translation of the Chinese 1950 Marriage Law, which sought to abolish patriarchal and co-

16 Tsegaye Beru, 'Brief History of the Ethiopian Legal Systems - Past and Present', International Journal of Legal Information, 41, 3, 335, (2013) at 351.

17 Arthur Schiller, 'Customary Land Tenure among the Highland Peoples of Northern Ethiopia: A Bibliographical Essay, African Law Studies, (1969), at 1- 2.

18 T.W. Bennett and T. Vermeulen, 'Codification of Customary Law', Journal of African Law, 24, 206, (1980) at 208. *See* also Liku Worku Haile, 'A critical analysis of Ethiopian Civil Code: in the light of the core features of Continental European Codification', (LL.M thesis on file Institute of Advanced Legal Studies School of Advanced Study University of London, 2012) at 36.

19 Norman J. Singer, 'The Ethiopian Civil Code and the Recognition of Customary Law', House Law Review, 9, (1971-1972), at 460.

20 Legal penetration refers to the reality which answers the question whether people have actually adjusted their behaviors to the prescriptions of the imported laws or not.

21 In the mid of 1960's, the prominent founders and leaders of the EPLF, Isias Afeworki and Romodan Mohammed Nur, was trained in China for more than a year. *See* Dan Connel, *Conversation with Political Prisoners of Eritrea,* (2011), at 27-73.

ercive marriage practices. Article 1 of the EPLF Marriage Law closely mirrors Article 1 of the Chinese Marriage Law, both outlawing feudal norms and emphasizing gender equality.

EPLF Marriage Law Article 1.: *'The feudal marriage norm based on the supremacy of men over women which entails arbitrary and coercive arrangements and which did not safeguard the welfare of children shall be banned.'*[22]

Chinese Marriage of 1950, Article 1.: *'the law proclaimed that the arbitrary and compulsory "feudal marriage" system which was based on superiority of man over woman and which ignored the interests of the children was to be abolished.'*[23]

EPLF also introduced a constitutional framework inspired by the Chinese Communist Party and the Palestine Liberation Organization. To ensure legitimacy, the Front undertook public consultation efforts involving fighters, diaspora Eritreans, members of clandestine organizations, and community organizations. This public participation program was similar to the customary style of law-making process in Eritrea and was a highly contributory factor to strict observance of the foreign-based law among the freedom fighters.

At its second Congress in 1987, the EPLF established a High Court and a Prosecution Office under the Department of Justice. The growing need to manage conflicts between fighters, civilians, and bandits (*shiftas*) led to the drafting of comprehensive substantive and procedural codes. These were heavily influenced by socialist legal systems of countries such as Romania, Yugoslavia, and China, and also by socialist-oriented Ethiopia. This was due to the EPLF's Marxist ideology and the availability of related literature. Ethiopian legal codes were also referred to, largely because most of the drafters were educated under that system at Haile Selassie I University, and also because of the cultural and socio-economic similarities between Ethiopia and Eritrea. These codes had been in use in Eritrea for decades, making them familiar to the population. Additionally, the EPLF's wartime legal practices, especially in criminal matters, were incorporated. All drafts were ultimately approved by the EPLF Central Committee in line with its constitutional procedures. Despite thorough preparation, these codes were never officially enacted or published, underscoring the limitations of wartime legal reform.

Transitional Codes of Eritrea

After de facto independence in 1991, the newly formed Eritrean Department of Justice was tasked with issuing transitional legal codes. The codes prepared after EPLF's Second Congress were deemed unsuitable for national application, as they had been designed only for the period of armed struggle. Eight transitional codes were introduced in 1991, mainly adaptations of Ethiopian codes, with minimal modifications. These were meant to serve until a permanent legal framework was established. The reforms prioritized consistency with EPLF's values. The test was Eritrean transitional laws should not be in conflict with the EPLF Constitution and should not contradict the values and guiding principles for which the EPLF fought during armed struggle for independence.[24] This major legal transplantation undertaken in September 1991 can be regarded as a pragmatic and strategic response to

22 Crish Silkin, 'The 1997 Marriage Law Arrangements within the EPLF' (Paper presented at the Conference of the Research and Information Centre on Eritrea, Eritrea, July 1988).

23 Roxana NG, 'The Marriage Law and Family Change in China With Special Reference To Kwantung Province, 1950 -1953', (MA thesis in file of University of British Colombia, 1975).

24 Yoel Kidanemariam, 'Legislative Process In Eritrea,' (LL. B Thesis in file College of Business and Social Sciences, 2022), at 34-38.

transitional challenges. At the time, the Provisional Government of Eritrea (PGOE) faced a significant dilemma about how to administer justice in the newly liberated Eritrea. Legal transplantation offered an immediate and functional solution. By adopting the transplanted transitional codes, the government was able to preserve legal continuity and maintain public order, thereby ensuring institutional stability and safeguarding the status quo of the Eritrean legal system during a critical period of transformation.

Although the transplantation of the codes was intended only for the transitional period, their implementation faced significant challenges, many of which stemmed directly from their transplanted nature and lack of contextual adaptation. First, the judiciary lacked trained personnel; many judges were former fighters with only six months to one year of legal training. Second, cultural mismatches hindered legal observance. For instance, the committees failed to distinguish between legal domains that require careful contextual adaptation—such as family law, the law of persons, and internal administrative matters—and those of a more universal character, like contract, commercial, and maritime law. This undifferentiated approach to transplantation resulted in several cultural and legal disconnects.

A notable example in this context is the treatment of dowry and material contributions in marriage celebrations. Under the Transitional Civil Code, both the request and acceptance of dowry are prohibited and deemed illegal. This provision was directly transplanted from the Chinese Marriage Law. In the Chinese context, such legal reforms were introduced to eliminate what were referred to as "feudal elements" in marriage, such as the exchange of bride price and ceremonial gifts. The Chinese law sought to prevent women from being commodified in marriage and aimed to dismantle traditional family structures altogether.[25] By contrast, the EPLF Marriage Law focused primarily on safeguarding women's rights within marriage, grounded in the belief that some customary marriage practices were inconsistent with the EPLF's principles of gender equality. However, the objective was not to abolish the entire Eritrean marriage system.

As some argue that dowry practices were not akin to bride selling in Eritrean context. They were seen as compensation for the bride's loss of inheritance and often matched the expenses incurred by the groom. Typically, the amount of dowry provided by the bride's family is proportionate to the expenses borne by the groom, including jewelry, clothing, and other personal items that ultimately remain the property of the bride.[26] Given this cultural context, the provisions of the Transitional Civil Code concerning dowry and marriage gifts have become practically unenforceable relegated to "dead wood" due to their lack of resonance with Eritrean social norms and practices.

Overall, the transitional codes, initially intended to serve only during the interim period, continue to function as the principal legal instruments in Eritrea. Although the government has undertaken a comprehensive law reform process aimed at replacing these codes with newly drafted Eritrean laws, the outcome of that reform remains incomplete. The draft versions of the new codes were circulated to the public in 2015 as part of a participatory effort to gather feedback. However, to date, they have not been officially promulgated and therefore do not have legal force.

---

25 See Roxana above n 23.

26 Mulubrhan Berhe Hagos, 'Customary Versus Modern Laws of Eritrea and Intangible Cultural Heritage', (Paper submitted to the National Consultation Meeting in Eritrea on the 2003 Convention of Intangible Cultural Heritage, April 2009), at 13.

Legal Transplantation in Contemporary Eritrea

Motivations for Legal Transplantation

Legal transplantation may involve the transfer of a single legal rule or the wholesale adoption of an entire legal system. As Valderrama (2006) explains, there are several motivations behind a state's decision to transplant foreign laws. In some cases, transplantation occurs involuntarily, most notably through colonization, where legal systems are imposed by external powers. In other instances, it is undertaken voluntarily, either in admiration for the donor system or in pursuit of legal modernization. Transplantation may also serve as a tool for unifying or harmonizing divergent legal regimes. Additionally, it can offer a solution to a legal reform problem shared by both donor and recipient jurisdictions. Finally, states may adopt foreign laws simply because those solutions have proven effective in practice and are perceived as functionally successful.

In Eritrea, legal transplantation continues to serve multiple purposes depending on the legislative context. A significant motivation has been the need for an immediate and functioning legal order, especially evident during the adoption of the transitional codes in 1991. At that time, the government adopted Ethiopian laws—originally based on Western legal systems—to meet the urgent need for legal stability. Legal transplantation is also driven by the need for harmonization with regional and international legal norms. In areas like marine transportation, Eritrean laws follow standards developed by international organizations such as the International Maritime Organization (IMO). Requirements like sea worthiness in maritime operations, for example, demand the adoption of universally recognized standards.

Another key motivation is the need to fulfill Eritrea's treaty obligations. The Ozone-Depleting Substances (ODS) Regulation (Legal Notice No. 117/2006) was enacted to domesticate Eritrea's responsibilities under the Montreal Protocol. Similarly, legal transplantation has also occurred to support institutional goals. The Cultural and Natural Heritage Proclamation (Proclamation No. 177/2015), for instance, was enacted primarily to facilitate the inscription of Asmara in the World Heritage List maintained by UNESCO.[27] Finally, legal transplantation in Eritrea has been ideologically motivated. Gender equality provisions in Eritrea's marriage law were influenced by socialist legal models, aiming to reflect the EPLF's commitment to women's emancipation and equality.

Legal Transplantation Outcomes: Does it Work?

Legal transplantation is a multifaceted process, and its success cannot be guaranteed by a single set of conditions. Rather, the determinants of effective transplantation are diverse and context-specific, often differing widely from one jurisdiction to another. Among the most critical factors are the intrinsic transferability of the transplanted law—whether it can operate effectively in a new cultural or institutional setting—and the extent to which the law is adapted to local conditions during both the adoption and implementation phases. Equally important is the legal and societal demand for the law in the recipient country; laws that address real and recognized needs are more likely to be accepted and enforced. Public awareness, understanding, and perception of the transplanted law also play a vital role in shaping its legitimacy and practical observance. Without institutional support and societal engagement,

27 Senai W. Andemariam, 'Eritrea's New Heritage Law: Drafting and Implementation Issues', International Journal of Cultural Property, 26, 167, (2019), at 169.

even technically sound legal transplants may remain ineffective in practice.[28]

However, despite the complex and varied factors that influence the success or failure of a transplanted law, a common denominator remains: the extent to which the law achieves the objectives for which it was originally promulgated.[29] In other words, a successful legal transplant is one that fulfills its intended function within the recipient legal and societal framework, while a failed transplant is marked by ineffectiveness, non-enforcement, or social resistance. To assess the conditions that contribute to success or failure in the Eritrean context, this paper employs a comparative analysis of two notable pieces of legislation: the Regulation on Ozone-Depleting Substances (ODS Regulation) and the Tobacco Control Proclamation. These case studies offer valuable insight into how contextual alignment, institutional capacity, and public engagement influence the outcomes of legal transplantation in Eritrea.

The Tobacco Control Proclamation (Proclamation No. 143/2010)

The Tobacco Control Proclamation (Proclamation No. 143/2010) was drafted using the WHO Framework Convention on Tobacco Control (FCTC) as its model. It addresses a wide range of regulatory concerns including designated smoking and non-smoking areas, bans on the advertisement and sponsorship of tobacco products, and the prohibition of tobacco sales to minors. The law also requires that point-of-sale provisions and other detailed enforcement mechanisms be clarified through subsequent regulations. Despite its adoption, however, it remains difficult to affirm that the proclamation has been fully implemented in practice.

Several factors have contributed to its poor enforcement, beginning with the absence of a strong institutional framework. Although the proclamation designates the Ministry of Health as the principal enforcement authority, the specific department assigned to this role—the Department of National Medicines and Food Administration—has declined to assume responsibility. The department has argued that tobacco control falls outside its mandate, considering smoking a broader public health issue rather than a regulatory concern.[30] As a result, the department has made no meaningful efforts to enforce the law. This institutional gap is evident in several areas: the department has not issued secondary enforcement regulations as required by Article 6(4), has not established monitoring bodies, and has failed to appoint inspectors to oversee compliance. These lapses clearly illustrate the challenges of implementing a transplanted law in the absence of institutional readiness and commitment, a condition essential for the success of any imported legal framework.

A second major barrier to implementation lies in the government's failure to ratify the WHO Framework Convention. Ratification would have enabled Eritrea to access technical support and financial resources for public awareness and enforcement initiatives, which are critical for operationalizing the proclamation. Given that Eritrea does not have a significant

28 Fei Deng, 'Legal Transplant as a Device of Legal Change in Transitional Economies: The Case of Importing Common-Law-Style Corporate Fiduciary Duties into Contemporary China' (PhD thesis on file School of Law College of Social Sciences University of Glasgow, 2021).

29 Michal S. Gal, 'The Cut and Paste of Article 82 of the EC Treaty in Israel: Conditions for a Successful Transplant Special Issue Approximation to EU Law' European Journal of Law Reform, 9(3), 467, (2008), at 471.

30 Ahmed Mussa, 'Critical Review Of The Implementation Of The Eritrean Tobacco Control Proclamation No 143/2004', (LL. B Thesis on file School of Law, College of Business and Social Sciences, Eritrea, 2014).

tobacco industry or commercial interest in tobacco exports—unlike major producers such as Brazil or countries with dominant industry players like the United Kingdom—ratification would pose minimal economic risks. Yet paradoxically, despite adopting the content of the WHO Framework in its national law, Eritrea has refrained from joining the convention. This disconnect highlights a critical lesson in legal transplantation: effective implementation often requires adopting not only the text of international frameworks but also the accompanying support structures, such as treaty ratification and donor alignment.

A third factor limiting implementation is the lack of contextual adaptation to Eritrea's tobacco landscape. The WHO FCTC, and by extension the proclamation, focuses primarily on smoking products such as cigarettes. However, in Eritrea—especially in the lowland regions—smokeless tobacco products, known locally as *safas*, are far more prevalent. These products are often produced and distributed informally, packaged in small plastic wraps without labeling or health warnings. Because the proclamation's provisions were largely adopted wholesale from the WHO model, they fail to address these local practices. The absence of regulatory provisions for smokeless tobacco significantly undermines the law's relevance and enforceability in much of the country. This lack of adaptation has further contributed to the law's limited observance and, in effect, rendered key parts of the proclamation inoperative.

Ozone Depleting Substances Regulation [ODS Regulation]

The ODS Regulation (Legal Notice No. 117/2006) stands out as one of the most effectively implemented pieces of legislation in Eritrea. Its successful application offers valuable lessons on how legal transplantation can be carried out in a practical and context-sensitive manner. Several factors contributed to this success. First, the drafting process of the regulation was grounded in empirical research. A comprehensive national study was conducted to assess the types and annual consumption of ozone-depleting substances (ODS) present in Eritrea. Based on this data, a technical committee was formed to oversee the drafting, and major stakeholders affected by the regulation—drawn from both the public and private sectors—were actively involved. This inclusive approach enabled the identification of realistic phase-out targets aligned with the requirements of the Montreal Protocol and ensured that the law was tailored to Eritrea's specific environmental and industrial context.

Second, the nature of the law itself contributed to its successful transplantation. Unlike culturally embedded legislation such as the Tobacco Control Proclamation, the ODS Regulation is primarily technical and falls under the category of instrumental law. As a relatively new regulatory field even at the global level, ODS regulation had no cultural or historical antecedent in Eritrean society. This absence of cultural entrenchment meant that the law did not conflict with traditional norms or societal practices. For example, ODS-related substances in Eritrea are primarily found in refrigerators used for domestic purposes or in industrial cold rooms. Crucially, the regulation does not require the replacement of existing refrigeration systems unless the gas they contain leaks during maintenance. As such, the regulation posed no economic burden on the public and was unlikely to provoke resistance.

Third, institutional commitment has been central to the regulation's success. The Ministry of Land, Water and Environment established a specialized body known as the Ozone Protection Unit to oversee implementation. This dedicated unit performs regular monitoring, organizes training programs for refrigerator and air conditioning technicians, and conducts workshops with stakeholders such as the Ministry of Marine Resources, Asmara Brewery, and Coca-Cola Eritrea. This level of organizational focus contrasts sharply with the Tobacco Control Proclamation, where enforcement responsibility was vaguely assigned to the Ministry of Health without a designated implementing body.

Finally, international support has played a decisive role. Eritrea's ratification of the Montreal Protocol enabled the Ministry of Land, Water and Environment to receive substantial technical and financial assistance from the United Nations Environment Program (UNEP). According to officials from the Ministry, almost the entire operational budget of the Ozone Protection Unit is financed through UNEP subsidies. In addition to funding, UNEP has provided training opportunities for staff members and supplied modern equipment used to detect ODS at ports of entry. This international backing has been instrumental in ensuring the law's practical effectiveness and long-term sustainability. In sum, the ODS Regulation exemplifies a successful legal transplant. Its success stems from a combination of contextual research, technical relevance, institutional dedication, and international support—factors that can serve as benchmarks for future legal reform efforts in Eritrea.

Conclusion and Recommendation

Legal transplantation has long been a feature of Eritrean legal development, dating back to the evolution of customary laws and continuing through colonial, federation, and post-independence periods. Foreign laws, concepts, ideologies, and institutional models have played a central role in nearly every major legal reform effort undertaken in the country. These transplanted laws, which now constitute the majority of Eritrea's formal legal framework, have entered the system either involuntarily—imposed during periods of colonization and federation—or voluntarily, adopted by Eritrean lawmakers in the pursuit of modernization, harmonization, or international alignment.

The experience of Eritrea demonstrates that legal transplantation is neither inherently successful nor necessarily doomed to fail. Rather, outcomes vary significantly based on a range of factors that differ from one legislative context to another. Among these, two stand out as particularly decisive: the degree to which the transplanted law aligns with the cultural, social, and institutional context of Eritrean society, and the extent to which supporting legal and non-legal mechanisms—such as enforcement institutions, public awareness efforts, and financial resources—are established alongside the transplanted law.

In light of these observations, the issue of designing and executing a proper legal transplantation process becomes essential for any future legal reform in Eritrea. This involves not only identifying functional legal models abroad but also rigorously assessing their suitability within the Eritrean context. By prioritizing adaptation, stakeholder engagement, and institutional readiness, Eritrea can ensure that transplanted laws are not only adopted but also effectively enforced and socially embraced.

References

Andemariam, S. W. (2019). Eritrea's new heritage law: Drafting and implementation issues. *International Journal of Cultural Property, 26*, 167–169.

Bennett, T. W., & Vermeulen, T. (1980). Codification of customary law. *Journal of African Law, 24*, 206–208.

Connel, D. (2011). *Conversations with political prisoners of Eritrea.*

David, R. (1963). A civil code for Ethiopia: Considerations on the codification of the civil law in African countries. *Tulane Law Review, 37.*

Deng, F. (2021). *Legal transplant as a device of legal change in transitional economies: The case of importing common-law-style corporate fiduciary duties into contemporary China* (PhD thesis). University of Glasgow.

Duncanson, D. J. (1949). Sirat Adkeme Mlgae – A native law code of Eritrea. *Journal of the International African Institute, 19*(2), 141–149.

Ewing, W. H., & Abdi, B. (Eds.). (1972). *Consolidated laws of Ethiopia* (Vol. I). Haile Sellassie I University.

Favali, L., & Pateman, R. (2003). *Blood, land, and sex: Legal and political pluralism in Eritrea.* Indiana University Press.

Feyissa, H. G. (2020). Complexities of legal codification in Imperial Ethiopia. *Third World Approaches to International Law Review, 1*, 152–153.

Gal, M. S. (2008). The cut and paste of Article 82 of the EC Treaty in Israel: Conditions for a successful transplant. *European Journal of Law Reform, 9*(3), 467–471.

Hagos, M. B. (2009). *Customary versus modern laws of Eritrea and intangible cultural heritage* (Paper presented at the National Consultation Meeting on the 2003 Convention of Intangible Cultural Heritage, Asmara).

Haile, H. S. (1996). The roots of the Ethiopia-Eritrea conflict: The erosion of the federal act. *Journal of Eritrean Studies, 1*(1), 1–18.

Haile, L. W. (2012). *A critical analysis of the Ethiopian Civil Code: In the light of the core features of continental European codification* (LL.M thesis). University of London, Institute of Advanced Legal Studies.

Kahn-Freund, O. (1974). On uses and misuses of comparative law. *The Modern Law Review, 37*, 1–27.

Kidanemariam, Y. (2022). *Legislative process in Eritrea* (LL.B thesis). College of Business and Social Sciences, Eritrea.

Kemink, F. (1991). The Tegreñña customary law codes. *Paideuma: Mitteilungen zur Kulturkunde, 37*, 55–67.

Mussa, A. (2014). *Critical review of the implementation of the Eritrean Tobacco Control Proclamation No. 143/2004* (LL.B thesis). School of Law, Eritrea.

NG, R. (1975). *The Marriage Law and family change in China with special reference to Kwantung Province, 1950–1953* (M.A. thesis). University of British Columbia.

Schiller, A. (1969). Customary land tenure among the highland peoples of northern Ethiopia: A bibliographical essay. *African Law Studies*, 1–2.

Sedller, R. A. (1967). The development of legal systems: The Ethiopian experience. *Iowa Law Review, 53*, 562–582.

Semere, H. (1996). The roots of the Ethiopia–Eritrea conflict: The erosion of the federal act. *Journal of Eritrean Studies, 1*(1), 1–18.

Silkin, C. (1988). *The 1997 marriage law arrangements within the EPLF* (Paper presented at the Conference of the Research and Information Centre on Eritrea).

Singer, N. J. (1971–1972). The Ethiopian Civil Code and the recognition of customary law. *Hastings Law Journal, 9*, 460.

Tsegaye, B. (2013). Brief history of the Ethiopian legal systems—past and present. *International Journal of Legal Information, 41*(3), 335–351.

Watson, A. (1976). *Legal transplants and law reform*, 79–84.

Mining-induced Displacement and Resettlement in Eritrea: A Socio-legal Analysis

Yanet Rezene Abay

Ministry of Justice
Emial: yanetrezene@gmail.com

Abstract

Contrary to expectations, where mining happens as a development undertaking, it often causes impoverishment of affected communities because physical displacement and resettlement pose social risks: productive system is dismantled; income or productive assets are lost; social networks are weakened; kin groups are dispersed; and cultural identity is diminished or lost.

International financial institutions and development banks have been making efforts to address the issue and provide strategies towards relocation and resettlement of displaced people that guarantee livelihood restoration but have been criticized for their limitations and shortcomings.

With regards to Eritrea, the critical issue of providing assistance and compensation for the livelihood re-establishment of those displaced involuntarily through mining activities is not adequately addressed in its national laws. While certain laws make provision for compensation, it involves a narrow definition which does not include an obligation to restore people's well-being and ability to sustain themselves. Interviews have been conducted with experts from mining companies and consultants regarding the nation's resettlement practice policy. According to such interviews, experience with recent resettlement scheme has shown difficulties and delays in proper implementation of the process causing frustration for the parties involved.

There is no denying that the mining sector greatly contributes to a nation's economy and presents significant opportunities for development. However, there is a risk that community benefits will not be realized if the industry fails to mitigate the negative impacts that would have the effect of jeopardizing the livelihood of the affected people and their subsequent generation. Hence sustainable development – the ultimate goal - which in this case involves not only relocating and rehabilitating displaced people but assuring that they are better off than before and that they are beneficiaries of the project that was responsible for their displacement in the first place finds itself in a threat.

Keywords: mining, displacement and resettlement, livelihood restoration, national law, sustainable development.

Introduction

The mining industry is a 'double-edged sword': on the one hand, it brings socio-economic development by creating jobs, improving infrastructure and yielding benefit to countries when mining companies pay taxes.[1] On the other hand, the industry leads to socio-economic challenges which include forced displacement of communities.[2] The growing number of displaced or project affected people remains a particularly relevant issue here making mining-induced displacement an even greater problem in developing countries.[3]

Eritrea is rich in mineral resources. Substantial investment in mineral exploration has been going on, resulting in the discovery of important copper, zinc, gold and silver, whose extraction and production has started to play a major role in the economic growth of the country.[4] Currently five mining companies are operating in Eritrea: Bisha Mining Share Company (BMSC), Zara Mining Share Company (ZMSC), Asmara Mining Share Company (AMSC), Colluli Mining Share Company (CMSC) and Kerkebet Mining Share Company (KMSC).

At the national level, Eritrea lacks laws which regulate the critical aspects of involuntary resettlement of communities suffering from mining-induced displacement. There are some provisions in Proclamation 58/1994, Proclamation 68/1995, Proclamation 179/2017, the Transitional Civil Code of Eritrea (TCCE) and the National Environmental Assessment Procedures and Guidelines (NEAPG) that remotely address the issue, but they all point to one conclusion: whenever the government expropriates property from usufructuaries, adequate compensation must be paid. The ambiguity of the term compensation and the absence of further elaboration call into question the sustainability of compensation as a solution to restore the livelihood of displaced people.

The researcher's concern on resettlement is triggered by Eritrea's first and relatively medium-scale experience in mining-induced displacement and resettlement[5] which was undertaken in a Zara Mine project area.[6] The project directly impacted 154 households in the Koka area in Anseba region as their houses and rangelands were included in the mining land.[7] Resettlement programs were carried out and the residents became relocated to a new site.[8] The resettlement process encountered drawbacks including delays of putting in place various essential services that were promised.[9] Taking into consideration these facts, this research paper attempts to address the following research questions: What is the extent to which the issue of resettlement is addressed in both international and domestic laws? and What are the legal consequences that could follow with respect to failure of mining companies and competent authorities to be bound by resettlement action plans?By exploring the above listed research questions, the aim of this study is concerned with addressing the inadequacies of resettlement, in both law and practice. It is intended to contribute to the understanding of the effects of mining induced displacement and resettlement (MIDR), the

1 Lewis Tumbama, A thesis on 'An exploration of lived experience of 11 resettled families in Makambuza district, Zambia by a Nickel mine project', University of Cape town, at 1

2 Ibid.

3 Borgul Terminski, 'Mining-Induced Displacement and Resettlement: Social Problem and Human Rights Issue', research report, 2012, available at https://nbn-resolving.org/urn:nbn:de:0168-ssoar-327774.

4 Seife M. Berhe and Tedros Habtegiorgis, *Mining and Sustainable development: implication to social and environmental policy issues in Eritrea*, at 1.

5 In this case, MIDR is used to refer to both economic and physical.

6 Interview with Mebrahtu Ogbazghi, Director of Mineral Resources Management at the MoEM, conducted on 8th September 2021, at 10:30 a.m.

7 Ibid.

8 Interview with Alem Kibreab, Acting Minister of MoEM, conducted on 10 March 2023, at 9:30 a.m.

9 Interview with Dr. Seife Berhe, Executive Director of Andiamo Exploration and Resettlement and Consultant to ZMSC, conducted on 8 September 2021 at 3:30 p.m. and on 19 October 2024 at 12:25 a.m.

problems faced by displaced people. and advocates for fair and sustainable compensation of affected people through meaningful resettlement programs that aim to restore and/or improve their standard of living. Its significance becomes greater in Eritrea since the country is rich in mineral resources and future mining ventures' extraction of minerals could lead to further displacement of communities residing in mining project areas:

In carrying out this study, the following means of data collection were employed Interviews with:

- competent authorities within concerned ministries in charge of displacement of legal occupants in mining project area;
- superintendents and experts employed in mining share companies; and persons engaged in consultations to formulate resettlement action plan of Zara Mine project
- Published works that will elucidate the theories and impacts of involuntary resettlement
- Case study and follow up report on displacement and resettlement plan implemented by the local government of Zoba Anseba and ZMSC in Eritrea.
- Questionnaires distributed to the Ministry of Energy and Mines (MoEM) and mining share companies.

The research is divided into four sections. The first section introduces the concept of MIDR, its effects and related factors, with some basic international standards that address the issue of displacement and resettlement. The second section is dedicated to Eritrean mining industry: it analyzes the legal background of the mining sector and assesses the actual and potential problems faced therein by discussing practical resettlement experiences of the mining share companies. The third section concludes the research and provides some recommendations.

Mining Induced Displacement and Resettlement (MIDR)

MIDR is the extensive process of planning and implementing the relocation of people, households and communities from one location to another because of mining projects, and all the associated activities such as the provision of compensation and restoring and improving the social condition of the community.[10] Although no global survey has assessed the scale of MIDR, available evidence suggests that the scale is significant enough causing social disruption that necessitates policy attention and response.[11]

National governments justify displacement and resettlement by invoking larger goals of national growth and development.[12] MIDR frequently provides opportunities such as employment, infrastructure and education.[13] On the other hand, project land acquisition and the consequent displacement can cause much hardship and impoverishment[14] resulting from:

10 Sigsmond A. Wilson, 'Mining-induced displacement and resettlement: The case of rutile mining communities in Sierra Leone, Journal of Sustainable Mining', available at www.elsevier.com/locate/jsm.

11 See Tumbama above n 1.

12 Dolores Koeing, 'Towards Local Development and Mitigating Impoverishment in Development-Induced Displacement and Resettlement', final report organized by the Refugee Studies Centre, University of Oxford, July 2001, at 2.

13 William Rook, 'Responsible Resettlement: Implementing the Corporate Responsibility to Respect Human Rights in the context of Mining Induced Displacement and Resettlement', a thesis submitted to the University of Greenwich, September 2017, at 7.

14 Frank Vanclay, 'Project-induced displacement and resettlement: From impoverishment risks to an opportunity for development?', Impact Assessment and Project Appraisal, available at https://doi.org/10.1080/14615517.2017.1278671, at 3.

landlessness[15], joblessness[16], homelessness[17], marginalization[18], food insecurity[19], increased morbidity and mortality[20], loss of access to common property[21], social disintegration[22], loss of access to community services[23] and violation of human rights.[24] These impoverishment risks extend to not only communities and households affected by mining projects but also apply to host populations.[25]

In order to prevent and overcome the patterns of impoverishment of MIDR, it is necessary to develop reconstructive strategies taking the form of comprehensive and systematic resettlement program.[26] To formulate this idea, two basic messages must be conveyed: a policy message and a strategy message.[27] The policy message is that the general risk pattern inherent in displacement can be controlled through a policy response that mandates integrated problem resolution.[28] The strategy message is that specific resettlement programs are required each time resettlement is to take place to describe the relevant national legal framework, the characteristics of the affected population, the anticipated loss and the mitigating measures to be implemented.[29] The purpose is to prevent impoverishment risks and restore the project affected communities' livelihoods.[30] While livelihood restoration has been a stated goal of resettlement practice in the past, contemporary thinking has become more focused on improving their livelihoods enabling them to transition to new circumstances.[31] This is thought to be more consistent with human rights enjoyment and achieving sustainable development.[32]

International Standards

International financial institutions and the large development banks have developed resettlement policies, procedures and guidelines for public and private sector projects for which

---

15 Theodore E. Downing, 'Avoiding New Poverty: Mining-Induced Displacement and Resettlement', report n.58, to International Institute for Environment and Development, April 2002, at 6.

16 Michael M. Cernea, 'Impoverishment Risks, Risk Management and Reconstruction: A Model of Population Displacement and Resettlement', George Washington University, working paper presented to the UN Symposium on Hydropower and Sustainable Development, at 2.

17 Ibid.

18 See Downing above n.15, at 11.

19 See Cernea above n.16, at 22.

20 Ibid.

21 Ibid. See also W. Courtland Robinson, 'Risks and Rights: The Causes, Consequences and Challenges of Development-Induced Displacement', The Brookings Institution - Sais Project on Internal Displacement, occasional paper, May 2003, at i.

22 Ibid.

23 Ibid, at 13.

24 Ibid.

25 See Cernea above n 16, at 28.

26 Michael M. Cernea and Christopher McDowell, Risks and Reconstruction, Experiences of Resettlers and Refugees, the World Bank, Washington, March 2000, at 34.

27 Ibid.

28 Ibid.

29 Lidewij van der Ploeg and Frank Vanclay, 'A human rights-based approach to project induced displacement and resettlement', Impact Assessment and Project Appraisal, 35:1, 34-52, available at https://doi.org/10.1080/14615517.2016.1271538.

30 See Cernea above n 16, at 31.

31 See Vanclay above n 14, at 6.

32 Ibid.

they provide finance.[33] Key documents include the World Bank's Operational Policy on Involuntary Resettlement, and the International Finance Corporation Performance Standard 5 on Land Acquisition and Involuntary Resettlement with its guidance notes.[34]

World Bank's Operational Policy on Involuntary Resettlement

The World Bank is known to be the first major development agency to formulate an international standard on involuntary resettlement for those projects In which it is involved.[35] It lends money to national governments and in any situation where its finance is involved in a project, its Environmental and Social Standard 5 are required to be applied. These standards are largely similar to the International Finance Corporation's Performance Standards and modelled thereon.[36]

International Finance Corporation (IFC) 's Performance Standards

The IFC is an international financial institution that offers investment, advisory and asset-management services to encourage private-sector development in less developed countries.[37] For any project with some proportion of IFC funding, the IFC Performance Standards apply. The IFC standards provide guidance and framework on how to identify risks and impacts and avoid them in a sustainable way.[38] The IFC Performance Standards encompass eight topics.[39] Among these, Performance Standard 5 is concerned with Land Acquisition and Involuntary Resettlement and is considered to be international best practice regarding displacement, involuntary resettlement and livelihood restoration and improvement.[40] It recognizes that unless properly managed, involuntary resettlement may result in long-term hardship and impoverishment for the affected communities as well as environmental damage and adverse socio-economic impacts in areas to which they have been displaced.[41] Therefore, the IFC expects that its clients design and implement projects in accordance with the following objectives.

a. Avoid or minimize displacement by exploring alternative project designs;
b. Avoid forced eviction;

---

33 Ibid, at 5.

34 Ibid, at 10.

35 Ibid.

36 Natalie Bugalski and David Pred, 'Reforming the World Bank Policy on Involuntary Resettlement', International Accountability Project, submitted to the World Bank Safeguards Review, April 2013, at 2.

37 International Finance Corporation, https://www.ifc.org/wps/wcm/connect/corp_ext_content/ifc_external_corporate_site/about+ifc_new/, (visited on February 2, 2022).

38 International Finance Corporation's Performance Standards on Social and Environmental Sustainability, https://firstforsustainability.org/risk-management/implementing-ifc-environmental-and-social requirements/establish-and-maintain-an—esms/ifc-environmental-and-social-performance, (visited on 13 September, 2021).

39 Ibid. The seven topics excluding Involuntary Resettlement and Land Acquisition include:
1. Performance Standard 1: Assessment and Management of Environmental and Social Risks and Impacts
2. Performance Standard 2: Labor and Working Conditions
3. Performance Standard 3: Resource Efficiency and Pollution Prevention
4. Performance Standard 4: Community Health, Safety, and Security
6. Performance Standard 6: Biodiversity Conservation and Sustainable Management of Living Natural Resources
7. Performance Standard 7: Indigenous Peoples
8. Performance Standard 8: Cultural Heritage

40 See van der Ploeg and Vanclay above n 29.

41 Ibid.

c. Anticipate and avoid, or where avoidance is not possible, minimize adverse social and economic impacts from land acquisition or restrictions;
d. Improve or restore the livelihoods of displaced persons;
e. Improve living conditions among physically displaced persons through the provision of adequate housing.[42]

To achieve the above listed objectives, the IFC Performance Standard 5 requires every project proponent to stipulate: (i) alternative feasible project design to avoid or minimize physical or economic displacement;[43] (ii) a consultation framework with project stakeholders;[44] (iii) compensation and benefits for displaced;[45] (iv) grievance mechanism to address specific concerns;[46] (v) resettlement planning and implementation.[47]

Where involuntary resettlement becomes unavoidable, appropriate socio-economic data will be collected to identify the persons which will be displaced by the project and determine who will be eligible for compensation and assistance.[48] Based on a social and environmental assessment, the project proponent will develop a resettlement action plan or framework designed to mitigate the negative impacts of displacement, identify the development opportunities, develop a resettlement budget and schedule, and establish the entitlements of all categories of affected persons.[49]

MIDR in the Eritrean Context

Modern mineral exploration in Eritrea started in 1996 and issuing of mining license for modern mining commenced in May 2008 which led to commercial mineral production in January 2012.[50] The entity responsible for the administration, regulation and coordination of all types of activities in the energy and mining sectors of Eritrea is the MoEM. Eritrea is now known to host significant Volcanic Massive Sulphate deposits which comprise mainly of copper, zinc, gold and silver.[51] Five mining companies have thus far been issued with mining licenses: BMSC, ZMSC, AMSC, CMSC and KMSC.

Legal Framework of MIDR

Eritrea does not have laws or regulations which govern displacement and resettlement processes due to mining activities. When people happen to be economically or physically displaced, resettlement is carried out in relation to the existing general policies and laws.[52] It is usually left up to the concerned mining company and the local government where the

42 See International Finance Corporation's Performance Standards on Social and Environmental Sustainability above n 38.

43 International Finance Corporation's Performance Standard 5 on Involuntary Resettlement and Land Acquisition, available at https://firstforsustainability.org/risk-management/implementing-ifc-environmental-and-social-requirements/establish-and-maintain-an—esms/ifc-environmental-and-social-performance, (visited on 13 September, 2021)

44 Ibid.

45 Ibid.

46 Ibid.

47 Ibid.

48 Ibid.

49 Ibid.

50 Seife Berhe, 'Mineral and Energy Resources Opportunities in Eritrea', power point presentation, Andiamo Exploration Ltd., April 2021, at 30.

51 'Eritrea, Mining Journal', Asmara Mining Conference, Ministry of Energy and Mines, State of Eritrea, https://www.tesfanews.net/zara-mining-share-company-developing-eritrea's-next-gold-mine/ (visited on 22 February, 2022).

52 See interview with Dr. Seife Berhe above n 9.

displacement is about to take place to reach an agreement on determining and defining the displacement and resettlement process.[53] The existing Eritrean laws which address the question of displacement and resettlement are: Proclamation 68/1995, Legal Notice 19/1995, Proclamation 58/1994 and the NEAPG.

Proclamation 68/1995 and Legal Notice 19/1995

Proclamation 68/1995 and Legal Notice 19/1995 regulate the overall development of mining production. The preamble to the Proclamation states that all mineral resources are public property and that the State shall ensure the conservation and development of the resources for the benefit of the people.[54]

The only provision of Proclamation 68/1995 applicable to land acquisition, displacement of legitimate occupants and compensation issue is Article 22 (2). Pursuant to Article 22 (1), a mining licensee has the duty to take proper precaution not to interfere with other legitimate occupants of the license area, the land covered by a lease and adjacent land. Nevertheless, if displacement of the legitimate occupants is unavoidable, i.e. if the mining operation requires that the other occupants be displaced for safety reasons, the licensee shall attempt to negotiate the compensation payable to such occupants.[55] If the occupants refuse to be displaced or to agree on the compensation, the Licensing Authority may cause the expropriation of immovable property, if any, and the eviction of such occupants from the area of the licensee's mining operations.[56] Taking this scenario into account, the Proclamation neither addresses nor regulates how the displacement and resettlement of the evicted occupants will take place.

Legal Notice 19/1995 further regulates mining operations and stipulates certain requirements that every application for a mining license must fulfil.[57] Any particulars on displacement and resettlement issues which the project may face is not among the requirements to be included in the application.

Proclamation 58/1994

The reason why Proclamation 58/1994 is relevant to MIDR discussion is because it contains provisions of compensation for land expropriated by a governmental authority and must, therefore, be closely applied with the two mining laws.[58]

Proclamation 58/1994 bestows the government the right of ownership to all land in Eritrea.[59] The government retains the right to distribute land to citizens who are entitled to a piece of land with a usufruct right for a lifetime.[60] Thus, land cannot be sold or transferred and cannot be mortgaged. Under this Proclamation, the government or the competent author-

53 The MoEM does not get involved in the negotiations process. See interview with Mebrahtu above n 6.

54 Government of Eritrea, Gazette of Eritrean Laws, Volume 5/1995, No.2, Proclamation 68/1995, Preamble.

55 Ibid, Article 22 (2).

56 Ibid.

57 Government of Eritrea, Gazette of Eritrean Laws, Volume 5/1995, No.2, Legal Notice 19/1995, Article 5.

58 Proclamation 68/1995 and Legal Notice 19/1995 do not have a *lex specialis* character vis-à-vis Proclamation 58/1994 on the subject of compensation/displacement because they do not contain provisions that enable them to supersede over the latter.

59 Government of Eritrea, Gazette of Eritrean Laws, Volume 4/1994, No.6, Proclamation 58/1994, Article 3 (1).

60 Ibid, Article 4.

ity may, in the national or public interest, expropriate land from usufructuaries with appropriate compensation and in accordance with due process of law.[61] Activities in the national or public interest consist those that are carried out for development and capital investment project purposes aimed at national reconstruction, including mining activities.[62] The compensation must be either in the form of cash or substitute land and must be commensurate to the usufructuaries' loss of land.[63] The type and/or amount of compensation must be subject to negotiation between the competent government authority and the usufructuary; however, where agreement cannot be reached and a dispute arises between the parties, the case will be sent to the High Court by means of a petition or a suit, and the relevant provisions of the Transitional Civil Code of Eritrea (TCCE) will apply.[64]

Transitional Civil Code of Eritrea (TCCE)

Ultimately, the TCCE remains the body of law that regulates compensation for land expropriation due to the reason that the *lex specialis* (Proclamation 68/1995 and Proclamation 58/1994) require that recourse be made thereto whenever a dispute arises.[65] The fact that Proclamation 58/1994 makes no specific mention of the exact provisions of the TCCE that would regulate compensation allows a certain discretion in choosing the applicable articles the civil code.

Accordingly, the relevant provisions of TCCE concerning compensation for expropriated land can either be the provisions on extra-contractual liability[66] outlined in Articles 2027-2161 or the provisions of compensation for expropriation of immovable property stipulated in Articles 1470-1477 of the Code.[67]

The rules on extra-contractual liability apply on three bases of liability: fault-based liability, strict liability and vicarious liability.[68] The duty to pay compensation by the government or a competent authority for expropriating land corresponds to strict liability whereby liability is determined based on the damage caused to another by any activity the government or authority is engaged in. The reason is because Article 51 (1) of Proclamation 58/1994 requires that compensation be assessed on the basis of the loss or damage suffered by the usufructuary. With regard to the modes of compensation, material damages and restitution in kind would be appropriate to expropriation. Nevertheless, article 2119 of the TCCE forbids courts to order restitution in kind as a mode of compensation where the duty to compensate falls on the State.[69] So, going back to the MIDR discussion, the rules on extra-contractual liability allow compensation to be paid only in the form of material damages (i.e., cash) to

61 Ibid, Article 50 (1).

62 Ibid, Article 2 (9).

63 Ibid, Article 51 (1).

64 Ibid, Article 51(3) and (4).

65 In other words, we see a law of general application taking a prevalent stance over the special laws on a particular subject matter the standards of which has not been covered by the latter. Interview with Sium Tekle, former judge and current Director-General of the Ministry of Justice, conducted on March 10, 2022, at 7:00 p.m.

66 Interview with Ataklti Habtemariam, Former judge in the High Court and Part-time lecturer at the School of Law, conducted on May 18, 2022, at 6:40 p.m.

67 Interview with Zenawi Haile, Former public prosecutor and Part-time Lecturer at the School of law, conducted in March 2024, at 5:30 p.m.

68 Government of Eritrea, Gazette of Eritrean Laws, Volume 1, No.1, Transitional Civil Code of Eritrea, Proclamation 2/1991, Article 2027.

69 Ibid, Article 2119.

occupants displaced through expropriation. Replacing damaged or destroyed property is out of the question.[70]

As a second option, the provisions on compensation for expropriation of immovable property in the TCCE may be relevant to regulate compensation for expropriated land. In order to reconcile the respective provisions of TCCE and Proclamation 58/1994, the researcher has employed a constructive interpretation and arrived at the following reasoning:

a. Keeping in mind the differences of applicability of the two legislations, (i.e., the provisions on compensation of TCCE apply to persons whose immovable property owned by them have been expropriated, while those of Proclamation 58/1994 apply to persons whose land – over which they have usufructuary rights only – has been expropriated), and focusing exclusively on the scheme of compensation stipulated in the TCCE, the basis of compensation with respect to expropriated immovable property can be applied to compensation regarding expropriated land; and/or
b. By widening the scope of immovable property and considering land as part thereof, compensation on expropriation of land would be covered under the same provisions of that of immovable property.

Moving on to the provisions on compensation for expropriation of immovable property, Article 1470 of the TCCE imposes the duty upon any person who benefits from such immovable to inform the competent authority of the amount of compensation claimed by them within one month from having been served with the expropriation order. Where the competent authorities do not accept the amount of compensation claimed and dispute arises as a result of that, the amount will be fixed by an arbitration appraisement committee.[71] The amount of compensation or the value of the land must be equal to the amount of the actual damage caused by the expropriation.[72] In addition, Article 1473 states that the sole duty of the arbitral committee is to fix the amount of compensation to be paid by the competent authorities and precludes the committee from deciding disputes regarding the rights giving rise to the compensation.

Therefore, just like the rules on extra contractual liability dictate, the type of compensation that the arbitral committee is allowed to stipulate for expropriation of immovable property is only material damages. This means that the committee is excluded from calculating or determining other damages of non-material nature that displacement due to expropriation may cause, including provision of services, benefits and support necessary for restoring and enhancing the social functioning of the affected people. Moreover, the fact that arbitral committee does not have the power to decide by law any rights that displaced people are entitled as a result of expropriation puts into question the competency of the arbitral committee as the entity that should entertain resettlement matters. Resettlement process is by its nature a complex subject that requires extensive provisions and approaches the implementation of which should not be left to an arbitral committee whose sole duty is limited to fix monetary compensation which does not find ways to value and redeem the intangible loss suffered by the affected people.

Hence, the discussions made in the current and previous sections clarify that not only is there an absence of a compensation standard throughout the discussed laws but also the

70 Ibid.
71 Ibid, Article 1473.
72 Ibid, Article 1474.

existing provisions are not adequate to encompass resettlement of project affected communities as a means of compensation for displacement.

Proclamation 179/2017 and the NEAPG

Article 20 of Proclamation 179/2017 recognizes the NEAPG as an instrument to be mandatorily applied for every proposed project and activity that is likely to have a significant adverse social and environmental impact.[73] The NEAPG is a set of procedures and guidelines developed for the purpose of assessing environmental and social aspects of a project.[74] It prescribes that all development projects should pass through a screening process in order to establish their appropriate level of environmental assessment.[75] Based on the outcome of the screening process, the proposed project will be assigned into one of three categories A, B or C depending on the impact that the projects will have on the environment.[76]

All mining projects are categorized under category A as they are likely to cause huge impact on the environment. Accordingly, each applicant for a mining license must prepare and submit a comprehensive Environmental Impact Assessment (EIA) as part of the project feasibility study report. Since the EIA that is required to be undertaken encompasses both biophysical and social aspects,[77] resettlement would belong in the latter. The NEAPG however, does not provide specifics on how the resettlement process should take place, the entitlements of the project affected people and above all, the role of the ministries and the mining company involved in carrying out the resettlement process.

MIDR Practice in Eritrea

Out of the five mining companies, actual physical displacement and resettlement program was carried out only in the Zara Mine project area.[78] Although Bisha Mine has not experienced resettlement issues, the company had been involved in displacing cemeteries, livestock rangeland and farmland throughout its mining operations in the northern part of Bisha main area, Harena and Northwest zone areas.[79] The government provided the community with land replacement and the company was made to compensate them for their economic loss and provide them with social services including water supply, education and electricity.[80]

As for the AMSC[81], the community's farmlands and rangeland in the company's li-

73 Government of Eritrea, Gazette of Eritrean Laws, Volume 25/2017, No.2, Proclamation 179/2017, Article 20.

74 Ibid, Article 2 (14).

75 National Environmental Assessment Procedures and Guidelines, Procedures and Guidelines for Environmental Clearance of Projects, Monitoring and Evaluation of Projects, DoE, MoLWE, March 1999, at 8.

76 Ibid, at 12.

77 Interview with Kubrom Asmerom, Director General of DOE at the MoLWE, conducted on 14 March 2022, at 12:16 p.m.

78 See interview with Mebrahtu above n 6.

79 Girmay Iyasu Mekonnen, 'The Mineral Potential and Mining Activities of Eritrea, and Fe-oxide and Sericite alteration mapping using the remote sensing and GIS techniques', an internship program at the Japan Space Systems, March 2021, at 2.

80 Interview with Bereket Berhane, Supply and Logistics Superintendent at BMSC, conducted on 14 March 2022, at 9:44 a.m.

81 AMSC's license areas are three and are found in Debarwa, Emba Derho and Adi Nefas. The mine deposits, however, are four as two mining operations are being conducted in the Adi Nefas area, namely Adi Nefas and Gupo deposits. See interview with Kuflom Tekle, Geologist and Exploration Superintendent at AMSC, conducted on 12 March 2022, at 10:40 a.m.

cense areas have been occupied by the project mine causing thereby economic displacement.[82] The compensation scheme is still ongoing: displaced farmers are in the process of moving to a valley to conduct their farming activities and have been given financial assistance to set up shops and businesses.[83] The company has indicated in its Social and Environmental Management System that it would implement agricultural, rangeland and water supply improvement programs.[84] With regard to CMSC, no prediction of physical resettlement issues has been made.[85]

Following is ZMSC taken as one example to illustrate the practical impacts that MIDR creates:

The Case of ZMSC

The Zara Mine project area directly impacted 154 households in the Koka gold deposit area including their residential houses and rangelands.[86] Hence, for the mining operation to take place, the residents of upper Koka valley area had to be relocated to a new site.[87] The resettlement program was conducted by ZMSC with the assistance of Global Resources Development and Management Consultants who were involved in the development of a resettlement action plan.[88] In developing the plan, a study team was in charge of making studies on the socio-economic conditions of the households that would be affected by the mining project.[89] The team first reviewed the institutional and legal frameworks in Eritrea that concern resettlement matters. The relevant laws included: Proclamation 86/1996 (a Proclamation to Establish the Local Governments), based on which studies were conducted on the administrative structure of the site where the resettlement project would be located;[90] Proclamation 58/1994; Proclamation 68/1995 and the NEAPG. Since those were not enough, recourse was made to the IFC Performance Standards to guide the resettlement program.[91]

The study team identified the people affected by the project and the adverse impacts on their livelihoods associated with the project's land acquisition.[92] All 154 households were registered as eligible for compensation and relocation assistance.[93] The impacts identified as eligible for compensation were:

---

82 See interview with Dr. Seife above n 9.

83 Ibid.

84 Interview with Dr. Mussie Fissehaye, Environmental and Social Superintendent at AMSC, conducted on 11 March 2022, at 9:00 a.m.

85 The Colluli Mine is one of the world's shallowest potash deposits open pit mine located in the Danakil Depression region of Eritrea, approximately 75 km away from the Red Sea coast. See 'Colluli Potash Project, Danakil - Mining Technology', available at https://colluli-danakali.com.au. (visited on 11 March 2022). See also interview with Awet Solomon, Environmental Technician at CMSC, conducted on 15 March 2022, at 9:00 a.m.

86 Global Resources Development and Management Consultants, 'Resettlement Action plan, Zara Project Area, Operational Follow Up Plan for Geftu Resettlement Area', Final Report, February 2016, Asmara, Eritrea, at 9.

87 Ibid.

88 Ibid.

89 Ibid.

90 Ibid, at 3.

91 Ibid.

92 Ibid.

93 Letter correspondence with Gebru Haile, Director General of Infrastructural Unit of Zoba Anseba, conducted on 5 April 2022, at 9:00 a.m.

- Loss of land use for mine development (agricultural land/cropland);[94]
- Loss of structures, dwellings and animal sheds resulting from physical displacement,[95]
- Loss of access to livestock production[96] and woodland which the local village depended on for shade, livestock feed and energy source for cooking[97]
- Loss of livelihood due to reductions in pasture supplies and livestock production in the local area.[98]

Following the assessment, a consultation process was carried out with the residents of the Koka valley and the local administration to discuss the issue of relocation of the residents to an alternative place.[99] It was then agreed that the community would be relocated to Geftu, a site located at about 2 km south of Rikeb sub-town across the southern bank of Zara River.[100] The relocation process took place during March and April 2013.[101] The infrastructure plan for the resettlement site was designed by the Infrastructure Unit of Anseba region and approved by the Ministry of Land, Water, and Environment (MoLWE) which was the responsible government regulating agency.[102] ZMSC provided the budget for the implementation of the resettlement action plan based on the Infrastructure unit's expenditure assessment.[103]

Positive Impacts

Throughout the preparation and implementation of the resettlement action plan, ZMSC was committed to complying with IFC Performance Standards provided that the Infrastructural Unit of the local government would implement the action accordingly under the supervision of the MoLWE. Some of the positive impacts of the resettlement process were:

Compliance with Rights to Participation and Information

Stakeholder engagement was relatively present during the resettlement process.[104] The project affected community played a role in choosing the site to which they would be relocated and had constant communication with the local government and the regulating agency through a representative committee.[105] This shows that their right to participation and right to information throughout the resettlement process were duly respected.

Mitigating Measures

To mitigate the potential impact of the residents' relocation, the company developed a

94 Ibid.

95 See Global Resources Development and Management Consultants above n 86, at 3.

96 Interview with Mussie Tesfahiwet, Environmental Supervisor at ZMSC, conducted on 22 March 2022, at 9:00 a.m.

97 See Global Resources Development and Management Consultants above n 86, at 15.

98 Ibid, at 16

99 Interview with Aster Redaizghi, Director of Environmental Management of DoE at the MoLWE, conducted on 14 March 2022, at 12:40 p.m.

100 See Global Resources Development and Management Consultants above n 86, at 14.

101 Ibid.

102 See Global Resources Development and Management Consultants above n 86, at 13.

103 Ibid.

104 See interview with Aster above n 99.

105 Ibid.

Social and Environmental Management plan[106] where it promised to conduct a Community Development Program to address local priorities, including water supply, education, health facilities, rangeland and agricultural improvement program to mitigate any loss of livelihood and develop alternative livelihood opportunities.[107] The company had also developed a Reforestation Management Plan that outlines strategies and actions to be taken to ensure a successful planting program.[108]

Provision of Improved Housing

The shelters where the project affected households lived before they became resettled were huts (Agnets). The local government constructed houses for each household[109] in addition to transferring the original shelters to the Geftu resettlement site.[110]

Job Opportunities

Zara Mine project had created several job opportunities for the residents of the Koka valley and subjected them to training to acquire the expertise and skill required to become employed.[111]

Negative Impacts

Implementation Problems

The negative impacts of the resettlement process mostly lie in the implementation process as there were some discrepancies between what was stipulated in the resettlement action plan and what was provided.[112] This becomes a legal issue if we compare some provisions of Proclamation 68/1995 that outline certain expectations to be fulfilled when conducting mining operations, with the implicit conduct of ZMSC and the MoLWE during resettlement planning and implementation.

Proclamation 68/1995 requires any exploitation of mineral resources to be carried out in accordance with appropriate technology and sound principles.[113] Furthermore, Article 24 imposes an obligation for mining operations to be carried out in a prudent, diligent, efficient manner in accordance with good practices accepted in the international mining industry. So, keeping in mind that resettlement is an inherent component and impact of mining activity, the fact that ZMSC has shown allegiance to comply with IFC Performance Standards and that the MoLWE has given its approval for the resettlement process to be implemented accordingly shows a legal commitment on their part to be bound by such international practice. Failure to be bound by their commitments can be raised as a legal ground for holding them accountable under the Proclamation. The same applies to the MoEM, who has the duty to inspect and ensure that mining operations are carried out in compliance with the above stated provisions of Proclamation 68/1995.[114]

---

106 See interview with Mussie above n 96.

107 See interview with Aster above n 99.

108 See Global Resources Development and Management Consultants above n 86, at 16.

109 Ibid, at 13.

110 See interview with Mussie above n 96.

111 See interview with Dr. Seife above n 9.

112 Despite the agreement of building the houses' ceilings with grass, they were built with steel.

113 See Proclamation 68/1995 above n 54, Preamble, second paragraph.

114 Ibid, Article 43 (2) (d).

Other Negative Impacts of the Resettlement Process Included:

a. Delay of access to basic social services

Although resettlement commitments of social services for the resettled community was high among the list of priorities during the construction period,[115] the community was faced with delay of permanent water system supply.[116] Despite the need for re-afforestation for cattle shade and the fact that the community was articulated in a community meeting as a means to mitigate the adverse resettlement impacts, the promised agricultural project support program was delayed as well.[117]

ii. Mismanagement of funds

The consultants of the local government used a disproportionate amount of money while analyzing and identifying an efficient source of water system.[118] As a result, the funds required for the purpose of conducting infrastructural works of water pumps remained insufficient.[119]

iii. Lack of cooperation between the local government of Anseba and the mining company

ZMSC had undertaken socio-economic studies and an impact assessment that identified the potential impact of the project on the community and the environment. The local government, on its part, had also appointed its own study group.[120] Such group, however, did not conduct an extensive socio-economic impact assessment nor had it used the information of the studies undertaken by the company in preparing a resettlement plan.[121] Such absence of cooperation between the local government and the mining company reflected on the outcome of the resettlement process.

iv. Monitoring issues

The MoLWE was the agency in charge of monitoring post-resettlement process.[122] The monitoring schedule was however planned by the local government. So, the planners knew when the monitors would show up and they would always be prepared for them. Moreover, monitoring takes place twice a year and requires lots of expenses.[123]

Potential Impacts

Although the existing mining companies do not expect to face with resettlement issues in the short run, the uncertain nature of mining projects may lead to displacement and resettlement issues in the future. Moreover, the emergence of new mining companies further complicates the situation. A notable example is Kerkebet Mining Share Company which has recently been endowed with a preliminary mining license.[124] Mining operations have begun around Mier area and 150 hamlets are expected to resettle, provided the environmental clearance will be given green light.[125] Another factor to be taken into consideration is the scarcity of land. It is not news that displaced people wish to be resettled to free and fertile land, how-

115 See letter correspondence with Gebru above n 93.

116 Ibid.

117 Ibid.

118 The study group took time to decide where to place the borehole for water supply. See interview with Aster above n 99.

119 Ibid.

120 See interview with Dr. Seife above n 9.

121 See letter correspondence with Gebru above n 93.

122 See interview with Aster above n 99.

123 Ibid.

124 See interview with Dr. Seife above n 9.

125 Ibid.

ever the unavailability of such exacerbates the situation and calls for stringent measures and attention.

Conclusion and Recommendation

Resettlement is a glaring impact of mining development, yet it is an inevitable phenomenon as prioritizing national interest for the benefit of the majority is the main legal policy of every sovereign nation. While there will always be some level of residual nostalgia amongst resettled people with respect to the land originally occupied by them, it is possible for the opportunities created by the project to increase individual and collective well-being of resettled communities.

The concept of MIDR in Eritrea is guided by the piecemeal application of mining law, land law, the TCCE and environmental law and guideline which have not been specifically designed to cover resettlement nor are they sufficient to regulate it. While certain laws make provisions for compensation, it involves a narrow definition which does not include an obligation to restore people's well-being and ability to sustain themselves. Furthermore, experience with recent resettlement scheme has shown difficulties and delays in proper implementation of the process causing frustration for the parties involved.

The effects of mining-induced displacement and resettlement are a major issue of social justice and an area of policy and practice that requires attention. Therefore, government authorities are called upon to understand the gravity of such issue and to act thereupon by creating a balance between economic development and social development needs for a sustainable future life. Accordingly, a resettlement model that outlines the procedure, defines the responsibility of the involved stakeholders and ensures proper implementation through timeline schedules, supervision and cooperation is vital. Creating an environment that fosters knowledge exchange, increases awareness and improves practice within the mining industry is highly recommended as well.

Reference

'Eritrea, Mining Journal', Asmara Mining Conference, Ministry of Energy and Mines, State of Eritrea, https://www.tesfanews.net/zara-mining-share-company-developing-eritrea's-next-gold-mine/.

Bogul Terminski, 'Mining-Induced Displacement and Resettlement: Social Problem and Human Rights Issue', research report, 2012, available at https://nbn-resolving.org/urn:nbn:de:0168-ssoar-327774.

Colluli Potash Project, Danakil - Mining Technology, available at https://colluli-danakali.com.au.

Dolores Koeing, 'Towards Local Development and Mitigating Impoverishment in Development-Induced Displacement and Resettlement', final report organized by the Refugee Studies Centre, University of Oxford, July 2001.

Frank Vanclay, 'Project-induced displacement and resettlement: From impoverishment risks to an opportunity for development?', Impact Assessment and Project Appraisal, available at https://doi.org/10.1080/14615517.2017.1278671.

Girmay Iyasu Mekonnen, 'The Mineral Potential and Mining Activities of Eritrea, and Fe-oxide and Sericite alteration mapping using the remote sensing and GIS techniques', an internship program at the Japan Space Systems, March 2021.

Global Resources Development and Management Consultants, 'Resettlement Action plan, Zara Project Area, Operational Follow Up Plan for Geftu Resettlement Area', Final Report, February 2016, Asmara, Eritrea.

Government of Eritrea, Gazette of Eritrean Laws, Volume 1, No.1, Transitional Civil Code of Eritrea, Proclamation 2/1991, done in Asmara, on 15 September 1991.

Government of Eritrea, Gazette of Eritrean Laws, Volume 25/2017, No.2, Proclamation 179/2017, done in Asmara, on 26 January 2017.

Government of Eritrea, Gazette of Eritrean Laws, Volume 4/1994, No.6, Proclamation 58/1994, done in Asmara, on 24 August 1994.

Government of Eritrea, Gazette of Eritrean Laws, Volume 5/1995, No.2, Proclamation 68/1995, done in Amara, on 20 March 1995.

Government of Eritrea, Gazette of Eritrean Laws, Volume 5/1995, No.2, Legal Notice 19/1995, done in Amara, on 20 March 1995.

International Finance Corporation, https://www.ifc.org/wps/wcm/connect/corp_ext_content/ifc_external_corporate_site/about+ifc_new.

International Finance Corporation's Performance Standard 5 on Involuntary Resettlement and Land Acquisition, available at https://firstforsustainability.org/risk-management/implementing-ifc-environmental-and-social-requirements/establish-and-maintain-an—esms/ifc-environmental-and-social-performance.

International Finance Corporation's Performance Standards on Social and Environmental Sustainability, https://firstforsustainability.org/risk-management/implementing-ifc-environmental-and-social-requirements/establish-and-maintain-an—esms/ifc-environmental-and-social-performance.

Lewis Tumbama, A thesis on 'An exploration of lived experience of 11 resettled families in Makambuza district, Zambia by a Nickel mine project', University of Cape town.

Lidewij van der Ploeg and Frank Vanclay, 'A Human Rights Based Approach to Project Induced Displacement and Resettlement', Impact Assessment and Project Appraisal, 35:1, 34-52, available at https://doi.org/10.1080/14615517.2016.1271538.

Michael M. Cernea and Christopher McDowell, *Risks and Reconstruction, Experiences of Resettlers and Refugees*, the World Bank, Washington, March 2000.

Michael M. Cernea, 'Impoverishment Risks, Risk Management and Reconstruction: A Model of Population Displacement and Resettlement, George Washington University', working paper presented to the UN Symposium on Hydropower and Sustainable Development.

Natalie Bugalski and David Pred, 'Reforming the World Bank Policy on Involuntary Resettlement, International Accountability Project', submitted to the World Bank Safeguards Review, April 2013.

National Environmental Assessment Procedures and Guidelines, Procedures and Guidelines for Environmental Clearance of Projects, Monitoring and Evaluation of Projects, DoE, MoLWE, March 1999.

Seife Berhe, 'Mineral and Energy Resources Opportunities in Eritrea', power point presentation, Andiamo Exploration Ltd., April 2021.

Seife M. Berhe and Tedros Habtegiorgis, *Mining and Sustainable development: implication to social and environmental policy issues in Eritrea.*

Sigsmond A. Wilson, 'Mining-induced displacement and resettlement: The case of rutile mining communities in Sierra Leone', Journal of Sustainable Mining, available at www.elsevier.com/locate/jsm.

Theodore E. Downing, 'Avoiding New Poverty: Mining-Induced Displacement and Resettlement', report n.58, to International Institute for Environment and Development, April 2002.

W. Courtland Robinson, 'Risks and Rights: The Causes, Consequences and Challenges of Development-Induced Displacement', The Brookings Institution - Sais Project on Internal Displacement, occasional paper, May 2003.

William Rook, 'Responsible Resettlement: Implementing the Corporate Responsibility to Respect Human Rights in the context of Mining Induced Displacement and Resettlement', a thesis submitted to the University of Greenwich, September 2017.

Interviews

Interview with Alem Kibreab, Acting Minister of MoEM.

Interview with Aster Redaizghi, Director of Environmental Management of DoE at the MoLWE.

Interview with Ataklti Habtemariam, former judge in the High Court.

Interview with Awet Solomon, Environmental Technician at CMSC.

Interview with Bereket Berhane, Supply and Logistics Superintendent at BMSC.

Interview with Gebru Haile, Director General of Infrastructural Unit of Zoba Anseba.

Interview with Kubrom Asmerom, Director General of DOE at the MoLWE.

Interview with Kuflom Tekle, Geologist and Exploration Superintendent at AMSC.

Interview with Mebrahtu Ogbazghi, Director of Mineral Resources Management at the MoEM.

Interview with Mussie Tesfahiwet, Environmental Supervisor at ZMSC.

Interview with Dr. Mussie Fissehaye, Environmental and Social Superintendent at AMSC.

Interview with Dr. Seife Berhe, Executive Director of Andiamo Exploration and Resettlement and Consultant to ZMSC.

Interview with Sium Tekle, Former judge and current Director of the Ministry of Justice.

Interview with Zenawi Haile, Former public prosecutor and current attorney at law..

Politics and Regional Affairs

The Neoliberal Policies and Development Challenges in Africa: Dependent Economies, Neocolonial Education and Moral Poverty

Ali Ibrahim Mohammed-Ali

College of Business and Social Sciences, Adi-Keiyh
Email: alibrahim1192@gmail.com

Abstract

After two decades of winds of hope toward progress and development, in 1980s neoliberal policies dismantled the *'infant'* development structures of the post-colonial Africa. Yet, after more than four decades with the neoliberal policies, Africa's development has remained dismal. Adverse impacts of these policies are devastating, more so on the essential development sectors such as the economy and education. Besides, in the Social Science discourses, development theories which once were considered as keys to understanding of how developed societies got developed and why underdeveloped societies remain underdeveloped have disappeared. Instead, in Africa, preaching of the neo-liberal policies suffocated the development paradigms. Hence, the causes responsible for Africa's development impasse are still largely baffling. Therefore, it is against these backdrops that this paper revisits the African development discourses, and in the context of the neoliberal policies, critically examines the existing challenges with special emphasis on the state of education and its effects in Africa. Importantly, the paper demonstrates how these policies promoted neo-colonial education and its vices, and how these policies are today exposing the African societies into the worse human obscenities including '*moral poverty*'. By doing so, this paper unveils the fact that the 1980s neoliberal policies were 'predaceous neo-colonial recipes' designed to prolong the state of underdevelopment in Africa. Generally, the study concludes that the external imperialist states, their foreign economic commanders-in-chief such as the IMF and the World Bank, and the predaceous neoliberal policies on one hand, and remotely-controlled governments (acting as surrogates or puppets), states without political sovereignty, subordinate economies and neo-colonial education on the other, are among the major challenges prolonging Africa's development deadlocks. Therefore, in this paper, a briefly elucidated policy roadmap has been proposed as the way outs for the African continent.

Keywords: Africa, development challenges, moral poverty, neo-colonial education, neoliberal policies,

Introduction

After two decades of winds of hope towards progress and development in the post-colonial Africa, in 1980s the neoliberal policies took no time to dismantle the *'infant'* development structures (Gumisiriza, 2019; Nkongho, 2014; Heidhues & Obare, 2011). Yet, after more than four decades with the neoliberal policies, Africa's development has still remained dismal. Adverse effects of the policies are devastating, more on the essential development sectors such as the economy (Nkongho, 2014) and the education (Nyoka, 2014; Mlambo, 2006). According to Karl Marx, economy and education are the 'superstructures' and the 'infra-structures' respectively, by which a developing society is built, progressed and transformed. However, economically, since then, "Africa is being pushed back to where it is desperately trying to run away from, namely, to the production of a few raw materials for the Western markets in return for manufactured products in accordance with the principles of comparative advantage that are championed as good economic policy" (Mlambo, 2006:153). Reports show that economic poverty and inequality are higher today than in 1990s Africa (Reinders et al., 2019). Moreover, today, Africa's child finds himself not only in *material poverty* (i.e. lack of money, food or shelter), but also he/she is equally suffering from what is referred to as *learning poverty* (World Bank, 2019; UNICEF, 2022). Worse, the ineffectually prepared child in Africa is at risk of exposure to a third but most terrible kind of poverty, which the author refers to in this paper as the *moral poverty.* However, the causes responsible for the above development challenges in Africa are still largely baffling. On the other hand, theories of development which once were considered as keys for the understanding of how developed societies got developed and why underdeveloped societies remain underdeveloped have disappeared (Reinert, 2008). Instead, the neoliberal gospels' preaching of the so-called free trade and privatization in untamed markets have suffocated the development paradigms in Africa.

Therefore, it is against these backdrops that this paper revisits the African development discourses in the context of neoliberal policies. The argument of the paper is developed through deductive reasoning approach, as it is largely guided by Erik Reinert's Development Theory of developing societies. Thus, few empirical narrations have been drawn only from two African countries, Eritrea and Cameroon, where the author has/had personal experiences, in one as a residing citizen and the other as a student of two years' fellowship studies respectively. In his development theory, exploring the lost macro-historical development wisdom of the past 500 years, Reinert demonstrates how all historically successful development policies have depended on *'governing the market'* and *'getting the prices wrong'.* He argues that, the wind of privatization, free trade or free market blows towards progress "only when a high level of development has already been reached" (Reinert, 2008:15). Based on this theoretical development perspective, the poorer the society or the infant the development structure of the society, the less the winds of neoliberal policies blow towards progress, inclusive growth and sustainable development.

In this paper, the author attempts to be attentive to Xavier Renou's critiques that "there is a permanent reluctance among academics to call a spade, a spade and a predatory (or imperialist) policy as such" (cited in Mbeki, 2012:6). In line with this critique, Thabo Mbeki sets us the challenge that "in the interest of the peoples of Africa, we should have the courage to frankly confront the African reality, therefore dare to call a spade, a spade" (*ibid*). Indeed, an academician should not act as a diplomat or a politician that needs to be 'politically correct', but based on empirical evidence and/or logical reasoning, he has the *'intellectual responsibility'* (to borrow Noam Chomsky's phrase) to reveal the truth and call a spade, a spade.

Generally, the paper aims to demystify the impacts of the neoliberal policies in Africa's development with emphasis on the case of education, economy and culture of the African society. Also, the paper is expected to contribute its part in 'deconstruction' of the 'underdevelopment' theories and the neoliberal policies which has been carried out in Africa. Therefore, the paper is organized into three major parts. Following the introduction, the second part of the paper challenges the neoliberal policies that have reigned the post-colonial Africa. Subsequently, with emphasis on the case of education and economy as essential pillars of development, the effects of the neo-liberal policies in Africa are critically examined. Finally, the paper concludes with a policy roadmap for social justice, inclusive growth, peace and sustainable development in Africa.

Neoliberal Policies and the Macro-Historical Development Paradigm

Neoliberal policies encouraged free trade, privatization, commodification of the public goods such as education and health, removal of public subsidies, curbing public spending and in effect protecting the few elites or the riches while exposing the disadvantaged citizens, their families and children (Gumisiriza, 2019) into the reign of what Jonathan Sacks refers to as the *"markets without morals"* (Sacks, 2020:91). As argued by the Norwegian development economist, Reinert, development of any society however, is a result of "conscious policies of emulation where the market is a force tamed, like the wind. ... [and] the issue of free market and other policy decisions is one of *context* and *timing"* (Reinert, 2008:15). So, when was the neoliberal policies imposed to Africa? And how mature were, then the development structures of the newly born post-colonial states?

When the neoliberal policies came into effect in the 1980s, almost all African post-colonial nations' development was at a state of a *'society-in-transition'* from agricultural to manufacturing industrial sector (Nkongho, 2014; Heidhues & Obare, 2011; Gumisiriza, 2019). Moreover, the policies were put into effect amidst global market crises related to the 1973 hike in oil prices (Nkongho, 2014; Gumisiriza, 2019). Therefore, it was in such an unfortunate context and time that the paid-up apostles of the neo-liberal policies theoretically proselytized that Africa's development and progress would be a default condition so long as African states submit their *'infant'* development institutions to the market forces. As there is a direct relationship between a development theory, development policy and development structure of a nation however (Reinert, 2008), if a theory on which a policy is built up on is distorted, the policy takes the development of a sector or a whole nation in a wrong direction. The realities of Africa's subordinate economic (Nkongho, 2014) and dependent education (Nyoka, 2013; Mlambo, 2006) structures can be illustrated as the cases in point. The untimely enforced neo-liberal policies and the development deadlocks in Africa are closely related. The policies pushed Africa back into a subservient development position merely depending on raw materials or what Reinert (2008) refers to as the "diminishing returns activities".

Indeed, the neoliberal policies were 'ahistorical', artificially prescribed for development of the African societies. The patterns of all historically successful development structures of modern nations up to the recent development successes of South Korea, Singapore, China ... etc. have been developed "in exactly the same way, through policies steering them away from raw materials and diminishing returns activities into manufacturing" (Reinert, 2008:11). Moreover, all developed nations had a time in history when they were the bastions of economic protectionists (Gumisiriza, 2019; Reinert, 2008; Chang, 2007) and all were grown "because for decades, *often centuries*, their states and ruling elites set up *subsidized* and *protected* dynamic industries and services" (Reinert, 2008:7). However, despite this irrefutable macro-historical evidence of development routes of the developed society, in 1980s the apostles of neoliberalism have outlawed these historical development norms, paradigms and policies in Africa.

Neoliberal Policies and the Post-colonial State in Africa

Neoliberal policies in Africa failed not only in 'timing and context' but also masking its repercussions and preaching the ahistorical development theories to distort Africa's development, the 'economic commanders-in-chief' such as the IMF, the World Bank and their compradors in the West weakened the role of the post-colonial state. A sovereign state is a corner stone for ensuring inclusive socio-economic progress of a developing society. It has the ability to design and own its development policies and strategies, to maintain peace and security and to protect its infant development institutions from external exploitative market forces. However, with the neo-liberal policies, the *'newly-born'* post-colonial states in Africa were paralyzed. As a result, for more than four decades since the 1980s, majority of the African post-colonial states have been in a state of *'political independence without political sovereignty'*, leaving the peoples of Africa under the control of the markets without morals. A state 'without political sovereignty' cannot decide whether and when to close or open its doors. As a powerless entity, it simply awaits to get instructions from its colonial or neo-colonial masters. Unfortunately, since 1980s, the post-colonial state in Africa can typically be characterized as such a powerless entity.

A brief case of Cameroon may suffice to reflect how neo-liberal policies detached the state from the society. In 1977, the then middle-income Cameroon discovered offshore oil reserves, thus sparing from the economic difficulties other states in Africa were facing during that time (Nkongho, 2014). However, Cameroon did not take time to become a powerless entity, not more than a 'surrogate' for its colonial and neo-colonial masters, and a tax collector for the local political elites (*ibid*). With the neoliberal policies, in the words of Nkongho, "the entire population consuming amenities were plunged into immeasurable hardship. In the process, poverty of great proportion was instituted. The more the policies were implemented, the deeper the level of misery and Cameroon remained in the abyss" (2014:2). Since, the state in Cameroon become mounted robot incapable of protecting neither its own people nor the economy (Nkongho, 2014). One of the major consequences of the neoliberal - neo-colonial policies is that, for ten years now, the people of Cameroon is paying the price of a 'divided nation' with armed group conflicts ruining the society in the South West regions of the country (Bate & Yenshu, 2022). During his study time in Cameroon, the author realized how and why for more than four decades the colonial master (France) was able to closely manage the Cameroonian economy and the state itself (Nkohgho, 2014) and yet failed to help the nation in solving the decade long crisis of the on-going armed conflicts in the country (Bate & Yenshu, 2022).

Neoliberal Policies and Neo-colonial Education

Relevance and quality education are at the hearts of development sectors of a society. In Africa, however, the neoliberal policies paralyzed not only the post-colonial states and its economies (Nkongho, 2014) but also worsened the state of underdevelopment in Africa by promoting a 'neo-colonial education' (Mlambo, 2006, Nyoka, 2013, Ibrahim, 2015). In colonial Africa, education that was being provided to the African child was designed for the colonialists' vested interests and not for the peoples of Africa (Matasci et al., 2020). In the neo-colonial Africa, the case is not much different. Since 1980s, the neo-liberal apostles of the imperialist states propagated that developing societies' best returns is from basic education, and not from spending on the higher education and scientific research (Gumisiriza,

2019). For example, the World Bank consistently insisted that universities in developing societies to slash subsidies to students and instead dramatically increase tuition fees (*ibid*). Education aids or loans were also attached to the fundamentalist neo-liberal preconditions where national interest and priority of the developing nation is overlooked. To date, this has "decapitated African universities' ability to carry out autonomous indigenous research" (Gumisiriza, 2019:10) and contribute to a fairly balanced knowledge production. Therefore, like the colonial Africa, in the neo-colonial Africa also, from the production of empirical knowledge to what is being taught in the education classes in Africa continues to be based on the "Western epistemological schema and theories, deeply rooted in and informed by [the] colonial thought" (Mlambo, 2006:166). Africa's higher education and research institutions are made to collect data and just export it as raw as the African economic raw resources to the Western research laboratories for theory production in the Social Sciences (Ndlovu-Gatsheni, 2021, Nyoka, 2013). This is how Africa consumes not only the economic material products of the Western industries, but also consumes every Social Science theory or paradigm that is produced in the Western universities. Worse is that Africa still consumes even those out-of-date theories of the18th or 19th century, constructed in a different European time and context. Also, these theories, models and practices are 'uncritically adopted' to dictate Africa's development policies. For the developing societies of Africa, importing such distorted and superficial theories and/or uncritically consuming them however, is worse than just importing the manufactured economic materials.

Karl Mannheim (Mannheim, 1964:13) argues that "in our age, it is not enough to say that this or that educational system or theory or policy is good. We have to determine for what it is good, for which historical aims and whether we want the educational result". Indeed, for the imperialists, in order to prolong Africa's structural economic dependency and subjugation, instituting the neo-colonial education – it will be shown that it is the prevalently dominant content of education in Africa – was, and continues to be, an essential weapon. In a book co-authored by more than 40 scholars concerning African education, the authors admitted that "education in African countries is largely unproductive; [as] it enlightens by distancing and estranging Africans from their life circumstances" (Nsamenang & Tchombe, 2011: xxvii). Through the neo-colonial curriculum contents, education divorced African children from their indigenous environment. Of course, when education curriculum's content is disconnected epistemologically from the social, cultural, political and economic environments, worldviews and social values close to the learner, education and its outcome undoubtedly become unproductive (Chipango & To, 2024; Idang, 2015; Nyoka, 2013; Mlambo, 2006). Such an education has been estranging the African student from his primary worldviews and cultural values. As a result, when the student tries to look to his African continent through the uncritically adopted and decontextualized West-centric theoretical perspectives, he remains helpless to fairly understand and appreciate the continent. In this paper, the author argues that there are four alarming costs of the neoliberal-neocolonial education in Africa. These are: *axiological crises, xenocentrism, uncritical learning* and *learning poverty*.

Axiological-crises: In extreme instances, "an axiological crisis may be a reflection of *value-nihilism,* of the anarchic cultural state in which there are no established, socially recognized standards [values], and all parties act without scruples [moralities], and with a view solely to their own immediate advantage" (Bidney, 1945:543). Social values are standards by which members of a given society are guided by to collectively distinguish the right from the wrong, and the moral from the amoral. Indeed, in the animal kingdom, values and the capacity to think are two qualities that make human beings to be different from the other animal species. In other words, there is no specie, other than the human being, who has such a moral force. Values are passed from one generation to the next through different means. In contemporary societies, it is through education that a society safeguards the continuity of *one's-own* nationally cherished cultural values (Mannheim, 1964; Durkheim, 1961). Unfortunately, while for all other developed and developing societies, the goal to prepare the children for other world cultures and economies is secondary, in African schools, colleges or universities, oddly, it is a primary goal (Nsamenang & Tchombe, 2011). In the book co-authored by more than 40 African scholars, the authors concluded that hitherto, the "African children are the only ones in the world whose socialization begins with *acculturation"* (i.e. learning about other worlds history and cultural values more than ones' own) "instead of beginning with *enculturation"* i.e. being deeply imbedded primarily into one's own immediate world, one's own society, one's own cultural legacies and history (Nsamenang & Tchombe, 2011: xxv).

To the worse, in curriculums' teaching materials, empirical examples and cases are usually uncritically drawn from other cultures and merely adopted. These examples and cases, at times, even contradict some commonly cherished values of the African nations. A case in a course reading material, I reviewed in 2023 in one of our Colleges in Eritrea, may suffice to explain the problem. In the reading material, *"redistribution of wealth through socialist political structure"* is demonstrated as an example of a 'deviant behavior'. From the perspective of the writer of the textbook (who is American) which the readings were compiled from or even from the perspective of the audience i.e. American students and readers, whom the writer has written the textbook for, we may say that the example was right. This is because, for the capitalist-plutocratic America (Parenti, 2008), the principles that value equity and social justice are *'political taboos'*. However, how empowering or mind-liberating do you think when such an example is drawn from the books produced by the Western social scientists, in a copy-paste form, for the African or Eritrean student?

A neo-colonial education that devalues the values of the African society will only expose African peoples and the continent's young generations into the risk of value decadence, self-belittling and institutional decays. Africa can modernize its economic, political, education or even the religious institutions. However, this should not mean through uncritically adopting or copy-pasting the 21st century ideals of the depraved elites and the decaying institutions (Sacks, 2020) of the 'postmodern' societies. If African development is meant to adopt and imitate the Western postmodern *economic values* (e.g. social Darwinism), *political values* (e.g. the plutocracy in which democracy is without justice) (Parenti, 2008), *family values* (e.g. same-sex marriage) (Sacks, 2020) and so on and so forth, Africa will go nowhere except backwards.

Xenocentrism: The prominent Kenyan critical thinker, Ngugi wa Thingo, argues that domination of the mental universe of the colonized is "through control of culture of how people perceived themselves and their relationship to the world" (Ngugi, 1986:16). This is why, since the colonial times the imperialists have tarnished Africa depicting "by what it lacks or what it is not" (Lushaba, 2006:3). Africa is portrayed as 'a society with no history, as peoples who did not contribute anything to the Civilizations of the mankind at all, and thus has nothing to give but only to receive'. To legitimize these erroneous depictions, they produced West-centric Social Sciences where no standard literature confirms or contains African perspective, and used these as imperial instruments of brainwashing, xenocentrism and mental colonization of the generations in the postcolonial Africa (Amusan, 2016). Xenocentrism is a social illness that leads peoples towards cherishing others' cultures as superior to one's own cultural-historical legacies. With the neo-colonial education, by looking at everything African only from the Euro-American ethnocentric worldviews that cherishes the Western and tarnishes what is African, the African children have been developing the attitude of be-littling one's own peoples' cultural legacies. Therefore, the neo-colonial education in Africa has contributed its part not in liberating the mind-sets, but in exposing the generations into such a social illness.

Uncritical learning: For the African student, the neo-colonial education became more *abstract, alien, rote* and *superficial.* What is being learned in school has become hard to be comprehensible, relevant and meaningful to the African child and the African society (Nyoka, 2013). In the African higher learning institutions, especially in the stream of business and social sciences, such a miseducation is producing graduates who have no 'critical understanding' of their own societies, economies and cultures (Nyoka, 2013; Mlambo, 2006). These graduates are becoming typical of what Herbert Marcus refers to as the *'one-dimensional man'* (Marcus, 1964). In other words, the neo-colonial education systems are depriving the generations of their rights to get quality and relevant education that liberates the minds and stretches the streams of their consciousness. To be liberated from a colonial dependent mind-set means "to assume full responsibility for one's actions, use reason to understand your own world, and refuse other's tutelage" (Lushaba, 2006:46). Rote learning however does not equip a learner with critical thinking skills and competencies and therefore can never liberate the minds nor empowers the learner to open their eyes. This is why, in the post-colonial states of Africa, the neo-colonial education curriculum has been designed to produce a one-dimensional, uncritical generation. Basically, it is a neo-colonial in nature as it is aimed to control the minds and neutralize any potential resistance against the neoliberal (neocolonial) ideals in Africa. Indeed, when mind is neutralized and controlled by an imperialist, the eyes become blind and therefore useless. The mind disengaged from 'critical thinking' and the eyes are unable to see behind 'veils' of the imperialist forces. This is why, stressing on how fatal are such weapons of colonization, a South African revolutionist, Stephen Biko, left us a powerful maxim: *"the most powerful weapon on the hands of the oppressor is the mind of the oppressed".*

Learning poverty: The last but most disturbing effect of the neo-colonial education is the prevalence of high rate of 'learning poverty'. In Africa, neoliberal/neo-colonial education has served the forces of hegemony as a means to their ends: controlling minds of the oppressed and recolonization of the Continent. Learning poverty is defined as a child's inability to read and understand a simple text by the age of 10 (World Bank, 2019; UNICEF, 2022). In Africa, it is estimated that almost nine out of ten (90%) children are suffering from learning poverty (UNICEF, 2022). Such a high rate of learning poverty is a grieve violation of the children's right to enjoy quality education. In Africa, the rate of learning poverty increased during the global campaign around the MDG-2 (Van der Berg, 2021; Chataa & Nkengbeza, 2019). Now, the question is why has the rate of learning poverty been increasing since 2000 across Africa?

Though the factors are manifold, the hypothesis of this author is that such a learning poverty is more associated with the so-called *automatic promotion policy* or *normal progression* (the latter is a term used in Eritrea) which is widely practiced in the education system in

Africa. Majority of the African countries merely adopted and hence implemented the practice since 2000, and this was in line with the Millennium Development Goal (MDG, 2000-2015) number 2. The target for MDG-2 is read as "ensure that all boys and girls complete primary school" level. For example, Eritrea 'adopted' the normal progression practice in its education sector policy document as early as 2003 (MoE, 2003) and implemented in 2013. The findings of National Surveys Monitoring Learning Achievements (MLAs) in Eritrea, the children's ability to master the foundational literacy and numeracy skills has been deteriorating, starting from the third MLA conducted in 2015. In the third-round MLA (2015), only 25.4% of students at grade 5 attained the Minimum Mastery Level (MML) compared to 50% in 2008 (MoE, 2018). In three years after the normal progression practice is implemented in Eritrea, the third-round MLA conducted in 2015 shows a deterioration in the MML of the students by about 50%. Despite it is clear that such "high rates of learning poverty are an early signal that education systems are failing to ensure that children develop *critical foundational skills*" (UNICEF, 2022:8, emphasis added), yet normal progression practice is pandemic across Africa (Van der Berg, 2021). The normal progression practice obliges a teacher, a director or a school to allow the student to pass to the next grade regardless of his/her mastery of the foundational skills on numeracy and literacy or the expected outcomes of the learning in each grade (Chataa & Nkengbeza, 2019), mainly in the elementary level. Advocates of this ill-armed-practice in education disregard not only student's rights to quality education but also disregards the parents, teachers and other public outcry against it (Chataa & Nkengbeza, 2019).

In Africa, the agenda behind the push for automatic promotion practice seems to hardly been understood. Basically, the practice is in compliance with the neoliberal education policies whereby access to education has become a 'private commodity' that should be paid by the child as a consumer (Gumisiriza, 2019). In the past two decades, just increasing enrolment at primary levels and privatization of education services become the two standards by which the profile of a government rises high in the eyes of the IMF, the World Bank and their comprador states. For advocates of the neoliberal education policy, the automatic promotion practice is found to be cost-effective. This is why, despite public concerns (Chataa & Nkengbeza, 2019) of the learning poverty in which the child remains suffering from, in the past two decades, African states have been simply propagating the 'enrolment and literacy numbers' while the 'literacy rate' is meant to be measured by the number of grades the student has been automatically promoted to, not necessarily based on the foundational skills and competencies expected to be acquired by the student in each grade. This indicates that the automatic promotion practice is for the neoliberal markets, and not in the best interest of the African children and their families. Therefore, in Africa, if we do not want to lose this and future generations of children and youth, both ending the automatic promotion practice and fighting the learning poverty are inseparable challenges of our times.

Moral Poverty: The New Challenge for Africa

In the existing literature, the term 'moral poverty' is barely conceptualized. Therefore, in this paper, moral poverty is defined as a state of being in human institutional decadence and death of courtesy which resonate humanity's highest order of social values, virtuous life and social responsibilities. Jonathan Sacks' phrase; *"the death of civility"* (Sacks, 2020:204) may suffice to precisely describe the state of moral poverty in a given society. Nowadays, the neoliberal/neo-colonial policies expose Africa to the risk of *'moral poverty'*. A mind liberating education, both at schools and at home, is essential to 'arm the minds' and 'immune the hearts' of children from such a moral poverty (Bennett et al., 2010; Durkheim, 1961). According to Bennett and his colleagues, moral poverty is defined as the by-product of the state of being without responsible parents at home and responsible teachers in schools that can teach children to "feel joy at others' joy; pain at others' pain; satisfaction when you do right; [and] remorse when you do wrong" (Bennett et al., 1996:14). In other words, moral poverty of a society is the outcome of children's lack of responsible adults in functioning families and schools where the young is guided to differentiate the right from the wrong. The American children, for example, have been exposed to axiological crises, largely since 1960s, and today the society is paying the price with an 'abject moral poverty' (see Sacks, 2020; Bennett et al., 2010). Since 1960s, the American families left without moral codes and the schools and universities with value-empty education (Sacks, 2020, Stone, 1994). These expose the American families, children and the society into the present state of moral decadence (Bennett et al., 2010).

As every culture rise and fall in human civility and the moral force (Botta, 2016; Wei, 2011), for the decaying institutions of the Western Civilization (Sacks, 2020; Fukuyama, 2015), the social pathologies such as the so-called 'same-sex marriage' or 'LGBTQ+s' are age-related natural by-products. However, what makes these *'subhuman'* practices more appalling is that the morally bankrupted plutocratic regimes in the West have been busy in exportation of such social pathologies to spill over the yet-to-grow-and-develop Africa. Indeed, this is a subtle war. However, such a war is more terrible than the war declared on the post-colonial state, the economy and the education in Africa, as this is targeting more the intangible cultural heritages. Therefore, this time, Africa should not make another grave mistake by compromising with such neo-imperialistic agenda. The first mistake, committed during the colonial and neo-colonial times, weakens Africa by leading into abject but 'material poverty'. Since Obama's naked campaign for promotion of 'uncultured' subhuman practices however, more terrible weapons of moral decadence is looming over the children and cultural heritages of the peoples of Africa. If the moral poverty is natural for the decaying institutions of the societies in the West, for the yet-to-grow Africa however, it is a *'premature death'*.

Conclusion

The paper demonstrates that the ahistorical neoliberal development policies were a neo-colonial recipe, manufactured to perpetuate Africa's underdevelopment and subordination. This is why, tossing the continent back into the curse of the material poverty, political instability, armed conflicts, powerless states and development deadlocks, the Western neoliberal apostles with the help of certain local elites as 'surrogates' or 'puppets' reinvaded the society and weakened the development pillars in Africa. Moreover, the neoliberal policies now exposed the African society into the worse human obscenities including moral poverty. Therefore, the paper concludes that the challenges for Africa's development are set in two sides of the same coin. The external imperialist states, their foreign economic commanders- in-chief and the predaceous neoliberal policies on one hand, and the remote-controlled surrogates and puppet governments, subordinate postcolonial states, dependent economies and neo-colonial education on the other, are the challenges prolonging African development deadlocks. Based on the development challenges presented in this paper, therefore, below are some broadly elucidated policy roadmap proposed for the ways outs.

Recommendations

In his recently published book (2020), Ndlovu-Gatsheni, a leading African thinker and decolonial theorist, calls up on African leaders to "implement radical transformative policies to liberate the continent from all forms of oppressions, domestic and external, which made the lot of the African masses unbearable". Therefore, the following 4-point recommendations are in line with Ndlovu-Gatsheni's call for the African young generation to *'turn over a new leaf'* for the good of the peoples of Africa and for the common good of peoples of the world.

Defiance through United African Fronts: The Africa we all want cannot be achieved from our weak position. Divided we are weak and therefore we always fall into the trap of imperialists; and united we are strong and therefore we stand as equally respected partners of the world. Africa should choose now, from the position where it deserves to be respected as equal global partner or yet to languish into another century of humiliation and heartless imperialists' exploitations. In 1961, Kwame Nkrumah of Ghana advised every independent African nation to rededicate in the struggle to emancipate other members. The revolutionary Nkrumah wisely advised the independent nations of the time that their independence is 'meaningless unless it is coordinated with other members for total liberation of the African continent'. Nkrumah was right. As a chain cannot be stronger than its weakest link, without 'true independence' of all African nations from the imperialist chains, sovereignty and peace of any African nation is incomplete, and therefore unsustainable. However, no one will offer Africa the sovereignty it deserves, and the peace, the stability, social justice and the inclusive development it aspires. It will come nothing good for us, without us. Also, a true right is taken; it is not given. Therefore, the paper calls up on all African sons and daughters for heroic defiance against all kinds of neo-colonial machines through coordinated 'African Fronts'.

People-Centered and Inclusive Development Policies: African development policies should be anchored on 'people-first-principle' of social justice. Social justice is not alien to the African cultural values, as the concept is not different from the African '*Ubuntu*' values, except that the term sounds more modern. Ubuntu is a 'people-based' African socialist continental values that cherishes equality, equity, justice, fraternity, community, cooperation, caring and sharing among citizens (Chipango & To, 2024). Also, the need for social justice and equity is more echoing in international development agendas such as the 2030 Sustainable Development Goals. In the 21st century's modern societies, no justice not only breeds no peace but also impedes inclusive growth and sustainable development. Therefore, in the development endeavors, true partners of African governments should be the African people, not merely the markets. Ngugi (2018:103) argues "if Africa is to make progress, the African elite must return to its real base: *the people*". It is in a similar vein that this paper is calling up on the African leadership to reconnect itself with what he describes as the revolutionary capital - *the people*.

Decolonization of Education: As discussed above, one of Africa's major development challenges is the neoliberal-neo-colonial education. Such a miseducation will take Africa nowhere except to prolong its development deadlocks, and take the peoples of Africa from bad (material poverty) to the worse: *moral poverty* and its vices. As discussed above, Africa is now exposed into the new challenge. The author holds the view that the imperialists have reigned Africa with their predaceous neo-colonial policies, for more than four and half decades now, largely because they succeeded in 'recolonizing the minds' through the neo-colonial education where the Western Ethnocentric Social Sciences 'constructed' dependent mind-sets which belittles what it has under their shadows, and dreams of what it wishes to get, through miracles, in an overnight. On my two years stay in Cameroon, I realized some young university students' view at people from France as *"superior race on the earth"*. When repeatedly heard such assertions, I asked myself; 'for such misperceptions in the minds of the postcolonial African youth, who is to blame?' Indeed, the author does not blame the generation but the education curriculum which cherish everything French or West-centric, in the cost of the African or Cameroonian cultural and historical heritages. Therefore, Africa should help its young generations yet born and yet to be born to first learn their own shared cultural-historical heritages, and only then that of others. This could enable the African young to understand who really they are and empower to fairly appreciate their African heritages and the world broadly. In other words, the African decolonized education curriculums should aim to produce *'national citizens'* who are proud proponents of their own cherished values as well as *'global citizens'* who are cross-culturally competent.

Values-Based Education: Last but most important, African education curriculums should value the human 'values' in general and African values in particular (Chipango & To, 2024; Idang, 2015). In November 2022, Head of the Political Affairs of the Eritrean Peoples Front for Democracy and Justice (PFDJ), Mr. Yemane Gebreab, had addressed the community of the College of Business and Social Sciences in Adikeih. In his seminar, Mr. Yemane listed 6 Eritrean long-term National Development Goals, of which the 6th but most important goal that cought my attention was that: 'through education Eritrea will *(*and should) work *to strengthen the values and bury the taboos': (*ክብረታት ምሕያል፡ ነውረታት ምምካት*)*. I squarely agree with the above wisdom.

Steven Muller, the ex-President of Johns Hopkins University in the US, used to say: "Universities are turning out highly skilled barbarians because we don't provide a framework of values to young people who more and more are searching for it" (Khera, 1994:17). With expansion of the markets without morals, this is more worse today than when Muller accurately noticed in America. For health and development of a society, education without social values is menace. Such education alone, influences the head and produces individuals who does not think by their 'minds' and 'hearts', but only their 'bellies'. These individuals cannot think of the society but their bellies. They are socially decadent and, therefore are socially irresponsible (Bennett et al., 2010). Though, the neoliberal- neo-colonial policies have attempted to spread such decadence across Africa, the author holds the view that it is not too late for Africa to heal its injuries and protect the African cultures and societies.

Therefore, this paper recommends that African education should *'value the values'*. In all levels of the African education institutions, when 2+3 (i.e. a technical skill) is taught in harmony with human social values, education can influence the minds and hearts of the learner and produce 'socially responsible' and 'culturally competent' citizens. Such values-rich education also would 'immune' the African child against the risk of the looming moral poverty. Human history tells us that citizens of such minds and hearts build virtuous societies, cultures, civilities and civilizations (Sacks, 2020). Those who lack such education thereof think only by their bellies and always attempt to destroy those of others' cultures, societies and civilizations. Therefore, African education must teach the young generation not only how to make living but also how to live together a virtuous life with civility, peace and development.

Finally, as concluding remarks of the paper, the author reitrates the Italian enlightenment thinker's, Giuseppe Mazzini, wisdom. In his book (1860) titled, *"The Duties of Man"*, Mazzini affirms that "*in working for our own country on the right principle, we also work for humanity"*. The author holds a similar view. If Africa takes and protects its sovereignty, fights back the moral poverty, defends its cultural heritages and holds its right position in the world stage, then Africa could be one of the greatest forces for good of the humanity on the world.

References

Amusan, L. (2016). Social Sciences as Imperialism: Analysis of the Global Economic Crisis of 2008 and Development Gaps in the Third World States. North-West University, South Africa. *Journal of Administrative Sciences*. Vol. 14. No. 27. pp.13-31.

Bate, NA., Balgah, RA., & Vubo, EY. (2022). Conflict and Livelihood Changes in the South West Region of Cameroon: A Social Capital Perspective. *International Journal of Social Sciences Perspectives, 12(1).*

Bennett, WJ., DiIulio, JJ., Jr., & Walters, JP. (1996). *Body count: Moral Poverty Theory and How to Win America's War against Crime and Drugs*. New York: Simon & Schuster. DOI: http://dx.doi.org/10.4135/9781412959193.n22

Bidney, D. (1946). The Concept of Cultural Crisis. *American Anthropologists*, 534-550.

Botta, M. (2016). Neo-collectivist Consciousness as a Driver of Transformative Socio-cultural Change. *Journal of Futures Studies*, 21(2): 51-70.

Chang, H. (2007). *The Bad Samaritans. The Guilty Secrets of Rich Nations and the Threat to Global Prosperity.*

Chataa, B. M. & Nkengbeza, D. (2019). Challenges Faced by Primary School Teachers in Implementing the Automatic Promotion Policy at a School in the Zambezi Region. *Creative Education*, 10, 1731-1744. https://doi.org/10.4236/ce.2019.107123

Chipango, EF, & To, LS (2024). When Sustainable Development Competes with African Ubuntu: A Case Study. *Geoforum,* 154(2024)10407

Durkheim, E. (1961) [1902-1903]. *Moral Education: A Study in the Theory and Application of the Sociology of Education*. Translated by Everett K.Wilson and Herman Schnurer. New York: Free Press.

Fukuyama, F. (2015). *Political Order and Political Decay: From Industrial Revolution to the Globalization of Democracy.* London, Profile Books.

Gumisiriza, P. (2019). Neo-liberal Policies and Impact on Developing Countries. A. Farazmand (ed.), *Global Encyclopaedia of Public Administration, Public Policy, and Governance*, https://doi.org/10.1007/978-3-319-31816-5_3782-1

Heidhues, F. & Obare, G. (2011). Lessons from Structural Adjustment Programs and their Effects in Africa. *Quarterly Journal of International Agriculture* 50 (1): 55-64

Ibrahim, Z.(2015). From a 'World-system' to a Social Science Knowledge 'Scape' Perspective: Anthropological Field working and Trans-nationalizing Theory-making in the 'Periphery'. *Journal of Glocal Studies*.

Idang, GE. (2015). African Culture and Values. *University of South Africa Press*, 16(2).

Khera, S. (1994). *You Can Win.* The eEconomics Press.

Lushaba, LS. (2006). Development as Modernity, Modernity as Development. ASC Working Paper 69/2006. African Studies Centre, Leiden, The Netherlands.

Mannheim, K. (1964). *An Introduction to the Sociology of Education.* (London: Routledge and Kegan Paul).

Marcus, H. (1964). *One-Dimensional Man: Studies in the Ideology of Advanced Industrial society.* With an introduction by Douglas Kellner in 1991.

Matasci, D., Dores, HG & Jeronimo, MB. (eds.) (2020). *Education and Development in Colonial and Postcolonial Africa: Policies, Paradigms, and Entanglements,* 1890s–1980s. https://doi.org/10.1007/978-3-030-27801-4_1

Mbeki, T. (2012). Is Africa there for the Taking?
Ministry of Education (2003). National Education Policy.
Ministry of Education (2018). Education Sector Plan. Asmara, Eritrea.
Mlambo AS. (2006). Western Social Sciences and Africa: The Domination and Marginalization of a Continent. *African Sociological Review*, 10(1), 161-179.
Ndlovu-Gatsheni S. (2021) The cognitive Empire, Politics of Knowledge and African Intellectual Production: Reflections on Struggles for Epistemic Freedom and Resurgence of Decolonization in the Twenty-first Century. *Third World Quarterly* 42(5): 882–901.
Ndlovu-Gatsheni, SJ. (2020). *Decolonization, Development and Knowledge in Africa: Turning Over a New Leaf.* Routledge, Taylor & Francis Group.
Ngugi, WT. (1986) *Decolonizing the Mind.* Woodbridge: James Currey.
Ngugi, WT. (2018). *Ngugi at 80: Selected Essays.* In Digest, Research and Documentation Center (RDC), Asmara, Eritrea, 22 (95), July 2018.
Nkongho, T. (2014). *Explaining the State Crisis in Sub-Saharan Africa: A Critical Analysis of State Transformation Through Privatization.* 2nd ed. Lambert Academic Publishing.
Nsamenang B. & Tchombe, T. (2011). *Handbook of African Educational Theories and Practices: A Generative Teacher Education Curriculum.*
Nyoka, B. (2013). Negation and Affirmation: a Critique of Sociology in South Africa. *African Sociological Review,17*(1).
Parenti, M. (2008). *Democracy for the Few.* 8th ed.
Reinders, S., Dekker, M., Kesteren, F., Oudenhuijsen, L (2019). Inclusive Development in Africa: Synthesis Report Series.
Reinert, E. (2008). *"How Rich Countries Got Rich and Why Poor Countries Stay Poor".* In Digest, Research and Documentation Center, (RDC), Eritrea, No. 89, July, 2017.
Sacks, J. (2020). *Morality: Restoring the Common Good in Divided Times.* 1st ed. ISBNs: 978-1-978-1-5416-7532-2 (ebook).
Stone, L., (1994). Family Values in a Historical Perspective. In The Tanner Lectures on Human Values, delivered at Harvard University November 16-17, 1994.
UNESCO (2017). Education for Sustainable Development Goals: Learning Objectives.
UNICEF (2022). The State of Global Learning Poverty: 2022 Update. *Conference Edition* June 23, 2022.
Van-der, SB. (2019). Education Access and "Learning Poverty" in Seven Southern African Countries.
Wei, R. (2011). Civilization and Culture. *Globality Studies Journal.* Issue, 24.
World Bank (2019). Ending Learning Poverty: What Will it Take?

Eritrea's Counter-Hegemonic Foreign Policy in the Changing International System

Biniam Berhe

Eritrean Mission to the African Union, Addis Ababa
Email: biniber@gmail.com

Abstract

This paper argues that the primacy Eritrea gave to independence of choice and aversion to domination in crafting and implementing its foreign policy was born out of Eritrean people's long and arduous struggle for independence which triumphed against all odds. The implementation of this policy direction was enabled due to the country's ownership of its development agenda and the self-reliant approach it pursued. The independent path Eritrea chose was a pragmatic choice to safeguard the country's sovereignty and advance its interests. Eritrea's resistance to the unipolar international system and its advocacy for the creation of a just and inclusive international system by working with like-minded countries are natural extensions of its foreign policy objectives.

Eritrean foreign policy has been the subject of a few articles and books. Apart from some exceptions, most academic writings on Eritrean foreign policy summarily ridicule and demonize Eritrea's distinct foreign policy thought and praxis, without making serious attempt to understand the underlying causes and the interlink with the country's history and its guiding principles.

The paper adopted Neo-Gramscian framework of hegemony and counter-hegemonic struggle for change, as developed by Robert W. Cox.[1] The methodology used includes reviewing relevant first-hand resources and personal observation as well as academic writings on the matter.

Keywords: Eritrea, Foreign Policy, International System

Introduction

The paper analyzes the factors that contributed to the primacy given to independence of choice and resistance to domination in crafting and implementing Eritrean foreign policy. The Eritrean people's struggle for independence which triumphed against all odds served as an inspiration to the country's post-independent foreign policy. The independent path Eritrea took, enabled by its self-reliant development path, was a pragmatic choice to safeguard the country's sovereignty and advance its interests. In the same vein, Eritrea's firm resistance to the existing unipolar international system and its advocacy for the creation of a just interna-

1 Cox, Robert W. "Gramsci. 'Hegemony and International Relations'". *Millennium Journal of International Relations* 12, no. 2 (1983): 162-175

tional system are consistent with its foreign policy objectives.

The paper based itself on primary and secondary sources as well the writer's observation as a practitioner in diplomacy. The paper adopted Neo-Gramscian concepts and perspectives to look into the nature of hegemony and counter-hegemonic movement.

The paper seeks to provide academic and practical perspectives on the nature of Eritrean foreign policy as an alternative to the prevailing narrative in a number of academic circles that simplifies and denigrates Eritrean foreign policy.

The paper has three main parts. The first part discusses the theoretical perspectives on foreign policy of states and surveys some representative writings on Eritrean foreign policy. The following section provides a bird's eye view of Eritrea's recent history and its impact on the foreign policy choices of the Government of Eritrea. This leads to an analysis of the path chosen by independent Eritrea. The third section illustrates that soon after achieving its independence Eritrea resisted the imposition of liberal political and economic policies on it. This part also portrays the challenges Eritrea faced arising from the US' execution of antagonistic policies in the Horn of Africa region, and Eritrea's resistance to safeguard its sovereignty and territorial integrity. This section illustrates that Eritrea's effort to articulate a vision for a new international system emanating from its counter-hegemonic foreign policy. Finally, the main findings of the paper are summarized and a recommendation is made for academicians and practitioners on the subject to undertake further research to better understand the nature of Eritrean foreign policy within the international system.

Literature Review

The major International Relations theories give little or no consideration to the international relations of African countries or other non-hegemonic countries. The prominent Neorealist scholar Kenneth Waltz, contended that "it would be...ridiculous to construct a theory of international politics on Malaysia and Costa Rica...[A] general theory of international politics is necessarily based on the great powers."[2]

Neoliberal theory does not fare any better. Neoliberal International Relations scholars do not consider Africa as an important actor in International Relations. Dunn observed that "when neo-liberals have paid attention to Africa, they have been motivated by development theories aimed at reproducing Western economic, political and cultural ideas."[3]

The Neo-Gramscian approach this paper has adopted provides important concepts and analytical tools to look into the interlink between domestic policy and foreign policy, the constraining effect of the international system on foreign policy choices, the process of establishing hegemony, the mechanisms of co-option and punishments utilized by the hegemonic power, resistance to the order, and nature of counter-hegemonic movements.[4]

Academic writings on Eritrean foreign policy are few in number. In surveying the writings in preparing this paper, it was noticeable that the publications were predominantly by non-Eritreans. The academic papers discussed below were written between 2007 and 2019. Their approaches vary from International Relations perspective, to that of journalism/advocacy.

A political scientist, Tanja Müller analyzed Eritrean foreign policy making within the context of the Horn of Africa and Red Sea countries. Her findings show that Eritrean foreign

2 Dunn, Kevin C. "Introduction: Africa and International Relations Theory". in Dunn, Kevin C. and Shaw, Timothy M (ed). (Palgrave Publishers Ltd. 2001) p. 2

3 Ibid. p. 2

4 Cox. "Hegemony and International Relations" pp 162-175

policy, contrary to the popular disparaging narrative, was "embedded within the wider political geography of the Horn as a region, as well as underpinned by Eritrean identity politics and long-term survival strategies in a contested geopolitical environment."[5] Her quest to understand the rationale for Eritrean foreign policy led her to the conclusion that Eritrean foreign policy had a "clear logic behind it and ultimately follows common foreign policy patterns in the Horn."[6]

Interestingly, Muller draws lessons from Eritrea's case on how "developments within the global system not only relegate African states to peripheral players but at the same time offer new opportunities to reorient their foreign policies in positive and constructive – or indeed destructive ways."[7] The effort made by the writer to understand Eritrea's foreign policy from an academic angle is a rarity.

The perspective on Eritrean foreign policy propagated by a big portion of Western 'experts' on the Horn of Africa is articulated in the report of a workshop held in London in 2007 by the Horn of Africa Group on '*Eritrea's Regional Role and Foreign Policy: Past, Present and Future Perspectives*[8]. The report relays that the participants of the workshop ascribed what they described as Eritrea's 'isolationist stance' to its 'sense of uniqueness' and 'frontier mentality'. They generalized that Eritrea has "a warrior culture that sees diplomacy as weakness".[9] These unsubstantiated generalizations can hardly be taken as academic assertions. Their historical perspective was that "diplomacy was not institutionalized in the formative years of Eritrean liberation struggle".[10] This statement ignores the diplomatic struggle Eritrean patriots waged at the United Nations since the late 1940s, the advocacy by Eritrean diaspora communities, the efforts of Eritrean Liberation Front and Eritrean People's Liberation Front's (EPLF) representational offices abroad and the sustained engagements it had with friendly countries and liberation movements. The little attempt made to understand Eritrea's view of the international system led the participants to the condescending generalization that "Eritrea tended to perceive and treat the international community as an undifferentiated hostile entity, disregarding how others saw the world."[11]

Should, at least in a footnote, a mention not be made of Richard Reid's edited volume on Eritrea's external relations?

A similar straw man fallacy is observed in a book by a former BBC journalist, Martin Plaut. The title of the book '*Understanding Eritrea: Inside Africa's Most Secretive State*' contained both a promise of intellectual curiosity and dismissive generalization. The chapter dealing with Eritrea's 'Foreign Friends' opened by stating that the adage "Stand alone, stand proud: we can do it on our own" summed up the Eritrean government's ideological position[12].

The writer concluded that it was difficult to describe Eritrea's foreign policy "when so much is covert, personal and subject to such swift alteration."[13] He followed this up by

5 Müller, T. R. "Assertive foreign policy in a 'bad neighbourhood': Eritrean foreign policy making". ed. Z. Tsighe, S. M. Idris, Y. M. Asfaha, S. W. Andemariam, R. K. Taddesse, & G. Ogubazghi (Eds.). p 632

6 Ibid. p. 632

7 Ibid. p.632

8 Healy, Saly. "Eritrea's Regional Role and Foreign Policy: Past, Present and Future Perspectives". (Chatham House, 2008)

9 Ibid.

10 Ibid.

11 Ibid.

12 Plaut Martin. "Understanding Eritrea: Inside Africa's Most Secretive State". (Hurst & Company, 2016), 69.

13 Ibid, p. 100.

blaming the ‘intransigence’ of Eritrean diplomacy for leaving “Brussels and Washington floundering about in an attempt to fashion a coherent foreign policy towards Eritrea: a minor but strategically significant state, in a troubled region.”[14] The book did not attempt to understand Eritrean foreign policy in an objective manner; rather it regurgitated the oft-repeated demonization of Eritrea while portraying the West as benevolent partner.

The Eritrean Government’s view on the international system and its critical view towards US policy towards Eritrea and the Horn of African region is well documented.[15] However, these factually corroborated grievances are not given much consideration in academic literature and the media.

Among the few attempts made by academicians and practitioners at understanding the impact of US policies on Eritrean foreign policy, are two almost identically titled articles published by Hank Cohen[16] and Bronwyn Bruton[17]. The authors took into consideration Eritrea’s opposition, among others, to the measures taken by the US to reverse the decision of the Eritrea-Ethiopia Boundary Commission (EEBC) to favor Ethiopia, the United Nations Security Council (UNSC) sanctions imposed on Eritrea in 2009 and 2011 at the behest of the US, and singling out Eritrea for human rights issues. Bruton warned that US policy aimed at isolating Eritrea would have unintended consequences, including Eritrea’s deepening relations with China and the Gulf, and drifting away from the West.[18] Consequently, they separately recommended the US Government to improve its relations with Eritrea by pushing for the implementation of the EEBC decisions, the lifting of the UNSC sanctions on Eritrea, and adopting constructive approach on human rights.

This brief survey of academic writing on Eritrea hints at that the majority of the writers used historical and journalistic perspectives that tried to make sense of Eritrea’s foreign policy choices vis-à-vis what they dismissively viewed as the Eritrean Government’s ‘undiplomatic’ character. The few exceptions used theories and approaches of International Relations to understand the issue.

In the next section, the paper attempts to look into the factors contributing to Eritrea’s foreign policy choices, by analyzing some milestones during the struggle for independence and the path charted in the early days of independence.

Eritrea’s Journey from Subjugation to Sovereign Statehood

As recognized by Neo-Gramscian perspective, a country’s domestic situation and the prevalent international system, among others, play a role in shaping its foreign policy.[19] In this vein, this section will provide a bird’s eye view on Eritrea’s history since Italian colonization. The documents containing the reports and outcomes of discussions held during the armed struggle and the early days of independence show that the lessons drawn from Eritrea’s subjugation by successive colonial powers for a century, led the EPLF and the Government of

14 Ibid. p. 101.

15 One such document is the ‘Chronology of unfriendly policies and acts pursued by previous US Administrations. 4 May 2023, https://shabait.com/2023/05/04/chronology-of-unfriendly-policies-and-acts-pursued-by-previous-us-administrations/

16 Hank Cohen. “Time to Bring Eritrea in from the Cold”. African Arguments, December 16, 2013, https://africanarguments.org/2013/12/time-to-bring-eritrea-in-from-the-cold-by-hank-cohen/

17 Bronwyn Bruton. “Eritrea: Coming in from the cold” Atlantic Council, 6 December 2016, https://www.atlanticcouncil.org/in-depth-research-reports/issue-brief/eritrea-coming-in-from-the-cold-2/

18 Ibid.

19 Cox. “Hegemony and International Relations” pp 162-175

Eritrea to act deliberately to end the victimization and put the destiny of the country on the hands of the Eritrean people.

Historical Basis

Eritrea's long held advocacy for the creation of a just global order in which all countries - big or small - contribute and benefit from peace, security and development emanated from its understanding that there cannot be peace in situations of domination, as domination leads to resistance. Due to its strategic location on the Red Sea and the Horn of Africa, Eritrea has borne the brunt of the expansionist agendas of and competitions between global powers. The country has been victimized by the insatiable appetite of hegemonic powers for dominance. Such powers have utilized their might to shape Eritrea's destiny to their liking.

In the interest of space, it suffices to mention modern-day examples. In the final decades of the 19th century, the colonial aspiration of then newly united Italy to have a colony by the Red Sea, supported by Britain which opposed France's expansion in the region, resulted in the formation of Eritrean polity with its current borders. Italy's 50 years of colonization ended in 1941 with its defeat by the allied forces led by Britain. Britain in turn administered Eritrea as its protectorate for 10 years. British rule was replaced with Eritrea's forced federation with Ethiopia in 1952 that was imposed on the Eritrean people by the UN at the behest of the US. The Ethiopian emperor abolished the federation in 1962 and annexed Eritrea as the 14th administrative region of Ethiopia. By then the Eritrean struggle for independence had evolved from a political movement into an armed struggle.

Eritrea was a battleground between the allies and axis powers during World II. The two superpowers that emerged after WWII put their weight, one after the other, to extinguish the Eritrean people's struggle for independence. The policies of global powers have caused enormous loss and suffering on the Eritrean people. These powers considered Eritrea just a pawn in their geopolitical games.

This traumatic experience led the EPLF and the Eritrean Government to give due consideration to the international system in their developing of world view/policy making and implementation. As briefly discussed hereafter, during the Eritrean people's struggle for independence, the EPLF, and since independence the Eritrean Government, strategically analyzed the nature of the international system, its impact on Eritrea, the country's neighborhood and globally. The outcomes of the 1977 and 1987 EPLF congresses have documented the strategic view of the Front. The National Charter of 1994 and subsequent policy documents regularly feature analysis of the international system and its impacts. These meticulous analyses are articulated succinctly and communicated to the public, and are known by friends and foes. [20]

The EPLF's Memorandum to all progressive forces in the world in April 1978[21] was aimed at explaining the struggle of the Eritrean people. The section of the Memorandum on 'Current political situation and the Eritrean national question' analyzes the global geo-political situation of the time, the competition between the two super powers and their allies, the geographical areas of contention, the strategy of the superpowers to control the sources

---

20 The thinking and practice during the armed struggle is captured in the 'EPLF; Documents of the Second and United Congress of the EPLF and ELF-Central Leadership held in 1987; 2007, Sabur Printing Press, Asmara' the National Charter which was adopted at the conclusion of the 3rd EPLF Congress held two years after Eritrea's independence is guided by the same principles.

21 EPLF; Memorandum: By the Eritrean People's Liberation Front submitted to all progressive forces in the world; April 1978, PFDJ Archive, 13-14 (*Written in Tigrigna, quoted part translated by the author*)

of oil and secure markets for their products in developing countries; and the interference by Western powers to suppress liberation and anti-imperialist movements in the Horn of Africa and Arabian Gulf while prodding client regimes who do their bidding in the two regions.[22]

The outcome of the Second and United Congress of the EPLF and ELF-Central Leadership[23], that was held in 1987, before delving into EPLF's foreign relations activities, analyzes the global and regional situation that EPLF operated in during the past decade, focusing on the role and influence of the then two super powers from economic, military and political aspects.[24] The National Democratic Program, that was adopted by the Congress outlined the foreign policy principles of the Front.[25]

The liberation of Eritrea was achieved on 24 May 1991. Eritrea's independence that was achieved after a long political and military struggle coincided with major regional[26] and global changes. The global competition for dominance between the US and its NATO allies, on the one hand, and the USSR and its Warsaw Pact allies, on the other, ended with the disintegration of the Soviet Union and the collapse of its alliance. These regional and global changes presented numerous challenges and opportunities.

Post-Independence: Charting the Path

Three years after liberating Eritrea and less than a year after the country's formal independence declared following an internationally supervised referendum, the EPLF held its 3rd Congress in February 1994 in Nakfa. The Congress adopted the People's Front for Democracy and Justice (PFDJ).[27] National Charter which outlined the vision, basis for political programs and organizing principles. In the midst of the major transformations at national, regional and global levels, the Front vowed to remain grounded in Eritrea's realities.[28] The Charter declared that "[f]or the first time, the people of Eritrea have become masters of their own country and hold their future in their own hands."[29]

The Charter underscored that Eritrea's foreign policy was the extension of its internal policy. The bases for Eritrea's foreign policy were listed as: "preserving [Eritrea's] national

---

22 EPLF, 1978, 13-14

23 EPLF. *Documents of the Second and United Congress of the EPLF and ELF-Central Leadership.* (Sabur Printing Press, 2007) (*Written in Tigrigna, quoted part translated by the author*)

24 Ibid. pp. 115-116

25 Follow peaceful and non-aligned foreign policy.

- Establish diplomatic relations with all countries of the world, regardless of their political and economic systems, based on: respect for independence, territorial integrity; non-aggression; non-interference in domestic affairs; equality and mutual interest; peaceful coexistence.
- As the Eritrean struggle is part and parcel of the struggles of peoples' of the world, especially that of the struggles against global domination, interference and racism of the peoples of Africa, Asia and Latin America, the EPLF pledged to support all just and democratic movements.

26 At the regional level, a new government led by EPRDF took power in Addis Ababa. The Somali state disintegrated leading to a bloody civil war. The coup d'état in Sudan that brought Al Bashir to power happened only two years before the transformation year that 1991 was. Hence, Eritrea emerged into regional and global environment that presented new challenges and opportunities.

27 At the Congress EPLF transformed into PFDJ.

28 The front stressed that the Charter was: "…not copied from books or from the charters of other countries. It starts from the realities of our country and society, and from our rich experience. It does not borrow wholesale any analyses or formulas that are fashionable in today's world either. Rather, by critically examining all ideas and relating them with the realities of our society and our experience, the purpose is to chart an independent line that works."

29 Ibid.

interest, working in the interests of peace and stability in [the] region and in the world, and enhancing cooperation in all areas with all governments and peoples."[30] The Charter summarized peace and non-alignment as being Eritrea's foreign policy. The document listed Eritrea's regional and international foreign policy objectives.[31] These foreign policy objectives are for the most part similar to the National Democratic Program adopted in 1987, with the addition of the responsibilities that come with Statehood.

The National Charter, which formed the bedrock of Eritrea's post-independence socio-economic and political agenda, was situated within Eritrea's history and prevalent domestic, regional and international realities of the time. In the same vein, the foreign policy objectives enshrined in the Charter took into considerations the country's experience during colonial occupation; its political and armed struggle for independence; the challenges and opportunities that came with statehood; the turbulence in the Horn of Africa and Red Sea regions; as well as the emergent post-cold war unipolar international system.

Walking the Talk: Foreign Policy in Action

Eritrea's interactions with the superpowers and their allies had prepared it well for the challenges of operating within the international system. Eritrea's decisive battlefield victory had by then softened the attitude of those who had condemned the people of Eritrea to colonial subjugation. Western press sang EPLF praises for its military might and discipline. There was massive support for the Eritrean people's referendum. A long list of countries recognized Eritrea's *de jure* independence even before independence was formally declared. Eritrea's application for membership in UN, Organization of African Unity (OAU) and Intergovernmental Authority on Development (IGAD) were swiftly accepted, and Eritrea assumed its rightful place in these organizations the same week as the declaration of formal independence.

Eritrea joined the international system with clear principles, and determined to be the master of its destiny. This approach reflected sovereign statehood; however, it was not the acceptable norm when developing states dealt with hegemonic powers. The discussion below gives a few examples of the difficulty Eritrea faced in implementing its self-reliant development path and independent foreign policy within the US dominated post-cold-war international system.

---

30 Ibid.

31

1. To follow a foreign policy of peace and non-alignment based on independence and national interest.
2. Starting with our neighbors, and our surrounding, to establish relations with all countries, regardless of their political and economic, systems, based on independence, respect of territorial integrity and national dignity, non-aggression, non-interference in internal affairs, equality and mutual interest.
3. To build friendship with all peoples, regardless of historical and cultural differences, based on universal brotherhood, equality, peace and understanding
4. To strive for regional as well as global peace and stability.
5. To develop economic and cultural cooperation with all countries in order to accelerate our country's economic and social development.
6. To abide by all international laws and agreements to which Eritrea is a signatory

Charting Self-Reliant Development and Resisting Policy Prescriptions

The external pressure put on Eritrea to toe the line, during the early years of independence and Eritrea's defiant position is captured in an unpublished report prepared in 1993 by the Foreign Relations Department of the Transitional Government of Eritrea. The document narrates that in their discussion with the Transitional Government, Western countries raised issues to do with the economic and political policies that Eritrea would follow. More specifically, the Western countries were keen to know what kind of political system Eritrea would establish after referendum; when and how elections would take place; how the constitution would be drafted and adopted, are if political parties were allowed to be established. The response given by the Eritrean side was that establishing democracy was a self-initiated process and would be implemented in accordance with the local situation and would not be based on the formulas set by external entities.[32]The report recommended that the new country had to prepare to reject attempts by the external entities to determine its policies and the imposition of preconditions; and in turn Eritrea had to encourage its counterparts to adopt positive and constructive approaches and to engage cooperatively.[33]

In the same vein, President Isaias Afwerki shared the attempts made to impose liberal policies on Eritrea using foreign aid.[34] The article narrates that in 1993, a high-level delegation from a friendly country visited Eritrea with the aim of supporting the country "make the difficult political and social transition to peace and salvage [Eritrea's] devastated economy."[35] The Eritrean government recognized that the offer was well intentioned, but had to decline it because "the aid package arrived with an appalling number of strings attached; chiefly those associated with privatization of Eritrea's public enterprises."[36]

The article also provided the case of a mission to Eritrea from another country that intended to support Eritrea in the energy and education sector. The offer was tied to the Government's efforts on women empowerment targets outlined in the Beijing Conference of 1995. The Government of Eritrea concluded that the condition was "apparently made without any knowledge or scrutiny of [Eritrea's] overall policy of affirmative action aimed at promoting the long-term equality of both genders."[37]

The article used these examples to demonstrate the dangers of the donor-recipient, pre-condition-laden, self-perpetuating model of aid. In a bid to mend the flaws of the model, the article proposes that "[t[he only yardstick for measuring the viability, effectiveness, and desirability of foreign aid, in the form of grants or concessions, is the extent to which it enables the beneficiaries to gradually reach the point where they no longer need it."[38]

The tug-of-war between Eritrea and development partners on ownership and intervention was acknowledged in a report prepared by the World Bank assessing the Bank's assistance to Eritrea from 1992-2000.[39] The Bank recognized the fact that "[t]he Government

32 Transitional Government of Eritrea - Department of Foreign Affairs, "Report submitted to the 6th meeting of the Central Committee of EPLF". February 1993 (Unpublished report in Tigrigna)

33 Ibid.

34 Isaias Afwerki, "Foreign Aid Works Best When It's Self-Limiting." Forum for Applied Research and Public Policy. Winter 1997; 12 no 4; ProQuest, pp. 68-70

35 Ibid.

36 Ibid.

37 Ibid.

38 Ibid.

39 World Bank. Operations Evaluation Department. Eritrea Country Assistance Evaluation. 2004, https://ieg.worldbankgroup.org/sites/default/files/Data/reports/cae_eritrea.pdf

of Eritrea (GOE) outlined its core development principles prior to independence in 1993. During the liberation struggle, the EPLF leadership, in consultation with Eritreans living abroad, evolved a political philosophy based on self-reliance, equity and social redistribution."[40] While noting the development partner's admiration of Eritrea's position, the report indicated that this same issue became a bone of contention: "the Government's aversion to any sign of outside intellectual or financial interference coupled with donor difficulty in adapting to Eritrea's ownership orientation led to many instances of misunderstanding. Early relations with donors, including the Bank, were often marked by strain and miscommunication."[41]

The development partners understood the validity of and logic behind Eritrea's commitment for a new approach guiding developmental cooperation. Their resistance emanated from the long-held approach. "Eritrea's insistence on control was unusual for a small less developed country (LDC). While this independent stance was admired, in practice many donors had difficulty adjusting to it and found the authorities to be inflexible or excessively rigid at times."[42]

The Foreign Relations Department report, the article by President Isaias Afwerki, and the assessment report prepared by the World Bank showcase the deliberate approach the young nation adopted to safeguard its independence of choice, not to lose the ownership of its socio-economic and political agenda, to resist compromising its principles for short term gains. This happened in an international system that was not amenable for developing states to exercise sovereign choice.

Confronting the Impositions of the New Security Architecture

The tension with development partners surrounding Eritrea's insistence to own its development agenda and to primarily rely on its own resources was mild compared with the challenges faced in the security area. As proclaimed in the National Charter, Eritrea pursued a policy of engaging with all countries regardless of past antagonisms, or current difference of views. However, the international security architecture that independent Eritrea found itself in was one that was dictated by a single superpower, overzealous to expand its 'democratic ideals'.

The plan was spelt out in the Defence Strategy written by Dick Cheney.[43] The Strategy briefly discussed US' plans for Sub-Saharan Africa. Among the core priorities were "... strengthening the security, stability and economic development of friendly states and supporting their democratic development."[44] The architecture provided clear roles: the victorious West led by the US wanted to dominate the international system unopposed, and use its dominance to push its political and economic views on the rest of the world.

Within this context, the initial peaceful years of independence that engendered significant developmental gains to Eritrea. This was disrupted in May 1998 by the war declared by Ethiopia on the pretext of a border dispute. The U.S. which got involved to broker peace between the two neighboring countries ended up escalating the conflict to a full-fledged war. Some observers pointed the blame at the then US Assistant Secretary of State, Susan Rice,

40 Ibid. p. 2
41 Ibid. p. 3
42 Ibid. p. 6
43 Cheney, Dick, Secretary of Defense. "Defense Strategy for the 1990s: The Regional Defense Strategy". January 1993
44 Ibid.

who was tasked to approach the two countries. In her dealings with the Eritrean Government "[s]he came in like a scoutmaster, lecturing them on how to behave and having a public tantrum when they didn't act the way she wanted."[45]

From that time on, the US policy towards Eritrea was seen in the context of Ethiopia's interests, retracing the same policy adopted to federate Eritrea with Ethiopia in 1952. This was taken to the extreme following the Bush era National Security Strategy of 2002[46] that strategized to deal with "countries with major impact on their neighborhood such as South Africa, Nigeria, Kenya, and Ethiopia [as] anchors for regional engagement"[47]

The disastrous nature of this approach was witnessed with US' attempt to undermine and change the rulings of the EEBC, in order to appease the Ethiopian government who refused to implement the final and binding ruling. The then US Ambassador to the UN, John Bolton, recounted that "Eritrea had a point. Ethiopia had agreed on a mechanism to resolve the border dispute in 2000 and now was welching on its deal in flat violation of its commitments."[48] His efforts to push for the implementation of the decision was thwarted after, "Frazer [then US Assistant Secretary State for Africa] reversed course, and asked in early February [2005] to reopen the 2002 EEBC decision, which she had concluded was wrong, and award a major piece of disputed territory to Ethiopia."[49] Eritrea resisted all forms of pressures aimed at reversing the EEBC decisions. Ethiopia finally accepted the decisions in 2018.

The Obama Administration continued the antagonistic policy with the imposition of UN Security Council sanctions on Eritrea in 2009 and 2011, accusing the latter of supporting the Somali group Al-Shebab, and using the dispute between Eritrea and Djibouti as a pretext. Eritrea rejected the allegations on both counts and characterized the imposition of the sanctions as "brazen act … neither based on fact nor on the provisions of international law...' [and that it] constitute[d] a travesty of justice and amplifies the dangers inherent in a unipolar world."[50] The UNSC sanctions were lifted in 2018.

These are just a few of the measures taken by successive US Administration targeting Eritrea through various means. In all instances, the Eritrean Government opted to resist the pressure. Eritrea's decision not to embrace the dictates emanating from Washington to reopen the EEBC decision was interpreted by some as intransigence[51]. This same characteristic is considered by Neo-Gramscian scholars as constituting the seeds of counter-hegemonic movement.[52]

Eritrea's Contributions towards the Shaping of a New International System

Articulating ideas for change and pushing the same with like-minded countries is at the core of the establishment of a counter-hegemonic movement that works to change the existing structure with a new one that provides opportunities for realization of the ideals of equality, peace, stability and development.

45 Rosenblum, Peter. "Irrational Exuberance: The Clinton Administration in Africa." *Current History*, vol. 101, no. 655, 2002, pp. 195–202. *JSTOR*, http://www.jstor.org/stable/45317205.

46 President of the United States, "The National Security Strategy of the United States of America", https://2009-2017.state.gov/documents/organization/63562.pdf, September 2002

47 Ibid. p. 11.

48 Bolton, John. *Surrender is Not an Option: Defending America at the United Nations and Abroad,* (Threshold Editions, 2008), 344

49 Ibid. p. 344

50 Ministry of Foreign Affairs of the State of Eritrea: "A Shameful Day for the United Nations", 23 December 2009, https://shabait.com/2009/12/23/press-release/

51 Plaut, Martin. *Understanding Eritrea: Inside Africa's Most Secretive State*, 101

52 Cox. "Hegemony and International Relations"

The Eritrean Government has been advocating in major international fora for the end of the international system dominated by a sole superpower and for the establishment of a just and inclusive system. In this vein, the President of Eritrea made powerful statements, to name a few, during the Africa-Russia Summit held in St. Petersburg, Russia in July 2023, and at the 2023 BRICS Summit. In his remark to the 2023 BRICS Summit, President Isaias outlined a vision "for a just and fair global order where justice and the rule of law prevail; where nations and peoples forge meaningful and symmetric ties of cooperation and partnership on the basis of respect for national independence and sovereignty; for societies anchored on compassion and justice."[53]

The statements resonated beyond the Summit halls and ignited discussions in the media and academic circles. The statements of Eritrean delegations at the annual General Debate of the UN General Assembly over the last several years contain similar messages. Eritrea is also a founding member of Group of Friends in Defense of the UN Charter which opposes the use of 'exceptionalism' by some powers and their disregard for sovereign equality of states and their attempt to establish a 'rule-based system'.[54]

Conclusion

The paper contended that the primacy given to independence of choice and resistance to domination in crafting and implementing Eritrean foreign policy was inspired by the country's struggle for independence. The choices were pragmatic in nature aimed at safeguarding the hard-won independence and sovereignty of the country and advance its interests. The implementation of this policy occurred in an international system dominated by the US that sought to implement its security architecture unopposed, and imposed its political and economic ideas on other countries. In this context, Eritrea's ownership of its development agenda and the self-reliant path it adopted, were foundation of and the enabling factor for the implementation of the independent foreign policy agenda.

As seen routinely, hegemonic powers do not tolerate other countries to pursue independent foreign policy or resist their hegemonic dominance. In such occasions they apply co-option or punishment. In the case of Eritrea, the punishment meted out by the hegemonic power included attempts to reverse the decisions of the EEBC and the imposition of UNSC sanctions. Eritrea's unflinching resistance and safeguarding its right to sovereign choice and not to compromise on its territorial integrity could serve as inspiration to other oppressed nations. Eritrea's standing against hegemonic powers that was once a rare exception in the African continent is being replicated in Mali, Niger and Burkina Faso.

The counter-hegemonic characteristic of Eritrea's foreign policy is further expressed in the efforts the country has made, in conjunction with like-minded countries, towards changing the existing unipolar international system with a just and inclusive system that can fulfil the aspirations of the majority of the people of the world.

Most of the literature on Eritrean foreign policy and its portrayal in the media is that of simplistic denigration. The paper's findings indicate that there is an acute need for serious analysis on the factors that led Eritrea to adopt an independent foreign policy and its consistent resistance to oppression; and to explore its contribution to the counter-hegemonic movement. This is especially relevant currently when there is wide ranging consensus that

53 Remarks by President Isaias Afwerki at the XV BRICS Summit, 26 August 2023. https://shabait.com/2023/08/24/remarks-by-president-isaias-afwerki-at-the-xv-brics-summit/

54 Group of Friends in Defense of the Charter of the United Nations. https://www.gof-uncharter.org/

the US-led unipolar World Order that came into being after the collapse of USSR in 1991 is unravelling, and that the outlines of a new international system with two or multiple poles can be traced in the on-going high-stake conflicts.

It is worth indicating that Eritrean academics and practitioners of International Relations have a responsibility to document and analyze Eritrea's foreign policy thoughts and practices. Equally, non-Eritrean writers interested on the subject are well advised to approach the issue objectively.

Reference

Bruton, Bronwyn, "Eritrea: Coming in from the Cold". Atlantic Council, December 7, 2016, https://www.atlanticcouncil.org/in-depth-research-reports/issue-brief/eritrea-coming-in-from-the-cold-2/

Bolton, John. Surrender is Not an Option; Defending America at the United Nations and Abroad. Threshold Editions. 2008

Cheney, Dick. Secretary of Defense. "Defense Strategy for the 1990s: The Regional Defense Strategy". January 1993

Cohen, Hank. Time to Bring Eritrea in from the Cold". African Arguments, December 16, 2013, https://africanarguments.org/2013/12/time-to-bring-eritrea-in-from-the-cold-by-hank-cohen/

Cox, Robert W. "Gramsci. 'Hegemony and International Relations'". *Millennium Journal of International Relations* 12, no. 2 (1983): 162-175

Dunn, Kevin C. "Introduction: Africa and International Relations Theory". in Dunn, Kevin C. and Shaw, Timothy M (editors). *Africa's Challenge to International Relations Theory*. Palgrave Publishers Ltd. 2001

EPLF. "Memorandum by the Eritrean People's Liberation Front Submitted to all Progressive Forces in the World". 1978. PFDJ Archives (in Tigrigna)

EPLF. Documents of the Second and United Congress of the EPLF and ELF-Central Leadership. Sabur Printing Press. 2007 (in Tigrigna)

Group of Friends in Defense of the Charter of the United Nations. https://www.gof-uncharter.org/

Healy, Sally. "Eritrea's Regional Role and Foreign Policy: Past, Present and Future Perspectives". 2008. Chatham House. www.chathamhouse.org/sites/default/files/public/Research/Africa/171207eritrea.pdf

Isaias Afwerki, "Foreign Aid Works Best When It's Self-Limiting." Forum for Applied Research and Public Policy. Winter 1997; 12 no 4; ProQuest, pp 68-70

Isaias Afwerki (President), Remarks at the XV BRICS Summit, 26 August 2023, https://shabait.com/2023/08/24/remarks-by-president-isaias-afwerki-at-the-xv-brics-summit/

Ministry of Foreign Affairs of the State of Eritrea: "A Shameful Day for the United Nations", 23 December 2009, https://shabait.com/2009/12/23/press-release/

Müller, T. R. "Assertive Foreign Policy in a 'Bad Neighbourhood': Eritrean Foreign Policy Making". In Z. Tsighe, S. M. Idris, Y. M. Asfaha, S. W. Andemariam, R. K. Taddesse, & G. Ogubazghi (Eds.), International Conference on Eritrean Studies, 20-22 July 2016, Proceedings (Vol. 1, pp. 631-651). National Higher Education and Research Institute. 2018

PFDJ, National Charter. 1994, https://ecss-online.com/data/pdfs/PFDJ-national-charter.pdf

Plaut, Martin. *Understanding Eritrea: Inside Africa's Most Secretive State*. Hurst & Company. 2016

President of the United States, "The National Security Strategy of the United States of America", https://2009-2017.state.gov/documents/organization/63562.pdf, September 2002

Rosenblum, Peter. "Irrational Exuberance: The Clinton Administration in Africa." *Current History*, vol. 101, no. 655, 2002, pp. 195–202. *JSTOR*, http://www.jstor.org/stable/45317205.

Transitional Government of Eritrea, Department of Foreign Affairs, "Report Submitted to the 6th Meeting of the Central Committee of EPLF". February 1993 (Unpublished report in Tigrigna)

World Bank Operations' Evaluation Department. Eritrea Country Assistance Evaluation. Report No. 28778. 22 April, 2004, https://ieg.worldbankgroup.org/sites/default/files/Data/reports/cae_eritrea.pdf

Eritrea's National Charter: Navigating through Contemporary Realities

Mehretab Medhanie[1] and Temesgen Tesfamariam[1]

[1]People's Front for Democracy and Justice, Research and Documentation Center (RDC)
Email: tesfam12@gmail.com

Abstract

The National Charter of Eritrea has been a cornerstone of the country`s development, shaping its transformation over the past three decades. This pivotal document has served as a guiding force. It has encapsulated the collective aspirations of the Eritrean people and provided a framework for state-building. Using primary and secondary sources, this article revisits the document to analyze its relevance in the contemporary Eritrean society. Accordingly, this article argues that the changing socio-economic and political landscapes, both domestically and globally, demands a re-evaluation of the Charter`s relevance and adaptability. Demographic shifts within Eritrea have introduced new challenges and opportunities. The younger, educated, and connected generations, with their distinct perspectives, require a more inclusive and participatory approach to governance. This requires embracing a dynamic and flexible approach, one that acknowledges the changing world while staying grounded in the country`s core values. Adopting this approach will make the National Charter remaining the symbol of the nation`s aspirations.

Keywords: Democracy, EPLF, Market Economy, PFDJ, Social Justice,

Introduction

In February 1994, the Eritrean People`s Liberation Front (EPLF) convened its third congress. The congress marked a new dawn in sovereign Eritrea's political history. This congress signified the end of an era defined by armed struggle and the beginning of a new chapter focused on state-building. Motivated by Eritrea`s hard-won independence, the EPLF—reconstituted as the Peoples' Front for Democracy and Justice (PFDJ)—embarked on a mission to build a prosperous and technologically-advanced society within two decades.[1] The mission was framed as a collective endeavor, supported by the Eritrean people. While the leadership acknowledged that state-building would be more challenging than achieving political independence, the realities of governance in the first three years of independence provided critical insights. The discussions and declarations from the congress, documented in various texts, reflect the depth of the leadership`s understanding and aspirations for Eritrea`s state-building project.

Among the key documents produced, the National Charter stands out as the most significant articulation of the PFDJ`s post-independence vision for state-building. The Charter

1 Government of the State of Eritrea (1994): The Macro-policy, November, 1994

outlines the Front`s vision, aspirations, and objectives, serving as its *raison d'être* and as a precursor for the macro-policy document which was issued a few months after the National Charter. While the vision itself is not new—rooted in the principles espoused during the three-decades-long struggle for independence—the Charter represents a formalization of these ideas. It acknowledges the contributions of the past and present generations to the nationalist struggle while emphasizing the need to empower future generations to address the challenges of state-building. In essence, the National Charter serves as the unifying agenda for Eritrean society, transcending diverse backgrounds and experiences.

Over the past three decades, the National Charter has remained a guiding framework for Eritrea`s state-building processes. However, significant socio-economic transformations, both domestically and globally, have altered the conditions that shaped the practicality of some of the Charter`s principles and mission. These changes are often the subject of public and intellectual debates about the Charter`s relevance. This paper seeks to critically examine the National Charter in light of these transformations, addressing a key question: What relevance does the National Charter hold under the fast-changing national and global dynamics? This article addresses the question by analyzing the Charter through post-development lens, assessing its alignment with domestic and global development dynamics since its adoption. While 30 years may seem a brief period in the history of state-building, the rapid pace of historical changes underscores the importance of critically evaluating the Charter by focusing on the political and economic systems that it envisioned.

Political System

The Charter raises two critical questions regarding Eritrea`s political system: What is the most appropriate political system for an independent Eritrea? And what type of government should be established? These questions reflect the Charter`s emphasis on aligning the political system with the aspirations of the Eritrean people. To understand the Charter`s political framework, it is essential to examine the historical and ideological context that shaped its drafting.

The contents of the Charter were intensely debated during the revolutionary movement and continued to be so after independence, including in formal conventions at the State Palace and *Balineki* where the third congress was held. Central to these discussions were the principles of national unity, nationalism, and social justice, which were identified as the ideological foundations of the new nation. The drafting committee also acknowledged the influence of global political and economic shifts in the early 1990s, a period marked by intense debates about the future of political systems across the globe. The rise of neoliberal capitalism, declared as the "end of history," contrasted with "clash of civilization" thesis, which emphasized cultural and geographical factors in shaping global politics, had become the chorus globally. Against this backdrop, the Charter sought to articulate a uniquely Eritrean vision of nationhood, rooted in the cultivation of cultural values and social cohesion.

The emphasis on nationalism and social justice reflects Eritrea`s unique historical context. Pre-colonial Eritrea lacked a hegemonic state structure, a vulnerability that often leads to national disintegration. This is further compounded by the rather negative experiences during the early period of armed struggle especially, that of the ELF leadership, which were marred by divisions along ethnic, religious, and regional lines. In this regard the Charter notes, "that it [the ELF] submerged itself in divisions in the Eritrean society along ethnic, religious and regional lines." However, sub nationalist thinking persisted despite 'advances in nationalist consciousness.' Hence, it is against the foregoing that the Charter emphasized the continued need for a unifying national identity. Mindful of the centrifugal tendencies in diverse societies such as Eritrea, the Charter proposed for the establishment of a democratic political system rooted in multiparty system, a functioning civil society and an independent media.

However, by invoking 'nationalism' as the organizing principle or legitimation of Eritrea's continued revival, how does the front foresee to balance the conundrum between cultural diversity and the demand for national unity? Are the more contentious political ideals of multi-partyism and cultural plurality the aspirations of the political elite or were part of the demands of the expansive social base that the EPLF/PFDJ has cultivated over the years? In what follows, the paper attempts to breakdown these two points.

The Conundrum of Cultural Diversity and National Unity

The tension between forging national unity and accommodating sub-national cultural attachments are vivid across multi-national societies. In the Eritrean case, key challenges include reconciling ethnic and religious affiliations with a broader concept of Eritrean nationhood and harmonizing cultural diversity with the demands for unity.[2] These issues were extensively debated during the armed struggle and in the post-independence period.

But the key ideas on how the front intends to confront these contentious issues is comprehensively captured in the seminar paper that the EPLF Secretary General (SG) presented to a gathering of the Front's top cadres in October 1988.[3] In a paper titled "Guideline for Political Education", the SG outlined 9 points that should be considered in charting out post-independence political system. The nine points were: i) generational gap; ii) geography; iii) urban vs. rural; iv) Eritreans in diaspora vs. in country; v) ethnic composition; vi) religious composition; vii) cultural diversity; viii) gender; xv) social class. Based on these points, the SG suggested Eritrea's political system be based on: a) nationalism; b) social justice; c) moral values (ስነ-ምግባር). These along, with additional three criterion, were adopted as the 'basic principles' of Eritrea's political system in the Charter—with some variation in language and articulations.

At independence in 1991, Eritrea faced glaring socio-economic inequalities. The drafters of the Charter recognized the dangers of failing to address these disparities, warning that 'perceptions of marginalization' by one social group or the other could undermine peace and stability as seen in many African countries. As put in the Charter, "if government is perceived as close to some but remote to others, let alone to build and develop Eritrea, we will not be able to preserve the peace and stability we now have." Various government and PFDJ members emphatically stress that it is only by narrowing the gap and invoking social and economic redress that Eritrea will be able to avoid the pitfalls of religious and ethnic sectarianism that have enmeshed many countries into conflict. In effect, the Charter espoused for the establishment of a national political system that is based on 'nationhood'. The Charter espouses to cultivate 'nationalism' based on 'citizenship' where the nation and the state are congruent.

Some of the critical voices against nationalism are that nationalism undermines multiculturalism and is inherently divisive. The liberal political tradition specially makes a pacifist

---

2 Fouad Makki (1996): Nationalism, State Formation and the Public Sphere: Eritrea 1991-96. Review of African Political Economy, Vol. 23, No. 70 (Dec., 1996), pp. 475-497

3 መደረ ብጸይ ኢሳይያስ ኣፈወርቂ ኣብ ሰሚናር ፖለቲካዊ ትምህርቲ (1988).

critique of the nation-statism project in Africa as a 'shackle' in itself. However, such assertions largely come from the experiences of Europeans in the wake of ultra-nationalist chauvinists that drove the entire continent of Europe to chaos and wars. Basil Davidson asserts that the [discourse against nationalism] is:

> easily acceptable in and for Europe in the wake of Fascism, Nazism, and Stalinism, under whose ideological impetus more crimes and horrors have been committed in the name of one or another kind of nationalist glory than will ever be counted or even remembered. But for the nationalists of Africa the matter had to look quite otherwise. [4]

This was exactly what the drafters of the National Charter seemed to have realized. They looked for alternatives that came from their own experiences. The kind of nationalism that is implored in the National Charter is a 'civic nationalism' which isn't assimilationist, but one that implores on the supremacy of national identity over all other primordial identities. Having enough experiences and knowledge, the drafters of the Charter also scrutinized the dangers that one group dominating another might bring, with the seriousness that it deserves. Because, in many ways, their armed struggle experience was equal to the task that nation building merits. The Charter also asserted the need to learn from others without losing a cultural balance and "without resorting to blind emulation and worshipping of foreign culture… to build [a] new culture on the foundation built during the revolutionary struggle." It is clear that the drafters of the Charter see the process of state-building largely as a cultural process that can be cultivated over a long period of time.

The process wasn't looked as only setting up the coercive state apparatuses. Nationalism was taken as an integrative ideology which will provide a moral framework for the enhancement of national cohesion among the different cultures of Eritrea. Eritrea's nation-building process largely emphasized "the primacy of 'national' loyalties over 'primordial' affiliations deemed divisive or prejudicial to the 'national interest' ". [5] Unlike in many developing countries, nationalism wasn't seen as a replacement of multiculturalism nor was there any attempt to form a singular identity.

There was almost a universalist appeal for nationalism and social justice as the basis of the political system for the new nation. This universalist appeal stems from their perceived capacity to transcend social divisions that have characterized the political environment of the early years of the armed struggle. Divisions along the nine points mentioned above were deemed as potential factors that might potentially undermine national cohesion. A consciousness of national unity and undiluted loyalty to the state would be cultivated by ensuring the fair distribution and ownership of national resources and respect for one's cultural identity. The state-building process was thus premised to be based on the supremacy of the 'national identity' but without undermining the cultural values and social markers outlined above.

However, the expansion of a globally interconnected world – hence the high probability for the percolation of external culture into the national culture – might be deemed as a challenge sustaining a national culture and identity. The question is then whether Eritrea's model of 'nationhood' will adapt to the emerging global trends. The Charter's cautious approach insinuating on the need to make cultural balance, by preserving Eritrea's identity heritage, while at the same time adapting to evolving global trends indicates that the two aren't necessarily antithetical. Bhabha`s work of 'hybridity' reminds us that national identity

4 Basil Davidson (1991), The Blackman's burden

5 Fouad Makki (1996): Nationalism, State Formation and the Public Sphere: Eritrea 1991-96. Review of African Political Economy, Vol. 23, No. 70 (Dec., 1996), pp. 475-497

is neither fixed nor monolithic.[6] The Charter has addressed the issue of hybridity by allowing the expression of multiple affiliations but pivoting national identity as the supreme marker.

Multipartyism and Democracy

The National Charter exhorts that Eritrea's political system must be based on a "multiparty system in which political parties legally participate, and compete among themselves in a peaceful and democratic way." This stance was taken during the EPLF's second and unity congress in 1987 which declared commitment to multiparty and market economy system. We will revert to discussions of the economy at the later section of this paper, but the commitment demonstrated a shift from an earlier adherence to Marxist-Leninist ideology to a more pragmatic position, asserting commitment for a democratic political system.

The drafters of the Charter were very cautious that, the kind of democratic political system should echo the historical and political developments and actual conditions of the Eritrean society. But the question is whether the social, economic and institutional condition was favorable for a 'democratic' and multiparty systems to take immediate root and sustain. The Charter acknowledges the absence of positive experience of multiparty system in Eritrea's history, and stresses on substantive rather than procedural democratic exercise, invoking the "existence of democratic institutions and culture, broad public participation in decision-making and a government that is accountable to the people."

The Charter addressed the conceptual and practical challenges of democracy, that democracy is a process that takes years to be cultivated and developed. Nor did it accept democracy as a panacea to all societal problems. Yet the Charter believed that democratic government, an active civil society and an independent and reliable mass media are the three pillars of democratic political system. Nonetheless, the drafters recognized that realizing such a system right in the aftermath of a war would be a daunting task. Hence, the commitment for multipartyism and popular democracy was just an aspiration that would be implemented in the long-term, phase by phase. One of the key informants stressed "we need multiparty systems to have a healthy political environment. But it must be associated with building strong institutions, peace and stability'. Still there are others who believe, there was nothing wrong with that aspiration as the political process to transition to democracy and multipartyism was only disrupted due to the 1998-2000 'border war' with Ethiopia.

That the formation of multiparty system delayed is framed as a gap between policy and practice. The inherent problem with popular democratic and multiparty system for a new nation that never had 'a positive experience of multiparty system', the plausibility of that system for the socio-economic conditions of the Eritrean people doesn't seem to be questioned. Yet, this might not be only a manifestation of a gap between policy (the Charter) and practice. Drawing from different development experiences, it is observed that what makes for a good practice isn't necessarily driven by good policy and good policies doesn't necessarily lead to implementable practices.[7] Often, good policies are the result of good practices accumulated over long period of time. The Charter expressed clearly that there hasn`t been 'positive experience with multipartyism' in the Eritrean society. This calls for a critical analysis of the pertinence of the multiparty system to the socio-economic and institutional condition of Eritrea; as the dangers of adopting a system that Eritrea never had a 'positive

6 Bhabha, H. (1994). The Location of Culture, London and New York: Routledge.

7 David Mosses (2005): Is Good Policy Unimplementable? Reflections on the Ethnography of Aid Policy and Practice. Development and Change 35(4): 639–671 (2004). Institute of Social Studies 2004. Blackwell Publishing.

experience' about can result in perpetuating unimaginable difficulties hidden behind a cloak of 'promises of the Charter'.

Historically, the PFDJ is known for its pragmatic views on any policy measures that it adopts. For instance, when the Front declared that it was abandoning its aim of forming a 'classless society', President Isaias justified the decision as "our ideology of cultivating a society that is free of exploitation of man by man was utopian. The strategies and procedures that are drawn and the decisions that are taken to achieve this utopian thought will require that we work out of established norms. This will cause a moral conflict and lead to unnecessary frustration and confusion among the society."[8] This shows the Front's pragmatic approach driven by practicality and considering the real conditions of the Eritrean people.

The reports of the Charter's drafting committee show that the issue of democratization and multipartyism was deeply debated with a few divergent views, especially on the adoption of multiparty systems. Even prior to the preparation of the Charter, the issue was deeply debated during the armed struggle by members of the Central Council of the EPLF. The 1991 January issue of the internal magazine of the EPLF, *Harbegna*, hosted a debate among four EPLF central council members under the title "What political system should Eritrea adopt in the future?"[9] Although all of them except one agree for the adoption of a multiparty system, they all warn of the dangers that could come with it in a society that still has to consolidate its state structure. There were those that contended that "multipartyism can be adopted as a right" as political parties are just instruments of political organization where society ensures participation of its representatives. But, experiences with multiparty system in post-colonial societies haven't been encouraging. In a society that still sticks to its social base the lofty ideal of thinking about 'national interest' is faint. First society needs to be educated, reach a certain degree of social organization and consolidate the things that it shares together under the 'nation'. Hence, it is one thing to adopt multiparty system as a 'right of every citizen' another thing to encourage and allow it occupy the political thinking of the citizenry. For instance, "it is one thing to state divorce as a right of every citizen, another thing to allow it to be practiced with ease."

The arguments against multiparty seemed to be popular and stronger than those supporting it, which revived during the constitution drafting process. For those who went for one-party system, if bourgeoisie democracy has any great success, it was its ability to create a forum of tolerance for contending views, and not of political parties.[10] The establishment of democratic political systems isn't necessarily contingent upon the presence of political parties and democratic political systems aren't solely about political parties as also alluded in the Charter. Indeed, the party system is only one among many systems for democratic governance.[11] To the contrary, it is argued that multipartyism in societies that haven't built well-functioning bureaucratic institutions might lead to unnecessary compromises and a 'coalition government' that is internally weak."[12]

---

8 saias, Akieba Balineki 1993

9 Harbegna (1991): What political System Should Eritrea Adopt in the Future. January, 1991. (ሓርበኛ፡ መጽሔ. ፖለቲካዊ ስርዓት ኣብ ኤርትራ፡ ጥሪ 1991)

10 Gelawdios Araya (1991): The Political Economy of Transition in Eritrea. Conference paper presented at the Economic Policy for Eritrea Conference (July 15-19, 1991).

11 Samuel P. Huntington (1994): Political Development in Ethiopia: A peasant based dominant party democracy. Report to USAID/Ethiopia. Consultations with the Constitutional Commission 28 March-1 April 1993; Addis Ababa, Ethiopia

12 Robert Kaplan (2010): Was Democracy Just a Moment? in "The State and Democracy- An Anthology" P-214. Published by Hidri Publishers (2010), Asmara, Eritrea

Key interlocutors of this paper believe that the PFDJ didn't see itself as the only hegemonic political organization to operate in Eritrea. It wanted to open up the political arena to diverse views and organizations as long as they hold a nationalist structure and prioritize national interest over any other interest. The question then is, can transition to a party-based political system be complete without the dangers of weakening the national unity that has been cultivated over the years? Can the Front`s popularity and hegemonic existence as broad-based mass movement and the characteristics of multiparty system be compatible? The Charter itself is wary of these risks, stating "during peacetime, unlike in wartime, it is evident that certain tendencies and thoughts that gradually weaken unity may emerge...... as in other Third World countries, this process [of forming a nationhood] is not yet concluded, we should consider the development and strengthening of Eritrean nationalism and the unity....". This kind of apprehension is evident in the experiences of other 'third world' countries, where initially popular revolutionary political organizations rush to establish a political system that is based on party politics and evidently fail to galvanize and consolidate internal cohesion and national political consciousness.

The Charter`s adoption of constitutional democracy reflects the long-term aspiration of the Front's leadership for a stable and harmonious society. The Charter acknowledges that democracy based on factitious political parties and elections alone doesn't foster democratic development in state or society. The attempt here isn't to question the importance of political parties and regular elections as essential instruments of a healthy and an accountable political system. But with all its characteristics and the social institutions that it demands, the paper askes, was the time ripe for multipartyism to be declared as the desired political system for Eritrea? As with the aspiration of 'classless society' that the front had adopted in its 1st and 2nd congress, which it later abandoned as mere 'utopian' in the lead-up to the third congress. Doesn't multiparty democracy also lead to unnecessary expectations, uncertainty and frustration? Wouldn't it be best, to first cultivate some of the requisite collective characteristics before declaring a policy choice such as the need to develop a critical middle class, efficient bureaucracy, and addressing the primary issues of consolidating sovereignty.

Since 1998, several changes have emerged in the socio-economic and political atmosphere of the nation. As one interlocutor from the PFDJ Executive body noted "a fresh political programme is in the need. One that factors the experiences and changes that have emerged since 1998 and one that takes into account global changes." The Front continues to enjoy substantial popularity among the Eritrean masses. It is considered not just as a political movement, but as a true reflection of the collective wishes of the Eritrean people to have an organization that binds them together. Yet, for any political movement the size of the PFDJ (600,000 members) the difficulty remains in staying as monolithic. The Front's adoption of the market economy also has a political implication that it might result in social differentiation potentially conceding the fervent support that it now enjoys. The Charter's analysis of developing a democratic political system that is rooted in the socio-economic condition of the Eritrean people didn't sufficiently factor the potential social differentiation that would come with the adoption of the market economy. It appears that it is of great essence to explore other systems of political organization such as the Chinese System, where the Communist Party of China (CPC) has adopted a political system that is different from the Western model, and is claimed to be based on the Chinese characteristics.

Political parties emerged as solutions to problems of social cleavages that Europe experienced in the 17th-18th centuries in the event when suffrage came in place of the monarchic system particularly those related to center-periphery or centrifugal/centripetal forces within the different nationalist groups.[13] They were only vehicles of social organization in a society that was organized along nationalist (often ethnic), religious and ideological bonds. The reality in Eritrea as in most Africa is different. The core issue is therefore: Can political parties serve the lofty ethos of social justice and national unity principles enshrined in the Charter?

Economic System

National Charter outlines that Eritrea's economic system must be based on the principle of social justice. The economic essence of social justice is explained as narrowing the socio-economic gaps, creating essential conditions for people to be productive, and developing economic democracy. The Charter emphasizes the need to balance between the role of the government in the interim period and with the critical role of the private sector in the longer term. Hence it stated that Eritrea's economy would be based on free market, while the government would reserve its regulatory role in the interim period until the gaps between the different socio-economic settings in the country are narrowed and national economic recovery reaches a certain stage. The government's role is framed as a catalytic role of economic development and paving conducive environment for the private sector. This policy statement is a known neoliberal capitalist dictum. This is a new ideological cajoling with the West when compared to the political program of the EPLF as adopted after the first and second congress in 1977 and 1987 respectively.

Changes in EPLF/PFDJ economic policies

1977	1987	1994
- Non capitalist development - The main sectors of the Eritrean economy would be administered by the state - Internal and foreign commerce would be under the supervision of the State - The State would nationalize all the mineral wealth, exploited or in potential, and invest in the interest of the whole people.	- A command economy, based on the principle of self-reliance and is not controlled by foreigners. - The main economic resources should be state owned. - Capital investment- Citizens in the private sector; while foreigners shall be encouraged and allowed to invest in the government sector.	- market economy, with government playing a catalytic and regulatory roles - a mixed economy in which both public and private sectors exist.

The table above shows how EPLF/PFDJ economic policies have shifted at different times. Despite slight differences, the economic policy in 1977 and 1987 remained consistent—command economy. After independence, a major shift has occurred towards adopting a market economy. This pragmatic decision was influenced by the socio-economic condition internally and global dynamics of the time externally. A report presented by the SG to the fifth regular meeting of the EPLF central council in 1992 highlighted the grim reality.

> The economy that we have inherited is an economy of drought and relief aid. To depend on relief aid is to live under slavery. In this kind of unconventional situations, all the issues that are related to our economic policies doesn't give us the flexibility to look for other alternatives. Situations like this expose you to undue political and economic pres-

13 Carles Boix (July 2009). "The Emergence of Parties and Party Systems". In Carles Boix; Susan C. Stokes (eds.). The Oxford Handbook of Comparative Politics. Oxford University Press. doi:10.1093/

> sure. As a result, one is compelled to make compromises at the expense of core interests, finally instilling a mindset of dependency.[14]

By all accounts, Eritrea was a country in ruins, with well above 70 percent of the population dependent on food aid. Hence, the Charter struck a balance between the need for a quick economic recovery and an economy be led by a vibrant and responsible private sector in the longer term. Many pointed out that the leadership was not free of influences from the neoliberal tendencies of the time in the wake of the disintegration of the USSR, believing in the neoliberal capitalist policies as the most appealing. The adoption of the market economy and limited role for the front are a clear break from its previous stance on the central role of the state and emblematic of the neoliberal discourses of the time. However, given the state of economy, it seems plausible that the government decided to limit its role on the broader objective of macroeconomic stability, while espousing for a market economy that will eventually lead to 'economic democracy'.

One PFDJ member who participated in the various discussions preceding the release of the Charter noted "we didn't have a clear orientation at the time. The focus was more on recovering the devastated economy. The Western economy was also much more dominant at the time, making the pursuit of an 'autonomous economic policy' difficult. But we were as well aware of the need to control on national resources and avoid exploitations that we saw in so many other countries. Hence, later on, we made pragmatic choices and negotiated our interest to the maximum, as seen in deals with mining companies where we insist on having significant shares 40-50%.

The Charter was also wary of the implications of liberalizing the economy without laying the necessary infrastructure for sustainable economic take-off. It outlined the need to develop the industrial base and agriculture (rural development). In this respect, self-reliance was taken as the organizing principle in all fields. Self-reliance wasn't adopted as just an economic principle but the Front also wanted to make it as part of the thinking of every Eritrean. By adopting self-reliance, the Charter aimed to avoid cyclic dependency in all its forms—economic, political, education, security and other social services. Interestingly, the Charter prioritizes economic development over political development, suggesting that the type of political system the front wanted to build would be determined by the type of economic system adopted.

This topic has sparked debates among various scholars. One PFDJ member said "the assumption at the time was that we no longer would go to war, that peace will prevail in the region. Our internal stability wasn't an issue, as is today. We were looking for a broader regional integration." It is clear that the conditions at the time dictated the adopted policies. The need to win in the economy was equated with winning the war of liberation. In 1994 PFDJ's Political Affairs Head said:

> Unless we won the war, we risked losing everything. At this stage, it is the economy. Unless we develop the economy in a way that does not create social dislocations, we risk losing everything again. So, we don't want the party branches to be places where people just come to talk and talk. I don't doubt that Eritrea will succeed economically. The challenge is whether we can do this and really have social justice, whether we can maintain our own identity and our core values.[15]

---

oxfordhb/9780199566020.003.0021.

14 A report presented by the Secretary General to the fifth regular meeting of the EPLF central council in 1992. **ኣጠቓላሊ ጸብጻብ ዋና ጸሓፊ ህ.ግ.ሓ.ኤ ናብ 5ይ ኣኼባ ማእከላይ ሽማግለ** (1992)

15 Dan Connel (2001): The Importance of Self-Reliance: NGOs and Democracy-Building in Eritrea;

The assumptions that furnished this belief aren't in place today. The region has entered into a different dynamic relations between countries continuing to be unstable, mostly due to external interference, and the global economic system undergoing unprecedented changes. Under the circumstances, the need to balance between political stability and economic development are of profound importance. The Charter identified agriculture, industry, and commerce as priority sectors. Yet, in a country that is endowed with over 1200kms of sea coast and vast mineral resources, one would expect the Charter provide an outline of the long-term vision in the fishery, blue economy and the extractive sectors at least broadly. Going forward, the Charter might need to emphasize these important sectors of the economy.

Today, with changes in the global economic system, developing countries have more options of adopting autonomous economic policies. The global economy is more decentralized, given that it is competitive, interdependent and interlinked. Global financial and trade regimes are entering a new phase with the rise of other nations and the formation of new economic blocks, including the BRICS+. The dominance of the dollar economy and international financial systems is being challenged as major powers such as China, India, Russia provide more options.

Despite these promising changes, the competitive, interdependent and interlinked nature of the global financial and trade regimes have become more binding on national governments, making the 'autonomous' pursuit of economic policies considerably more difficult for developing countries. The Charter might need to include factors for Eritrea's macroeconomic stability as the competing sectors of development are numerous. Although, policies related to the sectors of the economy need to be prepared separately, the Charter need to provide essential ideological guidelines for those sectors as well. For instance, the Macro-economic policy that was prepared following the adoption of the Charter is criticized by many as essentially neoliberal and informed more by the monetarist belief that growth in the economy is best achieved through the free operation of market forces and the removal of state interference.[16] The government has indeed adopted the neoliberal 'grab-bag' and initially accepted the Washington Consensus, establishing the National Agency for the Supervision and Privatization of Public Enterprises in 1995 (NASPEE), laying-off public service employees…etc. the NASPEE slated 39 enterprises for privatization at the time.

Now, with changes in the global economic order, more developed and emerging economies are turning towards Africa for resources, markets and industrial bases. This requires a high level of discipline of the leadership for macro-economic stability, which Eritrea isn't lacking. The fact remains that, unlike in the 1990s there are quite a number of possibilities to take advantage of and many ways to navigate revisiting the National Charter. Lessons from many African countries indicate that they rushed to liberalizing their economies without laying the necessary self-supportive economic infrastructure and neglecting their rural economies, resulting in outsourcing most of their much-needed resources to foreign actors and rendering their economy stagnant or even deteriorate.

The evolution of Eritrea`s economic policies, as outlined in the National Charter, reflects a journey from a command economy to a market-oriented one. This transformation was driven by the need to recover from devastation left by prolonged conflict and to adapt the global economic dynamics of the time. The shift from a state-controlled economy to a

http://www.merip.org/mer/mer214

16 Fouad Makki (1996): Nationalism, State Formation and the Public Sphere: Eritrea 1991-96. Review of African Political Economy, Vol. 23, No. 70 (Dec., 1996), pp. 475-497

market economy was a pragmatic decision aimed at ensuring quick economic recovery while fostering long-term economic democracy and social justice. The challenge however, was, how to strike balance between government intervention and the promotion of a vibrant private sector. The emphasis on self-reliance, industrial development, and agriculture highlights the Charter`s focus on creating a sustainable economic foundation. However, the economic policies adopted in the early 1990s were heavily influenced by neoliberal ideologies, which called for reduced state interference and a greater role for market forces.

In fact, in practice, the economic model that Eritrea has followed since 1998 resembles more like what the Keynesian school of economic thought call the "Employer of Last Resort-ELR" model, a system that emphasizes dominant role of the government in driving the public employment sector. This economic model is thought to be superior to the traditional export and FDI-led development model emphasized in the Charter, especially for small developing countries such as Eritrea. The rationale behind the ELR is that the government turns into an employer of last resort, to sustain labor demand elasticity and serving as a stabilizer where the public sector would absorb the employable work force during difficulties, and it would constrict during relative stability allowing the private sector access to labor force at a premium above the ELR-wage.[17]

As the global economic landscape continues to evolve, Eritrea must navigate these changes by adopting flexible and adaptive economic policies. Internally Eritrea`s economic strategy must consider the lessons learned from past experiences and the current global economic environment. By emphasizing the development of key sectors such as fisheries, blue economy, and extractive industries, Eritrea can harness its natural resources to drive sustainable growth. Externally, the rise of new economic blocs and the challenges to traditional global financial systems presented both opportunities and risks for developing countries like Eritrea. Balancing political stability with economic development will be essential in achieving long-term prosperity and social justice for the Eritrean people.

The National Charter and Contemporary Realities

The relevance of the National Charter in the contemporary nature of Eritrean society remains essential. First, as a guiding document where the collective aspirations of Eritrean society are expressed and outlined, and as an extract of past and present political and economic experiences of Eritrean society, the National Charter stood a witness to the PFDJ leadership's proactive approach to present day dynamics. Secondly, the principles of the charter are the outcome of blood and sweat of the Eritrean society. The principles owe to the people of Eritrea, for they themselves are an end attained through heavy price. However, thirty years have manifested several changes that the Charter should consider during its revision.

The Charter`s pledge for democratic political and multiparty system, and market economy reflects the front and leadership`s commitment to a nation premised on lasting peace, justice, and prosperity. Given the previous stance of the EPLF and the global context of that period, this was a very pragmatic change. However, it also reflects the drafters` entanglement in what Park termed as 'the trap of hybrid systems.'[18] In their immediate post-independence periods, many African countries attempted to mix liberal democracy with neo-patrimonial culture that existed within society. In the process, they neither successfully formed democrat-

17 Kaboub, Fadhel (2007) : ELR-led economic development: a plan for Tunisia, Working Paper, No. 499, Levy Economics Institute of Bard College, Annandale-on-Hudson, NY

18 Jong-Dae Park (2019): Re-Inventing Africa's Development Linking Africa to the Korean Development Model. P-130

ic governments nor preserved their indigenous governance structures, essentially becoming status quo-oriented. The Charter`s adoption of free press, civil society, and democratic government, as the three pillars of democracy show its embrace of liberal democracy. However, its emphasis on empowering people`s assemblies and associations, people`s army, and a dynamic organization based on the people' also indicates an attempt to preserve indigenous social structures.

With the end of the Cold War, neoliberalism became the dominant framework in political and economic development processes. Non-democratic states were perceived as threat to their own people and the world, making liberal democracy appear as the panacea for the economic and political challenges of war-torn society.[19] However, this was essentially a reconfiguration of conflict-ridden and fragile society into a Western social mold. [20] In effect, it was an "updated version of *mission civilisatrice,*" which justified a perceived moral responsibility of Western countries to civilize other societies.[21] This assumption that neoliberalism constitutes a superior agenda for humanity leads to a top-down approach that often labels local cultures as obstacle to liberal and national governance. Consequently, neoliberal state-building has been ethically bankrupt, subject to double-standards, coercive, unconcerned with social welfare, and unfeeling and insensitive towards its subjects. [22] As a result, it inflicts more harm than it achieves peace.

As a foundational document for state-building, the National Charter remains crucial for Eritrea`s state-building efforts. However, the rapid socio-economic and political changes of the past three decades necessitate a critical reappraisal of its relevance. While the Charter`s emphasis on nationalism, social justice, and democracy reflects the aspiration of the Eritrean people, its implementation has been shaped by practical challenges and shifting global dynamics. As Eritrea continues to navigate these complexities, the Charter`s principles must be adapted to contemporary realities while remaining rooted in the nation`s unique historical and cultural context.

Internally, demographic dynamics of the Eritrean society are substantial. Over the last three decades, Eritrea`s demographics have shifted significantly. The majority of Eritreans now belong to the independence generation, thanks to the post-independence investment in education, it is educated and skilled with only vague memories of the armed struggle. Growing up in a more interconnected world, this generation is well-informed and their consciousness extends beyond national borders. Due to immense access to digital information, neither culture nor state can guide it effectively via old methods. This generation has new economic demands and political aspiration. Addressing these new demographic realities and their associated needs should inform the revision of the Charter.

Externally, the last three decades of a unipolar world have seen the universalization of guided identities largely construed by Western traditions, leading to societal confusion and conflicts. Under the guise of neoliberal human right, societies fell into divisive identities and developed weak state institutions incapable of conceptualizing national identity, let

19 Curtis, Devon and Dzinesa, Gwinyayi A. *Peacebuilding, Power and Politics in Africa*, Ohio University Press, Athens, 2012, 9

20 Bereketeab, Redie, Alternatives to Neoliberal Peacebuilding and State building in Africa, (London & New York, Routledge, 2021) 25

21 Paris, Roland "International Peacebuilding and the 'Mission Civilisatrice'", *Review of International Studies,* Oct., 2002,28 (4), 651

22 Richmond, Oliver P. "A post-liberal Peace: Eirenism and the Everyday" *Review of International Studies,* July., 2009, 35, (3), 557-8.

alone realizing it, pushing societies towards subnational identities. It has been long already since neoliberal principles are transformed from colonial language to exploitation practices. What has been evident in the last three decades under the unipolar world is that democracy of neoliberalism has been another form of divide and rule tactics in a more elusive way. It created an elite with exploitative capitalism, that can be explained as the Fanon`s term of "comprador class." Disconnected and illusioned, while they enjoy consumptive lifestyles and practices, the public endures destitution and desperation. The so-called democratic practices, free market, free and fair election, equality, women`s right e.t.c, beautifully decorated the thuggery nature of the process.

Under such reality, the National Charter has to devise ways to avoid such catastrophes. One important way is to vouch for home-grown and context-driven democracy. The Charter recognizes the dynamism of society and believes that the document itself must evolve with the dynamism in order to establish the best possible political and economic systems in Eritrea. To this time, the historic democracy relevant to the Eritrean experience is the participatory democracy that evolved from the Eritrean armed struggle, where society through the local and regional assembles, participate in the national issues without discrimination.

Conclusion

The National Charter of Eritrea has served as a pivotal document, guiding the journey of political and economic transformations in Eritrea. Over the past three decades, it has encapsulated the collective aspirations of the Eritrean people and provided a framework for state-building. However, the evolving socio-economic and political landscape, both within Eritrea and globally, necessitates a critical reassessment of the Charter`s relevance and adaptability.

Internally, demographic shifts have introduced new challenges and opportunities. The younger, highly educated, and interconnected generations, comparing to the armed struggle generation, demands a more inclusive and participatory approach to governance. Externally, the influence of neoliberalism and the unipolar world order has highlighted the limitations of imposed democratic models, underscoring the need for home-grown context-specific solutions.

As Eritrea navigates these complexities, the principles of the National Charter must be adaptive to contemporary realities while remaining true to the nation`s historical and cultural context. By embracing a dynamic and flexible approach, Eritrea can continue to strive for a future marked by lasting peace, justice, and prosperity for all its citizens. This evolution will ensure that the Charter remains a living document, reflective of the people`s will and responsive to the changing world.

References

Davidson, Basil, The Blackman's Burden: Africa and the Curse of the Nation-Sate, New York: Times Books, 1992.

Bhabha, H. (1994). The Location of Culture, London and New York: Routledge.

Bereketeab, Redie, *Alternatives to Neoliberal Peacebuilding and State building in Africa,* (London & New York, Routledge, 2021) 25

Carles Boix (July 2009). "The Emergence of Parties and Party Systems". In Carles Boix; Susan C. Stokes (eds.). The Oxford Handbook of Comparative Politics. Oxford University Press. doi:10.1093/oxfordhb/9780199566020.003.0021.

Curtis, Devon and Dzinesa, Gwinyayi A. Peacebuilding, Power and Politics in Africa, Ohio University Press, Athens, 2012, 9

Dan Connel (2001): The Importance of Self-Reliance: NGOs and Democracy-Building in Eritrea; http://www.merip.org/mer/mer214

David Mosses (2005): Is Good Policy Unimplementable? Reflections on the Ethnography of Aid Policy and Practice. Development and Change 35(4): 639–671 (2004). Institute of Social Studies 2004. Blackwell Publishing.

Fouad Makki (1996): Nationalism, State Formation and the Public Sphere: Eritrea 1991-96. Review of African Political Economy, Vol. 23, No. 70 (Dec., 1996), pp. 475-497

Gelawdios Araya (1991): the political economy of transition in Eritrea. Conference paper presented at the Economic policy for Eritrea conference (July 15-19, 1991).

Government of the State of Eritrea (1994): The Macro-policy, November, 1994

Jong-Dae Park (2019): Re-Inventing Africa's Development Linking Africa to the Korean Development Model. P-130

Kaboub, Fadhel (2007) : ELR-led economic development: a plan for Tunisia, Working Paper, No. 499, Levy Economics Institute of Bard College, Annandale-on-Hudson, NY

Paris, Roland "International Peacebuilding and the 'Mission Civilisatrice'", Review of International Studies, Oct., 2002,28 (4), 651

People`s Front for Democracy and Justice, National Charter, 1994

Report presented by the Secretary General to the fifth regular meeting of the EPLF central council in 1992. ኣጠቓላሊ ጸብጻብ ዋና ጸሓፊ ህ.ግ.ሓ.ኤ ናብ 5ይ ኣኼባ ማእከላይ ሽማግለ (1992)

Richmond, Oliver P. "A post-liberal Peace: Eirenism and the Everyday" Review of International Studies, July., 2009, 35, (3), 557-8.

Robert Kaplan (2010): Was Democracy Just A Moment? in "The State and Democracy- An Anthology" P-214. Published by Hidri Publishers (2010), Asmara, Eritrea

Samuel P. Huntington (1994): Political Development in Ethiopia: A peasant based dominant party democracy. Report to USAID/Ethiopia. Consultations with the Constitutional Commission 28 March-1 April 1993; Addis Ababa, Ethiopia

The Balineki Notes (1993): Unpublished notes of the Secretary General's Remarks at a special meeting to discuss the preliminary draft of the Charter. Minutes of Akieba Balineki (1993)

ሓርበኛ፡ መጽኢ ፖለቲካዊ ስርዓት ኣብ ኤርትራ፡ ጥሪ 1991)

መደረ ብጸይ ኢሳይያስ ኣፈወርቂ ኣብ ሰሚናር ፖለቲካዊ ትምህርቲ (1988).

Regional Security Challenges in the Horn of Africa and the Role of Eritrea

Metkel Atewebrhan and Sinit Niguse

College of Business and Social Sciences, Adikeih
Email: metkelabu93@gmail.com

Abstract

Since the end of the colonial period, the Horn of Africa has faced a series of multiple and complex security challenges. In addition to the existing different inter-state and intra-state conflicts and natural disasters, a new round of crises by the instability in Ethiopia, Sudan and the Red Sea is looming all around the region. In Ethiopia, a challenging internal conflict and an alarming humanitarian catastrophe have continued to strangle peace and development. In Somalia, a series of political crises and stalked by the specter of politically-motivated violence, against a backdrop of the increasing influence and capability of the al-Shabaab terrorist has remained the major challenge to the revival of the Somali state. In Sudan, a popular political reform descended to civil war that tore Sudan into politically inhabitable state. These crises have been aggravated by multiple forms of foreign interventions. Additionally, the Red Sea is increasingly becoming a conflict zone of competing global powers. Having stood as stable and strong state, Eritrea has remained firmed to make contributions in changing the conditions of the region. However, its role in the conflicts has been understood differently. While major Western actors view it suspiciously and have denigrated its role, others admire and encourage Eritrea's contribution. The purpose of this paper is to interrogate the role of Eritrea in escalating and deescalating the regional security threats using archival and oral source of data. Its central questions are: to what extent is the role of Eritrea an advantage and/or a threat to the regional peace and security? Why is its role perceived negatively by those who do so? Guided by these major questions that will be disaggregated to sub questions, the paper will unpack the complex interpretations on the role of Eritrea in the region.

Keywords: Security, Ethiopia, Sudan, Somalia, Red Sea

Introduction

The Horn of Africa, located at the easternmost tip of the African continent, has always been one of the critical regions within global geopolitics. From ancient time's right up to the present, the strategic location of this region has placed it in good favor among empires, colonial powers, and modern nation-states. As early as 1500 BCE, the Horn was an important hub for trade and commerce, with the Ancient Egyptians, Phoenicians, and Greeks all vying for control of the region's valuable resources (Kitchen, 1994). In the centuries that followed, the Horn would be shaped by the rise and fall of various empires, including the Axumite Empire, which at its peak in the third century CE was one of the most powerful states in the ancient world (Munro-Hay, 1991).

In the pre-colonial era, the region was marked by the rise and falls of various sultanates, kingdoms, and city-states, each vying for control and influence (Lata, 2003). The introduction of European colonialism in the late 19th century further exacerbated existing tensions, as colonial powers such as Italy, France, and Britain sought to swell their spheres of influence, often through the manipulation of local power structures (Woodward, 2013). It is during this period that the region was partitioned, arbitrary borders imposed, and seeds of discord sowed that would continue to plague the Horn of Africa long after independence.

The post-colonial era has been no less tumultuous, as the newly formed nation-states grappled with a range of security challenges, including interstate conflicts, civil wars, ethnic and religious tensions, and the rise of transnational terrorist organizations (Abbink, 2003) that matched De Waal's depiction as *"Political Marketplace"* (see Muller, 2016). Today, the region's strategic importance, marked by its proximity to the vital shipping lanes of the Red Sea coast and the Gulf of Aden, has made it a focal point for the interests of global powers, further complicating the security landscape (Mesfin, 2019). All these factors of the proliferation of military bases, involvement of external actors in local conflicts, and competition for influence have combinedly contributed to the fragility of the region and continuous challenges to regional security.

This article intends to explore the security challenges within this region while also highlighting Eritrea's contribution to regional security dynamics. Precisely, the research objectives will look into the historical and current security environment of the Horn of Africa, and determine how Eritrea, often perceived from a conflictual perspective as an instigator and a pariah state, can be considered a factor of stability. The central problem will be an understanding of Eritrea as an agent of only destabilization in the region, without regard for strategic alliances and possibilities for peacekeeping. Our central contention is that Eritrea has the potential to play a major role in the mitigation of security threats in the Horn of Africa by facilitating regional cooperation and various diplomatic enterprises toward stability in its bilateral relations. The thematic questions set for the research are: to what extent does the role of Eritrea remain an advantage and a threat to regional peace and security, and why is its role perceived negatively? Therefore, this article will make use of archival and oral sources of data that provide an in-depth understanding of the nuanced security landscape. The article is organized into four sections.

The first section is the introductory section, which presents the geopolitical significance along with the thesis statement. The second section examines the current security challenges in the Horn of Africa and the ongoing conflicts on a spectrum analysis of states of the region separately. The third part will analyze Eritrea's role in the Horn of Africa, its foreign policy, explore Eritrea's relationship with neighboring countries, the role of external powers, and international organizations in shaping regional security dynamics. The final section concludes the paper and offers recommendations for policymakers and scholars seeking to understand and address the complex security challenges facing the region.

Theoretical Framework

The research paper consists of two major theories as the framework of analysis for the collected data, Realism and Regional Security Complex Theory (hereinafter referred as RSCT). The realist perspective defines the security dynamics of the Horn, whereby states are conceptualized as the major actors in the international system driven by interests and struggle for power. According to Mearsheimer (2001), this framework could facilitate explanations of Eritrea's foreign policy and its interaction with its immediate neighbors, especially since the country may adopt strategies or power-balancing to ensure its survival and security.

RSCT provides a clear contextual framework within which to understand the regional dynamics in the Horn of Africa (Buzan &Waever, 2003). Accordingly, this theory recognizes that regional security is determined by interactions and interdependence among states that reside within a geographically defined area. The region can be considered a distinct regional security complex, where the security of individual states is closely linked to the security of the broader region (Buzan &Waever, 2003). In the context of the Horn of Africa, this theory underscores the complex web of political, economic, and sociocultural factors that contribute to the region's security dilemmas, including the legacies of colonialism, the propagation of armed conflicts, the struggle for scarce resources, and the influence of external powers (Mesfin, 2019,). RSCT can help analyze how the security concerns and policies of Eritrea are influenced and interwoven with those security issues of its neighbours. The realist theoretical framework and RSCT provide a broad approach for investigation into the regional security challenges of the Horn of Africa and the role of Eritrea. Using these theories, the researchers made an attempt to explore the security-seeking behaviour of the states in the region, the dynamics of regional power relations, and the interconnectedness of security issues within the region.

Security Challenges

Sudan: Unsettled Crisis

Sudan's past is intricate in cultures, kingdoms, and finally, colonization, which has led to its current crisis. At the juncture of sub-Saharan Africa and the Arab world lies the country of Sudan, upon which much of ancient Sudan was home to the mighty Kingdom of Kush, which existed from 1070 BCE to 350 CE, its capital at Meroe (Heidorn, 1994). The Ottoman Empire's incorporating of Sudan into its domain in the 19th century paved the way for the Mahdist Revolt, a significant uprising against Turco-Egyptian rule from 1881 to 1899 (Holt 1989, 110-112). After crushing the Mahdist state, Sudan came under Anglo-Egyptian Condominium rule in 1899, setting the stage for future political and ethnic tensions by delineating artificial boundaries and uneven development favouring northern regions (Collins, 2008, 156). The British administration fostered north-south divisions with policies that stunted economic development and education in the south, thereby sowing the seeds of hatred that boiled over in the years following its independence in 1956 (Woodward, 2011, 206).

Independence ushered Sudan into an uncertain future characterized by democratic governance and successive coups d'état that plunged the country into numerous civil wars. First and Second Sudanese Civil Wars (1955-1972 and 1983-2005, respectively) were characterized by retaliations against systematic marginalization and Islamization policy against the predominantly non-Arab southern population (Johnson, 2011). These conflicts created terrible humanitarian situations that attracted international interest and eventually gave way to the comprehensive peace agreement in 2005, giving autonomy to the south and paving the way for the independence of South Sudan in 2011 (Hilhorst, 2013, 147). Independence, however, has not solved the recurring wars in Sudan. The chronic Darfur conflict, which ignited in 2003, saw government forces and allied Janjaweed militia engage in what the ICC called a campaign of genocide. Flint & De Waal (2005, 21)

Still, there was political turmoil: the government of Omar al-Bashir, because of internal dissent and international sanctions, finally fell in 2019 after huge protests against his government. The ensuing transitional government struggled amidst deep economic woes and tribal conflicts, with negotiations towards democratic transition often uncertain (Gallopin, 2019). The October 2021 military coup shattered these fragile hopes by ending civilian rule, reverting to a military junta that intensified unrest and led to widespread protests (El-Tom, 2022, 53). Through these turmoils, the complex ethnicity of Sudan, along with the geopolitically strategic position, has often engaged neighbouring states, Middle Eastern powers, and international organizations whose interventions have ranged from humanitarian aid to political pressure. Efforts for reconciliation are complicated by the fragmentation of armed groups, the multiplicity of actors engaged in Sudan's ongoing conflicts, and deep-seated mistrust within the civilian population. The transitional government faced undefeatable challenges, as various factions vied for power, and the impacts of the COVID-19 pandemic further strained the already precarious humanitarian conditions (Pettigrew, 2020). As regional dynamics shift, with interest from international actors seeking influence in the Horn of Africa, Sudan's trajectory remains uncertain yet pivotal for the stability of the region (Khan, 2021).

Contemporary Sudan is a tapestry of historical grievances and unfulfilled socio-political aspirations, the crisis of which has become increasingly exacerbated by continued manifestations of climate change, intensifying the competition for resources and further displacement. The current peace processes have to grapple with long-standing cultural divisions and building just governance structures while trying to address legacies of violence and human rights abuses in their wake (Verjee 2020, 64). The understanding of Sudan's historical trajectories will therefore enlighten on root causes of conflicts today and a potential way toward sustainable peace in the future.

Ethiopia: The Consequences of Ethnic Mapping

Ethnic mapping and categorization can be traced back to centuries in Ethiopia, with quite intensive and far-reaching consequences that continue to shape the nation's political landscape, social dynamics, and relationships between groups. In ancient Axumite and medieval Ethiopian societies, distinctions between ethnic and linguistic groups were often fluid and overlapping, with identities shaped more by regional affiliations, religious practices, and socio-economic roles than rigid ethnic classifications (Marcus, 1994, 7-9). Nevertheless, the process of state formation and expansion under successive Ethiopian empires, especially during the rule of Emperors Menelik II and Haile Selassie I in the late 19th and early 20th century, saw the growing codification and institutionalization of ethnic categories as a means of imperial administration and control (Zewde, 1991).

This trend was taken to even more extreme limits during the Italian occupation of 1936-1941, when the fascist regime instituted divide-and-rule policies along lines of pseudo-scientific racial theories, exacerbating tensions between groups designated as "Amhara," "Oromo," "Tigrayan," and others (Sbacchi, 1985, 167-170). The restored monarchy of Haile Selassie during the post-liberation period sought a Pan-Ethiopian identity, but such a dream was always contradicted by the realities of ethnic hierarchies and regional disparities (Clapham, C., 1969). The revolution of 1974 and the subsequent Derg regime promised to resolve the "nationalities question" but then adopted policies of forcible villagization and cultural homogenization that further fuelled ethnically based grievances (Donham, 1999).

Yet the institution of ethnic federalism under the EPRDF in 1995 marked a dramatic break, inscribing ethnicity as the fundamental basis of political organization and territorial administration. Whereas proponents of ethnic federalism argue that such a system would grant greater autonomy and cultural rights to the diverse ethnic groups of Ethiopia, critics maintain that ethnic chauvinism, clientelism, and resource competition along ethnic lines have been the result (Aalen, 2011, 87-90). Thus, ethnic federalism had been designed to deal with historical grievances and self-determination of Ethiopia's diverse ethnic groups. This system of government, which segregates the country into ethnic enclaves in the form of regional states, has inadvertently inflamed ethnic tensions. As Abbink argues, "The ethnic-based federal structure has led to a strengthening of ethno-nationalist sentiments and a weakening of Pan-Ethiopian identity" (Abbink, 2011). About this statement, Dr. Temesgen quoted Michela Wrong's book "It's Our Time to Eat: The Story of Kenyan Whistle-Blower" (2009) (Tesfamariam, 2024) reflecting the scenario. It finds manifestations in the fragmentation of national identity, especially in areas where ethnic groups overlap along borders due to territorial and resource claims. For example, the protracted conflict between the Somali and Oromo communities in eastern Ethiopia has been partly attributed to the rigid ethnic boundaries imposed by the federal system. This has promoted competition and tensions between different ethnic groups, as well as the marginalization of minority groups, in the ethnically defined regions. According to Fiseha, "The politicization of ethnicity has turned ethnic identity into one of the most fundamental avenues toward the mobilization of people and resources, often at the expense of national unity and shared citizenship" (Fiseha, 2012).

Recently, ethnic-based conflicts and political mobilization, exemplified by the ongoing crisis in Tigray and inter-communal violence in [Amhara,] Oromia, Benishangul-Gumuz, and other regions have prevailed (Ylonen, 2021, 310-315). These developments highlight the enduring legacy of ethnic mapping and categorization in Ethiopia, raising critical questions about the viability of the current federal system and the prospects for fostering a truly inclusive national identity that transcends ethnic divisions (Abbink, 2021). As Ethiopia grapples with these challenges, policymakers and scholars alike must confront the complex historical legacies of ethnic mapping and seek innovative approaches to promoting unity in diversity without erasing the rich cultural tapestry that defines the Ethiopian experience (Fiseha, 2018).

Somalia: Prolonged Instability

Historically speaking, Somalia's journey towards instability can be traced back to the colonial period. British and Italian colonial powers carved the territory into separate regions, this division laid the groundwork for the clan-based conflicts that emerged post-independence (Lewis, 2002). The differing colonial policies, Italy's more integrated approach in the south versus Britain's neglect of the north further aggravated regional disparities, fostering economic inequalities that persist today (CJA, 2024). The political situation worsened with the emergence of Mohamed Siad Barre in 1969. Barre's regime tried to centralize power by undertaking a socialist agenda and suppressing clan identities-a situation that offended many groups. His government used a combination of heavy-handed oppression and political manipulation, deepening the divides within Somali society (Samatar, 1992) myopic policy of provoking the Isa clan to fight his regime (Dualeh, 1994, 104), attempts to centralize power while navigating clan dynamics resulted in violent repression and civil war following his ousting in 1991 (Magan, 2016). International efforts, including a 1992 United Nations humanitarian intervention under Operation Restore Hope, did little to bring stability in the long run.

The mission encountered violent resistance, with the notorious "Black Hawk Down" incident in 1993, even depicted in a Hollywood movie, embodying the perils of outside interference in a strongly fragmented society (Penketh, 2001). This led to an inability to manage subsequent operations and a withdrawal of international forces, further descending into chaos. Since then, various Islamist factions have emerged, most notably al-Shabaab, which gained heavy media attention for its violent tactics and flatly rejecting the Western-supported Transitional Federal Government (TFG). Al-Shabaab managed to have the political vacuum to its advantage by aligning itself with global jihadist movements and exercising terrorism to claim control over large strips of territory. As stated by Menkhaus (2014), this further complicates reconciliation and governance efforts as the successive Somali governments-often controversially supported by the African Union-struggle to regain footholds against an ever-adapting enemy.

Despite the establishment of the Federal Government of Somalia in 2012, challenges remain formidable. Political infighting, corruption, and the lack of centralized authority prevent effective governance (Shire, 2015). Attempts to stabilize the economy and rebuild infrastructure continue to be hampered by ongoing violence and the threat of al-Shabaab's insurgency. Not only that, the 2006 military intervention by the Ethiopian government added fuel to the flames of crises (Tesfamariam, 2024). Amongst all these phenomena, the role of international actors has remained crucial yet contentious throughout the Somali conflict. The role of the African Union Mission in Somalia (AMISOM hereinafter) in support of the Somali National Army against al-Shabaab has been highly instrumental and has therefore illustrated both the potential benefits and limitations of military intervention (Williams 2017, 112). Whereas AMISOM has been able to take back territory from extremists, the permanence of these gains remains questionable because of continued insurgency and a lack of locally driven Somali initiatives toward peace-making. Furthermore, international responses have also often been relegated to criticism of being reactionary versus proactive. Aid and military support at times have exacerbated situational dynamics rather than contributed to authentic processes for reconciliation (Crawford, 2016).

Maritime Geopolitics and its Influence

The first way in which geopolitics of the Red Sea has destabilized the region is through the proliferation of military bases and increased militarization of the area. In fact, the large numbers of foreign military outposts being set up in countries such as Djibouti, which currently hosts a number of international forces, increase the danger of confrontation and conflict spill over (Ash, 2017). This has raised concerns that the Horn of Africa may become a theatre for larger geopolitical rivalries. Besides, the economic and diplomatic influence enjoyed by the external actors in the Red Sea region has emboldened them to interfere in the internal affairs of the countries of the Horn of Africa. This has undermined the sovereignty and stability of these nations, as they are increasingly beholden to the shifting priorities and agendas of external powers (Cedric, 2019). This dynamic has made it more difficult for local actors to address pressing issues, such as ethnic tensions, political instability, and security challenges, without the interference of outside forces.

The competition for control over the Red Sea's maritime chokepoints has also had destabilizing effects on the Horn of Africa. The potential disruption of trade routes or the blocking of vital sea lanes could have severe economic and political consequences for the countries in the region, further exacerbating existing tensions and conflicts (Gartenstein-Ross & Barr, 2019). This has resulted in increased militarization and a risk of miscalculation with further destabilizing effects the wider Horn of Africa. Perhaps the most emblematic example is the race for influence in Djibouti, a tiny country that guards the entrance to the Red Sea. Djibouti's location has made it a sought-after destination for foreign military bases, with the United States, China, France, and Japan all maintaining a presence in the country (Cedric, 2019). This influx of foreign military forces has contributed to heightened tensions and the potential for conflict, as the various powers vie for control and influence in the region. Secondly, the Red Sea's role in global trade and energy security has also had political implications for the Horn of Africa. Countries like Ethiopia, which are landlocked and dependent on access to the sea, have had to navigate complex diplomatic and economic relationships with their coastal neighbours, such as Eritrea, Somalia and Djibouti, in order to ensure the uninterrupted flow of goods and resources (Gartenstein-Ross & Barr, 2019). This has at times led to border disputes, economic tensions, and even military confrontations, which have shaped the political dynamics in the region.

Additionally, agreements like Ethiopia's now-stalled pact with Somaliland for Red Sea access have heightened tensions with Somalia, reflecting broader geopolitical rivalries (CIDOB, 2024). The competition for resources and influence continues to shape the political landscape, often at the expense of local populations. Interestingly, the investment by external factors, such as the Gulf States, in the affairs of the Horn of Africa has been partly because of their strategic interests in the Red Sea. For instance, the Saudi-led intervention in Yemen has had spillover effects across the Horn of Africa, with the conflict spilling over and influencing the political-security situation in countries like Somalia and Eritrea (Ash, 2017). In addition, there has been some economic repercussions for the local populations in relation to the competition for control over the Red Sea's maritime chokepoints and trade routes. For instance, any disruption in trade or blockage of vital sea lanes could result in shortages of essential goods, erratic fluctuations in prices, and economic hardship for the communities reliant on such trade networks for their livelihood (Gartenstein-Ross & Barr, 2019). This may exacerbate poverty and inequality that already exist within the region. Transnational Threats within the Horn of Africa with all these intra-states and inter-states crises roaming, the region faces a complex array of transnational threats that have far-reaching consequences for regional stability and global security.

These threats, underpinned by a combination of factors that are political, economic, and social, pose significant challenges to the national governments and international institutions tasked with addressing them. One of the primary transnational threats in the region is the proliferation of terrorism and violent extremism (Bruton, 2010). The presence of groups such as al-Shabaab in Somalia and the potential for the radicalization of disaffected youth have created an environment rife with security concerns, not only threaten the lives of civilians but also undermine economic development and hinder the region's ability to address pressing social issues. These groups have taken advantage of the region's political instability, porous borders, and relatively weak governance structures to establish safe havens and new recruiting networks. The region is also grappling with the challenge of environmental degradation and resource scarcity, which have been aggravated by climate change (Gemeda & Sima, 2015). Droughts, desertification, and the depletion of natural resources have fuelled competition over scarce resources, heightening the risk of conflict and displacement. These environmental challenges have direct implications for food security, public health, and the overall well-being of the region's population.

The colliding of illicit trade and climate change creates a setting where each issue fuels the other. For instance, illicit economies often thrive on areas where climate change has brought instability. Bayo, for one, poses that "the illegal economy thrives in contexts of weakened governance, which is often the case in regions affected by climate disruption" (Bayo, 2023). Amplification of illicit trade in crisis situations shows how climate change acts as a catalyst for conflict. Climate-induced migration poses yet another challenge.

Global reports have pointed out that "displaced populations often move into areas already experiencing tensions further straining resources and leading to conflict" (United Nations Environment Programme, 2021). According to Kamal, "the movement of people can worsen existing ethnic and social tensions, creating fertile ground for conflict" (Kamal 2019). Such a situation can therefore weaken regional cooperation and stability, negatively impacting efforts to resolve the underlying drivers of conflict and insecurity.

Eritrea's Role in the Regional Security

Amidst narratives frequently overshadowed by conflict and tension, it is argued in this chapter that Eritrea has begun to assume a pivotal (albeit underappreciated) role in fostering regional stability. This section examines Eritrea's influence on these areas, highlighting both the challenges and opportunities it presents for regional stability. Historically, Eritrea's foreign policy has been shaped by its tumultuous history and complex relationships with its neighbouring countries. As a relatively young nation having gained its de facto independence in 1991 and de jure independence in 1993 from Ethiopia, Eritrea has had to navigate a delicate geopolitical landscape in the Horn of Africa region (Negash &Tronvoll, 2000). Its foreign policy has been characterized by a blend of pragmatism and assertiveness, often driven by a desire to assert its sovereignty and protect its national interests (see Muller, 2016). Zoubir (2018) discusses how Eritrea's strategic geographical location allows it to influence various diplomatic relations; thus, it plays a crucial role in mediating conflicts and fostering dialogue between neighbouring nations (Zoubir, 2018). However, the complexities of the region often present significant challenges that Eritrea must address.

Nevertheless, the intricate interplay of international relations particularly with regard to military intervention and unilateral sanctions has long been a subject of debate in global politics. Some nations face severe consequences while others seem to operate with impunity (Krasner, 2018). This dichotomy often leads to certain states, for instance Eritrea, being labelled as pariahs, rogue, hermit and even the North Korea of Africa (Tesfamariam, 2024) on the world stage, a designation that has had far-reaching implications for their economic and diplomatic standing (Armstrong, 2019). In this context, it is crucial to examine the role of countries like Eritrea, which have frequently been ignored or misunderstood in the broader geopolitical narrative (Connell, 2021). Eritrea's emphasis on self-reliance and its resistance to external pressures has, in some ways, positioned it as a unique case study in the pursuit of national sovereignty and development (Muller, 2020). Significantly, Eritrea strides in its developmental agenda particularly in education and health even under challenging conditions.

As noted by Mahteme (2020), Eritrea's investment in human capital has been crucial to its strategy for promoting social stability and economic resilience (Mahteme, 2020). Although the international community has often focused on Eritrea's challenges, it is equally important to acknowledge the country's positive contributions such as in regional peacekeeping and its commitment to sustainable development goals (Bereketeab, 2022). This more nuanced approach to understanding Eritrea's position highlights the need to avoid double standards in international relations and to recognize the diverse paths that nations may take in their pursuit of stability and progress (Weldemichael, 2023). It is imperative to examine whether Eritrea is a stabilizer or an instigator of regional crises. The researchers attempted to crystalize uncertainty within the scholarly rim of the international arena by elucidating the rationale behind Eritrea's actions. The array of data analyses is presented as follows:

One notable example is its efforts to address consequences of ethnic mapping in Ethiopia, which has led to augmented tensions and violence between different ethnic groups (Mengisteab, 2019). The complex interplay of ethnic, religious and political interests within this region has contributed to the spread of crisis in Ethiopia, as neighbouring states and regional powers have often intervened either directly or indirectly to pursue their own strategic interests (Clapham, 2017). Well, in the context of Ethiopia's ethnic federalism challenges, Eritrea has advocated for a more unified national identity approach; this potentially justifies the risks of balkanization that ethnic mapping policies have created (Ahmed, 2023). This perspective highlights the role of Ethiopian government in attempting to maintain its territorial integrity and political dominance, often at the expense of ethnic minority groups (Woodward, 2013). The government's heavy-handed approach to managing ethnic tensions has, however, fuelled further resentment and resistance from marginalized communities. Thus, tension escalated into violent conflict in late 2020, resulting in humanitarian crisis.

In the aftermath of the Algiers agreement in 2000, Eritrea has made a determined effort to normalize relations with Ethiopia (Negash &Tronvoll, 2000, 123). According to Iyob (2006), Eritrea's president Isaias Afewerki has since prioritized regional cooperation and stability, recognizing that "peace and security in the region are essential for economic development" (Iyob, 2006). Especially after the 2018 peace agreement between Ethiopia and Eritrea, which ended a two-decade standoff (Tesfamariam, 2024), Eritrea's engagement has facilitated dialogue and cooperation between the two nations, promoting regional stability and economic ties. Eritrea's President Isaias Afwerki has been instrumental in mediating talks between Ethiopian Prime Minister Abiy Ahmed and leaders of various ethnic groups, facilitating dialogue and promoting reconciliation (Eritrea Ministry of Information, 2020). However, challenges remain; the TPLF's exclusion from peace processes complicated lasting resolutions (Stigant & Phelan, 2019; Tronvoll, 2019; Reid, 2020). This was followed by two years of north Ethiopian war (hereinafter referred to as the Tigray war).

Theoretically speaking, anarchic nature of the international system compels states to prioritize their survival, which led Eritrea to intervene militarily in the Tigray war (to counter the Tigray People's Liberation Front (TPLF), as it historically posed a threat to its sovereignty (France 24, 2021)). According to an interview conducted with a couple Eritrean soldiers, the intervention was also an act of peacekeeping, because Eritrea soldiers play a key role in pacifying the instability and avoiding ethnic cleansing (Interview conducted with Tesfom, 2024 & Habtemariam, 2024). However, the historical conflict between Ethiopia and Eritrea has created a complex web of interdependencies and rivalries that significantly impact regional security. The interplay of military interventions, ethnic tensions and geopolitical ambitions continues to shape the stability of this volatile region; although the situation remains precarious, this dynamic underscore the challenges faced by both nations.

In reference to Sudan, Eritrea has advocated and acted for dialogue between conflicting factions to address the persistent crisis while also respecting Sudanese sovereignty

(Tribune, 2023). Eritrea's government has historically aimed to uphold a balance of power in the region, perceiving the turmoil in Sudan as a potential threat to its own security and regional influence (Habtom, 2016). This viewpoint is further emphasized by the RSCT; thus, instability in Sudan might generate spillover effects on Eritrea, especially regarding refugee flows and cross-border insurgencies (Buzan and Waever, 2003, 44).

During Sudan's political transition (which followed the ouster of Omer al Bashir in 2019), Eritrea expressed support for a peaceful resolution and offered to facilitate dialogue between various stakeholders (Vertin, 2019, 45). Moreover, Eritrea's emphasis on self-reliance and national ownership of development projects, highlighted by Habtom (2016), reflects a desire to avoid dependency on foreign aid. This approach not only aims to strengthen internal stability but also seeks to inherit it within regional platforms as a stabilizing factor against ongoing problems (Habtom, 2016). Recently, Eritrea had worked closely with the Transitional Military Council (TMC) and opposition groups to address the unsettled crisis in the country (International Crisis Group, 2020). As De Waal notes, "Eritrean diplomats have leveraged their ties to both sides to encourage compromise and prevent a further descent into violence" (De Waal, 2022, 95). Although Sudan's transition remains fragile, Eritrea's efforts have contributed to preventing the worst-case scenarios of state collapse or civil war.

Nahom Tesfamichael from the Foreign Affairs Office agrees with De Waal that Eritrea attempts to mitigate crises from their onset by presenting a peace roadmap, which is prepared and overseen by the president's office. This roadmap underscores the significance of a Sudanese-led political transition (Tesfamichael, 2024). However, Eritrean initiatives to engage with Sudan reflect an understanding of collective responses to crises, advocating for conflict resolution mechanisms that tackle root causes (Tesfamichael, 2024). In Somalia, Eritrea has played a role in efforts directed at state-building and countering extremism. The aftermath of the Eritrea-Ethiopian border war has left unresolved issues, which continue to fuel hostilities, because both nations have different approaches to the Somalia case. Ethiopia's military interventions aimed at countering perceived threats from Islamist groups have often been met with accusations of exacerbating instability (Council on Foreign Relations, 2020), which diminishes the nascent hope for peace that began to revive in Somalia.

On the contrary, Eritrea has leveraged its diplomatic and intelligence capabilities to support reconciliation efforts and counter-terrorism operations against Al-Shabaab. According to Healy: "Eritrean security assistance and training programs have enhanced the capacity of Somali government forces to combat extremist groups and assert control over larger swathes of territory" (Healy, 2023). This has helped chip away at Somalia's prolonged instability, however, significant challenges remain. The peculiarity is that training boosted counter-terrorism operations and national unity (Tesfamichael, 2024). Eritrea's support for the ICU, which briefly controlled Mogadishu and other parts of Somalia in 2006, was a strategic move aimed at reconciling Somalia's problems (Menkhaus, 2014). This alignment of interests illustrates the realist notion that states act primarily in pursuit of their national interests, although often through pragmatic alliances this notion is imprinted in Eritrean foreign policy.

What Eritrea believes is its genuine effort to secure peace and seurityin the Horn has, however, drew international condemnation too (Menkhaus, 2014, 45) and sanctions from the United Nations, which have labelled Eritrea as a "geopolitical pariah" because of its alleged destabilizing activities (World Politics Review, 2023). In recent years mainly following the restoration of diplomatic relations with Somalia in 2018 after a 15-year break Eritrea's renewed diplomatic engagement with Somalia marks significant shift in the geopolitical landscape of the region. The agreement to restore relations in July 2018 was not merely a bilateral gesture; it represented broader strategy to reshape regional alliances and address mutual security concerns. As noted in the analysis of Somalia's diplomatic manoeuvres, the restoration of relations was accompanied by Somalia's call for an end to international sanctions against Eritrea, a move that has drawn the ire of Djibouti, Somalia's long-time ally (Oxford Analytica, 2018). This diplomatic balancing act highlights complexities of regional politics, where alliances are often fluid and influenced by historical grievances and contemporary security challenges.

The Red Sea a crucial geopolitical arena has also witnessed Eritrea asserting its interests by maintaining sovereignty while collaborating with international partners to secure trade routes against piracy and transnational threats. These threats include organized crime and illicit trade networks that thrive in lawless environments; however, the impacts of climate change often exacerbate resource scarcity and contribute to migration and conflict. This has prompted Eritrea to engage in regional environmental strategies that address these vulnerabilities and foster resilience (Green, 2023). Although Eritrea's foreign policy principles emphasize a commitment to regional stability and security, they showcase its potential as a proactive actor in addressing the multifaceted challenges within the Horn of Africa, but the complexities involved cannot be overlooked.

Conclusion

Eritrea's experience underscores the significance of regional cooperation, dialogue and self-help in sustaining regional security. Through its endeavours to foster peace and stability in neighbouring countries combined with its own domestic stability and independent foreign policy Eritrea has assumed a pivotal role in stabilizing the region. Although the Horn of Africa continues to confront numerous security challenges, Eritrea's example offers an additional impetus to the efforts of other states aiming to enhance peace and stability within the region. By implementing the following recommendations and pursuing future research directions, scholars and policymakers can collaborate to devise effective strategies for upholding regional security in the Horn of Africa; however, this requires a commitment to mutual understanding and shared goals.

References

(CFR), C. o. (2020). *Ethiopia-Eritrea Proxy War in Somalia Risks Broader Regional Conflict.* 24, France . (2021, March 9). Eritrea's murky role in Ethiopia conflict.

Aalen, L. (2011). *The Politics of Ethnicity in Ethiopia: Actors, Power and Mobilisation under Ethnic Federalism* . Leiden: Brill.

Abbink, J. (2021). "The Atlantic Charter and Ethiopia's Sovereign Integrity: A View from the Longue Durée.". In N. A. Studies.

_______ (2003). Ethiopia-Eritrea: Proxy Wars and Conflict Resolution in the Horn of Africa. *Journal of Contemporary African Studies, 3* (21), 403-424.

_______ (2011). Ethnic-Based Federalism and Ethnicity in Ethiopia: Reassessing the Experiment after 20 Years. In *Journal of Eastern African Studies* (Vol. 5 , pp. 596-618).

Ahmed, S. (2023). Ethnic Federalism and National Unity in Ethiopia. Journal of African Studies.

Analytica, O. (2018). *Somalia's Missteps Will Resonate at Home and Abroad.*

Armstrong, D. (2019). The Evolution of International Society: A Comparative Historical Analysis. *Routledge* .

Ash, R. (2017). "The Red Sea: From the Periphery to the Centre of global geopolitics. *RUSI Journal 162, 32.*

B.,Buzan & O.Waever. (2003). *Regions and powers: The structure of international security.* Cambridge: Cambridge University Press.

Bayo, S. (2023). Illicit Economies in a Changing Climate. *Journal of Conflict Studies, 15*, 65-90.

Bereketeab, R. (2022). Eritrea's Foreign Policy: Explaining the Anomaly in Regional and International Politics. In *Journal of Eastern African Studies* (Vol. 16, pp. 123-142).

________. (2017). Eritrea's National Security Strategy: A Critical Analysis. *Journal of Security Studies, 6*, 1-12.

Brown, M. (2024). Geopolitics of the Red Sea: Security Implications. *Routledge* .

Bruton, B. E. (2010). Somalia: A new approach., : . *Council on Foreign Relations* .

Cedric, B. (2019). The Red Sea: From Backwater to Geopolitical and Economic Hotspot. *Chatham House* .

CIDOB. (2024). Scramble of External Powers and Local Agency in the Horn of Africa.

CJA. (2024). *Somalia: Colonial Legacy*. Retrieved from THE CENTER FOR JUSTICE AND ACCOUNTABILITY: http://www.cja.org/where-we-work/somalia/related-resources/somalia-colonial-legacy/

Clapham, C. (1969). *Haile Selassie's Government.* New York: Praeger.

________ (2017). *The Horn of Africa: State formation and decay.* Oxford: Oxford University Press.

Collins, R. O. (2008). *A History of Modern Sudan_.* Cambridge: Cambridge University Press.

Connell, D. (2021). *Eritrea: A Country Under Siege." In The Horn of Africa: State Formation and Decay.* (R. Bereketeab, Ed.) London: Hurst & Company.

Crawford, N. C. (2016). *Accountability for Human Rights Atrocities in Somalia: The Politics of International Intervention.* University of Pennsylvania Press.

D O Gemeda & A D Sima. (2015). The impacts of climate change on African continent and the way forward. In *Journal of Ecology and the Natural Environment* (Vol. 7, pp. 256-262).

Donham, D. L. (1999). *Marxist Modern: An Ethnographic History.*

Dualeh, H. A. (1994). *Somalia: The Agony of a Nation: from Barre to Adeed.* Nairobi, Kenaya: Stella Graphics Ltd.

El-Tom, A. S. (2022). The 2021 Sudan Coup and Its Aftermath. *Journal of African Studies, 43* (1), 51-67.

Fiseha, A. (2012). Ethiopia's Experiment in Accommodating Diversity: 20 Years' :Balance Sheet. In R. &. Studies, *Regional & Federal Studies 22* (pp. 435-473).

_________ (2018). Federalism and Development the Ethiopia Dilemma. International Journal of Minorityand Group Rights.

Gallopin, J.-B. (2019). "Sudan's Unfinished Transition.". *Middle East Report 290* , 28-31.

Gartenstein-Ross Daveed & Barr Nathaniel. (2019, March 20). The Horn of Africa's Faultlines," War on the Rocks.

Green, S. (2023). Climate Change and Security: The Horn of Africa's Vulnerabilities. *Environmental Policy Journal, 12*, 30-40.

Group, I. C. (2020). *The Horn of Africa and the Arabian Peninsula: Interconnected Crises.* Crisis Group Middle East Report N°210.

Habtemariam, S. T. (2024, July 26). Regional Security Challenges in the Horn of Africa and the Role of Eritrea. (M. Atewebrhan, Interviewer) Asmara, Eritrea.

Habtom, H. (2016). Foreign Aid Effectiveness and Development Strategies in Eritrea: A Lesson for Sub-Saharan African Countries. *Journal of African Studies and Development, 8*, 10-15.

Healy, S. (2023). Somalia's Security Landscape: New Actors, Old Challenges. *International Affairs 99* , 103-120.

Heidorn, L. A. (1994). The Meroitic State: Northeastern Africa, 2100/1900 BCE to Fourth Century CE. *Journal of African History, 1* (35), 1-27.

Hilhorst, D. (2013). *"The Politics of Peace Processes." In_Conflict and Peacebuilding in the African Great Lakes Region.* (T. R. Büscher, Ed.) London: Zed Books.

Information, E. M. (2020). *Eritrea's National Development Strategy (2020-2030).* Asmara: Eritrea Ministry of Information.

Iyob, R. (2006). Sudan: The Elusive Quest for Peace. *International Peace Institute* .

Johnson, D. H. (2011). *South Sudan: A New History for a New Nation.* .Bloomington: Indiana University Press.

Julie Flint & Alex De Waal. (2005). *Darfur: A Short History of a Long War.* London: Zed Books.

Kamal, F. (2019). Displacement and Conflict in the Horn of Africa. *Journal of Refugee Studies, 32*, 115-132.

kfcris. (April 2020). Red Sea Peace Initiatives: Saudi Arabia's Role in the Eritrea-Ethiopia Rapprochement. *King Faisal Center for Research and Islamic Studies* .

Khan, M. (2021). Geopolitics in the Horn of Africa: Sudan in the 21st Century. *Journal of International Relations, 17* (2), 30-45.

Kitchen, K. (1994). *Documentation for Ancient Arabia.* Liverpool: Liverpool University Press.

Krasner, S. D. (2018). *Sovereignty: Organized Hypocrisy.* Princeton: Princeton University Press.

Lata, L. (2003). The Ethiopian State: At the Crossroads: Decolonization and Democratization or Disintegration? *The Journall of Modern African Studies, 41*, 39-72.

Lewis, I. (2002). *A Modern History of the Somali: Nation and State in the Horn of Africa.* Athens: Ohio University Press.

Magan, A. (2016). *Somalia: Instability, Conflict, and the Role of Clan.* NMBU Brage.

Mahteme, T. (2020). Human Development in Post-Independence Eritrea. In *East Africa Journal of Peace Studies* (Vol. 6 , pp. 330-340)

Marcus, H. G. (1994). *A History of Ethiopia.* Berkeley: Univeristy of California Press.

Mearsheimer, J. (2001). *The Tragedy of Great Power Politics.* New York: W.W. Norton & Company .

Mengisteab, K. (2019). The Horn of Africa: A Region in Transition. *7* (2), 1-10.

Menkhaus, K. (2014). *The Crisis in Somalia: Tragedy in Five Acts.* United States Institute of Peace.

Mesfin.B. (2019). The Geopolitics of the Horn of Africa: The Red Sea and the Rivalry Between Saudi Arabia and Iran. *Monograph Series, Institute for Security Studies* .

Muller, T. (2016). Assertive foreign policy in a 'bad neighbourhood': Eritrean foreign policy making. *International Confernece of Eritrean Studies*, (p. 20). Asmara.

__________ (2020). Asserting Eritrea's Agency in Regional Politics and Post-Conflict Developments. In *African Affairs* (Vol. 22, p. 119 (474)).

Munro-Hay, S. (1991). *Aksum: An Afriacan Civilzation of Late Antiquity.* Edinburgh: Edinburgh University Press.

N.,Hirt & A. H. Mohammad. (2018). The Eritrean diaspora and its impact on regime stability: Responses to UN sanctions.

Niguse, M. A. (2024, October 7). Regional Security Challenges in the Horn of Africa and Ther Role of Eritrea. (D. T. Tesfamariam, Interviewer) Asmara, Eritrea.

Omer, A. (2022). Regional Integration and Cooperation in the Horn of Africa. *Journal of African Studies, 15*, 75-90.

P.M.Holt. (1989). *History of the Sudan from the Coming of Islam to the Present Day.* Stafford, United Kingdom: Longman Group United Kingdom.

Penketh, A. (2001). African Union Forces to Withdraw from Somalia. The Independent.

Pettigrew, J. (2020). Humanitarian Challenges in the COVID-19 Era: The Case of Sudan. *Global Health*

Action, 13 (1), 109-115.

Programme, U. N. (2021). Climate Change and Migration in Africa. *UN Environment Programme* .

Reid, R. (2020). Conflict between Tigray and Eritrea – the long-standing faultline in Ethiopian politics. *University of Oxford* .

Review, W. P. (2023). *Singled Out? Eritrea and the Politics of the Horn of Africa.* Retrieved from World Politics Review: https://www.worldpoliticsreview.com/singled-out-eritrea-and-the-politics-of-the-horn-of-africa/

Samatar, A. I. (1992). The Social Origins of Violence in Somalia. *The Journal of Modern African Studies* , 63-84.

Sbacchi, A. (1985). *Ethiopia Under Mussolini: Fascism and the Colonial Experience.* Michigan , United of States: Zed Books.

Shire, A. M. (2015). Challenges of State-Building in Somalia. *Horn of Africa Bulletin, 27*, 43-48.

Susan Stigant & Phelan Michael. (August 2019). A Year After the Ethiopia-Eritrea Peace Deal, What Is the Impact? *United States Institute of Peace* .

Tekeste Negash & Kjetil Tronvoll. (2000). *Brothers at War: Making Sense of the Eritrean-Ethiopian War.* Oxford: James Currey.

Tesfamichael, N. (2024, Octer 7). Regional Security Challenges in the Horn of Africa and The Role of Eritrea. (M. A. Niguse, Interviewer) Asmara .

Tesfamariam, D. T. (2024, October 7). Regional Security Challenges in the Horn of Africa and The Role of Eritrea. (M. A. Niguse, Interviewer) Asmara, Eritrea.

Tribune, S. (2023, June 15). Eritrea Calls for Peaceful Resolution to Sudan Crisis.

Tronvoll, K. (2019). After Making Peace, Ethiopia and Eritrea Now Focus on Development. *Africa Renewal* .

USIP. (2019). A Year After the Ethiopia-Eritrea Peace Deal, What Is the Impact?

Verhoeven, H. (2024). *The Red Sea: Geopolitics, Power Projection and the Scramble for the Horn of Africa.* London: I.B. Tauris.

________ (2020). Amid Red Sea Rivalries, Eritrea Plays for Independence. *United States Institute of Peace* .

Verjee, N. (2020). Climate Change and Conflict in Sudan: Climate-Related Displacement and Migration. *Journal of Peacebuilding & Development, 15* (3), 63-77.

Waal, A. D. (2022). *Sudan's Unfinished Democracy: The Promise and Betrayal of a People's Revolution.* London: Hurst Publishers.

Weldemichael, A. T. (2023). Eritrea's Foreign Relations: Understanding Its Regional Role and International Pursuits. In *International Journal of African Historical Studies* (Vol. 56 , pp. 27-49).

Williams, P. D. (2017). *War and Conflict in Africa.* Wiley-Blackwell.

Woodward, P. .. (2013). *The Horn of Africa: Politics and International Relations.* I.B. Tauris.

Zewde, B. (1991). *A History of Modern Ethiopia, 1855-1974.* . London: James Currey.

Zoubir, Y. H. (2018). Eritrea's Diplomatic Engagements: A New Approach for Regional Stability. In *African Security Review* (Vol. 27 , pp. 210-220).

Key Causes of Governance Failures and Challenges in Multi-Ethnic Societies: Ensuring Peace, Security, and Development in the Horn of Africa

Redie Bereketeab

Associate Professor and Senior Researcher, The Nordic Africa Institute,
E-mail- redie.berekcteab@gmail.com

Abstract

This chapter investigates the root causes of governance failures and challenges in multi-ethnic societies, with a focus on the Horn of Africa. It identifies ongoing conflict, insecurity, and underdevelopment in the region as consequences of an incomplete nation-state formation process. As the most conflict-prone area on the African continent, the Horn of Africa continues to suffer from both internal structural weaknesses and persistent external geopolitical interventions that have compounded its instability. At the core of these governance failures lies the unresolved and fragmented process of nation and state formation. Accordingly, this chapter engages with key concepts and theories of nation-state formation to explain how these underlying dynamics contribute to the region's lack of peace, security, and development. Methodologically, the chapter employs qualitative textual analysis. Through this approach, it examines how conceptual and theoretical frameworks on nation-state formation help explain both the process itself and the structural obstacles impeding its completion. The chapter concludes that a dialectical relationship exists between incomplete nation-state formation and the persistent challenges of governance, peace, security, and development in the Horn of Africa. It further argues that past and ongoing experiments with different models of the nation-state—namely, the proto-state, the colonial state, and the national liberation state—demonstrate the unfinished and evolving nature of state formation in the region.

Keywords: governance, multi-ethnic societies, Horn of Africa, peace, security, development

Introduction

The Horn of Africa (HOA)—comprising Sudan, South Sudan, Somalia, Ethiopia, Eritrea, and Djibouti—is widely recognised as the most conflict-prone region on the African continent. Multiple factors contribute to this persistent instability. Modern states in the region are largely the result of colonial designs or legacies of the colonial era, which helps explain their vulnerability to various structural and political challenges. Geographically, the HOA occupies an exceptionally strategic location through which major global trade routes pass. This has made it a focal point for the interests and interventions of global powers over centuries (Woodward 2006; Yordanov 2016; Schmidt 2013).

Successive external powers and traders—including Arabs, Ottomans, and Europeans—have left lasting imprints on the region. Modern political boundaries and institutions, in particular, are products of European colonial influence. The era of colonialism was followed by the Cold War, neo-colonial dynamics, the U.S.-led global war on terror, maritime piracy, and, more recently, intense competition for natural resources. These overlapping phases of foreign involvement have consistently undermined peace, security, stability, and development in the region. The Cold War, driven by the geopolitical and ideological interests of superpowers, heavily shaped the post-colonial environment in the HOA (Yordanov 2016).

Nations, as we understand them today, are modern constructs. In the HOA, most nations are colonial artifacts (Markakis, Schlee, and Young 2021; Smith 1983; Mazrui 1983; Davidson 1992; Bereketeab 2023), with fewer than a hundred years of statehood. Their independent histories are even shorter—typically around sixty years. These states are thus still in the early phases of nation-state formation. This relatively young age, combined with the region's diverse and often competing identities, has contributed to the incomplete nature of nation-state formation. This incompleteness presents ongoing challenges in managing pluralism and establishing effective governance.

This chapter examines the challenges and failures of governance in multi-ethnic societies. It explores the roots of ongoing conflict, insecurity, and underdevelopment in the HOA as outcomes of unfinished nation-state formation processes. The central argument is that a dialectical relationship exists between the incomplete nation-state formation on one hand, and governance, peace, security, and development on the other.

The chapter is structured into seven sections. Section One introduces the central theme. Section Two defines key concepts. Section Three explores the governance challenges in multi-ethnic contexts. Section Four addresses conceptual and theoretical frameworks. Section Five presents empirical case studies and outlines three models of nation-state formation. Section Six discusses external interventions and their impact on state-building processes. Finally, Section Seven concludes with a summary of key findings and reflections.

Defining Features of Insecurity, Instability, and Underdevelopment

The Horn of Africa (HOA) is arguably the most conflict-prone region on the African continent. Numerous studies point to a wide range of interrelated factors that contribute to the region's persistent instability. Key characteristics that define the insecurity, instability, and underdevelopment of the region include: (i) intra-state and inter-state conflicts, (ii) state fragility, (iii) identity-based conflicts, (iv) democratic deficits, (v) environmental degradation, (vi) extreme poverty, (vii) mass migration and displacement, and (viii) externally driven geopolitical interventions (Bereketeab 2013; Woodward 2013; Schmidt 2013; Mengisteab 2014). The complexity and interdependence of these crises make effective solutions extremely challenging. Limited state resources and capacity, combined with both regional and international interference, exacerbate the situation and render many problems intractable. The region's strategic location increases its significance to global powers, further compounding internal challenges.

The failure of states in the region to collaborate—often opting for violent conflict over peaceful resolution—reflects the dysfunction of regional integration. The weakness of the Intergovernmental Authority on Development (IGAD), the regional organization, further hampers efforts to address these issues. At the root of many of these problems is the incomplete process of nation-state formation.

The severity of the situation in the HoA is underscored by indicators such as the Fragile States Index and the Human Development Index (HDI), both compiled annually by the

United Nations Development Programme (UNDP). The HDI assesses levels of human development, while the Fragile States Index evaluates the resilience and performance of states using a variety of metrics. In both indices, most HOA countries consistently rank among the lowest globally, as shown in the table below.

Table 1: Human Development Index and Failed States Index

Country	HDI (2022)	FSI (2024)	Rank* (2024)
Somalia	193	111.3	1
Sudan	170	109.3	2
South Sudan	192	109	3
Ethiopia	176	98.1	12
Eritrea	175	92.1	26
Djibouti	171	81.6	48

*Ranks for failed States Index, Source: UNDP

Conceptualizing Governance in Multi-Ethnic Societies: A Brief Note

The concept of governance is highly flexible, ambiguous, and context-dependent. It is employed in diverse ways across various disciplines and contexts. The term is widely used in development literature, by donors, NGOs, UN agencies, businesses, and financial institutions. Often, it is conflated with the notion of "good governance" (Kooiman 1999), while "bad governance" is commonly viewed as a root cause of many societal challenges. For example, the United Nations Economic and Social Commission for Asia and the Pacific (n.d., 1) defines governance as "the process of decision-making and the process by which decisions are implemented (or not implemented)." Governance manifests in different forms, including international, national, local, and corporate governance. Furthermore, governance is increasingly recognized as being as much about the collective framing of problems as it is about formulating shared solutions (Leach et al. 2007, 28).

This chapter focuses on the political dimension of governance, specifically its political sociology. It examines governance as a process of decision-making and implementation, particularly in relation to nation formation, inclusion, participation, and representation. It addresses how the structure and exercise of governance can reflect and accommodate social plurality. In essence, this involves managing identity cleavages and fostering reflexivity in multi-ethnic societies. Put simply, governance here refers to the navigation and negotiation of identity and diversity within a pluralistic societal framework.

The Incomplete Process of Nation and State Formation: A Theoretical and Conceptual Approach

The complex and multifaceted challenges facing the Horn of Africa (HOA) stem from deeply unresolved issues, chief among them being the process of nation and state formation. In multi-ethnic societies, this process is especially difficult and prolonged. Unlike in mono-ethnic societies—where nation-building is assumed to be relatively straightforward—multi-ethnic contexts present significant obstacles.

A major challenge lies in creating mechanisms to unite diverse ethnic groups into a cohesive and harmonious society. This includes fostering a shared national consciousness and a collective will to coexist. Achieving this requires time, and a gradual transformation in people's views, attitudes, aspirations, interests, emotions, and identity—shifting away from primordial affiliations toward civic awareness and shared identity.

The intellectual history on the subject emphasizes just how complex, arduous, and extended the process of nation-state formation is (cf. Anderson 1991; Gellner 1983; Smith 1986). Compared to Europe—where the idea of the nation-state originated and has evolved over at least five centuries—Africa is a relative newcomer, with only a few decades of experience in this endeavour.

Thus, the incomplete nature of nation-state formation underlies many of the region's current challenges, and more broadly, those faced by multi-ethnic societies worldwide. But what, specifically, does nation and state formation entail in such contexts? Let us now briefly explore the technical aspects of this process.

State Formation

The concept of the state has been approached from a variety of perspectives, including legal, sociological, territorial, functional, and power-based frameworks. Broadly speaking, the state is understood as an institution—an abstract entity made tangible through the operations and presence of its executive organs. Thus, the process of state formation is fundamentally about the formation of institutions. In this context, the state is defined as the institution par excellence. Accordingly, when we refer to state formation, we are essentially referring to institution-building (Poggi 1978; Evans 1989; Kamrava 2000). Emphasizing this point, Poggi (1978, 1) writes:

> The Modern State is perhaps best seen as a complex set of institutional arrangements for rule operating through the continuous and regulated activities of individuals acting as occupants of offices.

Kamrava (2000, 2) defines institutionalisation as "the effective establishment of state authority over society through specially created political structures and organs." The state, as an institution, is an abstract entity whose foundation becomes tangible through the embodiment of its executive, legislative, judicial, and bureaucratic branches. These components are the facets of the state through which society confers legitimacy. This legitimacy, in turn, depends on the state's ability to deliver social service provisions (Schaar 1981). The interaction between legitimacy and service provision ultimately converges into a social contract, which facilitates both state consolidation and societal stability.

In multi-ethnic societies, the nature and structure of state institutions significantly influence governance patterns and behaviours. These patterns and behaviours reflect the institutional form and content of the state. A functional state institution must incorporate three core principles: (i) representativeness, (ii) inclusivity, and (iii) participation. Representativeness requires that state institutions reflect the full spectrum of society, including all ethnic and identity groups. It is crucial that these groups feel genuinely represented. Inclusivity is closely tied to representativeness; the former depends on the latter. Inclusion, in turn, fosters participation. Groups that are neither represented nor included are unlikely to engage meaningfully in governance and may resort to centrifugal activities that challenge the legitimacy of the central state.

The emergence of representativeness, inclusivity, and participation signals the maturation of state institutions. Conversely, institutions lacking these attributes indicate not only an underdeveloped state apparatus but also an incomplete process of state formation. A key indicator of this incompleteness—and a major threat to state stability—is the persistent presence of centrifugal forces. The operation and implications of these forces will be discussed in the following section.

Nation Formation

The other side of the coin is nation formation. At times, it can be difficult to distinguish between a nation and a state. However, from an analytical perspective, it is entirely possible to draw a clear distinction between the two. So, what is a nation, and what does nation formation mean? A nation is commonly defined as:

> A term often used rather vaguely, to mean any sovereign state with political autonomy and settled territory. That usage represents a confusion between country and nation state. The definition given here is not political but social:
>
> A nation consists of a people, sharing a common language (or dialects of a common language), inhabiting a fixed territory, with common customs and traditions, which may have become sufficiently conscious to take on the aspect of law, and who recognize common interests and a common need for a single sovereign (Scruton 1982: 312).

Different scholars define a nation in various ways. Max Weber (1948) describes it as a *community of sentiment*, while Ernest Renan (1882) refers to it as a *daily plebiscite*. In brief, a nation encompasses sentiment, consciousness, a shared sense of belonging, and the will to live together. Identity is multi-layered. In multi-ethnic societies, the two most significant forms of identity formation are commonly referred to as ethnic and civic identity (Bereketeab 2011). These criteria, as definitions and foundations of pluralistic identity, pertain specifically to multi-ethnic contexts. Within such contexts, two distinct levels of identity formation exist: sub-national (ethnic) and national (civic). The former relates to cultural or communal identity (often described as primordial), while the latter corresponds to political identity.

Multi-ethnic societies face the challenge of balancing these two identity formations during their nation-state building process. As a political construct, the nation must transcend cultural or primordial identities, which is especially crucial in multi-ethnic settings. The case of Somalia illustrates that even in a mono-ethnic environment, the political construction of society—leading to the establishment of a civic identity—is decisive. This demonstrates that a nation is the product of conscious, purposeful, and ongoing construction.

Political and Domestic Nature of Nation-State Formation

The project of nation-state formation by its very nature is domestic and political (is inherently domestic and political). In terms of domesticity, the process of nation and state formation presupposes continuous dialogue, negotiations, compromises and equitable governance among the multiplicity of identity formations. It requires ongoing dialogue, negotiation, compromise, and fair governance among diverse identity groups (Bereketeab 2023). The multi-ethnic groups need to put their feet down and engage in serious negotiations and chart a common and consensual road they would like to march through together. They must firmly engage in meaningful negotiations to establish a shared and consensual path forward. This can only be achieved when space, time and freedom prevail, there is sufficient space, time, and freedom. Such progress could not be achieved in a situation where rampant external interventions are prevalent, is impossible amid widespread external interference. External involvement could only complicate and distort the process without meaningfully contributing to the resolution of nation and state formation problematics. In other words, it tends to complicate and undermine the process without genuinely helping to resolve issues related to nation and state formation.

An ongoing external intervention—including colonialism, the Cold War, the war on terror, piracy, and resource competition—have disrupted and distorted nation-state formation in the HOA. This has contributed to persistent instability, insecurity, and underdevelopment in the region.

Another significant aspect of this process is its political embeddedness. I argue that the process of nation-state formation is by its very nature political. Being political by nature means two things. First, a nation-state is a political construct and should be consciously, purposefully, and intentionally built. This responsibility primarily lies with the political elite. Second, the project of nation-state formation is about politics, involves politics—specifically, questions of power: who gets what, when, and how. It concerns the allocation of power within society. Therefore, it is the exclusive prerogative of societal groups to ensure the fair distribution of power. At the core of political power lies the governance of identity cleavages, which entails ensuring inclusion, representation, and participation of societal groups—goals external forces cannot achieve.

State Emancipation and Societal Pacification

Two closely related concepts fundamental to societal development are state emancipation and societal pacification. These concepts serve as indicators of the level of maturity a society has reached in its historical process of state and nation formation and transformation. From a functionalist perspective, state emancipation is based on the emergence of three dialectically interconnected phenomena: (i) the autonomy of the state, (ii) the state's position above societal groups, and (iii) the supremacy of the state over societal groups, encompassing its hegemonic power (Chabal and Daloz 1999; Young 1994).

Autonomy grants the state the independence to exercise legitimate power and control over society. This implies the submission of sub-national groups. When sub-national groups challenge the state's monopoly on legitimate power, it evidences the immaturity or incompleteness of state formation. A mature state, as a fully developed institution, emerges as an autonomous entity distinct from societal groups. Without autonomy, the state cannot exert legitimate authority over society. Authority depends on consent, whereas mere coercion results in power but not authority. Authority carries a moral dimension. The state's hegemony over society entails a monopoly on the means of violence (Bereketeab 2021: 82).

State emancipation is particularly significant in poly-ethnic societies. If the state is perceived as representing only certain societal groups, the nation's survival is jeopardized. The rise of the state above societal groups often marks the beginning of the modern state (Christensen and Laitin 2019: 70). A state lacking emancipation is viewed as a sectarian or feudal formation. The fulfillment of the three criteria above epitomizes the evolution of the modern state. Ultimately, however, the state's hegemonic power should serve society, as power originates from society itself.

The evolution of the modern state symbolizes the pacification of society. Pacification denotes a state of peace and harmony between society and the state. It represents a condition where peace, harmony, and equilibrium prevail. Pacification has two dimensions. The first is the internal dimension, which concerns society's submission to state authority. The prevalence of domination and submission, as a voluntary historical and politico-cultural expression of the modern state, manifests in national institutions and structures. Personal power transforms into institutional and structural power, embodying and objectifying non-personal entities. The state-society relationship reflects amicable harmony. Historically, the state is understood as the product of a protracted pacification process. "In this historical context, it is society that pacifies itself and authorizes the state rather than the state arrogating to itself the means of coercion" (Bereketeab 2021: 87).

The external dimension of pacification concerns territorial integrity, sovereignty, security provision, and harmonious international diplomacy based on the Westphalian consensus (Coggins 2014: 8; Evans and Newnham 1990; Morgenthau 1985; Osiander 2000: 261). The Westphalian Peace Treaty, a response to prolonged inter-European wars, was designed to regulate interstate relations and promote pacification by alleviating the constant threat of conflict.

Western support for non-state or centrifugal forces with the declared aim of regime change undermines the pacification process that African societies must undergo. A pacified society entrusts the means of violence exclusively to the state. Without this, "the state may be viewed as a ruling organization that competes for power with other political, economic, and social organizations and groups" (Callaghy 1984: 90). Therefore, the contest over control of violence between state and society must be resolved in favour of the state. When this fails, society descends into war and conflict.

Centrifugal and Centripetal Forces

The emergence of centrifugal and centripetal forces clearly indicates the incomplete nature of nation-state formation. It also highlights the absence of three critical dimensions of state institutionalisation: inclusion, representation, and participation. The failure to fulfil these dimensions drive centrifugal forces. The concept of centrifugality varies depending on the context in which it is used. A common definition describes "centrifugal tendencies" as a developmental process that gradually weakens the political centre's power in favour of regional politicians (Belous and Tarasov 2020: 3). This means centrifugal forces contest the authority of the central state, effectively undermining it. At the core of this contestation lies the legitimacy of the state.

In many cases, centrifugal forces represent excluded, disenfranchised, and marginalized ethnic groups. This exclusion lies at the heart of successful nation-state formation. Being denied access to power and relegated to second-class citizenship within their own country motivates ethnic groups to challenge the central state through extra-legal and often non-peaceful means. The ultimate goals of centrifugal forces vary—from peaceful state reform to outright secession and the creation of sovereign states.

Counterbalancing centrifugal forces are centripetal forces, which originate from the centre of state power and represent privileged, enfranchised, and dominant ethnic groups. Centripetal forces are motivated by the desire to preserve the unity, integrity, and dominance of the central state. Their aim to maintain the status quo directly opposes centrifugal forces. The irreconcilable interests of these two groups often plunge the state into chronic crisis. For example, in Ethiopia, the Amhara have historically acted as a centripetal force, while various ethnonationalist insurgencies serve as centrifugal forces. The Amhara view themselves as the custodians of Ethiopian nationalism and nationhood. Similarly, in Sudan, the ruling elite from Khartoum since independence—largely from three riverine groups: the Ja'aliyyin, Shaygiyya, and Danagla (Sorbo and Ahmed 2013: 5)—contrast with other groups that have long challenged their dominance since 1955.

State Penetration

Another concept that complements state emancipation and societal pacification is state penetration. There is an inherent assumption that behind the ideas of emancipation and pacification lies state penetration. A further underlying assumption is that for emancipation and pacification to prevail, the state must first penetrate society. Without this penetration, meaningful emancipation cannot occur. The analytical state-society binary is based on fusion and rupture: while fusion represents unity, rupture signifies separation. The state is expected to penetrate society, but society retains the power to restrain the state; ultimately, power lies with society.

State penetration is secured through the widespread presence of state institutions at every level of society. However, mere presence is insufficient; these institutions should embody representativeness, inclusivity, and participation. Penetration depends on time, place, and capacity, and it also aligns with modernity. In this context, state penetration in Africa can be evaluated by considering historical trajectories, including pre-colonial, colonial, post-colonial periods, and the liberation movements. The latter arguably represents the most penetrating force. The rationale behind the national liberation state's penetrative capacity stems from its origins: born and nurtured in rural areas, it eventually conquers state power, expanding its influence from villages and towns to the capital.

Earlier, it was emphasized that the formation of a nation-state is inherently a domestic process. It is based on the people themselves building their own nation-state. However, Africans were deprived of this opportunity. The primary manifestation of this denial was colonialism. As Lewis (1983: 73) aptly describes, colonialism,

> It is remarkable irony that the European powers who partitioned Africa in the late 19th century when the idea of the nation-state was paramount, should have created in Africa a whole series of Habsburg-style states, comprising a medley of peoples and ethnic groups lumped together within frontiers which paid no respect to traditional cultural contours.

The fragmented entities created by colonialism continue to be a significant source of governance failures and challenges. In Africa, colonialism was succeeded by the Cold War and neo-colonialism, each bringing its own obstacles to the formation of nation-states. The global war on terror and the scramble for resources that followed the Cold War reintroduced interventionist Western powers—alongside those who never truly left (cf. Yordanov 2016; Schmidt 2013; Woodward 2006).

These complex external interventions have had corrosive and corrupting effects on the process of nation-state formation. As a fundamentally domestic process, nation-state formation requires inclusivity, representation, and participation from all ethnic groups. This necessitates negotiations, bargaining, compromises, and dialogue that lead to a consensus on how to build the nation-state. However, these efforts were undermined by persistent external interference. Furthermore, the process is inherently political. Politics concerns power, and there must be an equitable distribution of power among the constituent national units.

Widespread external intervention disrupts the inter-ethnic negotiations, bargaining, dialogue, and compromises essential for consensus-building. It also undermines the equitable distribution of power. The consequences of these disruptions are division, conflict, and societal fragmentation, which in turn make governance exceedingly difficult.

Experimentation in State and Nation Formation: Three Models/Cases

The entities known today as states and nations in Africa are largely creations of colonialism. However, the colonial creation of states and nations was, from the outset, fraught with complex challenges. One such challenge concerns the relationship between the state and society. A clear divide emerged between state structures and society, revealing multiple crises. The relationship between nation and state was characterized by a fundamental misfit. The enduring question is how to resolve this misfit and what the proper formula is for constructing a viable nation-state. For some time, there has been a cycle of trial and error in how to build such a nation-state.

The African context has witnessed several experiments in the nation-state formation process. Three models can be identified: notably, the proto-state, the colonial state, and the national liberation state. Analytically, each model exhibits distinct characteristics. With reference to the Horn of Africa, these models are exemplified by Ethiopia (proto-state), Sudan (colonial state), and Eritrea (national liberation state). Let us analyse each in turn.

Proto-state Formation Experimentation: Ethiopia

Ethiopia is the only country in Africa that claims to have evaded colonization, hence the label "proto-state." This label is often applied to societies that are not colonial creations or are perceived as first-generation states in the historical evolution of nation-states. In other words, they are not imitations of pre-existing models; rather, they are considered pioneers. Ethiopia is thus counted among those that pioneered the nation-state concept. Ethiopians claim over three thousand years of continuous existence, statehood, and nationhood. This longevity suggests a mature state and nation that have undergone the full range of transformations leading to the completion of their formation.

However, the Ethiopian case is riddled with paradoxes. The paradox stems from the deeply entrenched myth of an unbroken three-thousand-year history of nationhood and statehood (Levine 2000; Marcus 2002; Rubenson 1978). Despite this, Ethiopia today faces some of the most severe nation-state formation crises in Africa. It is paradoxical that a country claiming such an ancient history suffers more instability than others established less than a century ago. Modern Ethiopia, like any African nation-state, is a product of the colonial era. During the Scramble for Africa, Ethiopia expanded to include many ethnic groups, transforming it into a multi-ethnic nation-state. Its failure to be inclusive, participatory, and representative largely explains the predicament it faces.

The crisis of state and nation formation in Ethiopia has led to experimentation with three models over the past century. First, under the monarchy (1931–1974) (Zewde 2001), a feudo-bourgeois model of nation-state formation was tested. This model failed to resolve Ethiopia's fundamental problems. Under the monarchy, issues related to nations and nationalities, reflecting the failure of nation-state formation, reached a critical level, eventually leading to the monarchy's collapse in 1974 (Tareke 2009). Severe inequalities, often along ethnic lines, mobilized university students and leftist political groups advocating for the marginalized and disenfranchised, bringing the country to a boiling point by 1973 (Zewde 2010). The emperor not only lost power but also his life, triggering a radical societal transformation. Subsequently, the military seized power, ending the feudo-bourgeois nation-state experiment.

Second, the military regime that followed introduced a new nation-state formation model: the military-socialist state. This experiment, however, met the same fate as its predecessor. It failed to address Ethiopia's core problems and instead revived the issues that had contributed to the monarchy's downfall. Ethno-nationalist insurgencies engulfed the country, lasting seventeen years. In 1991, the military regime was replaced by a coalition of ethno-nationalist insurgent groups, the Ethiopian People's Revolutionary Democratic Front (EPRDF) (Kefale 2009).

Third, the EPRDF implemented yet another model: ethnic federalism. The state was restructured along ethno-territorial lines, creating ethno-national regional states (Kefale 2009). Ethnic federalism also failed to resolve the persistent problems of nation-state formation. The EPRDF collapsed in 2019 and was succeeded by the Prosperity Party (PP).

The current model pursued by the PP remains unclear. Conflicts in the Tigray, Amhara, Oromo, and Benishangul-Gumuz regions fundamentally concern national identity and state structures. The lack of consensus on these issues underpins ongoing wars. Ethiopia stands at a crossroads, and it is increasingly apparent that the survival of Ethiopia as a united nation is in grave jeopardy.

It can be argued that Ethiopia's predicament can only be resolved through the recognition and acceptance of the modern Ethiopian state's recent creation. The majority of people living in today's Ethiopia were incorporated during the 19th century. The myth of a three-thousand-year history does not represent them, nor do they identify with it. One of Ethiopia's primary problems is disagreement over history, identity, and belonging. A potential point of consensus among all Ethiopians could be acknowledging when modern Ethiopia was formed. This shared understanding could foster a sense of common history and belonging. The history and identity prior to the 19th century belong predominantly to a minority group, the Abyssinians. Reconciling with history would not only help Ethiopians address their problems but could also promote peace, security, and development in the region.

The ongoing conflicts in Ethiopia are driven by contested histories, disputes over national identity, and state configuration. All of these highlight the failure of previous nation-state formation experiments and the challenges of governance in a multi-ethnic Ethiopia.

Colonial and Post-Colonial Nation-State Experimentation: Sudan

Sudan serves as a significant case study in post-colonial nation-state formation. The primary challenge for Sudan was to transcend the inherited colonial structures and institutions. This transcendence required the construction of a nation-state that was genuinely representative, inclusive, and participatory—one in which all citizens of post-colonial Sudan could feel a sense of belonging. From the outset, the transition from a colonial state to a post-colonial nation-state was met with formidable obstacles.

These challenges largely stemmed from the refusal of the northern political elite to acknowledge the concerns of southern Sudanese. As early as the 1950s, southern Sudanese politicians began advocating for a federal system following Sudan's independence, which would allow them to exercise self-rule. This was intended to prevent the marginalization of southern Sudanese in the independent state (Wassara 2015). This fear catalyzed the Torit Uprising on 18 August 1955, when southern Sudanese soldiers rebelled on the eve of Sudan's independence (Oromo 2015). This incident triggered what would become a decades-long civil war, which finally ended in 2005 with the signing of the Comprehensive Peace Agreement (CPA) (Comprehensive Peace Agreement 2005).

The fears of southern Sudanese were ultimately realized. The process of Islamisation and Arabisation began, resulting in the marginalisation and disenfranchisement of the southern population. Two years after Sudan's independence in January 1956, a group of military officers led by Major-General Ibrahim Abboud overthrew the civilian post-colonial government (Theron 2022: 45). The military regime pursued a nation-state formation process that mirrored the colonial state's exclusionary policies. The emerging post-colonial state was characterized by sectarianism. A small elite group—composed of the Ja'alyyin, Shaygiyya, and Danagla—formed the ruling class. The civil war in South Sudan continued for over four decades, while conflicts in Darfur, South Kordofan, and Blue Nile further devastated the country. Since April 2023, Sudan has been engulfed in yet another violent conflict.

The failure of nation-state formation in post-colonial Sudan has been so pronounced that the country has yet to establish a stable government. It has experienced frequent transitions between civilian and military rule. Indeed, post-colonial governance was largely dominated by military regimes that came to power through coups (El-Battahani 2016). Both civilian and military administrations failed to address the core issue: the challenge of genuine nation-state formation. The fundamental cause of this failure lies in the flawed model adopted—one based on sectarianism, exclusion, and a lack of representation and participation. This approach has driven Sudan to the brink of collapse.

Abboud's aggressive policies of Islamisation and Arabisation initiated a flawed nation-state formation project from the very beginning. This misguided process has plunged Sudan into chronic conflicts and wars, reflecting the exclusion, marginalisation, and lack of representation experienced by the majority of its people. Consequently, the nation-state formation experiment and governance remain dysfunctional.

National Liberation and Nation-State Experimentation: The Case of Eritrea

The third model of nation-state formation experimentation is national liberation. This paper uses Eritrea as an example of the national liberation model of nation-state formation. The case of Eritrea is somewhat complex. Eritrea was originally a colonial creation, which situates it within the colonial model. However, it was denied the right to self-determination and decolonisation, which compelled a national liberation struggle to achieve self-determination, nationhood, and statehood. The national liberation movement (NLM) assumed the responsibility of nation-state formation in Eritrea, spanning both the liberation and post-liberation periods. The critical question is: how does the national liberation model differ from the other two models?

The NLM differs from the others in several respects. The seeds of deliberate nation-state formation were sown during the liberation struggle. Formation and transformation projects were diligently carried out from the grassroots level. The NLM systematically waged liberation war while mobilising and organising the civilian population in liberated areas under its control. Popular administrative organs were established at the village level and gradually expanded to district and provincial levels. As a result, the foundations of nation-state institutions were laid. Moreover, in the economic, social, and cultural spheres, the NLM provided health and educational services, supported cottage industries, trade, and cultural initiatives targeting both its fighters and the general population. In summary, by the end of the liberation struggle, the NLM was capable of performing the functions expected of a state, earning it the designation of a "state in waiting" (cf. Connell 1993; Pool 2001).

To achieve this, beyond building highly disciplined liberation forces, effectively fostering profound societal transformation depended on developing a conducive political culture and ideology. The chosen political culture and ideology were those of national liberation. By necessity, this political culture and ideology were leftist and progressive. Some tenets of liberation political culture are described by Amílcar Cabral (1980:152) as follows:

> The armed liberation struggle demands the mobilization and organization of a significant majority of the population, the political and moral unity of the various social categories, the efficient use of modern weapons and other means of warfare, the gradual elimination of remnants of tribal mentality and the rejection of social and religious rules and taboos contrary to development of the struggle (gerontocracy, nepotism social inferiority of women, rites and practices which are incompatible with the national character of the struggle, etc.). The struggle brings about many other profound changes in the life of the population. The armed liberation struggle implies, therefore, a veritable forced march along the road to cultural progress.

The political culture and ideology cultivated under the auspices of the NLM are characterised by specific qualities that ensured the success of the liberation struggle. These qualities primarily include high centralisation, absolute secrecy, a strict hierarchy, firm discipline, unwavering loyalty, and self-reliance, all encompassed within the principle of democratic centralism. These attributes were fundamental to the success of the liberation struggle.

Of critical importance to the post-liberation reconstruction effort—and with significant negative implications for the transition from a national liberation political culture and ideology—these qualities were transferred to the post-liberation societal reconstruction phase. It appears there was a lack of awareness that the demands and objectives of these two periods were fundamentally different. The political culture and ideology that served the liberation struggle well were insufficient for addressing the post-liberation realities. Consequently, the transition to a civic political culture and governance proved to be extremely challenging. In other words, the liberation-era political culture and ideology became a hindrance in the formation of the post-liberation nation-state (Bereketeab 2007).

The objectives of the post-liberation nation-state formation were narrowly defined and framed, focusing on nation-building, state-building, national unity, and socio-economic development—objectives that could be seen as reminiscent of the liberation struggle period. The conflation of national liberation objectives with post-liberation goals illustrates the challenges faced by all liberation movements during this transition. However, a unique feature of the Eritrean NLM is its use of national service as a cornerstone for nation-state formation, socio-economic development, and national unity in the post-liberation period. The 1994 proclamation establishing national service, along with the creation of the Sawa Institute for youth—where they receive military training followed by completion of their senior secondary education—became the central hub for national integration. Perhaps the most significant contribution of the Sawa Institute lies in nation formation, national unity, and socio-economic development (cf. EPLF 1994; Government of Eritrea 1994).

Echoing the liberation front, the Sawa Institute was intended to serve as a convergence point where youth from all ethnic groups, social backgrounds, and regions would be moulded into a unified nation. In the socio-economic realm, the youth constitute the backbone of the development agenda. Naturally, in matters of defense—securing sovereignty, territorial integrity, and national unity—the Sawa youth have played a decisive role.

The principle of self-reliance, materialized through the implementation of national service, is a unique feature of Eritrea's national liberation and nation-state formation experiment. It also partially fulfils the requirements of inclusion, participation, and representation, thereby facilitating solutions to governance challenges.

Conclusion

This chapter aimed to examine and analyse the sources and failures of governance in multi-ethnic societies. It also highlighted the implications for peace, security, and development in the Horn of Africa. The central premise supporting the arguments presented here is that the incomplete process of nation-state formation underlies the governance challenges these societies face. Furthermore, it is argued that there is a correlation between the incompleteness of nation-state formation and the lack of inclusivity, representation, and participation among societal groups. This, in turn, explains the absence of peace, security, development, and effective governance practices. The logical conclusion drawn from this argument is that a fully realized nation-state formation would demonstrate inclusivity, representation, and participation.

Additionally, this chapter contends that the various experiments with nation-state models provide clear evidence of this incomplete process. The three models discussed—proto-state, colonial state, and national liberation state—are still undergoing continuous transformation. While these models both converge and diverge, they represent distinct governance challenges. Although each model is intended to address specific socio-historical, political, and cultural realities, they have ultimately revealed significant failures. These cases serve as classic examples of the crisis of nation-state formation in Africa, resulting in chronic governance problems, which are further intensified in multi-ethnic societies.

References

Anderson, Benedict. 1991. *Imagined Communities: Reflection on the Origin and Spread of Nationalism*. London and New York: Verso

Belous, Yulia A. and Tarasov, Andrey S. 2020. *Factors of Centrifugal Political Tendencies in Spain: A Qualitative Comparative Analysis*, Working Paper. Series: International Relations WP BRB 36/IR/2020

Bereketeab, Redie. 2023. *Historical Sociology of State Formation in the Horn of Africa: Genesis, Trajectories, Routes and Consequences*. Switzerland: Palgrave Macmillan

Bereketeab, Redie. 2021. *Alternative to Neoliberal Peacebuilding and Statebuilding in Africa*. London and New York: Routledge

Bereketeab, Redie. 2013.*The Horn of Africa: Intra-state and Inter-state Conflicts and Security*. London: Pluto Press;

Bereketeab, Redie. 2011. 'The Civic and Ethnic Foundation of Citizenship and Identity in the Horn of Africa', *Studies of Ethnicities and Nationalism*, vol. 11, no. 1, pp. 63-81

Bereketeab, Redie. 2007. 'When Success Becomes a Liability: Challenges of State Building in Eritrea (1991-2005)', *African and Asian Studies*, vol. 6, no. 4, pp. 395-430

Callaghy, Thomas M. 1984. *The State-Society Struggle: Zaire in Comparative Perspective*. New York: Colombia University Press

Chabal, Patrick and Daloz, Jean-Pascal. 1999. *Africa Works: Disorder as Political Instrument*. Oxford, Bloomington and Indianapolis, IN: James Currey and Indiana University Press

Christensen, Darin and Laitin, David D. 2019. *African States Since Independence: Order, Development, and Democracy.* New Haven, CT and London: Yale University Press

Coggins, Bridget. 2014. *Power Politics and State Formation in the Twentieth Century: The Dynamics of Recognition*. New York: Cambridge University Press

Comprehensive Peace Agreement (CPA). 2005. *The Comprehensive Peace Agreement between the Government of the Republic of the Sudan and the Sudan People's Liberation Movement/Sudan People's Liberation Army*. Nairobi, Kenya, 9th January, 2005

Connell, Dan. 1993. *Against All Odds: A Chronicle of the Eritrea Revolution*. Trenton NJ: Red Sea Press

Davidson, Basil. 1992. *The Black Man's Burden: Africa and the Curse of the Nation State*. London: James Currey

EPLF. 1994. *A National Charter for Eritrea: For a Democratic, Just and Prosperous Future. Approved by the Third Congress of the Eritrean People's Liberation Front (EPLF)*. Nacfa: February 1994

Evans, Graham and Newnham, Jeffrey. 1990. *The Dictionary of World Politics: A Reference Guide to Concepts, Ideas, and Institutions*. Hemel Hempstead: Harvester Wheatsheaf

Evans, Peter B. 1989. 'Predatory, Development, and Other Apparatus: A Comparative Political Economy Perspective on the Third World State', *Logical Forum*, vol. 4, no. 4, pp. 561-587

Fragile States Index. 2023. Prepared by the Fund for Peace.

Gellner, Ernest. 1983. *Nations and Nationalism*. Oxford: Basil Blackwell

Government of Eritrea. 1994. *Macro Policy Document*. Asmara: Government Press

Human Development Index. 2023. Prepared by United Nations Development Programme.

Kamrava, Mehran. 2000. *Politics and Society in the Developing World*. London and New York: Routledge

Kefale, Asnake. 2013. *Federalism and Ethnic Conflict in Ethiopia: A Comparative Regional Study*. London and New York: Routledge

Kefale, Asnake. 2009. *Federalism and Ethnic Conflict in Ethiopia: A comparative Study of the Somali and Benishangul-Gumuz Regions*, Doctoral Thesis. Faculty of Social and Behavioural Sciences. Leiden: Leiden University

Kooiman, Jan. 1999. 'Social-political Governance: Overview, Reflections and Design', *Public Management: An International Journal of Research and Theory*. Vol. 1, issue 1, pp. 67-92

Leach, Melissa et al. 2007. *Understanding Governance: Pathways to Sustainability*. STEPS Working Paper2. Brighton: STEPS Centre

Levine, Donald N. 2000. *Greater Ethiopia: The Evolution of a Multiethnic Society*, 2nd ed. Chicago, IL and London: University of Chicago Press

Marcus, Harold. 2002. *A History of Ethiopia*, updated ed. Berkeley, CA, Los Angeles, CA and London: University of California Press

Markakis, John, Schlee, Gunter and Young, John. 2021. *The Nation State: A Wrong Model for the Horn of Africa,* Studies 14. Halle: Max Planck Research Library for the History and Development of Knowledge

Mazrui, Ali A. 1983.Francophone Nations and English-Speaking States: Imperial Ethnicity and African Political Formation, in Donald Olorunsola and Victor Rothchild (eds.), *State Versus Ethnic Claims: African Policy Dilemma*. Boulder, CO: Westview Press

Mengisteab, Kidane. 2014. *The Horn of Africa*. Cambridge and Malden, MA: Polity Press

Morgenthau, Hans J.1985.*Politics Among Nations: The Struggle for Power and Peace*, 6th ed. Revised and by Kenneth W. Thompson. New York: McGraw-Hill

Oromo, Sirisio L. 2015. 'Sudan's First Civil War for Self-determination', in Redie Bereketeab (ed.), *Self-Determination and Secession in Africa: The Post-colonial State*. London and New York: Routledge

Osiander, Andreas. 2001. 'Sovereignty, International Relations, and the Westphalian Myth', *International Organization*, vol. 55, no. 2, pp. 251-287

Poggi, Gianfranco. 1978. *The Development of the Modern State*. London: Hutchinson

Pool, David. 2001. *From Guerrillas to Government: The Eritrean People's Liberation Front*. Oxford, Athens: James Currey and Ohio University Press

Renan, Ernest. [1882] 1991. 'What is a Nation', in Homi K. Bhabha (ed.), *Nation and Narration*. London and New York: Routledge

Rubenson, Sven. 1978. *The Survival of Ethiopian Independence*. London: Heinemann

Schaar, John H. 1981. *Legitimacy in the Modern State*. New Brunswick and Oxford: Transaction Publishers

Schmidt, Elizabeth. 2013. *Foreign Intervention in Africa: From the Cold War to the War on Terror*. Cambridge, New York, Melbourne, Madrid, Cape Town, Singapore, Sao Paulo, Delhi and Mexico City: Cambridge University Press

Scruton, Roger. 1982. *A Political Dictionary of Political Thought*. Pan Books: The Macmillan Press

Smith, Anthony D. 1986. *The Ethnic Origin of Nations*. Oxford and Cambridge, MA: Basil Blackwell

Smith, Anthony D. 1983. *State and Nation in the Third World.* Hemel Hempstead: Wheatsheaf Books

Sorbo, Gunnar M. and Ahmed, Abdel Ghaffar M. 2013. 'Introduction: Sudan's Durable Disorder', in Gunnar M. Sorbo and Abdel Ghaffar M. Ahmed (eds.), *Sudan Divided: Continuing Conflict in a Contested State.* New York: Palgrave Macmillan

Tareke, Gebru. 2009. *Ethiopian Revolution: War in the Horn of Africa.* New Haven, CT: Yale University Press

Weber, Max. [1948] 1994. 'The Nation', in John Hutchinson & Anthony D. Smith (eds.), *Nationalism*. Oxford, New York: Oxford University Press

Wassara, Samson Samuel. 2015. 'Political History of Southern Sudan before Independence of the Sudan', in Redie Bereketeab (ed.), *Self-Determination and Secession in Africa: The Post-colonial State*. London and New York: Routledge

Woodward, Peter. 2013. *Crisis in the Horn: Politics, Piracy and the Threat of Terror*. London and New York: I. B. Tauris

Woodward, Peter. 2006. *US Foreign Policy and the Horn of Africa.* England and USA: Ashgate

Yordanov, Radoslav A. 2016. *The Soviet Union and the Horn of During the Cold War: Between Ideology and Pragmatism*. Lanham, Boulder, New York and London: Lexington Books

Young, Crawford. 1994. *The African Colonial State in Comparative Perspective*. New York and London: Yale University Press

Zewde, Bahru. 2010 (ed.). 2010. *Documenting the Ethiopian Student Movement: An Exercise in Oral History*. Addis Ababa: Forum for Social Studies

Zewde, Bahru. 2001. *A History of Modern Ethiopia, 1855-1991*, 2nd edi. Oxford, Athens. OH and Addis Ababa: James Currey, Ohio University Press and Addis Ababa University Press